SYNTHESIS AND PROPERTIES OF BORON NITRIDE

SYNTHESIS AND PROPERTIES OF BORON NITRIDE

Edited by

John J. Pouch and Samuel A. Alterovitz

NASA Lewis Research Center
Cleveland, Ohio 44135
USA

TRANS TECH PUBLICATIONS

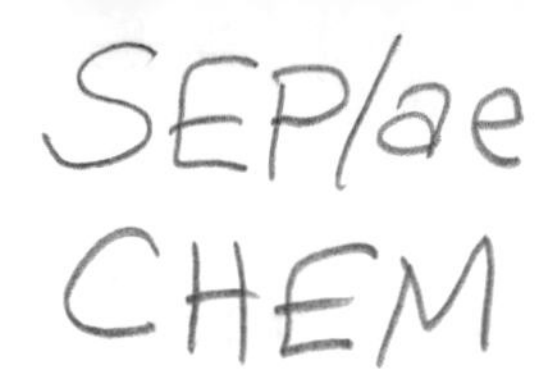

ISBN 0-87849-606-8

Volumes 54 & 55 of
Materials Science Forum
ISSN 0255-5476

Distributed in the Americas by

Trans Tech Publications
Old Post Road
Brookfield VT 05036
USA

and worldwide by
Trans Tech Publications
Segantinistr. 216
CH-8049 Zürich
Switzerland

INTRODUCTION

In the last 5 years, boron nitride (BN) films became more available. Renewed interest in this material has yielded new preparation methods and applications. A wide variety of synthesis methods are now being used and boron nitride films can be grown in many phases. The cubic phase, the most desirable phase for applications, is still hard to grow. A large number of applications will benefit from BN films. In particular, three main areas show the biggest immediate promise: thin membranes for x-ray masks, electronic applications for high temperature active devices and insulators, and friction reducing coatings. The low atomic number combined with mechanical strength give BN the edge for the x-ray masks, while the feasibility to dope the semiconducting cubic BN with both n and p type impurities makes BN a strong competitor to diamond-based electronics. It is safe to state that BN is competing with diamond and silicon carbide (SiC) in most applications. Therefore, most of the Materials Research Society and other meetings that included BN were part of a diamond and SiC symposium. Each proceedings volume includes a small number of short papers. There is no single book describing in detail the most important work done in the last several years in the field of BN films. In order to fill this gap, Dr. Fred Wohlbier (Trans Tech Publications) invited us to edit this book.

There are two main purposes to this book: first, to give a scientist starting in this field a way to know, in detail, the state-of-the-art; second, to give researchers already in the field a timely and more complete description of work done in laboratories all over the world, without looking for original publications. Most chapters contain original results that have not yet been published elsewhere. In addition, the book can serve as a reference book for people not working in the area.

The book is divided into three sections: synthesis, properties and applications. All the chapters in the book are invited chapters. Each section contains 5-7 chapters. Most of the chapters summarize the work that has been done over several years. In many cases, a chapter could easily be classified under more than one section. Each chapter contains synthesis, and/or properties and/or applications. Therefore, in these cases, the classification of the chapters may look somewhat arbitrary.

The first section is devoted to synthesis. This is a very important issue, as BN film growth is not a trivial endeavor. In the first chapter, Professor Aita is reviewing the sputtering method for BN deposition. Sputtering and ion beam methods are described in the second chapter (Guzman and Elena), while other modifications of the ion beam deposition technique are discussed by Fujimoto and by Halverson et al. in chapters 3 and 4 respectively. Chemical vapor deposition (CVD) methods are described in chapters 5 (Nakamura) and 6 (Sugiyama and Itoh). The pulsed laser evaporation method is described in the last chapter (Murray) of this section. Most chapters also give characterization results, some of them in great detail.

The section on properties starts with a theoretical paper (Lam et al.) devoted to the electronic and structural properties of several BN phases, while the next chapter (McKenzie et al.) reviews the work done analyzing the structure of BN films. A study of Auger transitions in the BN system is reported in the next chapter (Hanke et al.). The subsequent two chapters (Dana and Karnezos) describe the low pressure CVD process used to grow thin BN films for x-ray lithography masks. A wide variety of characterization methods are covered. Stress, dielectric properties and radiation effects are also described. The following two chapters (Osaka et al., Montasser et al.) focus on films made by plasma CVD. Results on the mechanical and optical properties are presented. As mentioned above, four chapters in this section also describe preparation methods, while some include applications.

The last section is devoted to applications. Growth and application of single crystal cubic BN, with both p and n doping, are described by Mishima. Although this achievement was done using crystal growth technology, we incorporated it in the book to show that this important application is within reach using cubic BN. Applications of BN films as a gate dielectric in field effect transistors are described in the next two chapters (Yamaguchi and Kapoor). The last two chapters of the book (Miyoshi and Kuwano) describe the applications of BN films in tribology for friction-reducing coatings. Similar to the other sections, several chapters in this section contain synthesis and characterization. Most notably, the chapter by Yamaguchi contains a theoretical study of the electronic structure of BN.

The book does not cover all the work and all the groups in the field of BN films. There is a limit to both the scope of this book and the ability to collect up-to-date material. However, we believe that we covered the most important aspects of preparation, characterization and applications.

The invited chapters are of high scientific quality and up-to-date results are presented. We believe that the book will be used extensively by researchers already in the field, newcomers and researchers that need a general reference to this new and developing area.

We would like to thank all authors for their contributions, without which this book could not be published.

John J. Pouch
Samuel A. Alterovitz
Cleveland, Ohio, USA

December 1989

TABLE OF CONTENTS

SECTION III APPLICATIONS

Materials Science Forum Vols. 54 & 55 (1990) pp. 1-20

SPUTTER DEPOSITED BORON NITRIDE: A REVIEW

C.R. Aita
Materials Department and the Laboratory for Surface Studies
University of Wisconsin-Milwaukee
P.O. Box 784
Milwaukee, WI 53201, USA

ABSTRACT

This paper addresses the growth of boron nitride thin films by reactive sputter deposition. First, the sputter deposition process is discussed as a method for growing nitrides in general, and the few published reports of sputter deposition of boron nitride are reviewed. Next, we discuss atomic order in the films and show in detail how ultraviolet-visible-infrared spectrophotometry was used to gain information about optical behavior, and hence about short range atomic order. Last, we suggest several starting points for future research leading to the development of process parameter-growth environment-film property relationships for the sputter deposited boron-nitrogen materials system.

figure 1. The chamber is evaculated to a base pressure typically in the 10^{-7} Torr range and backfilled with the sputtering gas in the 10^{-3} to 10^{-2} Torr range. A large negative voltage is applied to the cathode while the anode is kept near ground potential. The gas between the electrodes is ionized to form a self-sustained glow discharge. Radio frequency excitation allows the use of a target that is not an electrical conductor [1].

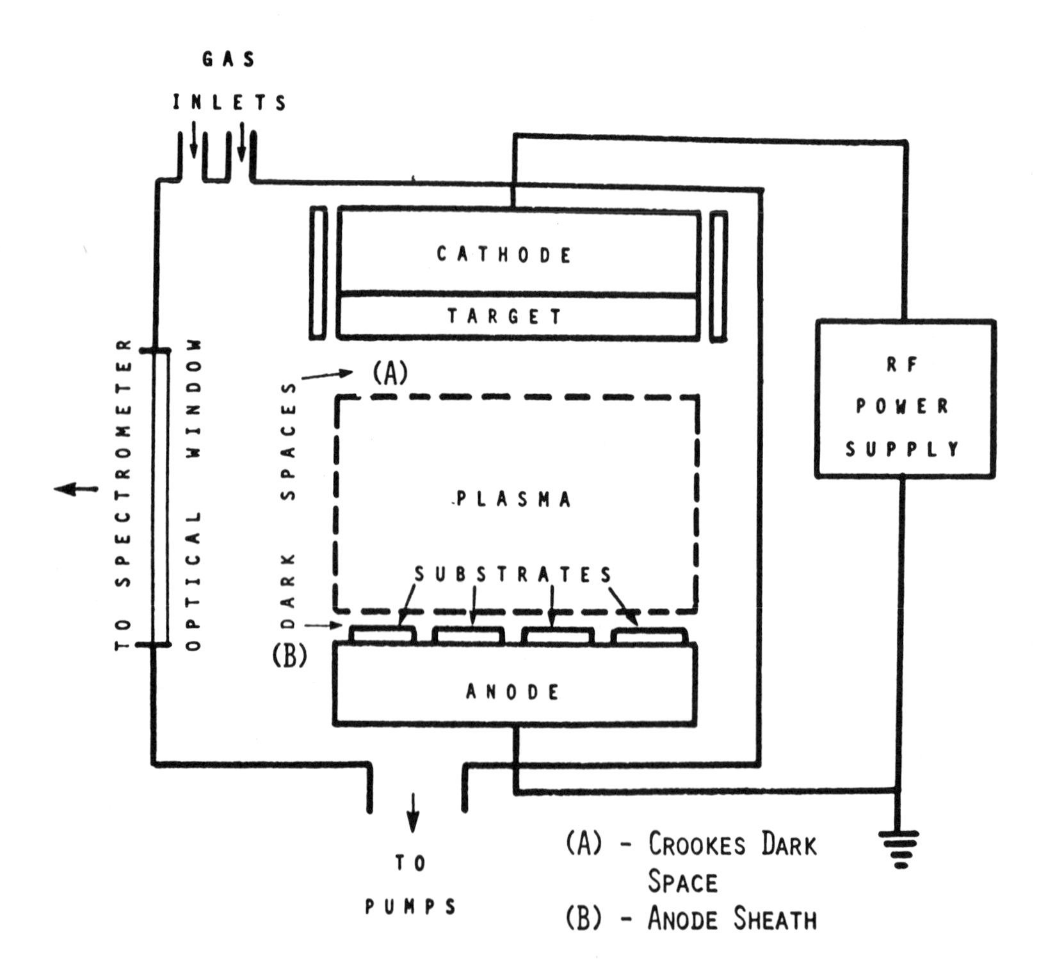

Figure 1: An rf diode sputter deposition apparatus.

I. OVERVIEW

The review that follows addresses the growth of boron nitride films by reactive sputter deposition. Section II contains an introduction to the sputter deposition process as a method for growing nitrides in general, and surveys the literature with respect to the growth of boron nitride by this process. Section III discusses the crystallography and short range atomic order of the films in relation to the bulk material. The term "short range order" is used here to denote parameters involving a central atom and its nearest neighbors: atomic species, coordination number, bond length, and bond angle. The application of ultraviolet-visible-infrared spectrophotometry to gain information about short range atomic order in boron nitride films is discussed in detail. Section IV contains suggestions for future research. We hope that this review is useful to scientists addressing basic questions involving process parameter-growth environment-film property relationships for the sputter deposited boron-nitrogen materials system, as well as those who wish to apply the answers to these questions to an emerging technology.

II. BACKGROUND

Reactive sputter deposition is widely used to grow alloy and compound films in which one or more of the constituent elements is volatile. Phase formation in the films is governed by kinetics. Low defect density films of high melting point materials can be therefore grown on unheated substrates. Metastable phases and multiphase structures with behavior not attainable in bulk material can be synthesized for specific applications. The principles of reactive sputter deposition applied to the growth of nitrides are reviewed in the first part of this section. The last part surveys the literature with respect to sputter deposition of boron nitride.

1. Reactive sputter deposition

Reactive sputter deposition is carried out in a chamber containing two electrodes attached to a power supply. A target is placed over the cathode. The substrates are placed over the anode. A typical sputter deposition apparatus is shown in

The distribution of electrical potential, light intensity, and space-charge for the rf diode system described above is discussed in detail in reference 2. To summarize here, referring to figure 1, a drop from the high negative cathode potential to ground occurs across the Crookes dark space. The potential acquires a small positive value in the plasma or negative glow. The plasma potential is strongly dependent upon the ratio of the cathode-to-wall area [3] and weakly dependent upon discharge power. A sheath separates the plasma from the anode surface, which floats at a slightly negative potential with respect to ground, on the order of several electron volts.

Physical sputtering of the target occurs when species that originate as positive ions in the plasma are accelerated across the Crookes dark space and, possibly after symmetric charge exchange collisions with less energetic neutrals [4], strike the target as high energy neutrals. Target material is ejected by a momentum transfer process [5]. Most of the ejected target species are uncharged, assume a random motion between the electrodes, and condense on any surface in their path. A film is deposited in this manner. However, in addition to thermalized target neutrals, other species impinge upon the growth interface. These species include: 1) high energy secondary electrons that cross the cathode-to-anode distance in one rf cycle, 2) high energy gas atoms that have either been elastically scattered from the target surface or created in the region near the target, and 3) positive ions created in the plasma.

The path of high energy secondary electrons emitted from the target can be deflected by an external magnetic field, usually accomplished by placing permanent magnets in the cathode assembly of a diode system. Secondary electrons that would be lost to the walls or the substrate in a single rf cycle in a diode system are constrained to remain in the plasma, as primary electrons, in a magnetron system. These electrons are now able to engage in ionizing collisions, and the power-voltage characteristics of the discharge changes considerably. There is also a reduction of substrate heating, the cause of which is primarily bombardment of the growth interface by highly energetic secondary electrons [6-9].

The effect on film growth of high energy gas neutrals that have been elastically scattered from the target surface or created in the region of the discharge near the target is not well understood. However, it is suggested that rare gas incorporation into the film results from these species [10].

Positive ions from the plasma are accelerated across the anode sheath and strike the growth interface with an average energy equal to the algebraic difference between the plasma potential and the potential at which the anode is floating. This effect can be deliberately enhanced by applying a voltage with a negative bias to the anode, and is used to modify film structure and chemistry [11,12]. It is possible to accelerate positive ions across the anode sheath with energies on the order of tens to hundreds of electron volts. In the case of a compound material, preferential resputtering of one component from the growth interface strongly modifies plasma characteristics in the region of the substrate, with resulting modification of the film characteristics [13].

A binary nitride film can be grown by sputtering an elemental target in a nitrogen-bearing atmosphere. A nitrided layer forms at the target surface once the reactive gas pressure has exceeded a critical value. It is important to note that this "critical value" depends upon other process parameters such as cathode voltage and rare gas type, as discussed with respect to other nitride systems in references 13-16. However, the sputtering process also can cause dissociation of the nitrided layer. Therefore, even when the target surface is fully reacted, the sputtered flux consists of both target atoms and nitride molecules sputtered without dissociation. There is currently no theory that allows prediction of the degree of dissociation of the nitrided layer at the target surface. Several trends that have been observed, and are in agreement with what is expected from binary collision theory, are that the molecular fraction of the total ejected flux increases a) with increasing bond energy, and b) if the atom struck can transfer enough energy to its molecular partner that both can be ejected [17,18]. However, other factors such as the dissociative effect of radiation from the discharge on the target surface have not yet been considered.

A binary nitride film can also be grown from a nitride target. However, in general sputtering dissociates the target material, as in the case of dissociation of the nitrided layer on an elemental target surface described above. Therefore, even though the as-received target material is stoichiometric, nitrogen must be added to the discharge to achieve stoichiometry in the film.

Process parameters that can be varied independently are the cathode voltage, anode bias voltage, anode-cathode spacing, total gas pressure, type of rare gas used, and partial pressure of the constituent gases. Other parameters are derived from these, such as the rf forward power which is coupled to the cathode voltage, and the deposition rate which is dependent upon the cathode voltage and the constituent gases.

The chemistry, short range atomic order, crystallography, and microstructure of a boron nitride film depend upon the relative flux of boron atoms, small molecules containing boron and nitrogen, and various types of nitrogen species that arrive at the growth interface from the plasma, are adsorbed and ultimately incorporated into stable nuclei that coalesce to form a continuous film. Factors controlling film chemistry and structure are:

1) Processes at the target surface affecting the balance between the rate of removal of boron atoms and the rate of formation and removal of nitrided boron molecules.

2) Processes at the substrate affecting the adsorption, surface diffusion, bulk diffusion, and desorption of boron atoms and nitrided boron molecules sputtered from the target surface, and nitrogen species from the plasma.

3) Plasma volume processes, that is, collisional and radiative processes affecting a) either the nitridation of sputtered boron atoms, or the dissociation of sputtered nitrided boron molecules in the plasma, and b) the creation of excited or ionized nitrogen species in the plasma [14,16,19] that ultimately engage in compound formation at either electrode.

2. Literature survey of sputter deposited boron nitride

There are few published reports of boron nitride grown by sputter deposition. In all cases, a boron nitride pressed powder target was used. An early report by Davidse and Maissel [1] demonstrated that dielectric targets could be sputtered using rf excitation and gave boron nitride deposition as an example. Noreika and Francombe [20] grew boron nitride and boron-aluminum nitride on silicon and fused silica, unheated and heated to 900°C, and studied the ultraviolet-visible optical behavior of the films. Wiggins et al. [21] studied the effect of discharge composition on the chemical composition, short range order, and ultraviolet-visible-infrared optical behavior of boron nitride grown on unheated silicon, glass, and sapphire. Rother and Weissmantel [22] discussed microcrystallinity in boron nitride grown on fused silica at an unspecified temperature. Goranchev et al. [23] investigated stress in boron nitride grown on silicon, sapphire, glass, and carbon at an unspecified temperature as a function of the total gas pressure and the type of rare gas used (argon or krypton) in conjunction with nitrogen. Siedel et al. [24] investigated the effect of substrate bias and discharge composition on phase content (hexagonal or cubic) of boron nitride grown on carbon, silicon, and tantalum substrates, unheated and heated to 350°C. Aita [25] modeled the optical behavior in the vicinity of the fundamental optical absorption edge of boron nitride grown by Wiggins et al. [21] using the coherent potential approximate assumption for the statistical independence of wavefunctions, and related the results to intralayer bond disorder of hexagonal boron nitride. The results of these studies will be presented and discussed in the next section.

III. <u>ATOMIC ORDER IN SPUTTER DEPOSITED BORON NITRIDE</u>

1. Review of bulk properties

Atomic order in bulk boron nitride will be reviewed next for reference before discussing the properties of the sputter deposited films. Bulk boron nitride crystallizes in two forms, hexagonal and cubic. The crystal structures are shown in figures 2 and 3. Hexagonal boron nitride (h-BN) is the stable polymorph at standard temperature and pressure, whereas cubic boron nitride (c-BN) is metastable. The two polymorphs are frequently compared to the graphitic and diamond forms of carbon, with which they are isoelectronic, and to the wurtzite and zinc blende forms of zinc sulphide [24]. However, if carried too far, these analogies may actually hinder our understanding of the properties of this unique material.

Hexagonal boron nitride has a layered structure, with B_3N_3 hexagons lying in the {0001} crystal plane and alternating boron and nitrogen atoms aligned along the <0001> crystal axis [26-28]. The h-BN structure consists of interlocking hexagonal boron and nitrogen sublattices displaced from each other by $c_o/2$. The bimolecular unit cell has the dimensions a_o=2.504Å and c_o=6.661Å. The B-N interlayer bond length is equal to 3.331Å. The B-N intralayer bond length is equal to 1.445Å. The geometric relationship between the lattice constant a_o and a B_3N_3 hexagon is shown in figure 2b. Hexagonal boron nitride is highly anisotropic because of the large difference in B-N bond length parallel and perpendicular to the {0001} plane. The compound has strong intralayer and weak interlayer B-N bonds. Misregistry between layers is observed. In the extreme, there is a random stacking of layers, resulting in a two dimensional variant of the h-BN structure called turbostratic boron nitride (t-BN) [29,30].

The c-BN structure consists of interlocking face-centered cubic sublattices displaced from one another along the cube body axis by a distance $(a_o/4)$<111>, where a_o=3.615Å [31]. The B-N bond length in the c-BN structure is equal to 1.565Å, or $(3^{1/2})(a_o/4)$. Each polymorph has a distinct short range atomic order. sp^2 intraplanar bonds are characteristic of h-BN, whereas sp^3 bonds are characteristic of c-BN. As a result, the boron nitride polymorphs are distinguished from each other by large differences in optical, electrical, mechanical, and chemical behavior.

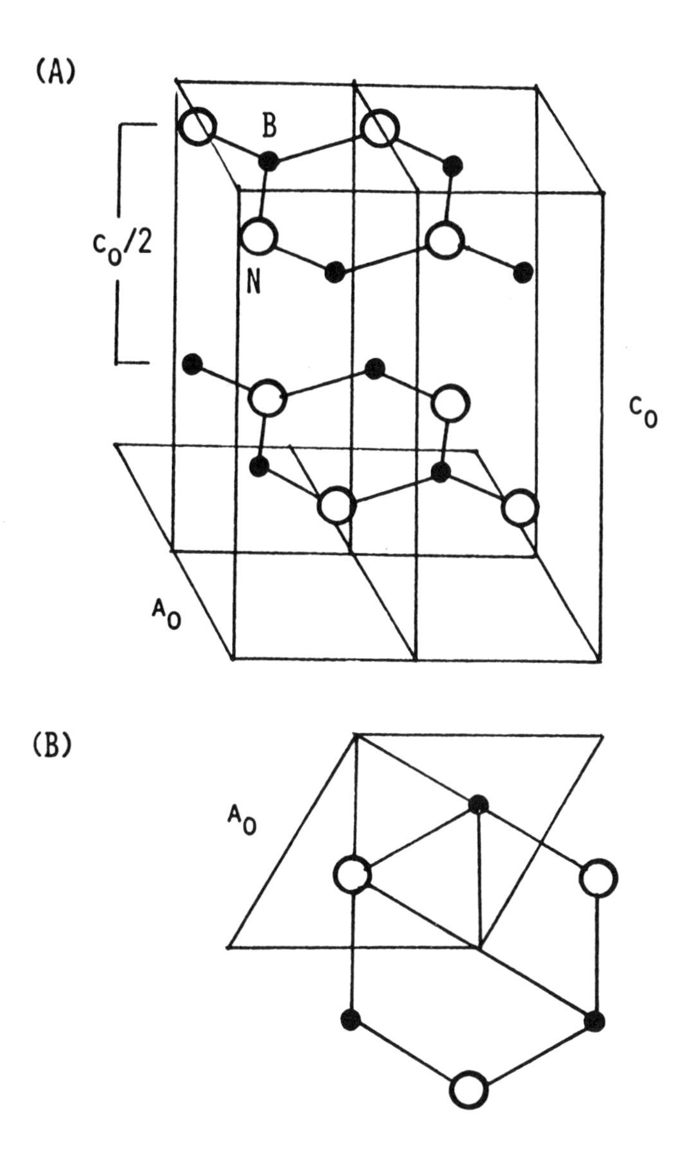

Figure 2: a) The unit cell of hexagonal boron nitride showing stacking of the B_3N_3 hexagons relative to the crystal axes. b) A projection of the {0001} crystal plane showing a B_3N_3 hexagon relative to the lattice constant a_o.

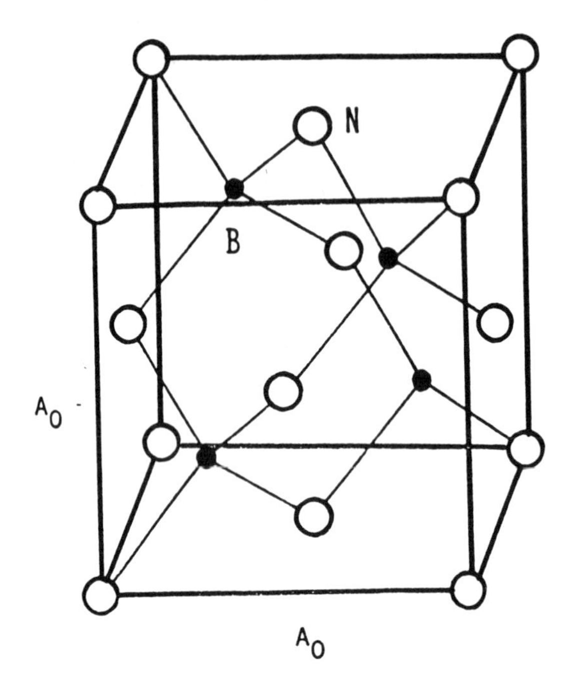

Figure 3: The cubic boron nitride unit cell.

2. Crystallography of sputter deposited boron nitride

The crystallography of sputter deposited boron nitride has been investigated by electron diffraction [20,22-24]. X-ray diffraction is not a good tool for investigating boron nitride thin films because both B and N atoms are poor x-ray scatterers. In fact, BN has been studied for use as a mask membrane material for x-ray lithography for this reason [32-36].

Films grown without applying an external bias voltage to the anode produce diffraction patterns attributable to the disordered h-BN lattice. The interlayer spacing, c/2, of these films is 5 to 15% greater than that of ideal h-BN, whereas the intralayer lattice constant, a, is within 2% of ideal h-BN [20,22]. t-BN is reported to have an interlayer spacing of up to 7% greater than ideal h-BN [29]. It is therefore possible that stacking disorder contributes to expansion of the interlayer spacing in the sputter deposited films. The large anisotropy with respect to the difference in inter- and intralayer lattice constants compared to ideal h-BN has also been observed for another layered material, sputter deposited vanadium pentoxide [37,38], and is not surprising considering the difference in inter- and intralayer B-N bond strength. Nor should it be surprising that the deviation in lattice constant from the ideal value is greater than is expected from a consideration of the elastic constants of the material. The quantity that is measured here is not the material's response to a long range elastic stress field, that is, not the elastic strain. Instead, it is the average value of the localized plastic strain due to isolated imperfections in the film that disrupt the lattice periodicity, such as displaced or missing B or N atoms, or rare gas atoms incorporated from the plasma [20].

Application of a negative bias voltage to the anode is reported to stabilize c-BN in sputter deposited films [24]. However, as we will discuss when presenting infrared spectrophotometry results, it is doubtful that c-BN is the sole phase present, although it may be the only phase with long range crystallographic order. The fact that the application of a negative bias voltage to the anode encourages c-BN growth is consistent with observations for the Ta-N system [13]: high density, more highly ordered phases are stabilized by the negative bias. A mass spectrometry study of the plasma showed that enhancement of the flux and energy of Ta^+ ions to the growth interface was responsible. A similar study of the B-N system would be interesting.

3. Short range order of sputter deposited boron nitride

Short range atomic order of sputter deposited boron nitride has been investigated by the interaction of the film with electromagnetic radiation in the optical spectral region [20,21,24,25]. In general, the transmittance, T, and reflectance, R, measured by ultraviolet-visible-infrared spectrophotometry, can be used to calculate the absorption coefficient, α, as a function of incident photon energy, E, for a film of thickness, x, using the relationship [39]:

$$T=[(1-R)^2\exp(-\alpha x)]/[1-R^2\exp(-2\alpha x)]. \quad (1)$$

Noreika and Francombe [20] measured the ultraviolet-visible transmittance of h-BN grown on fused silica, unheated and heated to 900°C, and found that the fundamental optical absorption edge was shifted to lower energy than that obtained for bulk h-BN [40]. Substrate temperature did not affect the position of the edge. Post-deposition annealing of the films in nitrogen shifted the edge to higher energy, coincident with bulk h-BN. Noreika and Francombe concluded that lattice disorder was responsible for both the displacement of the absorption edge in the as-grown material, and for the expansion of the interlayer lattice constant, discussed above in section III.2.

Wiggins et al. [21] showed that discharge composition had a strong effect on the ultraviolet-visible transmittance of h-BN grown on unheated alumina. Aita [25] calculated the absorption coefficient for these films, and analyzed the results using the coherent potential approximation with Gaussian site disorder in the valence and conduction bands [41,42]. According to this model for interband optical transitions in a disordered semiconductor, three energy parameters are defined: 1) the energy band gap of the virtual perfect crystal, E_g, 2) the energy band gap of the disordered crystal, E_x, and 3) the inverse slope of the exponential change of a as a function of E at the onset of the fundamental optical absorption band, E_o. These parameters are interrelated as follows:

$$E_x(W,T)=E_g-AE_o(W,T). \quad (2)$$

T is the absolute temperature, W is a measure of the structural disorder. For constant T, $E_o \propto (W^2/B)$, where B is the band half-width. A is a constant for materials with the same virtual crystal, hence the same value of B.

With respect to h-BN, figure 4 reproduced from reference 25 shows α as a function of E for three representative films whose growth conditons are recorded in Table I.

Table I: Growth Conditions, and Chemical and Infrared Optical Parameters of Sputter Deposited Boron Nitride (from reference 25).

Film	Gas $\%N_2$	B/N^a	Appear-ance	$\nu_s(cm^{-1})^b$	$\nu_b(cm^{-1})^c$	$\Delta\nu_s(cm^{-1})^d$
A	0	5.1	brown	1355	782	208
B	10	1.3	yellow	1380	810	167
C	50	1.2	clear	1380	810	130

a) Measured using wavelength dispersive spectroscopy to an accuracy of ±20%.
b) In-{0001} plane B-N bond stretch (E_{1U} mode), see figure 5.
c) Out-of-{0001} plane B-N-B bond bend (A_{2U} mode), see figure 5.
d) Full width at one-half peak maximum intensity.

Band-to-tail or tail-to-band transitions are responsible for the low energy region in which α varies exponentially with E:

$$\alpha^{BT}=\alpha^{*}\exp[(E-E_g)/E_o], \quad (3)$$

where α^* is the value of α^{BT} at $E=E_g$. The coordinates (E_g, α^*) define the focal point where curves with different values of E_o (different amounts of disorder but with the same band half-width) cross. The heavy lines in figure 4 were obtained by applying equation 3 to the data. Films B and C are within experimental error of stoichiometry and are assumed to have the same virtual crystal, and therefore the same valence and conduction band width. Solving graphically for E_g and α^*, equation 3 for Films B and C can be rewritten:

$$\alpha^{BT}=\{[7.5x10^5cm^{-1}]\exp[(E-6.88eV)/E_o]\}. \quad (4)$$

E_o, the inverse slope of α^{BT} versus E, is equal to 0.68 eV for Film B and 0.47 eV for Film C.

The absorption coefficient due to band-to-band transitions, when E is near to but less than E_g is given by

$$\alpha^{BB} \propto (E-E_x)^2. \tag{5}$$

The region over which equation 5 holds is shown as light lines through the data in figure 4. Solving for E_x by extrapolation of $(\alpha^{BT})^{1/2}$ versus E to $\alpha^{BT}=0$ yields 3.08 eV for Film B and 4.10 for Film C.

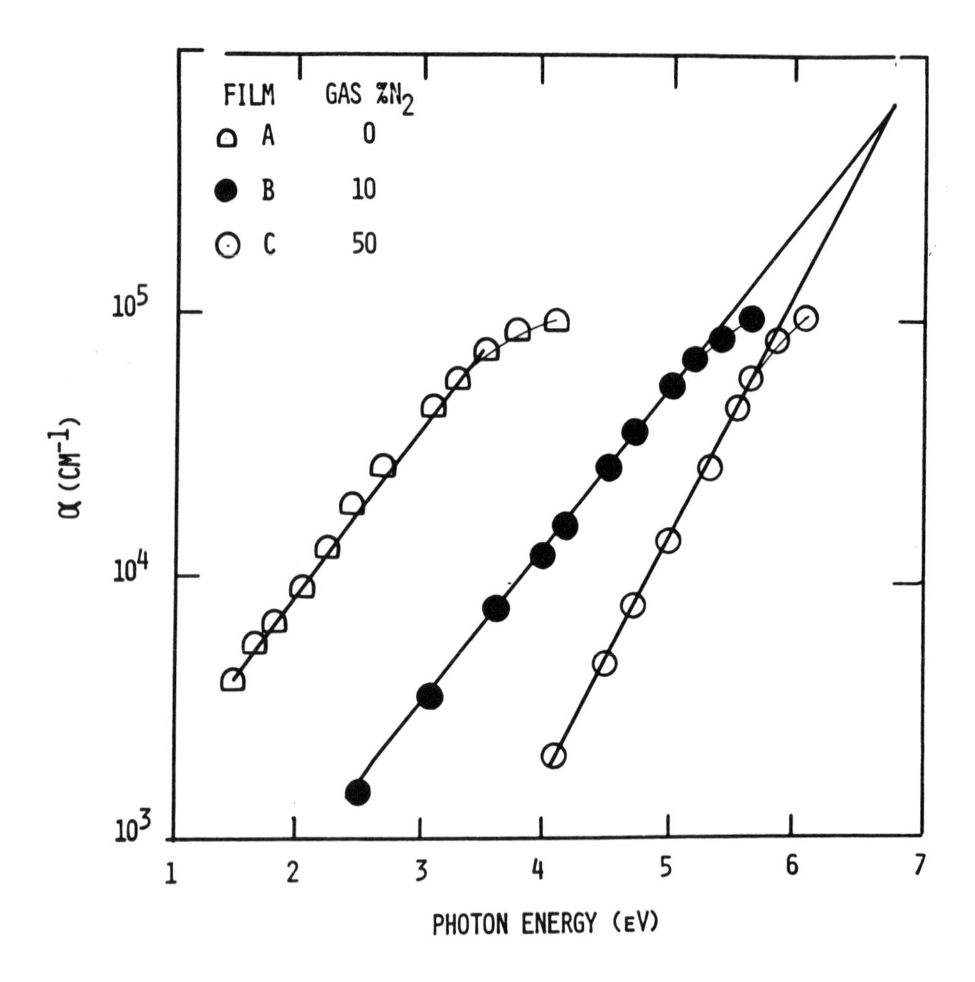

Figure 4: The absorption coefficient as a function of incident photon energy for three h-BN films. The heavy lines represent equation 3. The light lines are the best fit through the data in the region in which equation 5 applies.

The assumption made in the above analysis is that the virtual crystal is the same for Films B and C. Hence, a decrease in E_x and an increase in E_o are caused by a wider statistical distribution of B-N bond characteristics in Film B compared to Film C. The infrared spectrophotometry results discussed next indicate that the intralayer B-N bond length disorder is greater in Film B than in Film C. Atomic displacements associated with infrared-active normal modes E_{1U} and A_{2U} in h-BN are shown in figure 5 [40,43,44]. Wiggins et al. [21] used silicon substrates for infrared transparency, and deposited films under growth conditions identical to those used to grow Films A-C. These films are designated in the text as "Film A'...". It can be seen that both the average B-N bond stretch and B-N-B bond bend frequencies, ν_s and ν_b, are identical in Films B' and C', and identical to that for pyrolytic h-BN, indicating identical *average* intralayer B-N bond strength and hence equilibrium bond length. However, it can also be seen from Table I that the full width at one-half the maximum absorption band intensity for B-N bond stretch, $\Delta\nu_s$, is greater in Film B' than in Film C', indicating a greater statistical distribution about the average intralayer B-N bond length.

If ultraviolet and infrared spectrophotometry are measuring the same type of short-range atomic disorder and since $W \propto E_o^{1/2}$ for Films B and C, then we expect: $[\Delta\nu_s(\text{Film B'})/\Delta\nu_s(\text{Film C'}]=[E_o(\text{Film B})/E_o(\text{Film C})]^{1/2}$. Substitution of the appropriate values yields 1.28~1.20. Therefore, both ultraviolet and infrared spectrophotometry consistently show that the relative disorder is 1.2-1.3 times greater in h-BN grown in argon containing 10% nitrogen compared to h-BN grown in argon containing 50% nitrogen, holding other independent process parameters constant (section II.1). Infrared spectrophotometry enables us to make deductions about the nature of the disorder.

Equation 2 relates E_g, E_x, and E_o for each value of W and T at a fixed band width B. Therefore substitution of these energy parameters for either Film B or C should yield the same value for the constant A. Carrying out the substitution, we obtain A(Film B)=5.6~A(Film C)=5.9. Equation 2 can now be rewritten as follows:

$$E_x(W,T)=6.88eV-6E_o(W,T). \quad (6)$$

Equation 6 is applicable to all h-BN films in which the virtual crystal has the short range atomic order of a perfect h-BN lattice, that is, nearly stoichiometric films, and in which structural disorder is manifest by bond length randomness. E_o vanishes for the virtual crystal, and $E_x=E_g$. Physically, E_g is the zero-disorder limit (W,T=0) of E_x, the optical band gap, and the constant A, stated eloquently by Cody [42], "reflects conservation of states under the effect of disorder."

Film A has a B/N atomic concentration of five times that of stoichiometric h-BN. Unlike Films B and C, Film A has pronounced chemical disorder. The virtual crystal for Film A is unknown, but it is clearly not the perfect h-BN lattice. With respect to structural disorder detected by infrared spectrophotometry, Table I shows that both ν_s and ν_b are decreased for Film A' in comparison to Films B' and C', indicating an increase in average intralayer B-N bond length. In addition, $\Delta\nu_s$ for Film A' is increased above that for Films B' and C', indicating an increase in the statistical distribution of intralayer B-N bond length about the average value.

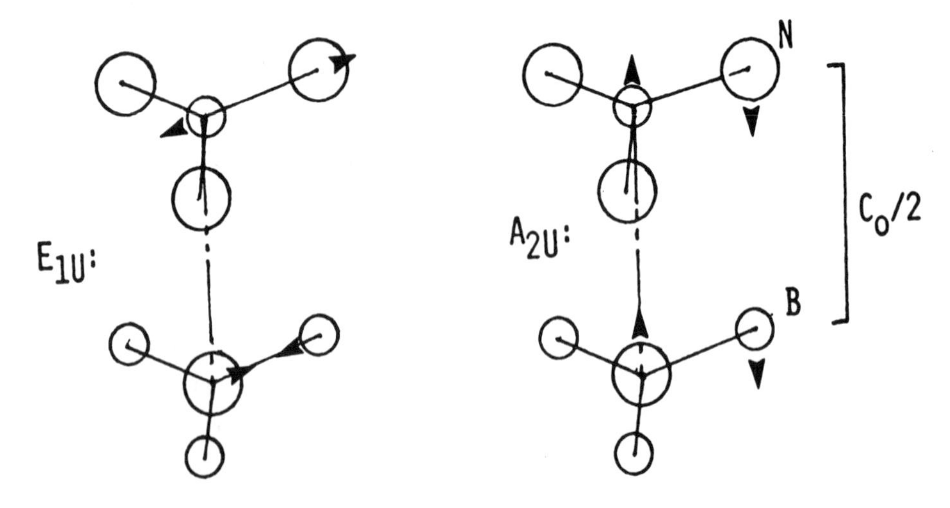

Figure 5: The atomic displacements associated with two infrared-active normal modes in h-BN. The optical phonons E_{1U} and A_{2U} are associated with the in-{0001} plane B-N bond stretch and the out-of-{0001} plane B-N-B bond bend, respectively.

With respect to the work of other researchers, features attributable to c-BN [45] were absent from the infrared spectra of all sputter deposited films. Recall that on the basis of electron diffraction, Siedel et al. [24] identified films as c-BN. Based on their infrared absorption spectra, we surmise that these films contain small regions of sp^3-bonded material embedded in a larger matrix of sp^2-bonded material with no long range crystallographic order, and hence yields no diffraction pattern.

IV. SUMMARY AND SUGGESTIONS FOR FUTURE RESEARCH

Considering the published research on sputter deposited boron nitride as a whole, two process parameter-film property relationships are observed:

1) A boron nitride target dissociates when sputtered. Nitrogen must be added to the discharge to achieve stoichiometry in the film and to minimize intralayer B-N bond length randomness in hexagonal boron nitride even after stoichiometry is achieved.

2) A negative bias voltage applied to the anode during deposition stabilizes cubic boron nitride in the film.

A study of the growth environment is needed to understand these relationships and to develop others. In order to achieve this goal, two types of *in situ* measurements are a good starting point for future research:

1) Characterization of the plasma volume using a) mass [2] and optical spectrometry [46] to determine the flux and energy of non-electronic species, and b) electronic probes to determine electron energy and density, as a function of the independent process parameters discussed in section II.1. Spatial and temporal resolution is desirable. For example, *in situ* plasma diagnostics have already been used to study the sputter deposition of two other group IIIA-nitrides, aluminum nitride [14,15] and indium nitride [47], with focus on the production of N_2^+ and the role this species plays in nitridation of metal atoms at each electrode. It would be interesting to compare these results to boron nitride growth. In addition, we mentioned in section II.2 that an enhanced metal ion flux at the plasma-anode sheath boundary, a consequence of resputtering of material from

the growth interface upon application of a negative bias voltage to the anode, results in denser, more highly ordered phases in the case of sputter deposited tantalum nitride films [13]. The literature shows that c-BN is stabilized by the application of a negative anode bias. It would be interesting to determine if B^+ ions are produced as a consequence of resputtering, and if their return to the growth interface with energy equal to the algebraic difference between the plasma potential and the anode bias is a determining factor for c-BN formation.

2) Characterization of the film using *in situ* surface analysis techniques including electron spectroscopies, electron diffraction, and optical index measurements, either continuously as the film is growing, or at various stages of growth after extinguishing the plasma if the measurement cannot be made in the plasma environment. The determination of film characteristics before exposure to laboratory air is important for basic nucleation and growth studies.

Acknowledgements

The author thanks C.-K. Kwok and R. Reeber for useful discussions concerning the material presented here. Preparation of this manuscript was supported under U.S. Army Research Office Grant No. DAAL-03-89-K-0022 and through a gift by Johnson Controls, Inc. to the Wisconsin Distinguished Professorship.

References

1) Davidse, P.D. and Maissel, L.I.: J. Appl. Phys., 1966, 37, 574.
2) Aita, C.R.: J. Vac. Sci. Technol. A, 1985, 3, 625.
3) Coburn, J.W. and Kay, E.: J. Appl. Phys., 1972, 43, 4965.
4) Davis, W.D. and Vanderslice, T.A.: Phys. Rev., 1963, 131, 219.
5) For a general reference on sputter deposition, see: Maissel, L.I. and Glang, R.: *Handbook of Thin Film Technology*, 1970, McGraw-Hill, New York.
6) Lamont, Jr., L.T., and Lange, A.: J. Vac. Sci. Technol., 1970, 7, 198.

7) Muth, D.G.: J. Vac. Sci. Technol., 1971, 8, 99.
8) Brodie, I., Lamont, Jr., L.T., and Myers, D.O.: J. Vac. Sci. Technol. 1969, 6, 124.
9) Ball, D.J.: J. Appl. Phys., 1972, 43, 3047.
10) Thornton, J.A. and Hoffman, D.W.: J. Vac. Sci. Technol., 1981, 18, 203.
11) Frerichs, R. and Kircher, C.J.: J. Appl. Phys., 1963, 34, 3541.
12) Krikorian, E. and Sneed, R.J.: J. Appl. Phys., 1966, 37, 3647.
13) Aita, C.R. and Myers, T.A.: J. Vac. Sci. Technol. A, 1983, 1, 348.
14) Siettmann, J.R. and Aita, C.R.: J. Vac. Sci. Technol. A, 1988, 6, 1712.
15) Aita, C.R., Gawlak, J.G., and Shih, F.Y.H.: unpublished.
16) Huber, K.J. and Aita, C.R.: J. Vac. Sci. Technol. A, 1988, 6, 1717.
17) Coburn, J.W., Taglauer, E., and Kay, E.: Jpn. J. Appl. Phys. Suppl. 2, 1974, 1, 501.
18) Coburn, J.W. and Kay, E.: CRC Crit. Rev. in Sol. State Sci., 1974, 4, 561.
19) Aita, C.R. and Gawlak, C.J.: J. Vac. Sci. Technol. A, 1983, 1, 403.
20) Noreika, A.J. and Francombe, M.H.: J. Vac. Sci. Technol., 1966, 6, 722.
21) Wiggins, M.D., Aita, C.R., and Hickernell, F.S.: J. Vac. Sci. Technol. A, 1984, 2, 322.
22) Rother, B. and Weissmantel, C.: phys. stat. sol. (a), 1985, 87, K119.
23) Goranchev, B., Schmidt, K. and Reichelt, K.: Thin Solid Films, 1987, 149, L77.
24) Siedel, K.H., Reichelt, K., Schaal, W., and Dimigen, H.: Thin Solid Films, 1987, 151, 243.
25) Aita, C.R.: J. Appl. Phys., 1989, 66, xxx, in press.
26) Wyckoff, R.W.G.: *Crystal Structures, Vol. 1,* 1963, Interscience, New York, pp. 184, 185, 187.
27) Geick, R. and Perry, C.H.: Phys. Rev., 1966, 146, 543.
28) Pearse, S.: Acta Cryst., 1952, 5, 536.

29) Thomas, J., Weston, N.E., and O'Connor, T.E.: J. Amer. Chem. Soc., 1963, 84, 4619.

30) Baronian, W.: Mat. Res. Bull., 1972, 7, 119.

31) Wyckoff, R.W.G.: op cit., pp. 109, 110.

32) Retajczyk, Jr., T.F. and Sinha, A.K.: Appl. Phys. Lett. 1980, 36, 161.

33) Johnson, W.A., Levy, R.A., Resnick, D.J., Sanders, T.E., Yanof, A.W., Betz, H., Huber, H., and Oertel, H.: J. Vac. Sci. Technol. B, 1987, 5, 257.

34) Levy, R.A., Resnick, D.J., Frye, R.C., Yanof, A.W., Wells, G.M., and Cerrina, F.: J. Vac. Sci. Technol B, 1988, 6, 154.

35) King, P.L., Pan, L., Pianetta, P., Shimkunas, A., Mauger, P., and Seligson, D.: J. Vac. Sci. Technol. B, 1988, 6, 162.

36) Duncan, T.M., Levy, R.A., Gallagher, P.K., and Walsh, Jr., M.W.: J. Appl. Phys., 1988, 64, 2992.

37) Aita, C.R. and Kao, M.L.: J. Vac. Sci. Technol. A, 1987, 5, 2714.

38) Aita, C.R., Kwok, C.-K., and Kao, M.L.: Proc. Mater. Res. Soc., 1987, 82, 435.

39) See, for example, Pankove, J.I.: *Optical Processes in Semiconductors*, 1971, Prentice-Hall, Engelwood Cliffs, New Jersey.

40) Hoffman, D.M., Doll, G.L., and Eklund, P.C.: Phys. Rev. B, 1985, 30, 6051.

41) Abe, S. and Toyozawa, Y.: J. Phys. Soc. Jpn., 1981, 50, 2185.

42) Cody, G.D.: *Semiconductors and Semimetals*, Vol. 21, Part B, edited by J.I. Pankove, 1984, Academic, Orlando, Florida, pp. 36-42.

43) Geick, R. and Perry, C.H.: Phys. Rev., 1966, 146, 543.

44) Nyquist, R.A. and Kagel, R.O.: *Infrared Spectra of Inorganic Compounds*, 1971, Academic, New York, pp. 114-115.

45) Kendall, D.N.: *Applied Infrared Spectroscopy*, 1966, Reinhold, New York, p. 442.

46) Greene, J.E.: J. Vac. Sci. Technol. 1978, 15, 1718.

47) Natrajan, B.R., Eltoukhy, A.H., Greene, J.E., and Barr, T.L.: Thin Solid Films, 1980, 69, 201.

Materials Science Forum Vols. 54 & 55 (1990) pp. 21-44

THIN BORON NITRIDE FILMS OBTAINED BY PHYSICAL VAPOUR DEPOSITION AND BY ION BEAM ASSISTED TECHNIQUES

L. Guzman and M. Elena
Istituto per la Ricerca Scientifica e Tecnologica
I-38050 Povo (Trento), Italy

ABSTRACT

The purpose of this paper is to present some novel preparation techniques, properties and applications of BN films.Boron nitride films were produced on several substrates using (i) nitrogen ion implantation of boron films; (ii) radiofrequency (r.f.) magnetron sputtering from a BN target, and (iii) ion-beam-assisted deposition of boron.

The films were characterized by scanning electron microscopy, Auger electron spectroscopy, secondary ion mass spectrometry, electrochemical and corrosion tests, micro-hardness, friction and scratch-tests in order to study the influence of the deposition process and parameters on the films properties.

1. INTRODUCTION

Boron nitride (BN) is one of the most interesting of the III-IV compounds, both from practical and fundamental viewpoints, due to its unique properties: low density, high thermal conductivity, exceptional strength and high chemical inertness, even at very high temperatures. BN may be obtained in at least two crystalline forms: a cubic zinc-blende, and a hexagonal structure. Much work has been done to synthesize this material in thin film form. However it is difficult to obtain at will a specific phase.

There have been some recent attempts to obtain BN coatings by using deposition techniques, such as plasma- or ion-beam-assisted processes [1-3], in which ionized reactive and/or inert species with an energy of about 1 keV are involved. These techniques have given promising results, although microanalysis has revealed that the films produced hitherto contained significant amounts of oxygen and carbon.

In order to produce compounds with predetermined (possibly non-equilibrium) composition, whose characteristics are interesting for the improvement of surface properties (wear, corrosion, thermal oxidation resistance), ion implantation has proved to be very effective [1]. However, the shallow thickness of the implanted layers constitutes a major limitation. In addition, surface sputtering effects drastically limit the concentration of implanted atoms.

The use of chemical vapor deposition (CVD) or of physical (PVD) methods such as ion plating or other ion-assisted processes is an important alternative. Hard and protective coatings are usually being obtained by such techniques.

The idea of enhancing the properties of a physically deposited film by ion bombardment is not a new one, and has been discussed, for example, by Weissmantel [5,6] and more recently by Dearnaley et al. [7]. In order to obtain thicker modified layer and higher implanted atom concentrations, new process which involve both film deposition and ion implantation are being developed. One such process is the ordinary ion mixing, in which a coating is first applied and subsequently bombarded with an ion beam. This process can be further extended by multiple sequential deposition/implantation steps, thus making the depth of the treated region nearly independent of the projected range of the implanted ions. Moreover, the ion beam plays a role in increasing the adhesion and homogeneity of the deposited film, and eventually in converting it into a chemically different material, e.g. a nitride.

2. PREPARATION TECHNIQUES

2.1. Formation of BN by nitrogen ion implantation of boron coatings [8].

Boron was evaporated at 10^{-4} Pa from an electron-gun-heated crucible and deposited onto iron substrates up to a thickness of 100-200 nm. The deposited boron films were then implanted with 100 keV nitrogen ions at a dose of 6×10^{17} ions/cm^2. The samples were then investigated by Auger Electron Spectroscopy (AES) combined with argon sputter etching; the analysed area size was about 100 μm x 100 μm.

Figure 1 shows the Auger depth profiles of boron-deposited sample, it is essentially free from contamination. The sputter etching in the Auger machine was carried out at a rate of 2.7 nm/min, as determined from the thickness of the B layer (140 nm, measured by a mechanical stylus).

Figure 2 shows the depth profiles of the implanted sample obtained using the same sputtering conditions as before. The majority of the nitrogen ions are observed to come to rest within the boron film. This confinement is unexpected because the projected range R_p for 100 keV N^+ ions in boron exceeds 220 nm. The flat-topped distribution of the implanted nitrogen is believed to be a dynamic effect of the high dose (6×10^{17} ions/cm^2) implantation. Furthermore, the interface between coating and substrate is shifted to a higher sputtering time.

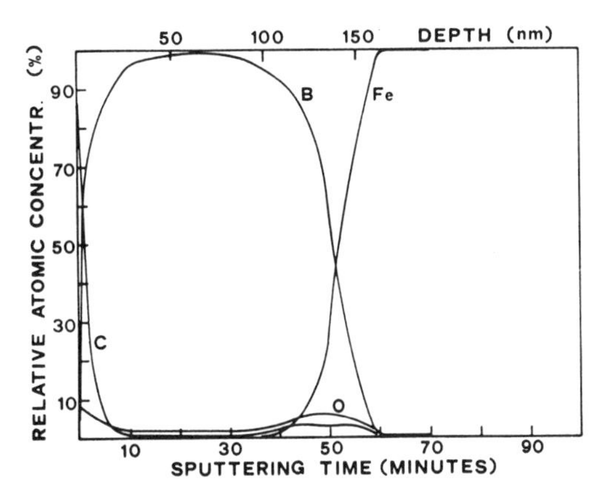

Fig. 1 - Auger depth profiles of the only-evaporated boron-on-iron sample.

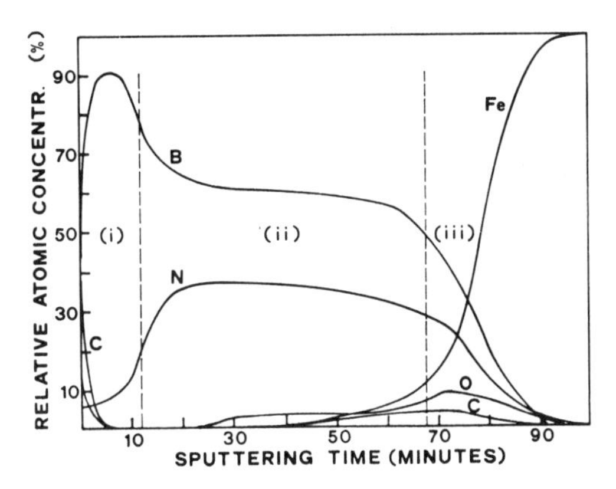

Fig. 2 - Auger depth profiles of the sample implanted with 100 keV N^+ ions.

This fact is quite surprising; because of the sputtering of the overlayer during implantation, a lower sputtering time is normally necessary to reach the substrate in the implanted sample with respect to the only evaporated one. Even making the unrealistic assumptions that no boron is lost by sputtering during implantation, that the retained nitrogen dose is equal to the nominal dose and that the density of the film does not change, we find a thickness expansion (by a factor 1.4) which, however, cannot account for the observed increase in the sputtering time (by a factor 1.5) Therefore the nitrogen implanted film has a sputtering yield lower than that of the only-evaporated film. This result suggests that the implanted surface layer might have better mechanical properties, see Section 5.

In the depth profile of Fig. 2, three regions are distinguished: (i) There is a near-surface region corresponding to a boron overlayer. Its Auger lineshape is similar to that presented in Fig. 3, spectrum a, for the boron deposit and is in agreement with that of elemental boron [9]. (ii) There is an intermediate region, in which boron and are present with a substantially constant ratio of their relative atomic concentrations of around 1.7, which is determined by using the elemental sensitivity factors method based on peak-to-peak heights [10]. This quantification must be taken with great care, particularly in this type of situation, the Auger lineshapes being strongly affected by the chemical environment. The boron Auger lineshape in this region is clearly different from that of pure boron. Both boron and nitrogen lineshapes (shown in Fig. 3, spectrum b) are in contrast very similar to those reported in the literature [11] and reproduced also in our laboratory for commercially available BN. We cannot rule out, however, the presence of some excess boron atoms not bonded to nitrogen atoms in the implanted film.

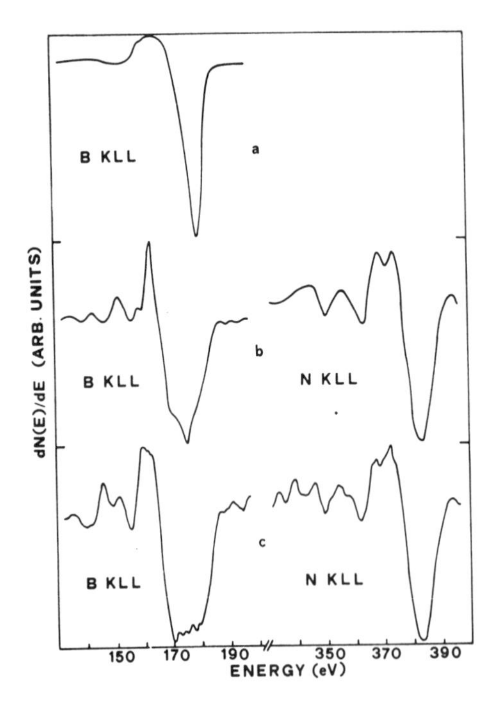

Fig. 3 - Auger lineshapes of boron and nitrogen: spectrum a, boron deposit; spectrum b, implanted sample; spectrum c, implanted and annealed sample.

Evidence of BN layer formation by nitrogen ion implantation of boron films was confirmed by IR spectroscopy [12]. (iii) A broad interface region exists between the film and the substrate. Ion-beam-mixing at the B-Fe interface is thought to be responsible for the observed adherence increase between coating and substrate after the implantation treatment.

The thermal stability of implanted surfaces may provide additional evidence for the presence of ion-induced phases. With regard to implanted nitrogen, out-diffusion from implanted iron samples [13] was observed at temperatures as low as 400° C. In order to check the thermal behaviour of our samples, we have carried out annealings at about 1000° C for 1 hour at a base pressure of 10^{-3} Pa. The Auger depth profiles of an annealed sample are shown in Fig. 4 (a). It is apparent that boron and nitrogen are left essentially in the same concentration ratio. The boron

KLL and nitrogen KLL lineshapes are shown in Fig. 3, spectrum c. The nitrogen Auger lineshape is the same as that of the implanted sample, whereas the boron lineshape changes and is more similar to that of oxidized boron [9]. Iron was detected from the very beginning of the sputter etching, this effect being mainly due to the presence of microcracks in the overlayer, as shown in Fig. 4 (b), caused by a great difference between the thermal expansions of coating and substrate. In spite of the presence of oxygen, the iron Auger lineshape never showed any evidence of oxidized iron [14]. We think therefore that the oxygen is bonded to the excess boron atoms in the implanted film. Surprisingly nitrogen is still present in the coating after such an extreme treatment. This is further evidence that the implanted nitrogen is strongly bonded to boron. The main effect of the 100 keV N^+ implantation of boron deposits is therefore the formation of a buried BN layer just below an "untransformed" surface boron layer; this BN is thermally stable up to at least 1000° C and acts as a protective coating against the oxidation of iron; see Section 4.

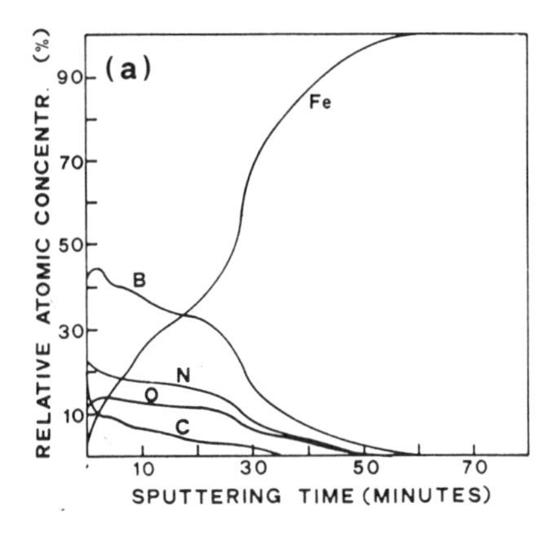

Fig. 4 - Characterization of an implanted sample after annealing at 1000° C for 1 h; (a) Auger depth profiles; (b) scanning electron micrograph showing thermal microcracks along the iron grain boundaries and within the grains.

2.2. Preparation of BN layers by r.f. magnetron sputtering

A possible application of BN is in the area of hard coatings, for instance for enhancing the performance of tools. In this sector, one can expect BN to represent a future alternative to well known materials like TiN or TiC. In order to investigate the possible role of BN as a tool coating, we produced BN thin films using the r.f. magnetron sputtering technique. Its main advantage is the capability of sputtering insulating targets. In addition one can obtain coatings covering in a uniform way substrates of even complex geometry.

BN coatings were prepared in an Alcatel SCM 440 machine by sputtering in Ar atmosphere, using a round BN target (crystalline structure: hexagonal) 10 cm in diameter. The substrates were hard metal (WC-Co) inserts characterized by a

Knoop microhardness value of 2600 (measured at 25 gf load), which were mechanically polished to a mirror-like finish using 3 μm diamond powder in the final step.

The samples were placed on a heating stage capable of reaching 400° C. Pre-sputtering was carried out in a 99.985 % pure Ar atmosphere at a pressure of 2 Pa. Sputtering was carried out at different Ar pressure values (from 0.3 to 0.8 Pa), as well as at different substrate temperatures (from 100 to 400° C). The base pressure before the start of the sputtering process was $5x10^{-5}$ Pa.

In Section 5 we report on the mechanical characterization (microhardness, scratch and friction) of boron nitride (BN) films deposited on WC-Co substrates. As we shall see, it is possible to obtain superhard coatings by r.f. sputtering, even starting from a hexagonal BN cathode.

2.3. Reactive ion beam assisted deposition (RIBAD)

A recently developed method consisting of simultaneous or sequential deposition/implantation steps, has proved to be very effective in the production of surface compounds with interesting surface properties [15-18]. This hybrid technique allows the depth of the treated region to be nearly independent of the projected range of the implanted ions and thereby to obtain thicker layers than by conventional ion-implantation. Moreover, with respect to unimplanted deposited layers, the ion beam increases the uniformity of the films and their adhesion to the substrate.

2.3.1. Apparatus

An electron gun was placed under the implantation chamber of a low energy (30 keV) horizontal ion implanter,in order to perform reactive ion beam assisted deposition (RIBAD) experiments under controlled conditions in a high vacuum environment (3.10^{-6} Pa base pressure). The samples could be rotated around an horizontal axis (perpendicular to the ion beam) in order (i) to expose their surface to deposition and implantation sequentially or (ii) to allow simultaneous deposition and implantation. More details about the apparatus can be found in [18].

Precalibrated film thicknesses and evaporation rates (kept fixed at about 0.1 nm/s) were monitored by a quartz crystal oscillator. The bulk temperature of the specimens was measured by a thermocouple and never exceeded 373 K.

2.3.2 Sample Preparation

All samples were mechanically polished and pre-implanted to sputter clean the surfaces and to create many favourable sites for deposition, enhancing adhesion.

RIBAD obtained BN films on copper, iron and WC-Co substrates, were obtained using the same technique by 4 to 10 sequential steps of a 27 nm B deposition, followed by a N_2^+ implantation to a dose of 1.4 x 10^{17} N atoms/cm^2 using a nominal 3:1 ratio of B:N. By partially screening the ion beam, some unimplanted evaporated samples were obtained in every run; in the following, we shall refer to them as to the reference samples.

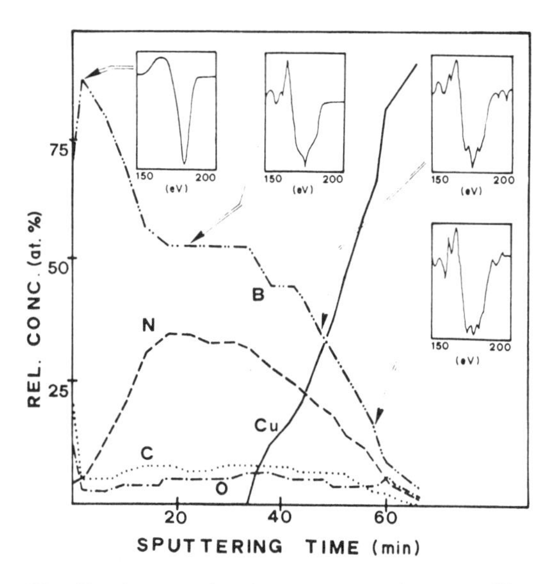

Fig. 5 - Auger depth concentration profiles for sequentially boron-evaporated and nitrogen-implanted copper substrate. The B KLL Auger line shapes corresponding to different regions of the profile are shown in the insets.

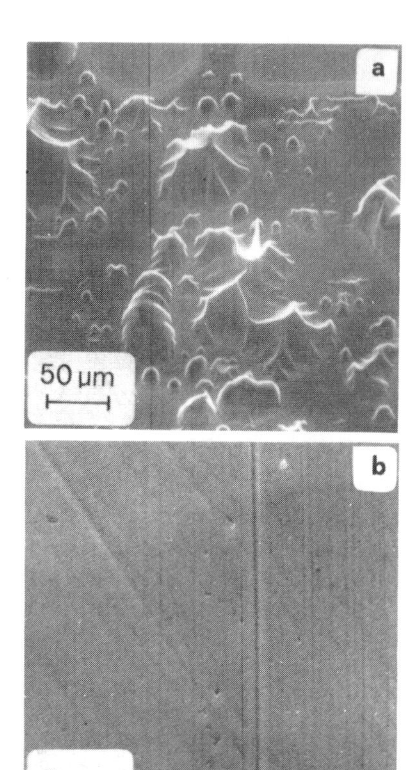

Fig. 6 - Scanning electron micrographs of: (a) a 100 keV nitrogen-ion-bombarded boron layer 140 nm thick on copper, showing extensive adherence failures; (b) well adherent homogeneous BN obtained by RIBAD on copper.

TABLE 1

Free corrosion potentials and instantaneous corrosion rates in 1 M NaCl solution, deaerated (sample area: 0.5 cm^2).

Sample	E_{corr} (mV)	R_p (kΩ)	β_a (V)	β_c (V)	i_{corr} (μA cm^{-2})
ARMCO Fe (standard ref.)	-780	24.5	0.05	0.165	1.35
Single B layer on Fe implanted with N_2^+ [20]	-810	178.7	0.08	0.14	0.25
Fe-B-Fe on Fe implanted with N_2^+ [21]	-820	20	0.04	0.16	1.39
Fe-B-Fe on Fe implanted with Kr^+ [21]	-790	14	0.065	0.10	2.44
RIBAD BN on Fe [23]	-720	650	0.065	0.16	0.062

2.3.3. Microstructural, structural and compositional characterization

Surface microstructure of the samples was examined by scanning electron microscopy (SEM); depth composition profiles were obtained by Auger Electron Spectroscopy (electron gun coaxial to a single pass cylindrical mirror analyzer with 0.6% intrinsic relative resolution), combined with 2 keV ion sputter etching. The AES depth profiles were elaborated with the sensitivity factors method [10]. The structure of the samples was studied by X-ray diffractometry (XRD), using Cu K_α radiation. Also secondary ion mass spectrometry (SIMS) was used to obtain qualitative sputter profiles as well as evidence of the RIBAD-induced phases.

Figure 5 shows the AES sputter profiles for a four-step sequentially boron-evaporated and nitrogen implanted copper substrate. Also the Auger lineshapes of boron and nitrogen recorded troughout the profile are shown in the insets. The multilayer coating obtained here shows features very similar to those observed previously [8] after a single 100 keV N^+ implantation step through a 140 nm thick boron film deposited on iron; i.e. the formation of a BN layer just below a surface layer of "untransformed" boron. Also the adhesion between the coating and the substrate appears to be highly enhanced as a result of more effective ion-mixing at the interface. (Surprisingly, a similar single 100 keV N^+ implantation through a boron deposit on copper never succeeded in enhancing the adhesion, as can be seen in Fig. 6a.) On the contrary, the BN film deposited by RIBAD and with same thickness on Cu appears to adhere well and to be homogeneous, Fig. 6b. The composition of the film, determined against suitable reference samples is nitrogen deficient; the Auger lineshape is representative of elemental boron in the near surface region while under the surface it is typically that of BN even if the N:B ratio attains only the value 2:3 over an extended depth. The coating appears slightly golden in colour. The presence of face centered cubic BN (lattice parameter 0.363 nm) was confirmed by XRD analysis of films deposited on silicon. It should be noted that diffraction peaks pertaining to BN thin films are difficult to observe, because of its very low electron density.

The corresponding SIMS sputter profiles have also shown the presence of a boron-rich BN layer. The SIMS spectra indicate the presence of Cu, BN, CuB and CuN in the interface region, although the last two compounds (CuB and CuN) are metastable and do not appear in the respective phase diagrams.

3. ELECTROCHEMICAL BEHAVIOUR AND CORROSION RESISTANCE

The electrochemical and corrosion behaviour of metal surfaces modified by direct energy beams is a research subject which has been widely studied in our laboratory during the last years [19-23]. We report first on the electrochemical characterization of iron samples sequentially treated with the two following processes: boron deposition and nitrogen ion implantation. The comparison with previously studied samples (e.g. alternate boron-iron multilayers deposited on iron and implanted with either Kr^+ or N^+ ions) [21] gives evidence of the effectiveness of the RIBAD technique in producing surface coatings with improved corrosion

TABLE 2

R_p (kΩ)measurements *vs.* immersion time in a 0.3 % Na_2SO_4 solution (sample area: 0.5 cm^2).

Sample	*Time (h)* 0	1	2	3	5	8	24
ARMCO Fe	5.06	5	4	3.84	-	-	3.1
Fe-B-Fe on Fe; N_2^+-implanted [21]	-	25	15.7	10.1	-	4.1	2.9
RIBAD BN on iron	90	67.5-	-	-	10	7	6

resistance properties. The electrochemical measurements were carried out on treated samples in the following environments: (i) 1M NaCl (pH=4) solution, deaerated, and (ii) 0.3% Na_2SO_4 neutral solution. Polarization resistance (R_p) measurements, potentiostatic I(t) and potentiodynamic polarization curves were recorded, using an AMEL (Milan) Metalloscan apparatus, at a voltage scanning rate of 720 mV/h.

The data from electrochemical measurements in 1M NaCl solution are summarized in Table 1. Free corrosion potential (E_{corr} and corrosion current (i_{corr})values are reported. The latter are obtained from Tafel lines (with slopes $ß_a$, $ß_c$) and R_p measurements, as described in [24]. For the sake of comparison, the values for pure Fe and for BN-coated samples [20] and for Fe/B multilayers implanted with N^+ [21] are also included. We first point out the noble free corrosion potential attained by the RIBAD boron nitride.

The corrosion current values observed are interesting if compared to the other values reported in Table 1. Indeed, they are about 2 orders of magnitude lower than that of pure Fe and about 1 order of magnitude lower than that of BN-coated Fe samples [10].The electrochemical parameters E_{corr} and i_{corr} clearly show the nobility and scarce reactivity of the nitride surface layers produced by RIBAD. Observation of the surface morphology of the samples after anodic polarization tests confirms indeed the above considerations.

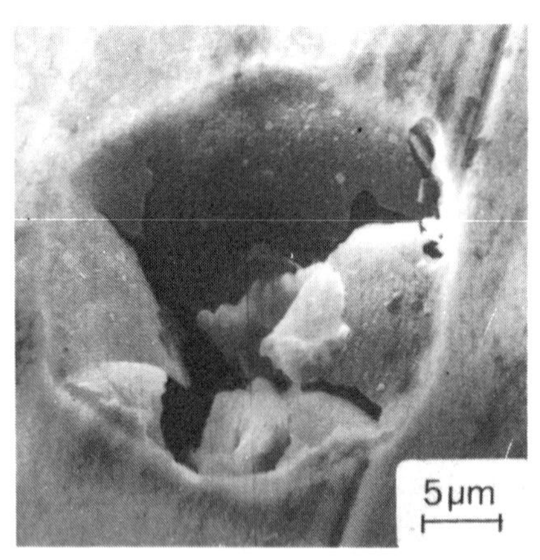

Fig. 7 - Scanning electron micrograph of a pit in otherwise chemically inert BN coatings obtained by RIBAD on Fe.

The corrosive attack is localized and several pits are present in some zones where the coating is defective, Fig. 7. The low reactivity and the cathodic behaviour of the BN layer is highlighted by the presence of undamaged coating partially covering the holes. An interesting feature of the pits is their "cubic" symmetry, which may give further evidence of the crystallographic structure of the boron nitride coating. Moreover, from a careful examination of the coating fracture in the pits, it is possible to learn something about the mechanical properties of the coating, which appears to be very hard and brittle.

R_p measurements were also performed in a 0.3% Na_2SO_4 solution as a function of time to study the behaviour of the material in a neutral environment (Table 2). All the samples show a better initial behaviour than iron, particularly the RIBAD-BN coatings, which remain effective over a long period. The fact that the surface attained by the RIBAD process is more homogeneous and compact than that obtained by single implantation or deposit confirms this.

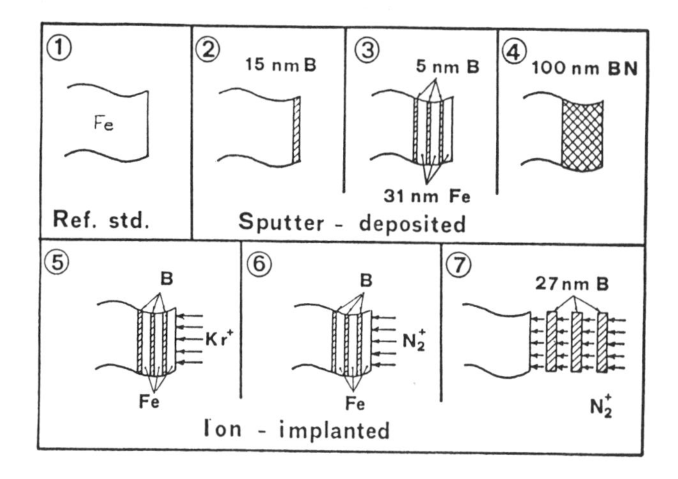

Fig. 8 - Scheme of the samples used in the oxidation study.

4. OXIDATION INHIBITION OF BORON-DEPOSITED AND ION-IMPLANTED IRON

The improvement in the oxidation resistance obtained by boron introduction in iron and iron alloys is now well established [25, 26]. Also several ion implantation studies have been performed to clarify the reasons for this behaviour [27, 28]. B^+-implanted iron shows excellent oxidation resistance at T< 600° C [26]. The lack of solubility of boron into iron and consequent segregation to grain boundaries and other fast diffusion paths is thought to inhibit the oxidation. In order to extend the protection to higher temperatures, however, it may be necessary to turn to other mechanisms such as blocking by particle agglomeration or by a diffusion barrier like BN.

BN appears to be extremely interesting for high temperature applications, in fact it is not easily oxidized , at least up to 1000° C. The inhibition resulting from

N^+ implantation across boron deposits is substantial [29] but the mechanisms involved need further clarification. It could result from the presence of both the BN barrier layer and the B/Fe ion-beam-mixed layer.

Here we report oxidation data of iron specimens treated by different deposition and implantation techniques, including reactive ion beam assisted deposition, in order to discriminate between the possible contributions from the respective layers.

Also the high temperature oxidation protection of titanium alloys with BN is currently being investigated [30] and has already shown promising results.

4.1. Experimental procedures

The following specimens were considered: iron coated with boron or BN, Fe/B/Fe/B... multilayers deposited onto iron, and analogous samples implanted with Kr^+ or N_2^+ ions at an energy of 160 keV.

Ion-assisted-coatings were also obtained on Fe using 4 sequential steps each consisting of 27 nm B deposition followed by N_2^+ implantation to a dose of $1.4x10^{17}$ N atoms/cm^2 [31].

All the samples used in this investigation were in the form of foils treated on both sides according to the procedures schematically described in Fig. 8.

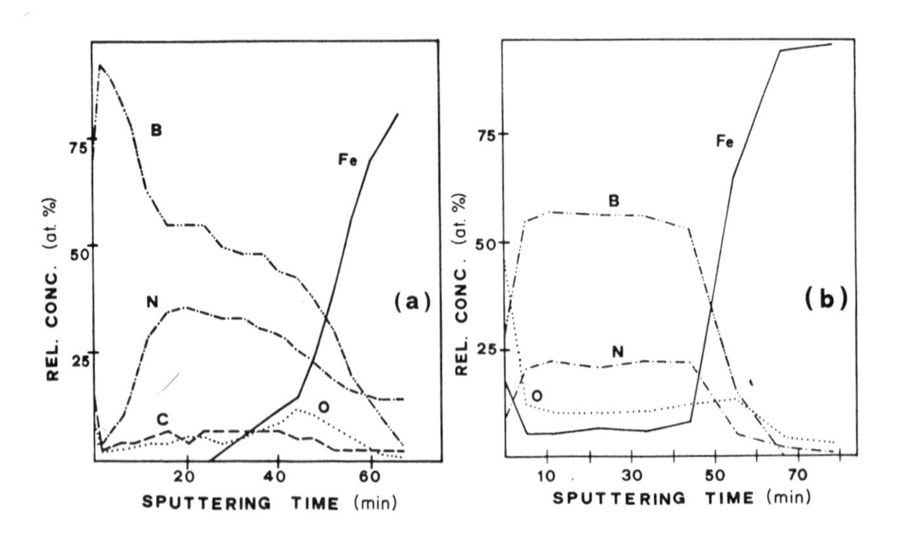

Fig. 9 - Auger depth concentration profiles of: (a) RIBAD obtained BN-on-Fe; (b) the same after oxidation at 500° C.

The specimens were cut in 1 cm x 0.5 cm pieces and suspended in a recrystallized alumina reaction tube which was part of a Mettler thermobalance apparatus. Oxidation was performed in air flowing at 10 l/hour and the specimen attained the test temperature (500° C, 650° C) with a rate of 10° C/min. At the end of the run, the specimens were allowed to cool at the same rate. Mass changes during oxidation were monitored continuously with an accuracy of ±0.1 mg, and checked after the oxidation with an analytical Mettler balance (accuracy, ±0.01 mg); loss of oxide during cooling was only observed occasionally.

4.2. Characterization of boron- and BN-coatings on iron

The B and BN deposited on iron substrates did not show any appreciable number of defects, the films being very fine grained and coherent. The RIBAD-obtained films also appeared to be homogeneous. Their Auger depth concentration profiles, shown in Fig. 9a, are very similar to those of previously studied samples, deposited with a single B layer and implanted with 100 keV N^+ ions, in which a buried BN layer is formed with highly enhanced adhesion to the substrate [8]. However, in the present case the uniformity of the RIBAD samples was better than that of the previous samples.

The SIMS spectra of the RIBAD-obtained films indicate the predominant formation of BN with some evidence of FeB, Fe_2B, FeN, Fe_2N and Fe_3N.

4.3. Oxygen uptake data

Plots of mass gain vs. time for unimplanted iron and for the different boron-treated samples during oxidation at 500° C and 650° C are presented in Fig.10. The kinetics can be fitted generally by two parabolic regimes: the first is valid for short oxidation times of less than 3h, while the second applies for times in excess of 6h.

In all cases, oxidation inhibition is observed. Thus even in the case of pure iron (sample 1) a single parabolic law is not observed and this is attributed to contamination in the oxidation chamber. In fact, SIMS analysis was able to find boron traces at the surface of oxidized iron.

Boron addition in the form of Fe/B multilayers on iron (sample 3) gives rise to an improvement visible only after reaction for a few hours. The oxidation is then no longer parabolic, but shows a slower dependence than $t^{1/2}$ on time, this being an indication that boron atoms enter progressively into the process. On the other hand, boron deposited on the surface of iron (sample 2) is immediately effective against oxidation. A substantial reduction in the oxidation rate at 500° C is also observed for the Kr^+-mixed surfaces (sample 5) as well as for the N^+-bombarded ones (sample 6). At this temperature the effectiveness of both implants is similar. The RIBAD-obtained BN-coatings (sample 7) behave even better, and are almost comparable with the sputter- deposited BN coatings (sample 4) which only react with oxygen within the measurement errors.

At 650° C the situation changes radically; the surface boron-layer (sample 2) is clearly inadequate, while the Kr^+-mixed boron-layer (sample 5) appears to be far less efficient than the iron-boron multilayer (sample 3), which probably is converted into an alloy layer at this temperature. A radiation damage effect can probably explain this trend inversion. For the N^+-implanted multilayer (sample 6), the BN phase is still very effective at this temperature. However, a coherent BN barrier layer (sample 7 or 4) performs better than a simple dispersed phase.

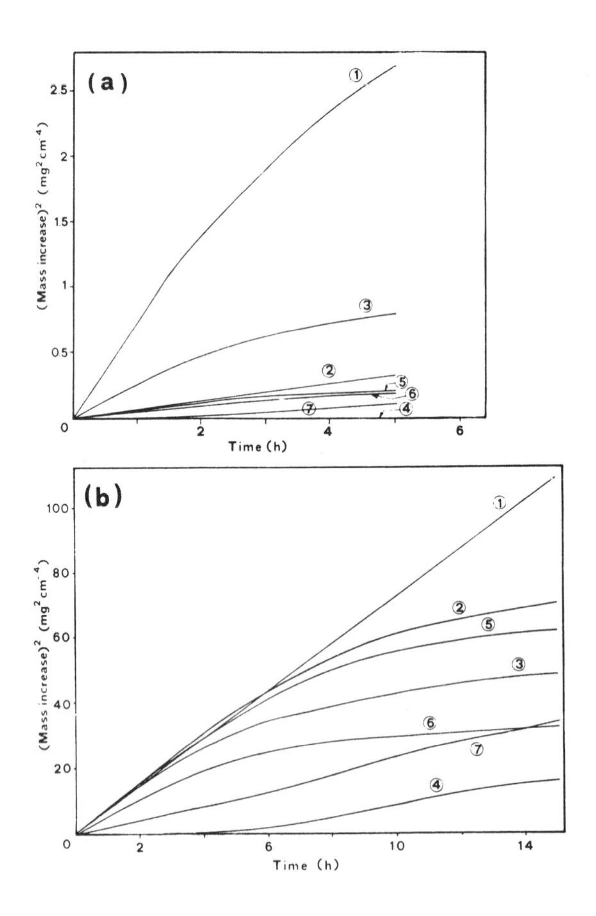

Fig. 10 - Square of the mass increase vs. time for samples 1-7, during oxidation in air at (a) 500° C and (b) 650° C. The samples are specified by the numbers on the curves (according to figure 8).

4.4. Oxide composition and structure

The samples oxidized at 500° C were also examined with SIMS and AES spectroscopies. A summary of the SIMS results can be found in Table 3.

Iron is present at the surface of all oxidized samples, as confirmed by AES. In sample 6 a 1.1 nm thick oxide scale is present; only a little quantity of boron is left at the surface while nitrogen is no more detected. In sample 7 (Fig. 9b), the boron rich BN layer still exists and only the near surface region appears oxidized. Nevertheless, it is important to note that, in spite of the presence of oxygen, the Auger lineshape analysis never showed any evidence of oxidized iron at the BN/Fe interface.

XRD analysis of the different oxidized samples, summarized in Table 4, only reveal a mixture of magnetite (Fe_3O_4) and haematite (Fe_2O_3) except for sample 4, where no oxide is present after oxidation at 500° C, in good agreement with the mass gain data. At 500° C, we observe clearly two different trends. In the first,in samples 3, 5 and 6 (the multilayer-coated series) haematite predominates, which suggests that the ion-induced phases act as a barrier against cationic diffusion. In the second group, in samples 2, 4 and 7 (B- or BN-coated samples) the inhibition mechanism derives from the sacrificial oxidation of boron. At 650° C, the barrier mechanisms still act for all samples but are less effective.

TABLE 3

Species present as found by SIMS.

Sample # (as in fig. 8)	*Prior to oxidation*	*After oxidation* (5h at 500° C)
1	Fe	FeO, Fe_2O,Fe_2O_3
5	Kr, FeB, Fe_2B	Kr, FeO, Fe_2O, B_2O_3, FeO . B_2O_3, Fe_2O_3
6	BN, FeB, Fe_2N, Fe_3N	BN, FeO, Fe_2O, B_2O_3, FeO . B_2O_3
7	FeB, Fe_2B, BN, FeN, Fe_2N, Fe_3N	FeB, Fe_2B, BN, B_2O_3 FeO . B_2O_3

TABLE 4

Peak intensities of phases present after oxidation at 500 and 650° C

Sample # (as in fig. 8)	*500°C* Fe_2O_3	Fe_3O_4	*650°C* Fe_2O_3	Fe_3O_4
1	w	s	m	s
2	m	s	m	m
3	m	w	m	m
4	-	-	m	m
5	m	w	m	s
6	m	w	m	s
7	m	m	m	s

s, strong; m, medium; w, weak.

Nitrogen ion implantation of boron deposits is extremely effective in inhibiting oxidation at 500° C and higher. BN formation is thought to be responsible for the significant effect observed. The presence of second-phase particles exercises some influence on the motion of defects across the scale. The greatest inhibition is provided by a continuous BN layer which provides a diffusion barrier. The magnitude of its influence depends on the coherency of the BN layer and also on its adhesion, which can be highly enhanced if the coating is produced by an ion-assisted technique.

5. MECHANICAL PROPERTIES OF BORON NITRIDE COATINGS

BN films were produced on hard metal (WC-Co) substrates using:(i) r.f. magnetron sputtering from a BN target and (ii) ion beam assisted deposition of boron. The deposition conditions are summarized in Table 5.

The films were characterized by SEM, AES, microhardness, friction and scratch-tests in order to study the influence of the deposition process and parameters on the mechanical properties of the films [32].

TABLE 5

Deposition conditions for BN films on WC-Co

Technique	*r.f. magnetron sputtering*	*RIBAD*
Target material	BN	B
Base pressure (Pa)	$5x10^{-5}$	$3x10^{-6}$
Sputtering atmosphere	Ar	-
Ar pressure (Pa)	0.3÷0.8	-
Substrate material	WC-Co	WC-Co
Substrate potential	grounded	grounded
Substrate temperature (°C)	100÷400	Room temp.
r.f. film growth rate (nm/s)	0.1	-
B deposition rate (RIBAD) (nm/s)	-	0.1
Coating thickness (μm)	≈1	≈0.3

5.1. Preparation of BN coatings

All our WC-Co samples, characterized by a Knoop microhardness value of 2600 (measured at a load of 25 gf) were mechanically polished to a mirror-like finish (3 μm diamond powder in the final step).

Some samples were prepared in an Alcatel machine by sputtering in Ar atmosphere, using a round BN target 10 cm in diameter. The samples were placed on a stage capable of being heated to 400° C. Variations in the Ar pressure (from 0.3 to 0.8 Pa) as well as in the substrate temperature (from 100 to 400° C) were made in order to study the variation of the mechanical properties of BN films.

The remaining samples were coated using reactive ion beam assisted deposition (RIBAD). BN was synthesized by using two different and independently controlled beams, one from a low energy (30 keV) horizontal ion implanter, the other from an electron beam evaporator installed in the implantation chamber.

The RIBAD process started with a N_2^+-implantation at a dose between 1 and $4x10^{17}$ N atoms/cm^2 intended to sputter clean the substrate surface and to create favourable sites for deposition. As an additional advantage, this procedure results in a strengthening of the WC-Co substrate [7]. The compound was then produced by first depositing a boron layer with a thickness of 27 nm. This value matched the projected range of the ions to be implanted, thus allowing good mixing with the underlying layer, resulting in enhanced adhesion. In order to achieve a stoichiometric composition we then implanted at a N dose of $3x10^{17}$ atoms/cm^2 equal to the expected number of atoms present in the deposited film. This dose may have to be corrected for the sputter losses, which have been evaluated to be less than

10%. To obtain the final coating thickness of about 300 nm, ten such steps (deposition followed by implantation) were made. It was not possible to have thicker films due to the limited volume of the electron-beam-gun crucibles used in this experiment.

5.2. Surface Microanalyses and Mechanical Tests

The microstructure of the samples was examined by SEM. The surface is rather smooth, and the film cross section shows a compact structure, Fig. 11.

Depth-composition profiles were obtained by AES combined with 2 keV Ar^+ sputter etching. The film composition was nearly stoichiometric all across the film thickness for the sputtered samples, whereas three clearly defined regions are present in the case of ion-beam-assisted coatings, see Section 4.2. The depth profiles indicate the formation of a BN layer just below a surface layer of "untransformed" B and just above a broad ion-mixed interface which should ensure a highly enhanced adherence. The coatings show a yellowish hue, similarly to other cubic nitrides which are golden in colour. XRD data are indicative of a structure at least partially cubic; this is in agreement with the high microhardness values observed.

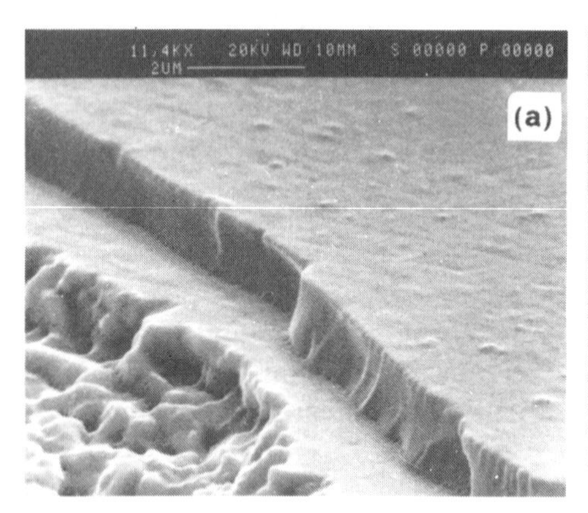

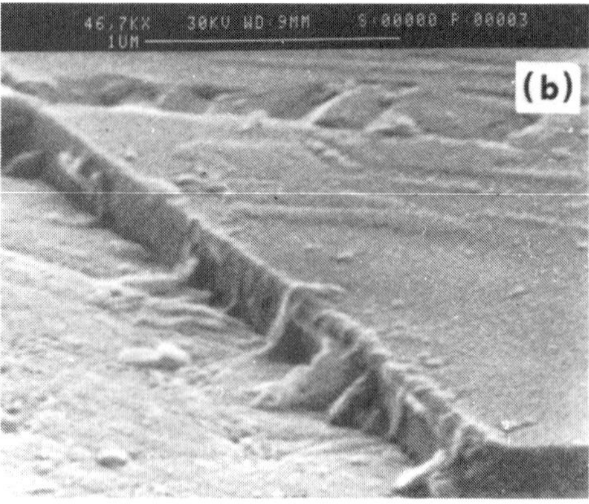

Fig. 11 - Scanning electron micrographs showing the fracture surface of (a) a r.f. magnetron-sputtered BN coating and (b) a RIBAD BN coating on WC-Co substrates.

5.3. Hardness testing

The BN coated samples obtained by r.f. sputtering were hardness tested, using a Knoop indenter with 25 gf load, giving an indent depth between 250 and 400 nm, i.e. well within the BN film (thickness ~1 μm). Figure 12 shows the hardness as a function of substrate tempebrature, for an Ar pressure of 0.5 Pa in the chamber. Six indenter readings were taken for each point on the graphs, which represent average values together with standard deviations. No strong dependence of hardness on substrate temperature was found. Figure 13 shows hardness as a function of Ar pressure in the vacuum chamber for a substrate temperature of 300° C. In this case an optimum pressure was found at 0.7 Pa, giving the highest hardness values.

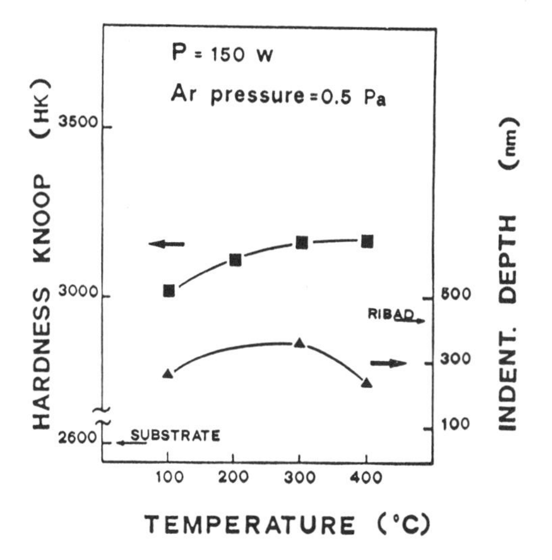

Fig. 12 - Knoop microhardness (squares) as a function of substrate temperature for r.f. magnetron-sputtered BN coatings. The indentation depths for scratches at a load of 2N are also shown (triangles).

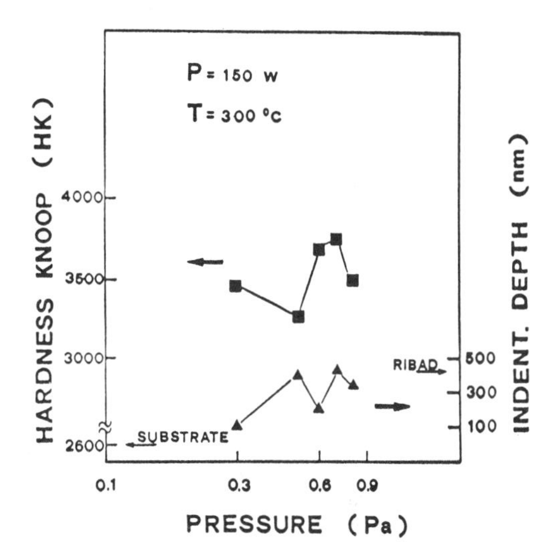

Fig. 13 - Knoop microhardness (squares) as a function of argon pressure in the vacuun chamber for r.f. magnetron-sputtered BN coatings. The indentation depths for scratches at a load of 2N are also shown (triangles).

5.4. Scratch- and friction testing

In the scratch-tests a loaded stylus was drawn repeatedly across the surface of the coated samples. The stylus radius was 200 μm and the following loads were applied: 2, 3, 4, 6, 8, 10 and 20 N. The depth of the scratches was measured by means of a profilometer; the data taken at very low loads (2 N) were found to simulate correctly the microhardness data (see Figs. 12 and 13) and have the advantage of not requiring the optical measurement of small indents. Moreover, the critical load for film detaching was deduced from optical microscopy as well as by an acoustic emission signal recorded during the scratch-test.

The most obvious difference between the sputtered and the RIBAD-obtained coatings is that, while a critical load L_c can definitely be measured in the former case (where the film suddenly detaches from the substrate), in the latter no such decohesion is present even if the substrate is heavily deformed at sufficiently high loads.

Figures 14 and 15 show both L_c and the friction coefficient μ as a function of substrate temperature and Ar pressure, respectively; they are a useful indication of the optimum deposition parameters: 300° C and 0.7 Pa. A minimum friction coefficient μ=0.05 is found, to be compared with the values 0.15 - 0.30 obtained for other samples.

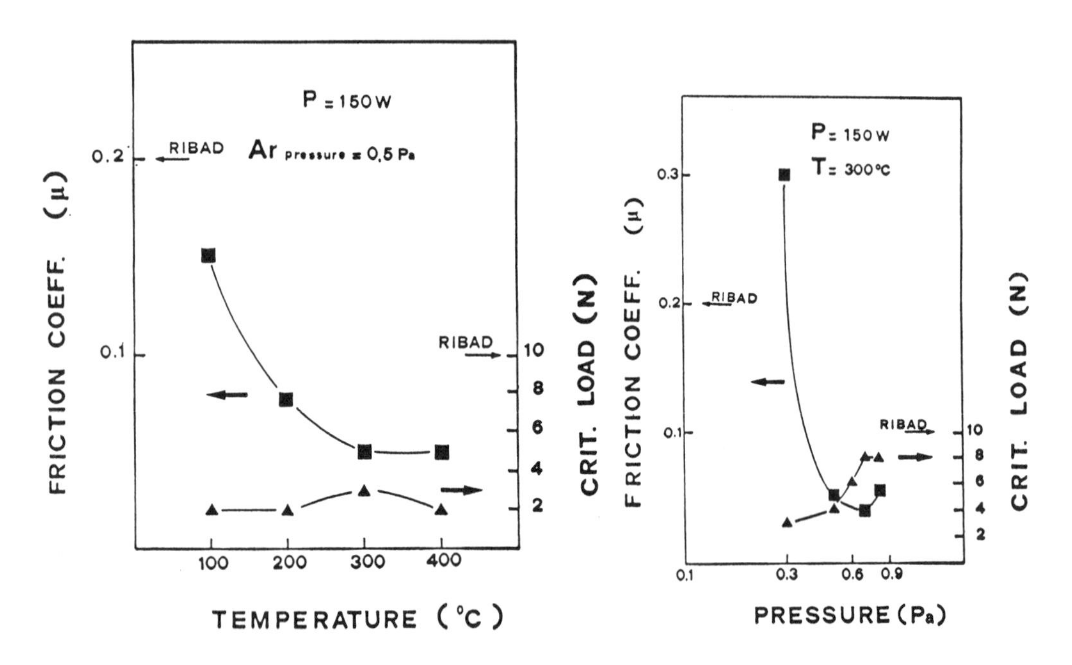

Fig. 14 - Friction coefficient (measured at a normal load of 2N; squares) and critical load (triangles) for film detachment from the substrate as a function of substrate temperature for samples obtained by r.f. sputtering. The values for a RIBAD film are also shown for comparison by arrows along ordinate axes.

Fig. 15 - Friction coefficient (measured at a normal load of 2N; squares) and critical load (triangles) for film detachment from the substrate as a function of argon pressure in the chamber for samples obtained by r.f. sputtering. The values of a RIBAD film are also shown for comparison.

5.5. Discussion

Both r.f. magnetron sputtering and RIBAD techniques are able to produce BN hard coatings. In the case of r.f. sputtered samples, hardness improvement with respect to the substrate is significant as values obtained are higher than 3500 HK. This very good value is confirmed by the very small scratch depths measured by means of profilometry. Moreover, the friction coefficients μ measured during the scratch tests were found to be of the order of 0.05, i.e. six times lower than the corresponding μ values of the substrates against the same diamond stylus. A possible explanation for this marked improvement may be the observed enhanced film roughness, which may reduce the contact area between the stylus and the coating. However these μ values are not especially low as could be expected on the basis of chemical inertness and hardness of BN. This complicated matter deserves further study in order to ascertain the role of the various mechanisms involved [32]. The critical load for film detaching shows a maximum value in concomitance with the best hardness and friction coefficient values. In fact, one of the most critical aspects of the coatings is their adhesion which is clearly related to the deposition technique.

Summarizing, we can indicate the best parameters for obtaining hard BN films by r.f. sputtering in the given system as 0.7 Pa argon pressure and 300° C substrate temperature during deposition. Moreover, the adhesion is considerably enhanced by ion-beam-assisted deposition, due to the absence of a sharp interface, as evidenced by the AES depth analysis. In this case BN coatings are observed still adherent even after substrate failure.

6. CONCLUSIONS

Various metallic substrates were coated with BN using different techniques. The hybrid technique (RIBAD) combines some of the best features of both deposition and implantation to produce BN layers with a high control of microstructure and of properties. Homogeneity of the formed films and their adherence to the substrate are noticeably improved with respect to those of samples obtained by conventional implantation or deposition alone. Relatively thick surface layers can be grown, depending on the deposition and implantation rates. Since the surface region is selectively treated these coatings can be used as finising steps, allowing surfaces to be engineered for optimum properties independently from bulk property requirements.

The BN films obtained are suitable for use in many metallurgical hard coating applications and particularly as protective coatings against high temperature oxidation of the base materials. The coatings are also chemically inert and thus corrosion resistant in different aqueous environments.

ACKNOWLEDGEMENT

The authors gratefully acknowledge the collaboration of their colleagues at I.R.S.T. and at the Department of Engineering, University of Trento.

REFERENCES

[1] Weissmantel, C.: J. Vac. Sci. Technol.,1981, 18, 179
[2] Weissmantel, C., K. Bewilogua, K. Breuer, D. Dietrich, U. Ebersbach, H.-J. Erler, B. Rau and G. Reisse: Thin Solid Films, 1982, 96, 31
[3] Shanfield, S. and R. Wolfson: J. Vac. Sci. Technol.,1983, 41, 323
[4] Dearnaley, G.: Thin Solid Films 1983, 107, 315
[5] Weissmantel, C.: Proc. of 3rd. Int. Conf. on Solid Surfaces (Vienna, 1977) p. 1533.
[6] Weissmantel, C. Le Vide et les Couches Minces 1986, 41, 45
[7] Dearnaley, G., P.D. Goode, F.J. Minter, A.T. Peacook and C.N. Waddell: J. Vac Sci. Technol, 1985, A3, 2684
[8] Guzman, L., F. Marchetti, L. Calliari, I. Scotoni and F. Ferrari: Thin Solid Films, 1984, 117, L63
[9] Joyner, D.J. and D.M. Hercules, J. Chem. Phys.:1980, 72, 1095
[10] Davis, L.E., N.C. Macdonald, P.W. Palmberg, G.E. Riach and R.E. Weber: Handbook of Auger Electron Spectroscopy, Phys. El. Div., Perkin Elmer, (Eden Prairie MN, 1978)
[11] Stulen, R.H. and R. Bastasz: J. Vac. Sci. Technol.,1979, 16, 940
[12] Singh, A., R.A. Lessard and E.J. Knystautas: Mater. Sci. Engineer. 1987, 90, 173
[13] Guglielmi, M., A. Oliani and S. Tosto: Appl. Surf. Sci., 1982, 10, 466
[14] Ertl, G. and K. Wandelt: Surf. Sci., 1975, 50, 479
[15] Calliari, L., F. Giacomozzi, L. Guzman, B. Margesin and P.M. Ossi, in NATO A.S.I. Series: "Erosion and Growth of Solids Stimulated by Ion and Atom Beams", G. Kiriakidis, G. Carter and J.L. Whitton Eds., (M. Nijhoff, Dordrecht, The Netherlands 1986), p. 316
[16] Guzman, L., B. Margesin, V. Zanini and F. Giacomozzi: Suppl. to Le Vide et Les Couches Minces 1987, 235, 285

[17] Guzman, L., F. Giacomozzi, B. Margesin, L. Calliari, L. Fedrizzi, P.M. Ossi and M. Scotoni: Mater. Sci. Engineer. 1987, 90, 349
[18] Guzman, L.: Advanced Materials & Manufacturing Processes: 1988, 3, 279
[19] Bonora, P.L., G. Cerisola, L. Fedrizzi and C. Tosello: Mater. Sci. Engineer. 1985, 69, 283
[20] Marchetti, F., L. Fedrizzi, F. Giacomozzi, L. Guzman and A. Borgese, Mater. Sci. Engineer., 1985, 69, 289.
[21] Elena, M., L. Fedrizzi, V. Zanini, M. Sarkar, L. Guzman and P.L. Bonora: Nucl. Instr. Methods, 1987, B19/20, 247
[22] Bonora, P.L., G. Cerisola, C. Tosello and S. Tosto, in "Ion Implantation into Metals", eds. V. Ashworth et al. (Pergamon, Oxford, 1982) p.1
[23] Fedrizzi, L., L. Guzman, P.L. Bonora and G. Cerisola: Surface & Coatings Technology, 1988, 35, 221
[24] Fontana, M.G. and N.D. Greene, "Corrosion Engineering", (Mc Graw-Hill, N.Y., 1978) p. 342
[25] Lea, C.: Metal Science, May 1979, p. 301
[26] Pons,M., M. Caillet and A. Galerie: Matériaux et Techniques, Dec. 1985, p. 699
[27] Pons, M., M.Caillet and A. Galerie: Nucl. Instr. Methods, 1983, 209/210, 1011
[28] Galerie, A., M. Caillet and M. Pons: Mater. Sci. Engineer., 1985, 69, 329
[29] Giacomozzi, F., L. Guzman, A. Molinari, A. Tomasi, E. Voltolini and L.M. Gratton: Mater. Sci. Engineer., 1985, 69, 341
[30] Elena, M., L. Guzman and A. Tomasi, work in progress
[31] Giacomozzi, F., L. Guzman, F. Marchetti, A. Molinari, M. Sarkar and A. Tomasi: Mater. Sci. Engineer., 1987, 90, 197
[32] Elena, M., L. Guzman and S. Gialanella: Surface Coatings & Technology, 1989, in press
[33] Rabinowicz, E., "Friction and Wear of Materials", J. Wiley & Sons, 1965, p. 82.

Materials Science Forum Vols. 54 & 55 (1990) pp. 45-70

COATING FILM FORMATION OF BORON NITRIDE BY MEANS OF DYNAMIC MIXING METHOD

F. Fujimoto
Institute of Scientific and Industrial Research
Osaka University, 8-1 Mihogaoka, Ibaraki
Osaka 567, Japan

ABSTRACT

One method of the coating film formation, the dynamic mixing or IVD one, using an ion beam of an element and simultaneous vapor deposition of another one, is introduced. Boron nitride films are prepared by nitrogen ions with the energy of 200 eV~ 40 keV and evaporation of boron. Their properties are studied by a transmission electron microscope (TEM), X-ray photoelectron spectroscopy (XPS), X-ray diffraction and infrared absorption spectroscopy (IR).

For samples formed by 25~ 40 keV ions, films consist of mixed phases of cubic (c-), wurtzite (w-), hexagonal (h-) boron nitride and amorphous phase. Percentages of the c-BN phase in films are not very high. However, grains of the c-BN phase with column and plate-like structures are observed by TEM and the ⟨ 100 ⟩ axis of c-BN is normal to the film surface. The w-BN phase is also observed as grains with column structure.

As the energy of nitrogen ions decreases, the ratio of c-BN in films increases. Films prepared by ions with lower energy than 5 keV show clearly a line due to c-BN in IR spectra. In X-ray diffraction patterns for films formed with lower energy than 10 keV, the diffraction peak from the (111) reflection of c-BN can be observed. Moreover, the diffraction patterns of films prepared at certain deposition rate by using 500 eV and 1 keV nitrogen ions show only a peak due to c-BN.

Vickers hardness of films with about 1μ m thickness made at 3 and 10 kV is about 1/2~ 1/3 of c-BN. Besides those at 1 kV and 500 V can not be tested, because the adhesion with a substrate is too weak for the films to be peeled off during the testing.

1. INTRODUCTION

Boron nitride has various kinds of crystal structures which are similar to those of carbon. Their crystalline phases are classified as cubic (zincblend, c-BN), wurtzite (w-BN), hexagonal (layer, h-BN), amorphous (a-BN) and rhombohedral (r-BN). The phase h-BN has similar structure and properties to the graphite, and is well-known as "white graphite". The c-BN phase is distinguished by its hardness and high thermo-conductivity which are very similar to that of diamond. Furthermore, c-BN is more stable at high temperatures than diamond and non-reactive with iron. These characters indicate that c-BN shows promising as a coating material for tools and machines. Besides c-BN is a kind of Ⅲ-V compound and has a large band gap; 8 eV. Accordingly, application of the c-BN coating film to a compound semiconductor with a wide range of operating temperatures can be considered.

The phase of c-BN does not exist in nature and is usually synthesized under high pressure and high temperature as well as in the case of diamond. Recently various methods have been attempted to prepare the coating film of c-BN, such as; the ion enhanced deposition [1], the ion beam deposition [2], the reactive sputtering [3] and the CVD [4] (chemical vapor deposition) like the case of diamond [5]. The principle of these methods is in how highly atoms of boron and nitrogen are excited. As a method to give rise to excited atoms, the condensation of energetic ionized particles can be considered. In this chapter, an ion-enhanced method for metastable film preparations, which was developed by the present author and collaborators and is called the dynamic mixing or the IVD (ion and vapor deposition) method, will be introduced and properties of the prepared boron-nitride coating films will be dealt with.

2. DYNAMIC MIXING METHOD (IVD METHOD)

As a method of surface modification, ion implantation is well-known and is widely used in the semiconductor industries in order to introduce various impurity elements into silicon and compound semiconductors. In this case the accelerating voltage of ions is usually 200 kV or higher and the amount of introduced impurity atoms is much smaller than it is for the case of the surface modification. For surface modification, the number of implanted atoms must be much higher than 10^{17} atoms/cm^2 because they form compounds with target atoms. Accordingly, if we use a conventional ion implantor for the surface modification, a long-term or high current implantation is necessary. In the latter case, the temperature increase cannot be avoided. Moreover, the thickness of the modified surface layer is limited by the range of the implanted ions, the density distribution of the implanted atoms in the layer is not uniform and a modified layer which does not include the substrate elements such as a boron nitride layer on steel cannot be prepared. However, the adhesion of the layers with the substrate is much stronger than those prepared by the electrochemical, vacuum evaporation, sputtering and chemical vapor deposition methods.

In order to utilize the characteristic of the ion implantation and to avoid its faults, the dynamic mixing method (or IVD method) has been developed [6]. This method is applied not only to the preparation of nitride films such as aluminum- [6,7], boron- [8,9,10,11], titanium- [12,13,14,15,16] and molybdenum [17], but also to the coating film formation of metal [18] and diamond [19]. The principle is that one element of the modified layer is evaporated on the substrate in vacuum and another one is simultaneously implanted as ions. The energy of ions is in the range of 200 eV ~ 40 keV which is much lower than the case of ion implantation. Thus it becomes easy to obtain an ion source with a high current.

2.1. APPARATUS

The first apparatus for surface modification by the dynamic mixing method consists of a PIG ion source, an evaporator with an electron gun and current and film thickness monitors [6] . The accelerating voltage was 10~ 40 kV and the maximum ion current was 1 mA. The mass of ions from the ion source was separated by an analyzing magnet and the ion beam converged by a lens system. The bombardment rate of ions was obtained by measuring the ion beam current. The deposition rate of the evaporated element was estimated by a thickness monitor. The bombarded area at the specimen holder was 10 cm^2. The second machine [18] was constructed for more practical use than the first one. The schematic diagram of the machine is shown in Fig. 1. The machine consists of a rectangular bucket type multiaperture ion source, an evaporator with an electron gun, a rotary specimen holder, film thickness and current monitors and a vacuum system. The temperature of the substrate was controlled by water cooling or electric heating of the specimen holder.

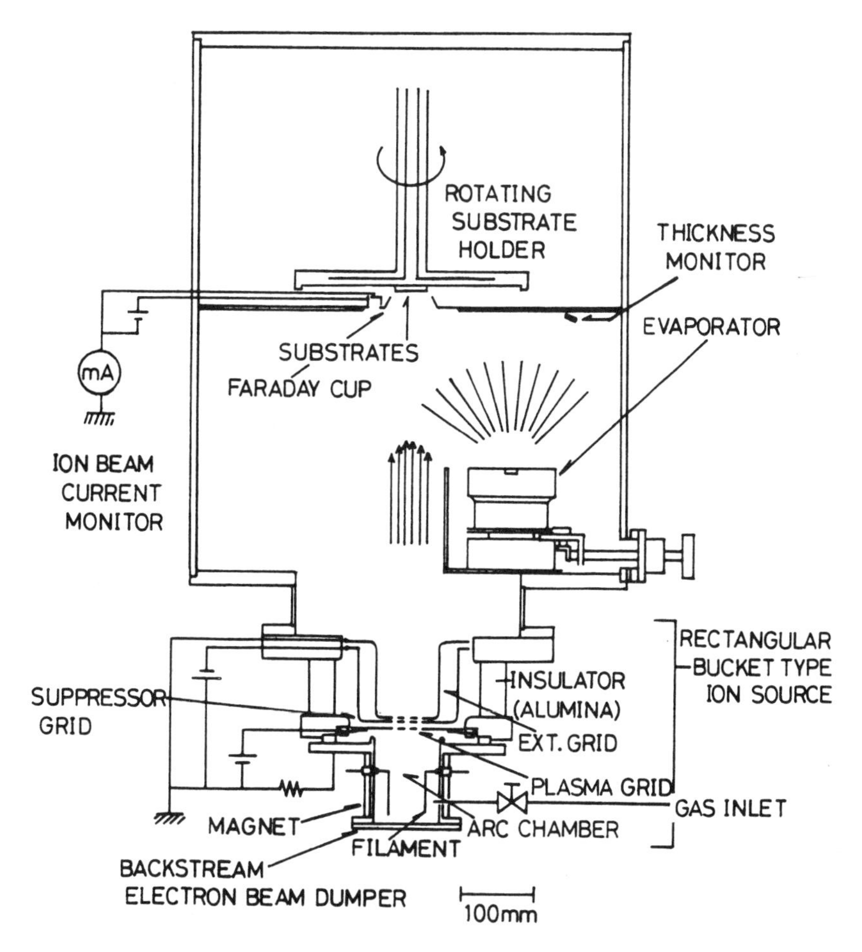

Fig. 1. A schematic diagram of the dynamic mixing (IVD) machine.

Performances are as follows,

1) Ion beam energy : 2 ~ 40 keV;
2) Maximum ion beam current : 100 mA for nitrogen ions;
3) Beam uniformity : smaller than ± 10% variation within 40 cm^2;
4) Base pressure of order of 10^{-7} Torr.

The machine shown in Fig. 1 have no analyzing magnet to bombard the ion beam on a large area of the substrate. The numerical ratio of atomic and molecular ions for nitrogen was nearly unity. After the second machine, various types of machine are used, such as a machine with the low accelerating voltage of 100 ~ 1000 V [19] and that with the high current of 500 mA [20] . These machines don't have the analyzing magnet like the second one.

2.2 PROPERTIES OF FILMS

Films prepared by the present method have various advantages such as (1) very strong adhesion between the film and the substrate, (2) shorter preparation time than the case of ion implantation, (3) the formation of a thick coating film, (4) the film formation at low temperature, (5) easy control of film components and (6) the formation of film composed of elements which are not included in the substrate. It was often observed that the films consisted of crystalline with a definite orientation, even if the substrate was polycrystallines, and the crystal orientation varies with the preparation condition and the ion beam direction [13,14,

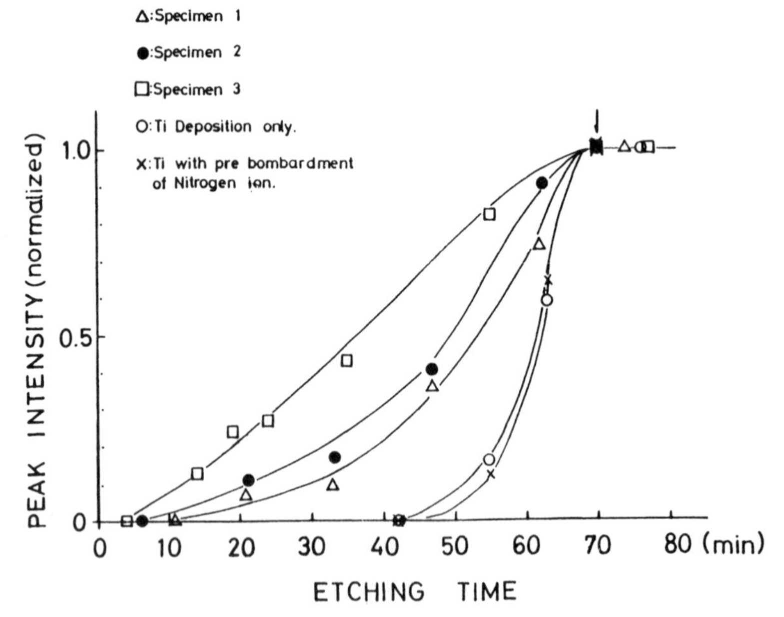

Fig. 2. Intensity variations of Fe-2p peak from stainless steel substrates observed in spectra of XPS from titanium nitride films with the sputtering time. Preparation conditions of specimens 1(△), 2(●) and 3(□) are shown in Table 1. Circles (○) and crosses (×) are shown results from samples of titanium films prepared by evaporation only and after bombardment with nitrogen ions, respectively, as references. Thickness of all films is about 1000 Å.

Table 1 Preparation condition of TiN films in Fig. 2

Sample No.	Evaporation rate of Ti (Å/s)	Current density of N ion beam (μ A/cm²)	Accelerating Voltage of ion beam (kV)
1	1.8	180	10
2	1.0	88	20
3	0.6	61	30

16,17,21,22] . The strong adhesion between the film and the substrate arises from the formation of an intermixed layer composed of elements from the target, evaporated and implanted materials at the interface.

In Fig. 2, the distributions of iron atoms from the substrate of stainless steel in the intermixed layers between the substrates and the films of titanium nitride are shown [12] . Here these films were prepared by the evaporation of titanium and the simultaneous bombardment of nitrogen ions with the energies of 10, 20 and 30 keV. The film thickness was about 1000 Å for all samples. The preparation condition of these titanium nitride films is given in Table 1 where nitrogen ion beams have the same power for different energy. The substrate temperature in these cases was about 300 ℃ . Distributions of iron atoms in three films in Fig. 2 were obtained by measuring the intensity variations of the Fe-2p electrons at 711.5 eV in the Al-Kα X-ray-induced photoelectron spectra (XPS) with the sputtering time by an argon ion beam. The energy and current of argon ions were 5 keV and 0.5 mA/cm^2 , respectively, and the beam was impinged at 45° to the substrate normal. The abscissa in Fig. 2 is the sputtering time which is approximately proportional to the etching depth. After the sputtering time of 70 min, the sputtering yield of iron atoms became constant. As references, observations on samples with titanium coating films prepared by evaporation only and after prebombardment with nitrogen ions are shown.

From Fig. 2, we can estimate the thickness of the intermixed layer. By defining the thickness where the intensity of the Fe peak varies from 20 to 80% of that at the substrate and by correcting the knock-on and surface roughness effects which are observed from two reference samples by means of the square sum rules, the intermixed layer thickness are found to be 440, 330 and 300 A, respectively, for specimens prepared by 30, 20 and 10 keV nitrogen ions.

In Fig. 3 (a) and (b), pictures of a scratch test for titanium nitride films prepared by the CVD method and the dynamic mixing one of a 30 keV nitrogen ion beam, respectively [23] . In each case, the film thickness was 2000 Å. The film prepared by the CVD method is cracked and stripped off from the substrate, while that by the dynamic mixing method is fixed and only a scratch trace can be seen.

The strong adhesion of films with substrates formed by the present method shown in Fig. 3 can be seen when the ion energy is higher than a few keV. However, if a thick film is prepared by lower energy than 1 keV, the film often peels off from the substrate as well as those prepared by other methods.

The crystallization of films has a close relation with the formation of the intermixed layer. A TEM (transmission electron

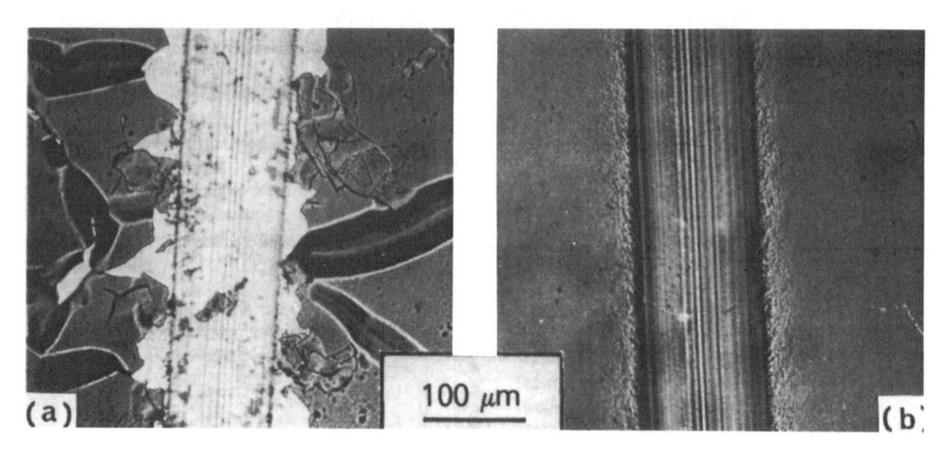

Fig. 3. Optical microgrphs of deformation produced during adhesion testing. Pictures show the case of a film prepared by the CVD method (a) and by the dynamic mixing one (b).

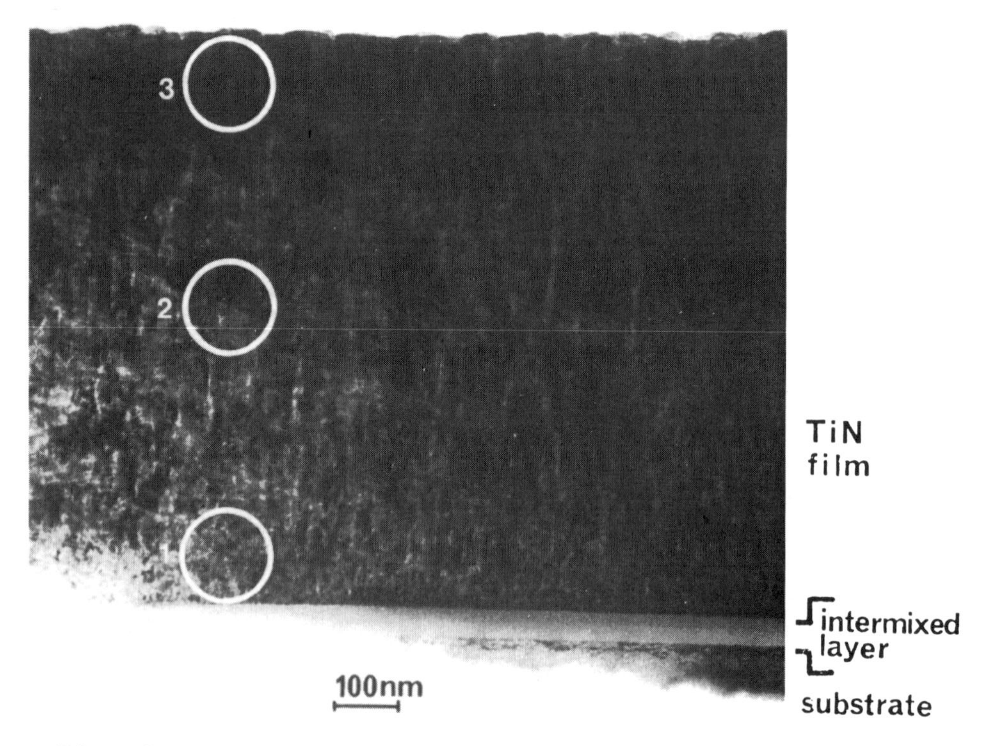

Fig. 4. Cross-sectional profile of TiN films on a silicn wafer observed by TEM.

microscopy) picture shown in Fig. 4 represents a cross section of a titanium nitride film prepared on a silicon wafer by a 30 keV

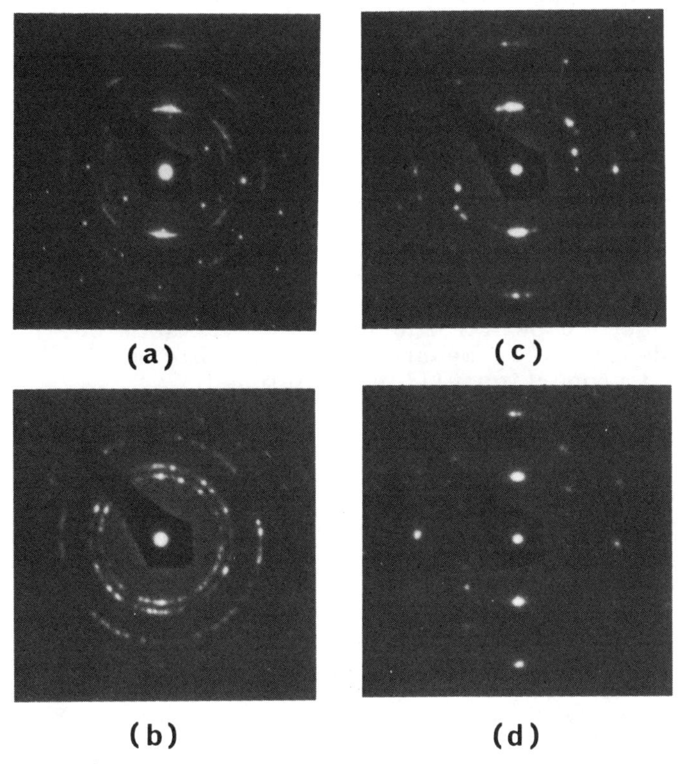

Fig. 5. Diffraction patterns obtained from parts indicatd in Fig. 4; (a) from total area, (b) from the part (1), (c) from the part (2) and (d) from the part (3).

nitrogen ion beam perpendicular to the substrate [13,14] . The film thickness is about 1μ m. We can clearly see the intermixed layer with the thickness of about 400 Å at the interface. This value agrees well with the result shown in Fig. 2. A high resolution image indicates that the structure of the intermixed layer is amorphous. Fig. 4 shows that the film has the column structure and the diameter of each column distributed in the range between a few hundred to over one thousand angstroms. The column direction which is that of the crystal growth, is normal to the substrate surface and parallel to that of the ion beam direction. Diffraction patterns at each part of the film can be seen in Fig. 5(a) ~ (d). These patterns explain that the film is polycrystal-like at the early stage of the growth and becomes single crystal-like with the progress of growth. The prepared film is the TiN crystal and its surface normal is the ⟨ 111 ⟩ direction.

In most cases, films prepared by the present method are crystallized and the direction of the crystallization is independent of substrate conditions such as crystal, polycrystal or amorphous materials because of the presence of the amorphous intermixed layer [14] . However, the crystal orientation depends on preparation conditions: ion beam energy [24] , the energy deposition per an evaporate atom [16] , the substrate temperature [17] and the beam direction [14] . Various problems concerning crystallization can not be explained and are important subjects to be studied in the future.

3. BORON NITRIDE FILMS

The main motivation for the boron nitride film formation is to prepare a c-BN film with its prominent properties already mentioned. However, coating films formed by these various methods have complicated structures which are not only the mixed phase of c-, w-, h-, and a-BN but also those not identified by each crystalline phase. The lattice constants for each phase of boron nitride are a = 3.62 Å for c-BN, a = 2.55 Å and c = 4.20 Å for w-BN and a = 2.50 Å and c = 6.66 Å for h-BN. In this section, preparation of boron nitride films by means of the dynamic mixing method, their structure and properties will be mentioned.

3.1. PREPARATION OF FILMS

Films were produced by the evaporation of boron and simultaneous irradiation of nitrogen ions on various kinds of substrate. Substrates were selected according to methods that diagnose films such as cleaved rocksalt for TEM observations, titanium plates for measurement of the electric resistance, aluminum foils for RBS (Rutherford backscattering spectroscopy) analysis and silicon wafers for IR (infrared absorption spectra), X-ray diffraction and XPS studies. The vacuum pumping system consisted of a turbomolecular pump and an oil rotary pump. The base and operating pressures were less than 5×10^{-7} Torr and $3\sim 8 \times 10^{-5}$ Torr, respectively. Purities of nitrogen gas for an ion beam and of metallic boron were 99.999 and 99.7 %, respectively.

The ion beam energy was widely changed from 200 eV to 40 keV. Beam currents at high and low energies were lower than 200 μ A/cm^2 and 500μ A/cm^2, respectively. The temperature of the substrates was lower than 300 ℃, although that was not always measured. The arrival ratio of boron and nitrogen atoms on substrates was varied in the range between B/N = 0.5~ 3. The arrival ratio is not the same as that of the composition, because the elements which composed films, mainly boron, were sputtered by the bombardment of nitrogen ions and implanted nitrogen atoms were sometimes diffused out from the film surface.

The composition ratio of films was measured by RBS spectra of 2 MeV protons from films on aluminum foils, comparing those from polycrystalline h-BN [9] . The other method used to obtain the composition ratio was to calculated from the area ratio of B-1s1/2 and N-1s1/2 peaks in XPS spectra considering the sensitivity of the detector, escape depth of photoelectrons and the photoionization cross section [10,11] .

Fig. 6 shows a variation of the XPS signal from B-1s1/2 state for the films prepared by the nitrogen ion beam of 3 keV and 40 μ A/cm^2 with a boron deposition rate 1.5~ 6 Å/s [10] . These spectra were obtained using ESCA 750 after 5 min etching by a 2 keV Ar ion beam. In the case that the boron deposition rate is 1.5 Å/s, the XPS peak is at about 190.3 eV. This value agrees with the value from boron bound with nitrogen [25] . However, the peak position gradually shifts to the low energy side as the boron deposition rate increases from 1.5 to 6 Å/s. Each peak can be separated into two Gaussian curves with their peak positions at 190.2 ~ 190.3 eV

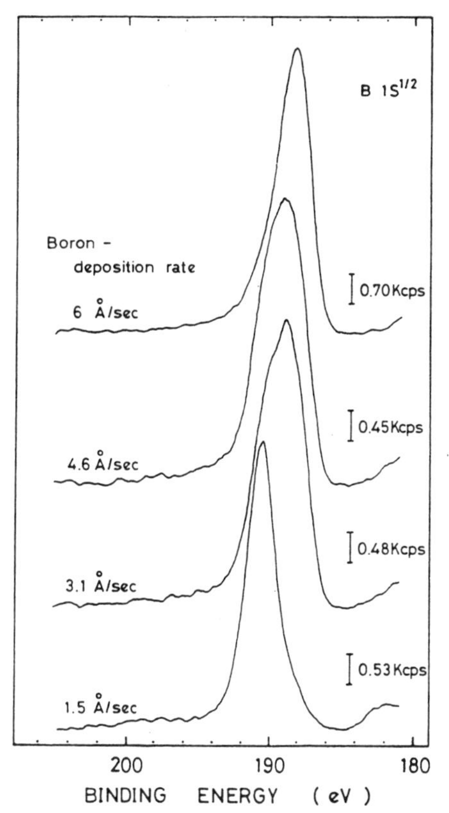

Fig. 6. XPS spectra of the B-1s1/2 state electrons for films prepared at 3 kV. Ion current dnesity was about 40 μ A/cm^2 and the deposition rate of boron 1.5 ~ 6 Å/s.

and 188.0 ~ 188.5 eV. The latter value of the peak position agrees with metallic boron. Concerning nitrogen, XPS signal in each film has a peak at about 398.5 eV.

Fig. 7 shows changes of the composition ratio B/N calculated from the area ratio of total boron and nitrogen peaks in XPS spectra (open circles) and the B/N ratio for boron bound with nitrogen (crosses) [10] . The figure indicates that the B/N ratio for total boron atoms linearly increases with the boron deposition rate, while that for bound boron is about unity. This fact indicates that all of the nitrogen atoms in the films bind with those of boron and the excess content of boron remains as metallic boron.

Similar measurement was carried out for films prepared by lower energy ions than 1 keV and the maximum current of 500μ A/cm^2 [11] . Fig. 8 shows that the concentration of boron increases linearly with the transport (arrival) ratio in the region B/N> 1 similar to the case of higher energy shown in Fig. 7. However, the composition saturates in the range of nitrogen-rich transport condition (B/N< 1), which is different from the high energy case. It seems that this disagreement is due to the difference in ion penetration depth. That is, irradiated nitrogen ions diffuse out from the near-surface because the penetration depth of low energy ion beams is very shallow. The problem of whether the implanted

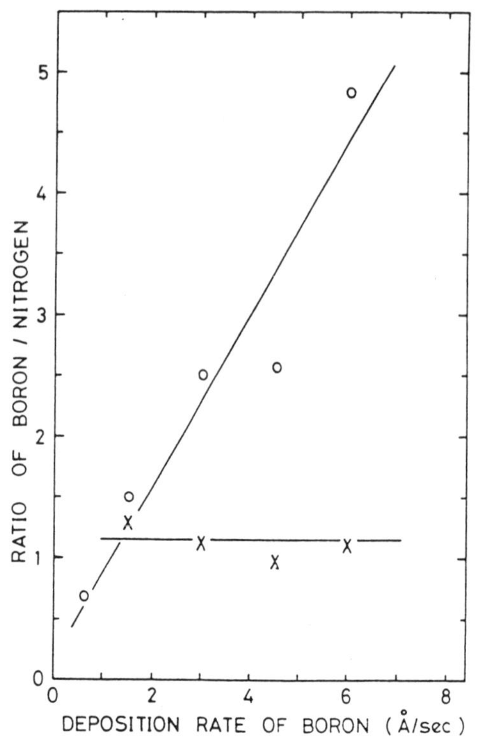

Fig. 7. Variation of the composition ratio of B/N calculated from the area ratio of each peak in XPS spectra with the deposition rate of boron. Open circles mean B/N ratios for all of boron and crosses for the boron bound with nitrogen.

atoms stay inside or diffuse out from the target closely correlates with that of the adhesion between the film and the substrate.

3.2. STRUCTURES OF FILMS

The structure of films was studied by means of TEM, X-ray diffraction and IR spectroscopy.

3.2.1. TEM STUDIES

The study by means of TEM and the microelectron diffraction is the best method to know the morphology and micro-structure of films. Samples for TEM were prepared by the first apparatus [9] . That is, N_2^+ ions with their energy 25 ~ 35 keV were bombarded on cleaved rocksalts and aluminum foils with 0.7 μm thickness and boron was simultaneously evaporated. The film thickness was 1000 ~ 6000 Å. Films formed on rocksalts were floated off on water and observed by a 200 kV electron microscope. Samples deposited on the aluminum foils were used to measure the composition ratio of film B/N by observing the RBS spectra of a 2 MeV proton beam. In Fig. 9 and Fig. 10, RBS spectra from the sample on a aluminum foil and from polycrystalline h-BN, which has the composition ratio is B/N = 1, are shown, respectively. The ratio of step heights for each

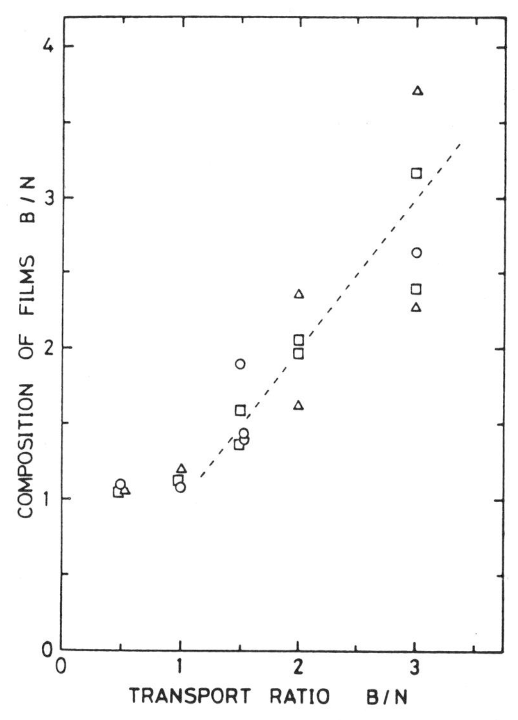

Fig. 8. Variation of the composition ratio with transport ratio of deposited boron atoms and irradiated nitrogen ions; ○ : 200 eV, △ : 500 eV and □ : 1000 eV.

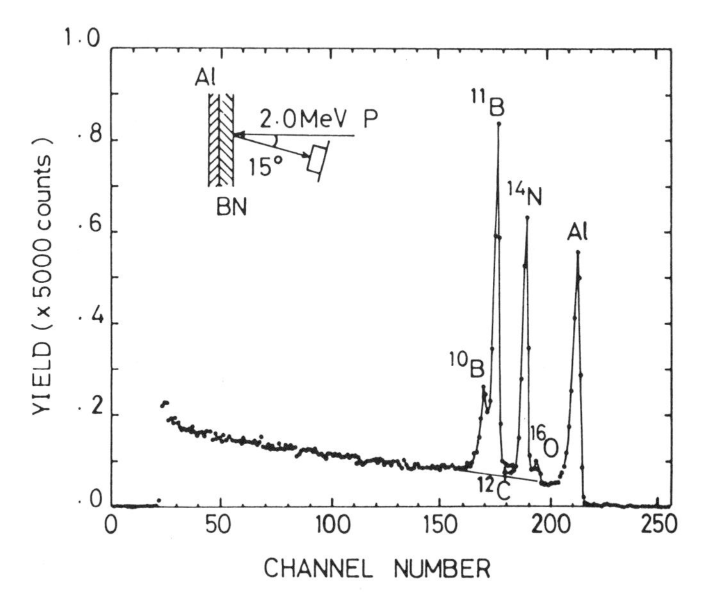

Fig. 9. RBS spectrum of 2.0 MeV protons from a BN sample with the composition ratio B/N = 0.75 on an aluminum foil.

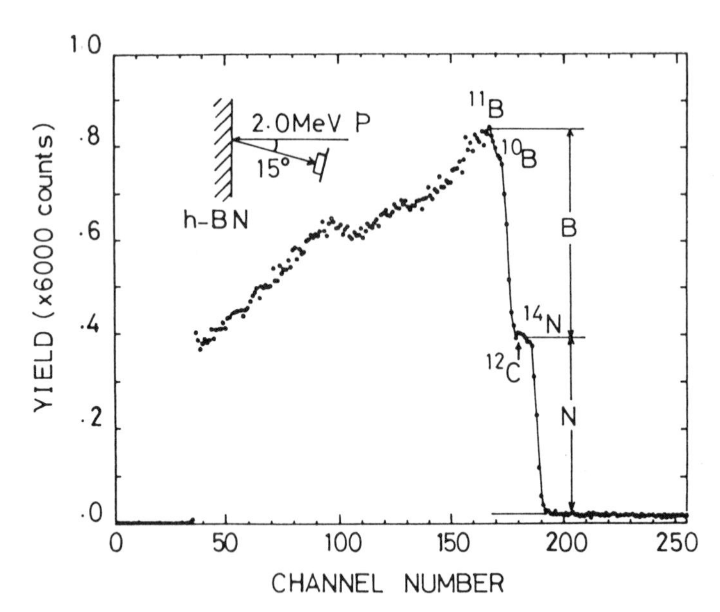

Fig. 10. RBS spectrum of 2.0 MeV protons from polycrystalline h-BN.

element is not the same value as the ratio calculated by using a formula of the Rutherford backscattering. This is the case because the nuclear reaction between film elements and proton affects on the spectra. However, the concentration ratio of the film can be obtained by comparing peak areas for each element in Fig. 9 with step heights in Fig. 10. On samples formed by 40 keV nitrogen ions, we did not identify the composition ratio, but the arrival ratio was indicated.

Figs. 11 and 12 show electron micrographs and their diffraction patterns. Composition ratios and ion beam energies are indicated in figure captions. In general, films prepared by ion beam energies in the region 25 ~ 35 keV can be morphologically classified into the three groups shown in Figs. 11(a), (b), and (c). Films belonging to a group shown in Fig. 11(a) are usually smooth and some of them are very uniform. The diffraction pattern consists of three fairly sharp rings which correspond to the lattice spacing 4.0, 2.4 and 1.4 A. These values suggest that films of this group have a structure close to c-BN and that the plane corresponding to (111) of c-BN is parallel to the film surface.

A group observed in Fig. 11(b) has a granular or a fiber structure and its diffraction pattern consists of fine rings and a broad background. It seems that the structure of films belonging to this group is a mixed phase of h-BN and a-BN. The third one shown in Fig. 11(c) is humpy and the diffraction patterns are quite diffuse. Atomic distances estimated from these diffuse rings suggest that the film is a-BN with the structure close to h-BN.

A sample shown in Fig. 12 consists of grains with the sizes equal to or smaller than 3000 Å. Their diffraction patterns indicate that these grains are crystallines with a definitely coherent structure. Films with a structure similar to those seen in Fig. 12 are very often observed in cases of titanium [13,14,21, 23] , aluminum [6] and molybdenum [17] nitrides in which grains

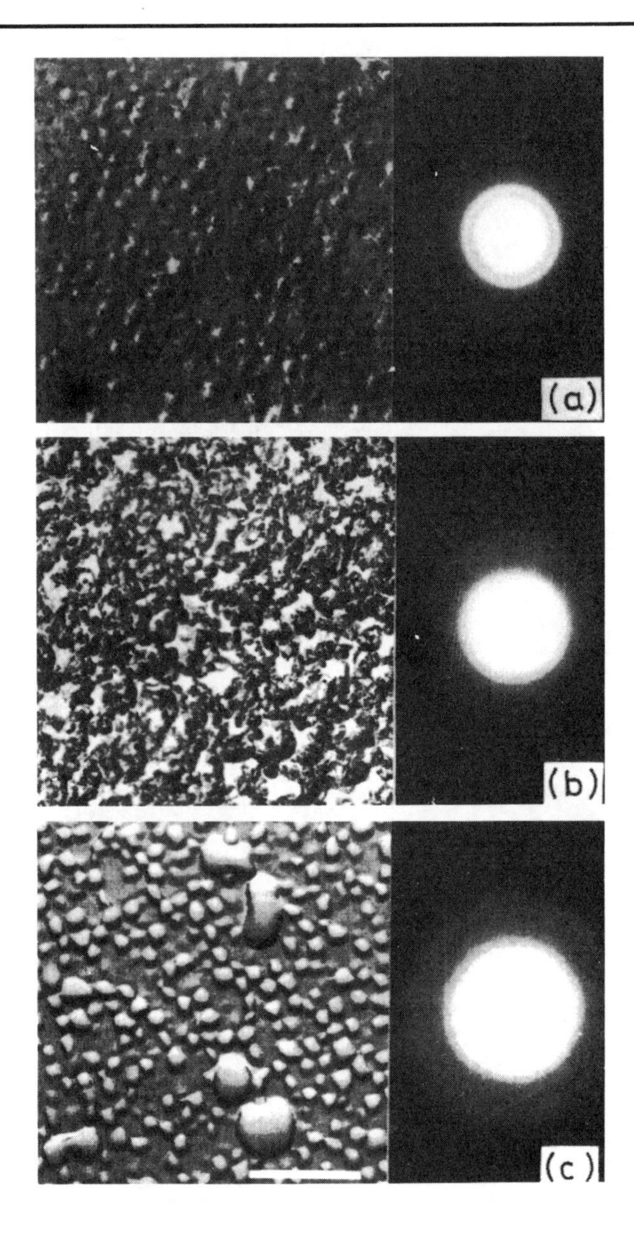

Fig. 11. Electron micrographs and their diffraction patterns. Composition ratio and ion beam energies are, (a) B/N = 1.1, 35 keV, (b) B/N = 0.95, 35 keV and (c) B/N = 0.95, 30 keV. The scale marker shown in the lowest micrograph = 2μ m.

have the column structure as shown in Fig. 4. The crystalline phase of the films seen in Figs. 12(a) and (b) is w-BN. The film of Fig. 12(a) partly includes the twin and polycrystallines and grains in Fig. 12(b) has the c-axis perpendicular to the surface. The structure shown in Fig. 12(c) is that of c-BN. Concerning the c-BN structure, we sometimes observed plate-like and rarely dentrite-like grains as seen in Figs. 13(a), (c) and (b), respec-

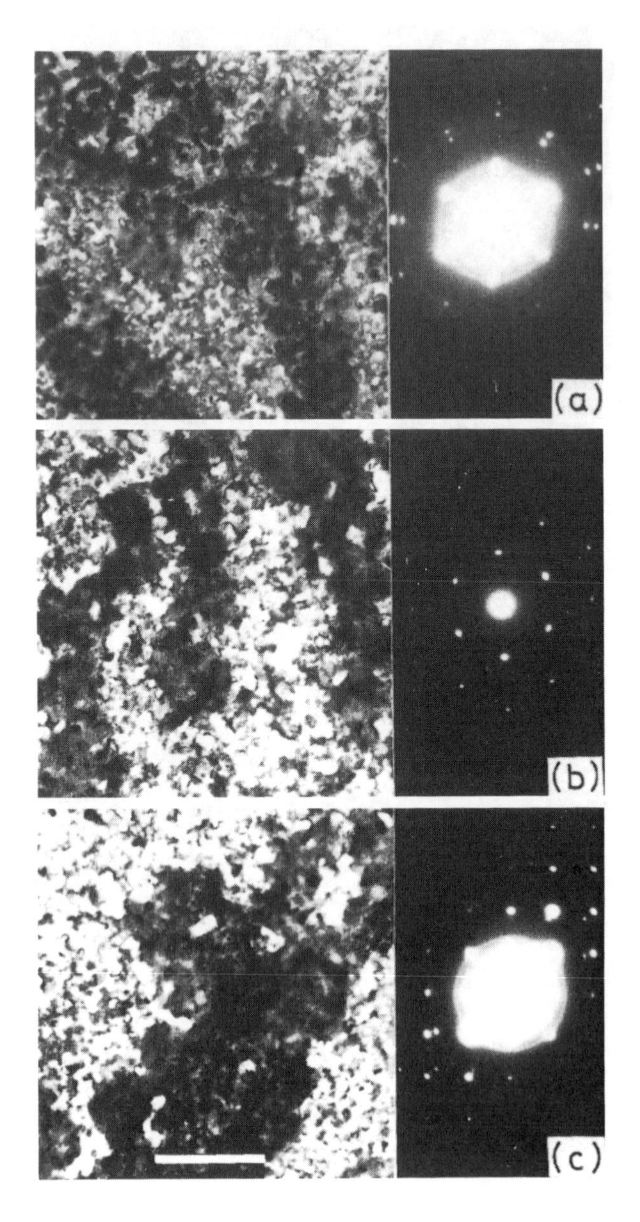

Fig. 12. Electron micrographs and their diffraction patterns of samples with B/N = 0.50 and prepared at 30 kV. (a) and (b) w-BN and (c) cubic. The scale marker shown in the lowest micrograph = 2 μm.

tively, in addition to the column structure, and some of them reaches a size of 10 μm [8] . The surface of plates is the (100) plane in most cases. The samples in Fig. 13 were prepared by 40 keV N_2^+ ion beam from PIG ion source (first machine). The plate-like c-BN grains were observed even in samples formed by nitrogen ion energy of 30 keV.

We confirmed that films formed by nitrogen ion beam of 25 ~

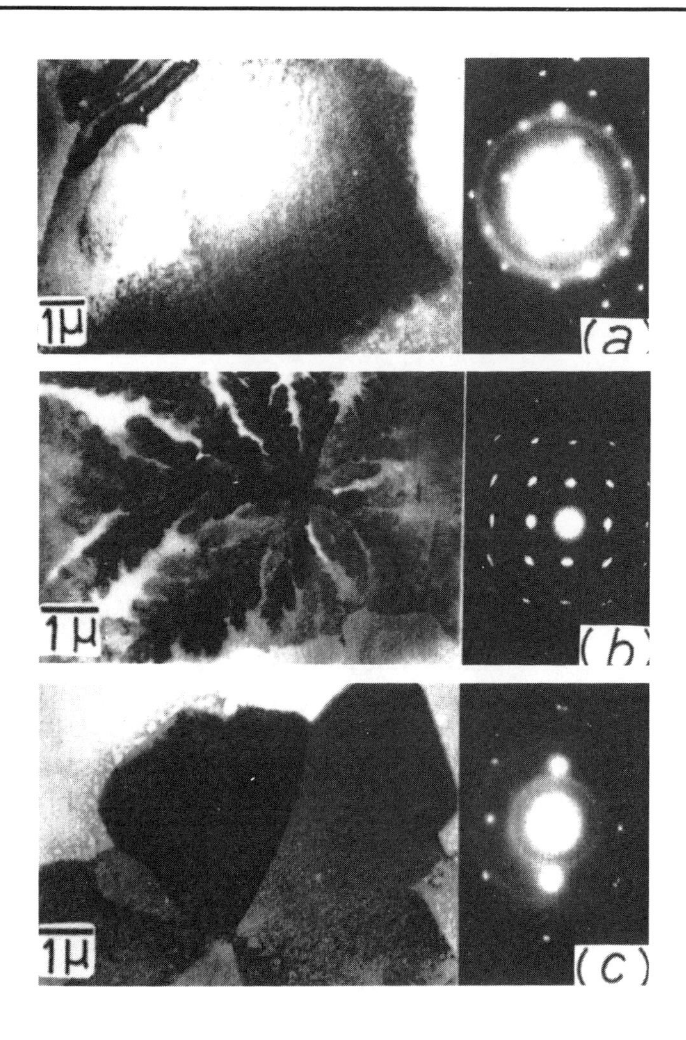

Fig. 13. Electron micrographs and their diffraction patterns of samples prepared with the deposition ratio B/N = 2.5 and at 40 keV N_2^+ ions. (a) a grain with the surface of (100) plane of c-BN, (b) dentrite-like grains of c-BN and (a) a grain with the surface of (110) plane of c-BN.

40 keV consist of a mixed phase of cubic, wurtzite, hexagonal and amorphous. However, results of the X-ray diffraction and IR absorption spectra measurements indicate that the percentage of c-BN phase in films is quite low as we can see in later sections.

3.2.2. X-RAY DIFFRACTION STUDIES

Samples manufactured by nitrogen ions with the energy of 25 ~ 35 keV were studied by the θ -2θ method of Cu-Kα X-ray diffraction for the surface, and those prepared by ions under 25 keV were analyzed by an X-ray diffractometer using Cu-Kα X-rays where the measurements of X-ray diffraction patterns were performed with an incident angle of the X-rays at 6.0° for the film surface by using

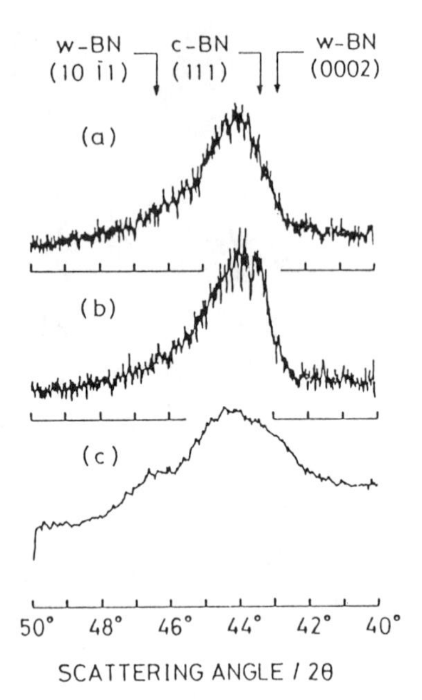

Fig. 14. X-ray diffractometer scans of boron nitride films prepared by 35 keV N_2^+ ions. Composition ratio of films are (a) B/N = 1.0, (b) and (c) B/N = 1.5.

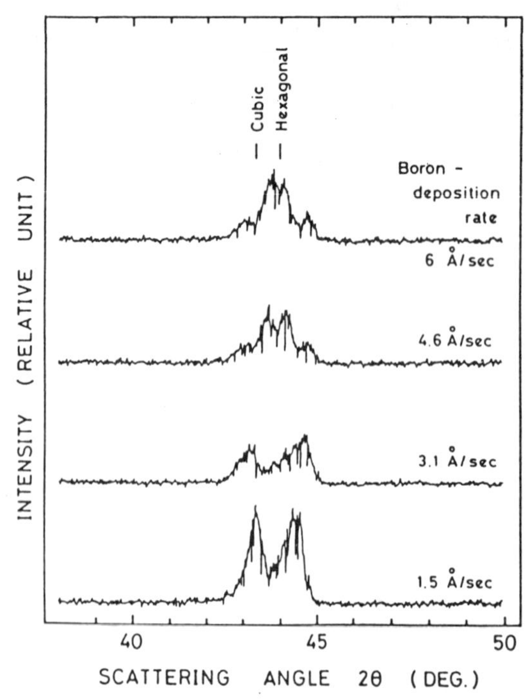

Fig. 15. X-ray diffraction patterns obtained from films prepared at 3 keV.

thin film attachment.

X-ray diffractometer scans of BN films formed at high energy ion beams between 25 ~ 35 keV are shown in Fig. 14 where only a broad peak can be observed in the vicinity of 44° [9] . This broad peak corresponds to the (111) reflection of c-BN (43.3°), the (0002) and ($10\bar{1}1$) ones of w-BN (42.8° 46.3°). The main peak at 44° is considered to be the broad peak from amorphous boron nitride. However, weak reflections from the w-BN phase can be seen in Figs. 14(a) and (c) and the reflection from the c-BN one in Fig. 14(b). It should be noted that no peak near 26.8 ° , which is the position of the (0002) reflection of h-BN, appears.

X-ray diffraction patterns can be obtained from quite a wide area of films, unlike the case of electron micrographs. The above observations suggest that films made by high energy ions (25 ~ 40 keV) composed mainly of amorphous and polycrystalline h-BN and the area of c- and w-BN phases is small.

For films prepared at low energy, a situation of their structure is very different from those at high energy. A variation of X-ray diffraction patterns in the range from 40° to 50° obtained from the films prepared at 3 kV and 40 μ A/cm^2 with the boron deposition rate is shown in Fig.15 [10] . In the case of a deposition rate of 1.5 Å/s, where this condition could yield a composition near to the stoichiometric one, the pattern consists of two dominant peak at 43.3° and 44.3° . The former angle agrees with the scattering angle from the (111) reflection of c-BN. By increasing the boron deposition rate, the peak at 43.3° decreases, while that in the vicinity of about 44° splits into a number of

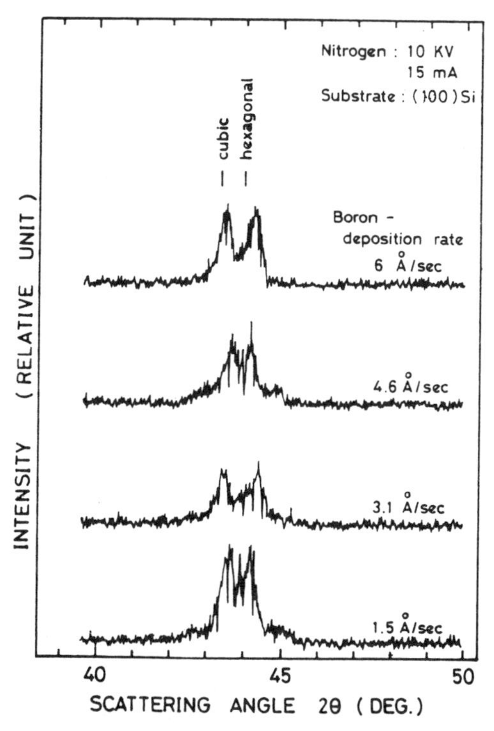

Fig. 16. X-ray diffraction patterns obtained from films prepared at 10 kV.

peaks. On thc films prepared at 10 keV, the diffraction patterns shown in Fig. 16 consist of two dominant peaks, where one of the two peaks approximately agrees with the scattering angle of c-BN and the other with that from h-BN [10] . However, the diffraction peaks in Figs. 15 and 16 are slightly shifted from any of the peak positions for the four kinds of boron nitride system; cubic, wurtzite, hexagonal and rhombohedral. Therefore, it seems that c-BN can be formed using the present method, while in the films excess damage is caused by high energy ion bombardment, and the crystal lattice becomes somewhat distorted.

Diffraction patterns formed by further low energy ion irradiation are shown in Figs. 17 and 18. The former and the latter samples were prepared by 500 eV and 1 keV ion bombardments, respectively, and the ion current was 2 mA/cm^2 in both cases [11] . The diffraction patterns of films formed by 500 eV ions at the transport (arrival) ratio B/N = 0.75 show a clear peak at 2θ = 43.4° which agrees with the (111) reflection of c-BN. The patterns of other films prepared at 500 eV shows two peaks at about 43° and 43.8° , which seem to be due to wurtzite and hexagonal structures, respectively. On the other hand, the films formed with 1 keV ion irradiation also show a peak at 2θ = 43.4° . For both conditions of ion energy, variation of X-ray diffraction patterns with the transport ratio shows similar tendency. The appearance of two peaks shown in Figs. 17 and 18 was also found in Fig. 16. However,

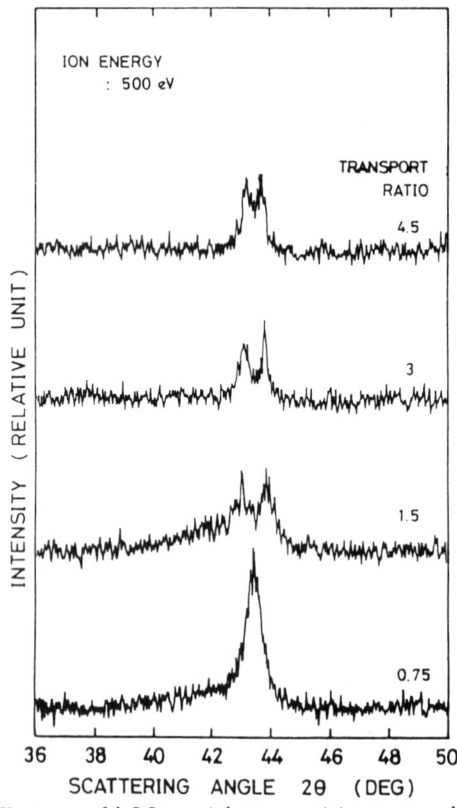

Fig. 17. X-ray diffraction patterns obtained from films prepared at 500 V.

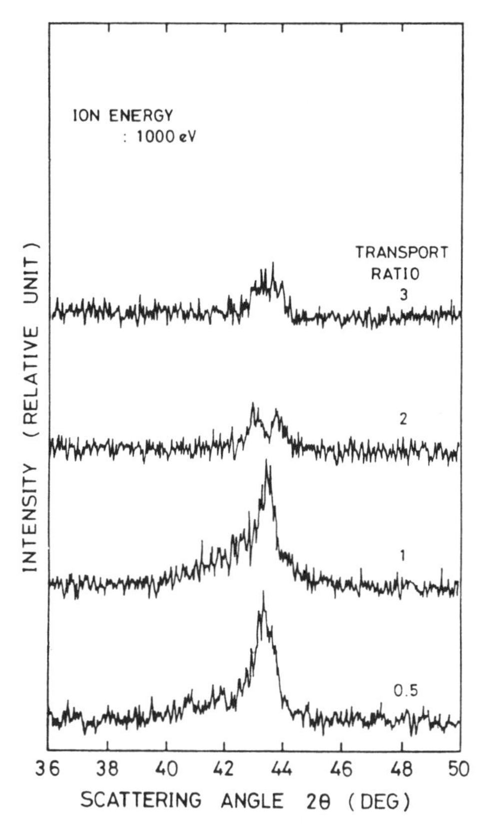

Fig. 18. X-ray diffraction patterns obtained from films prepared at 1 kV.

the diffraction angle of the peak at larger angle in Fig. 16 has a larger shifted angle side than the angle of the corresponding peak in Fig. 17.

From the above observations of the X-ray diffraction, we can conclude that films including the c-BN phase can be effectively manufactured by energy ion irradiation as low as 2 keV, and that the area of c-BN phase in films seems to increase with the decrease of the ion energy and to depend strongly on the composition ratio. The effects of the amorpharization and the shift of the diffraction line from any of the line positions for the definite phase of boron nitride are considered to be caused by damage due to ion irradiation and become large as ion energy increases. These conditions for crystallization of boron nitride are different from cases of titanium-, aluminum- and molybdenum nitrides as mentioned in the section 2.

3.2.3. IR STUDIES

A measurement of IR absorption spectra gives a useful information on the film structure, because each phase of boron nitride gives different absorption spectra. IR spectra were observed from samples formed on silicon wafers by irradiating infrared on a wide area. Accordingly, information on the average structure in a

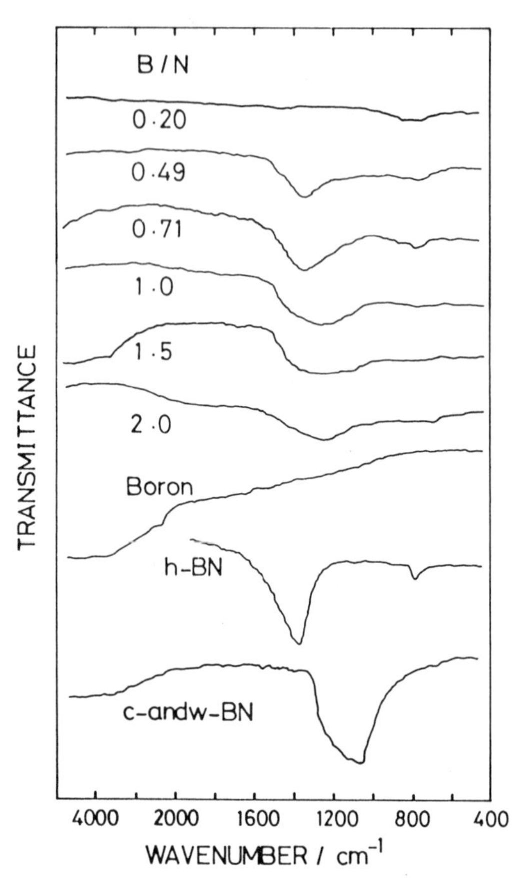

Fig. 19. A series of IR absorption spectra of films with various compositions together with those of boron, h-BN, c- and w-BN. Ion energy is 35 keV.

wide area of a sample can be obtained.

In Fig. 19, a series of IR spectra of films with various compositions and manufactured by using 35 keV N_2^+ ions is shown together with those of boron, h-BN and c- and w-BN [9] . The spectrum of h-BN shows two peaks at about 1350 cm^{-1} and 800 cm^{-1} [26] which are caused by the vibration modes in-plane and out-of plane, respectively. Spectra of c- and w-BN, which we observed, were essentially the same and had a large peak at 1080 cm^{-1} as seen in Fig. 19.

Spectra obtained from films with the composition ratio B/N lower than 0.4 did not show any dominant absorption peak and had only a small hump at 800 cm^{-1}. When the value of B/N increases, two peaks appear at about 1330 cm^{-1} and 800 cm^{-1} and their intensity increases. These peaks are considered to correspond to those of h-BN. In the region larger than B/N = 1, a new broad peak seems to appear at about 1100 cm^{-1} and the peak at 800 cm^{-1} disappears. This evidence indicates that the c- or/and w-BN phase was produced in the region B/N> 1.

Fig. 20 shows a variation of IR spectra from films prepared with the energy of nitrogen ions ($N^+/N_2^+ \sim 1$) in the range between

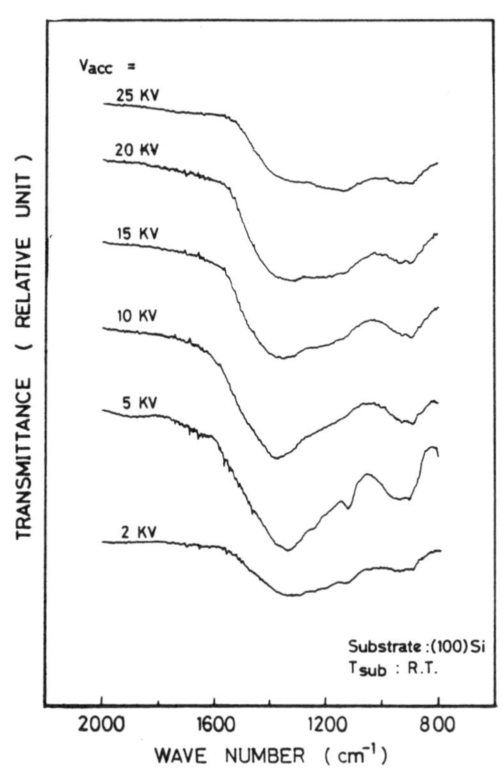

Fig. 20. IR spectra from films prepared with 2～ 25 keV ion beam. Ion current density was about 40 μ A/cm^2 and deposition rate of boron 1 Å/s.

2 and 25 keV [10] . The current density of the ion beam was about 40μ A/cm^2 and the deposition rate of boron about 1 Å/s. All of the spectra show not only absorption peaks at about 1350 cm^{-1} and 800 cm^{-1} from h-BN but also a small hump at 1100 ～ 1200 cm^{-1}, except for the spectrum from the film prepared at 5 kV in which a small peak was observed instead of the hump. The former spectra are similar to those observed from films with B/N = 1.0 and 1.5 in Fig. 19.

Spectra shown in Fig. 21 are those from films prepared by nitrogen ions with the energy 3keV and the current density 40μ A/cm^2 and the boron deposition rate 1 ～ 3 Å/s [10] . All of the spectra in this figure show a clear peak at about 1150 cm^{-1} due to c-BN [27] . This peak was usually observed from the films prepared with low energy ions and this result agrees well with that observed in the X-ray diffraction studies.

3.3. HARDNESS OF FILMS

Vickers hardness was measured only on films prepared on silicon wafers at 3 kV and 10 kV [10] . The film thickness was about 1 μ m. The result is shown in Fig. 22. The figure indicates that every film has the hardness of 3000 ～ 5000 Hv and is 3 ～ 5 times higher than that of the substrate silicon in spite of the

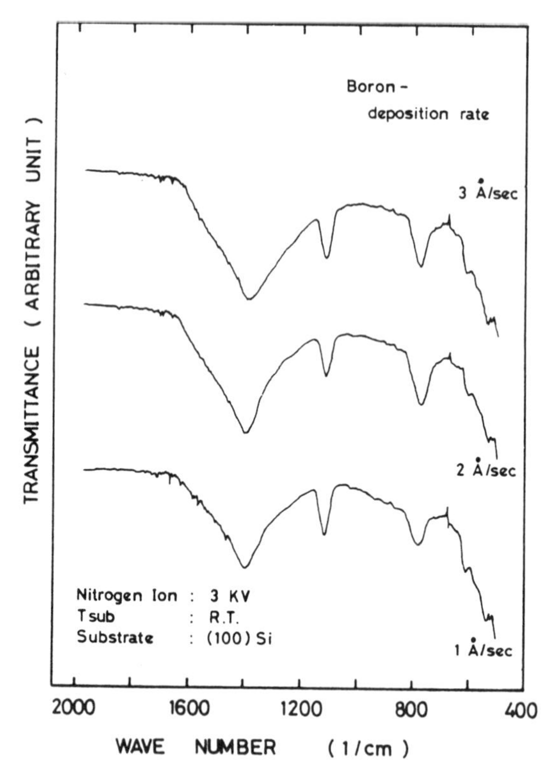

Fig. 21. IR spectra from films prepared at 3 kV. Ion current density was 40 μ A/cm^2 and deposition rate of boron 1 ~ 3 Å/s.

difference of the composition ratio B/N. This hardness is about 1/2 ~ 1/3 of the c-BN, 4700 ~ 8600 kg/mm^2. On films prepared at 1 keV and 500 V, the hardness testing was not successful because the adhesion strength was weak and the films were peeled off during the testing.

4. CONCLUSION

Boron nitride coating films have been prepared by the dynamic mixing (IVD) method which was developed by the present author and the collaborators. In this method, boron was evaporated on a substrate and nitrogen ions with their energy of 200 eV ~ 40 keV were simultaneously irradiated. For high energy regions, 25 ~ 40 keV, molecular ion beam N_2^+ were used and, for low region, 200eV ~ 25 keV, nitrogen ions mixed molecular and atomic ones of which the ratio was $N_2^+/N^+ \sim 1$ were bombarded. The ion beam current was lower than 200 μ A/cm^2 for 3 ~ 40 keV nitrogen ions and 500 μ A/cm^2 for 200 ~ 1000 eV ones. The substrate temperature was controlled at under 300℃ . The structure of films was studied by TEM, XPS, X-ray diffraction and IR absorption spectroscopy.

For samples prepared by 25 ~ 40 keV ions, films consisted of mixed phases of c-BN, w-BN, h-BN and amorphous. Percentages of the c-BN phase in the films were not so high. However, grains of the c-BN phase with column and plate like structures were observed

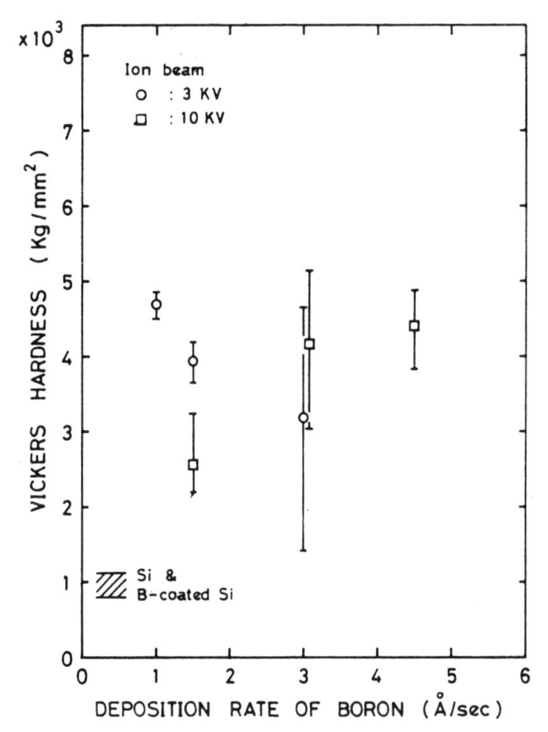

Fig. 22. Vickers hardness of films prepared with 3 keV and 10 keV nitrogen ions (load : 10 g).

by TEM in most of the samples where the ⟨ 100 ⟩ axis of c-BN was perpendicular to the film surface. The size of plate-like grains sometimes reached to 10 μ m and that of grains with the column structure distributed in the ranges 300 ~ 3000 Å. The w-BN phase was also observed as grains with the column structure and their c-axis was normal to the film surface.

As the energy of nitrogen ions decreases, the ratio of c-BN in the films increases. In fact, films prepared by ions with lower energy than 5 keV show a clear line in IR spectra at 1150 cm^{-1} due to c-BN together with two peaks at 1350 cm^{-1} and 800 cm^{-1} from h-BN. In X-ray diffraction patterns of films formed with lower energy than 10 keV, the diffraction peak from the (111) reflection of c-BN can be observed, though other peaks sometimes split into a number of peaks and shift from the (0001) reflection of h-BN. However, the diffraction patterns of films prepared at a certain deposition by using 500 eV and 1 keV nitrogen ions show a peak due only to c-BN.

Vickers hardness of films with about 1 μ m thickness manufactured at 3 and 10 kV was about 1/2 ~ 1/3 of c-BN. However, those at 1 kV and 500 V could not be measured, because the adhesion strength was weak and the films were peeled off during the testing. A similar effect can be seen in a relation between the deposition and the composition ratios. Films formed by 500 eV and 1 keV ion beams with the deposition ratio B/N > 1 had their composition ratio B/N = 1. This fact suggests that irradiated nitrogen ions diffuse out from the near-surface layer in the nitrogen-rich transport conditions. This is because the penetration depth is too shallow to keep nitrogen atoms inside films. From the results obtained by a series of experiments on the formation of boron nitride coating films, it can be expected that a coating film of c-BN with strong adhesion can be obtained by producing boron nitride film with high energy nitrogen ions at first and then, accumulating film with low energy ions.

REFERENCES

1) Weissmantel, Ch., Bewilogua, K., Dietrich, D., Erler, H.J., Hinneberg, H.J., Klose, S., Nowick, W., and Reisse, G.:Thin Solid Films, 1980, 72, 19.
Weissmantel, Ch.: J. Vac. Sci. Technol., 1981, 18, 179.
2) Shanfield, S. and Wolfson, R.: J. Vac. Sci. Technol., 1983, A1, 323.
Halverson, W. and Quinto, D.T.: J. Vac. Sci. Technol., 1985, A3, 2141.
3) Hovel, H.J. and Cuomo, J.J.: Appl. Phys. Lett., 1972, 20, 15.
4) Rand, M.J. and Roberts, J.F.: J. Electrochem. Soc., 1968, 115, 423.
Adams, A.C. and Capia, C.D.: J. Electrochem. Soc., 1980, 127, 251.
Takahashi, T., Itoh, H. and Kuroda, M.: J. Crystal Growth, 1981, 53, 323.
5) Weissmantel, Ch.: Thin Films from Free Atoms and Particles, ed. by Klabunde, K. (Acad. Press Inc.), 1985, p153.
6) Satou, M., Fukui, F. and Fujimoto, F.: Proc. Int. Workshop by Professional Groups on Ion-Based Techniques for Film Formation (Ionics Corp. Ltd.,Tokyo) 1981, p349.
7) Satou, M., Matsuda, K. and Fujimoto, F.: Proc. 6th Symposium Ion Sources and Ion-Assisted Technology, Tokyo, 1982, p425.
8) Satou, M. and Fujimoto, F.: Jpn. J. Appl. Phys., 1983, 22, L171.
9) Satou, M., Yamaguchi, K., Andoh, Y., Suzuki, Y., Matsuda K. and Fujimoto, F.: Nucl. Instr. and Meth., 1985, B7/8, 910.
10) Andoh, Y., Ogata, K., Suzuki, Y., Kamijo, E., Satou, M. and Fujimoto, F.: Nucl. Instr. and Meth., 1987, B19/20, 787.
11) Andoh, Y., Ogata, K. and Kamijo, E.: Nucl. Instr. and Meth., 1988, B33, 678.
12) Satou, M., Andoh, Y., Ogata, K., Suzuki, K., Matsuda, K. and Fujimoto, F.: Jpn. J. Appl. Phys., 1985, 24, 656.
13) Kiuchi, M., Tomita, M., Fujii, K., Satou, M. and Shimizu, R.: Jpn. J. Appl. Phys. 1987, 26, L938.
14) Kiuchi, M., Fujii, K., Tanaka, T., Satou, M. and Fujimoto, F.: Nucl. Instr. and Meth., 1988, B33, 649.
15) Kiuchi, M., Fujii, K., Miyamura, H., Kadono, K., Satou, M. and Fujimoto, F.: to be published in Nucl. Instr. Meth. B, 1989.
16) Satou, M., Fujii, K. and Kiuchi, M. and Fujimoto, F.: to be published in Nucl. Instr. Meth. B, 1989.
17) Fujimoto, F., Nakane, Y., Satou, M., Komori, F., Ogata, K. and Andoh, Y.: Nucl. Instr. and Meth., 1987, B19/20, 791.
18) Andoh, Y., Suzuki, Y., Matsuda, K., Satou, M. and Fujimoto, F.: Nucl. Instr. and Meth., 1985, B6, 111.
19) Ogata, K., Andoh, Y. and Kamijo, E.: Nucl. Instr. Meth., 1988, B33, 685.

20) Sato, T., Ohata, K., Asahi, N., Ono, Y., Oka, Y. and Hashimoto, I.: J. Vac. Sci. Technol., 1986, A4(3), 784.
21) Fujimoto, F. and Satou, M.: Energy Pulse and Particle Beam Modification of Materials, ed. by K. Herring (Akademie-Verlag, Berlin), 1988, p131.
22) Fujimoto, F.; to be published in Vacuum, 1989.
23) Kant, R.A., Sartwell, B.D., Singer, L.L. and Vardiman, R.G.: Nucl. Instr. and Meth., 1985, B7/8, 915.
24) Hentzell, H.T.G., Harper, J.M.E. and Cuomo, J.J.: J. Appl. Phys., 1985, 58, 556.
25) Carlson, T.A.: Photoelectron and Auger Spectroscopy (Plenum Press, New York, 1975)
26) Geik, R. and Perry, C.H.: Phys. Rev., 1966, 146, 543.
27) Gielisse, P.J., Mitra, S.S., Plenell, J.N., Criffis, R.D., Mansur, L.C., Marshall, R. and Pascor, E.A,: Phys. Rev., 1967, 155, 1039.

Materials Science Forum Vols. 54 & 55 (1990) pp. 71-110

ION ASSISTED SYNTHESIS OF BORON NITRIDE COATINGS

W.D. Halverson, T.G. Tetreault and J.K. Hirvonen
Spire Corporation, Patriots Park, Bedford, MA 01730, USA

ABSTRACT

Two methods for the production of boron nitride coatings with the aid of ion beams are presented. One technique, direct ion beam deposition, uses a Kaufman type ion source to accelerate a beam of ionized species derived from a borazine ($B_3N_3H_6$) plasma at various conductive and insulating substrates. Another technique, ion beam enhanced deposition (IBED), combines physical vapor deposition of boron (via electron beam evaporation) onto a substrate with simultaneous nitrogen ion bombardment from a similar Kaufman ion source.

The experimental methods and process parameters for each technique are discussed. Resulting coatings are characterized by a variety of analyses including Rutherford backscattering spectrometry (RBS), x-ray photoelectron spectroscopy (XPS or ESCA), x-ray diffraction (XRD), IR and Raman spectroscopy, transmission electron microscopy (TEM), and nuclear reaction analysis (NRA), among others. Mechanical properties such as microhardness, and tribological (friction/wear) behavior are also presented.

I. INTRODUCTION

Boron nitride is well known in its soft, graphite-like form. It is widely used as a lubricant and as a refractory dielectric material. Its molecular structure closely parallels graphite, and it was largely due to this similarity that attempts were made to synthesize a super-hard phase analogous to diamond. In 1957, R.H. Wentorf, Jr. was the first to develop a reproducible method for the formation of cubic boron nitride crystals [1].

Since that time, many methods have been investigated to produce boron nitride in both its soft (hexagonal structure) and hard (cubic structure) phases. Techniques have varied from the early high pressure/high temperature methods to those in high vacuum and near room temperature environments. Although hexagonal phase boron nitride is still widely investigated, recent years have seen an increase in interest in cubic phase BN due to its many unique electrical, thermal, chemical, and mechanical properties which closely resemble those of diamond.

Presented here are two relatively newer experimental techniques for making boron nitride thin films using ion beams. We will refer to them as i-BN denoting the role of ions in their formation. Also discussed are the analytical tools which are employed to characterize these coatings and the process parameters that are important to their production.

II. DIRECT ION BEAM DEPOSITION

Based on earlier work on direct ion beam deposition of diamond-like carbon films (i-C) [2,3,4] and ion-assisted deposition of boron nitride (i-BN) [5,6], Spire Corporation instituted a program in 1981 to investigate i-BN films formed by direct ion beam deposition. Borazine gas ($B_3N_3H_6$) was used as the source of boron and nitrogen for the i-BN coatings.

A. EXPERIMENTAL PROCEDURE

The experimental facility was described in more detail in Reference 7; the deposition process is illustrated schematically in Figure 1. Borazine gas ($B_3N_3H_6$) was injected into a Kaufman type ion source [8] in which a diffuse arc discharge ionized and partially dissociated the gas molecules. The magnetic field from an external solenoid provided some confinement of the plasma electrons. Ionized particles were accelerated by the potential difference between the grids and exit as a set of beamlets which mixed to form a uniform ion beam approximately 10 cm in diameter.

An electrically-heated tungsten filament passed through the beam could neutralize the ion beam space charge when its thermionic emission current was equal to the total ion current. The total current measured at the substrate holder during depositions could be positive, negative or zero, depending on the temperature of the neutralizer filament.

Substrates were mounted in the beam on an electrically isolated holder which was grounded through a milliameter. The substrate holder, located about 20 cm from the ion gun, could be heated to temperatures up to 850°C and be electrically biased with respect to ground potential.

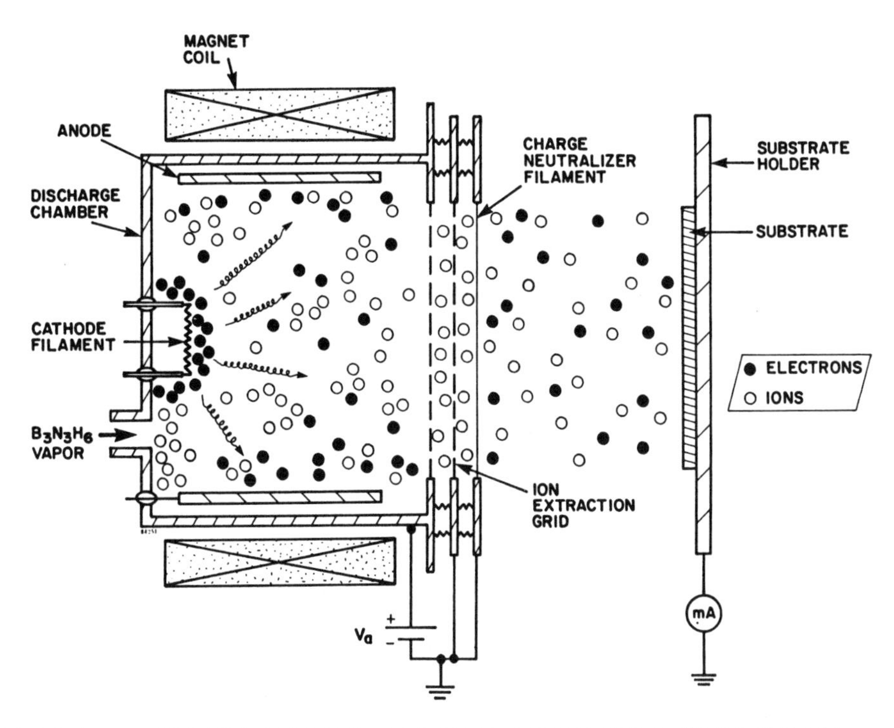

FIGURE 1. SIMPLIFIED SCHEMATIC OF ION BEAM DEPOSITION SYSTEM.

The depositions were conducted in a stainless steel vacuum chamber which was pumped by an oil vapor diffusion pump with an ultimate pressure of about 10^{-7} torr. During i-BN depositions, the pressure in the vacuum chamber was 5×10^{-4} torr or less. A stainless steel shroud on the internal wall of the vacuum chamber was cooled by liquid nitrogen to reduce contamination by water and other volatile vapors during depositions.

Ion beam depositions of i-BN films were performed on various substrate materials, over a wide range of beam voltage and current density, and substrate temperatures from about 200°C to 850°C. Additionally, ion beam deposition was performed with no space charge neutralization and compared with depositions with nearly complete neutralization. "Nearly complete" charge neutralization was indicated by a null (ion plus negative electron) current on the substrate holder. Under these conditions there may have been a small net space charge in the beam because secondary electron emission from the substrate holder is registered as a positive current on the milliameter.

Both electrically conductive and insulating substrates were used, including polished single-crystal silicon, tungsten carbide metal cutting tools, titanium nitride-coated tungsten carbide, stainless steels, tool steels, glass microscope slides, fused silica slides and SiAlON ceramic cutting tools. The substrates were mounted vertically on the holder by stainless steel machine screws. The heads of the screws obscured a small area of the substrates' surface and thus provided a convenient area for comparison between the i-BN-coated and uncoated surface.

The substrates were cleaned with a fluorocarbon solvent before mounting on the substrate holder. The substrates were then sputter-etched for 15 minutes by a 1000 eV, 0.2 mA/cm^2 argon beam from the ion source before commencing the i-BN deposition.

The beam energy was varied between 150 eV and 1800 eV and the ion current density between 0.01 and 1 mA/cm^2 at the substrate. Deposition rates of i-BN films were typically 0.5-1 nm/s at the highest ion currents and are nearly proportional to the beam current density.

Analysis techniques for the films included optical and electron microscopy, x-ray diffraction, and x-ray photoelectron spectroscopy (XPS). Auger electron spectroscopy (AES), which uses a probe beam of electrons, was unsuccessful for films thicker than 1 μm because of severe charging effects in the insulating BN. XPS uses an x-ray beam probe, hence only secondary electrons from the sample surface cause charging. In this case, charge neutralization was effected by controlling the thermionic emission of electrons from a tungsten filament near the sample, such that the energy of the O(1s) line was shifted back to 532 eV, which was used for reference in binding energy determinations. XPS analysis was performed in the as-received state after BN deposition and following sputter cleaning (3 kV Ar^+, 15 $\mu A/cm^2$, 60 min) to remove about 50 nm of the surface layers. In some instances the samples were electron-beam-heated (15 min, 100 W) to further clean the surface and then analyzed.

B. FILMS LESS THAN 1 μm IN THICKNESS

Films less than 1 μm in thickness had microhardnesses considerably greater than that of their substrates, were very adherent, and showed indications from X-ray diffraction analysis of containing cubic-BN. Figure 2 shows a rapid X-ray diffractometer

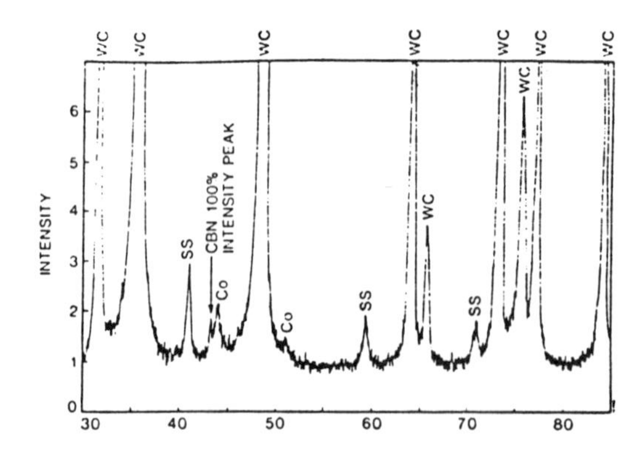

FIGURE 2. RAPID X-RAY DIFFRACTOMETER SCAN OF BORON-NITRIDE-COATED WC SUBSTRATE. The designation SS refers to (Ta,Ti)C solid solution. (From Ref. 7)

scan of an i-BN coating on a tungsten carbide substrate [7]. Because of the small coating thickness (less than $1\,\mu m$), diffraction peak intensities from the boron nitride coating would be expected to be substantially lower than those from the substrate. A small but distinguishable peak of cubic boron nitride appears in Figure 2 at 2θ = 43.25 degrees corresponding to a (111) lattice spacing of d = 2.09 angstroms. The standard powder diffraction file shows d = 2.085 angstroms for the (111) peak of cubic boron nitride at 100% relative intensity (most intense peak). The other cubic boron nitride peaks all have less than 6% relative intensity so there is little chance of detecting any other cubic boron nitride peak than 2θ = 43.25°.

Hexagonal phase boron nitride has a 60% relative intensity peak at d = 2.023 angstroms (2θ = 44.75°) which could interfere, in a rapid scan, with the main cubic boron nitride peak mentioned above. Additionally, h-BN also has a 100% intensity peak at d = 1.910 angstroms (2θ = 47.56°) and lesser peaks at d = 1.149 angstroms (2θ = 84.19°) and d = 1.083 angstroms (2θ = 90.67°). Since many of these peaks very nearly coincide with major WC peaks, the absence of hexagonal phase boron nitride cannot be conclusively proven.

Scratch hardness (Mohs scale) and Knoop indentation hardness tests were conducted on substrates with thin i-BN coatings. Table I summarizes some of the Knoop hardness measurements.

TABLE I. RANGE OF MEASURED KNOOP HARDNESS (25 g load) FOR SUBSTRATES BEFORE AND AFTER BORON NITRIDE COATING.

	Knoop Hardness (kg/mm^2)	
Substrate Material	Substrate	BN Coating
TiN-coated WC	1300-1650	2200-3100
Al_2O_3-coated WC	1200-1600	1600-2200
Ceramic	1400-1700	2100-2800
Glass	730- 770	1100-1300

Auger analysis of the thin coatings indicated that the B:N ratio was close to stoichiometric [7]. Principal impurities were found to be carbon (less than 10%) and oxygen (less than 5%).

C. THICKER i-BN COATINGS

Based on the encouraging preliminary results with thin films, thicker i-BN coatings were investigated. This required some modification of the ion source to allow operation for periods of more than one hour. It was found that the arc current in the ion source gradually decreased during deposition, because of the formation of an insulating film on the anode electrode. Installation of a cylindrical stainless steel mesh to increase the effective anode area allowed depositions of films of 3 to $4\,\mu m$ in thickness.

C.1 DEPOSITION WITHOUT NEUTRALIZATION

A striking feature during the operation of the borazine ion beam without neutralization was the appearance of small, bright flashes on the interior surfaces of the vacuum chamber and on the metallic substrate mounting structure [9]. We believe that these were "Malter" discharges [10] caused by dielectric breakdown between the charged surface of the i-BN coating and the conductive substrate. Surface charging can be caused by the incident positive ion flux, secondary electron emission, and photoemission. The Malter discharges occurred most frequently on surfaces directly in the line-of-sight of the ion source, although some flashes were visible in shadowed areas as well. These apparently were caused by secondary ion or electron emission, charge diffusion, and ion beam scattering.

The Malter discharges immediately ceased when the neutralizer filament was heated to the point where an apparent drop was observed in the beam current on the substrate holder. No discharges were seen during neutralizer operation either in the direct line-of-sight of the ion source or in shadowed areas. Electrical noise on the substrate holder also decreased substantially when the neutralizer filament was heated to incandescence.

A series of i-BN films deposited on various substrates without charge neutralization was examined by optical and scanning electron microscopy (SEM). The film surfaces on both conductive an nonconductive substrates were visibly rough and had indications of dielectric punctures and surface tracking. The visible damage, under magnification, shows clear evidence of melting and flaking, as illustrated in the SEM micrograph of Figures 3 and 4.

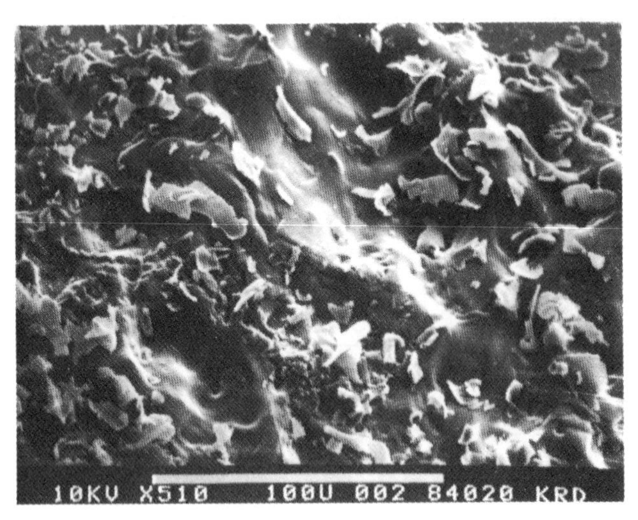

FIGURE 3. SEM MICROGRAPH OF i-BN FILM ON SILICON SUBSTRATE, DEPOSITED WITH UNNEUTRALIZED ION BEAM. The white bar indicates a distance of 100 μm.

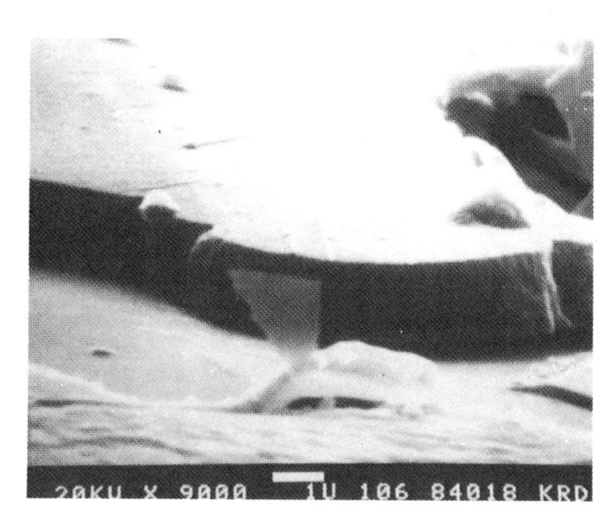

FIGURE 4. SEM MICROGRAPH (75° tilt) OF i-BN FILM ON SILICON SUBSTRATE, DEPOSITED BY UNNEUTRALIZED ION BEAM. The white bar indicates a distance of 1 μm.

Figure 3 shows an i-BN coating on a silicon substrate deposited for 85 minutes with ion beam energy of 600 eV and an average ion current density at the substrate holder of 0.1 mA/cm^{-2}. Melted zones, with surface roughness much larger than that of the polished substrate, underlie loose flakes of boron nitride. Figure 4 is a SEM micrograph from a 600 eV i-BN deposition with a current density of 0.4 mA/cm^2. Although the surface of the approximately 2 μm thick layer appears locally smoother than in the film of Figure 3, there are melted regions as well as delamination of the film from the substrate.

C.2 DEPOSITION WITH CHARGE NEUTRALIZATION

Depositions conducted with nearly complete beam neutralization produced i-BN films with smooth, uniform surfaces showing very little evidence of damage in the form of surface melting or cracking. There were, however, many instances of delamination and flaking due to internal compressive stresses in the films. Figures 5 and 6 are SEM micrographs of films formed by charge-neutralized deposition at a beam energy of 1000 eV and an average ion current density of 0.6 mA/cm^2.

The SEM micrograph of Figure 5 shows that the film was smooth and essentially featureless, with an occasional nodule projecting above the surface. These nodules may be initiated when the neutralizer filament was momentarily turned off during the deposition to measure the ion beam current on the substrate holer. Figure 6 shows the edge of a 1.8-μm-thick i-BN film from a silicon wafer that was scribed and fractured. The film has delaminated from the silicon substrate and the flakes bend upward, indicating residual compressive stress in the coating. There is no large columnar growth toward the top surface of the film, perhaps indicating continuous renucleation of the material during deposition. At this magnification, the i-BN material appears very similar to that formed with an unneutralized ion beam (see Figure 4).

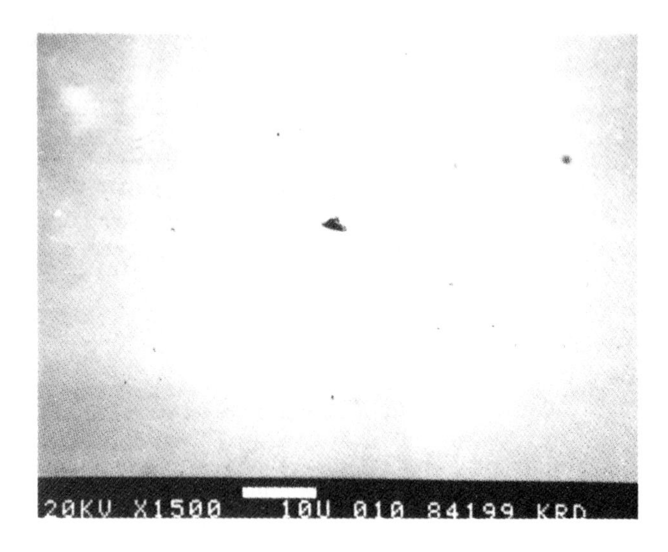

FIGURE 5. SEM MICROGRAPH (75° tilt) OF i-BN FILM ON SILICON SUBSTRATE, DEPOSITED BY CHARGE NEUTRALIZED BEAM. The white bar indicates a distance of 10 μ m.

FIGURE 6. i-BN FILM OF FIGURE 5 (75° tilt) AT SCRIBED EDGE. The white bar indicates a distance of 1 μ m.

D. COMPOSITION AND STRUCTURE

The composition of the deposited films, determined by XPS, was typically stoichiometric BN and varying amounts of oxygen and carbon. The binding energies and the ratio of peak heights of the B(ls) and N(ls) were coincident with published x-ray photoelectron data [11] for BN compounds. Table II shows the compositional analysis of the film (within an accuracy of about $\pm 5\%$ based on handbook relative sensitivity factors along with analysis of a film formed with an unneutralized beam but otherwise similar deposition conditions.

TABLE II. XPS ANALYSIS OF COMPOSITION OF i-BN FILMS DEPOSITED UNDER NEUTRALIZED AND UNNEUTRALIZED CONDITIONS.

Film		Composition (at.%)				
Deposition Conditions	ESCA Conditions	B	N	O	C	Na
Neutralized	As-received	42.7	46.3	4.4	4.5	2.0
	100 nm sputter etched and heated	46.9	49.4	2.0	1.6	-
Unneutralized	As-received	43.4	43.4	4.9	7.3	<1
	100 nm sputter etched and heated	47.1	45.3	3.6	3.2	-

Except for the general observation that the residual impurity levels tended to be as much as a factor of 2 higher in the i-BN films deposited without charge neutralization, there were no other systematic differences observed in the composition of the films.

E. DISCUSSION

Without beam neutralization, the surface of the boron nitride films can charge to a potential which is close to the beam energy. Secondary electron emission from other surfaces under ion bombardment and photoelectron emission from surfaces exposed to ultraviolet radiation from the ion source can reduce this potential somewhat, but these effects cannot compete against the positive charging by an ion beam of 0.1 to 1 mA/cm^2.

The dielectric breakdown strength of ion-beam-deposited BN is not known; BN films deposited by plasma-activated CVD [11] showed breakdown strengths as high as 3×10^6 V/cm. Based on this value, a 1-μm-thick i-BN film should charge to 300 volts before puncture to a grounded substrate. Thinner films will charge to lower levels before breakdown, although enough electrical energy must be available to ionize and excite the boron nitride molecules during the breakdown avalanche. The lower limit is probably a few tens of volts.

The energy released upon breakdown of a 1-μm-thick i-BN film at 300 volts is about 1.6×10^{-4} J/cm^2, based on a dielectric constant of 4 for i-BN [11]. At a beam current density of 0.1 mA/cm^2 this energy would be released about every 10^{-2} seconds, the time required to charge the thin-film capacitor structure. This energy and discharge rate, even if grossly overestimated, is quite sufficient to account for the damage observed on the i-BN films deposited without beam neutralization.

The increased oxygen and carbon impurity levels in the i-BN films deposited without charge neutralization can result from Malter discharges on the walls and internal structures of the vacuum chamber. The flashovers and punctures eject a considerable "plume" of ionized and unionized material from their discharge sites, and these materials can be redeposited on the growing films.

The apparent lack of differences between the bulk properties of films formed under neutralized and unneutralized deposition conditions indicates that the average surface charging was not severe enough to decelerate the incident ions to energies where other BN crystalline structures (e.g., hexagonal) could be formed. We have found that soft films with hexagonal-BN crystallography are deposited only when the ion source is operated with accelerating potentials less than about 150 V or when the substrate was heated to temperatures greater than 450°C.

Because of problems with adhesion and relatively low hardness values of the coatings [9], direct ion beam deposition of i-BN coatings was abandoned in 1985. Subsequently, we have found that ion beam enhanced deposition produces significantly better i-BN coatings, as discussed in Section III.

III. ION BEAM ENHANCED DEPOSITION (IBED)

A. BACKGROUND

The international ion beam R&D community has, in the past few years, given increased interest to the use of simultaneous ion bombardment and physical vapor deposition. Ion-assisted deposition has been reviewed by Harper et al. of IBM [12] and has been studied in several other laboratories [13-23] since the mid-1970's to study electronic materials, to understand the role of ions in plasma-assisted deposition processes, to prepare dense optical coatings as well as to prepare tribological coatings. It is to be noted that most work in this area has utilized much lower energy beams (100 eV-2000 eV) than used in ion implantation.

There are several aspects of film growth that are beneficially influenced by ion bombardment during thin film deposition including: (i) adhesion, (ii) nucleation and growth, (iii) control of internal stress, (iv) morphology, (v) density, (vi) composition, and (vii) possibility of low temperature deposition.

Obviously, this process allows thicker alloyed regions to be attained than by either direct ion implantation or ion beam mixing but still incorporates advantages attributed to ion beams, such as superior adhesion due to precleaning and ion mixing during the initial stages of deposition. As an example, the adhesion of IBED optical coatings onto substrates has been improved by over an order of magnitude compared to results for deposition performed without ion beam assistance. Several researchers have also observed a significant reduction in the stress and crystallite size for ion assisted films.

IBED processing can also change the structure of deposited coatings. Conventional low temperature deposition processes often result in porous, columnar microstructures. This is a concern for optical coatings, for example, whose transmission properties are sensitive to the presence of absorbed water. It is commonly found that their characteristics in air and vacuum are different and that the use of ion bombardment during the deposition can produce fully dense non-absorbing coatings whose characteristics are stable upon atmospheric exposure. Its use in depositing optical coatings is thus far the most advanced application area.

A number of studies [24] have demonstrated how the internal stress within deposited films can be changed by ion bombardment during deposition, being attributed to a number of factors including: (i) impurity incorporation, (ii) the preferential sputtering-out of impurities, and (iii) thermal spike events.

IBED processing also allows control over stoichiometry and structure. For example, Si_xN_y coatings prepared by the IBED technique using silicon evaporation and low energy nitrogen bombardment can be continually varied from pure Si to stoichiometric Si_3N_4 by varying the relative ion beam and evaporant flux ratios. It is this type of control which motivated us to explore its use in producing (ion-assisted) i-BN.

B. EXPERIMENTAL PROCEDURE

Figure 7 shows a schematic of Spire's Ion Beam Enhanced (or Assisted) Deposition (IBED) system combining electron beam evaporation with simultaneous (nitrogen) ion bombardment. A Kaufman type ion source as described in Section II.A operates from 200 eV to 1000 eV providing ion current densities up to 1000 microamps/cm^2.

In order to produce compounds combining evaporated material (B) with elements in the form of an energetic ion beam (i.e., nitrogen ions) one has to provide enough ions to ensure stoichiometry. Another consideration is that the non-mass-analyzed beam consists of a diatomic (N_2^+) component (approximately 90% abundant) as well as a monatomic (N^+) component (approximately 10% abundant). The diatomic component dissociates immediately up striking the substrate with each nitrogen atom carrying away one-half of the initial beam energy. Both components will tend to sputter away surface atoms in competition with build-up from deposition. The ion-assisted deposition process is thus very complicated and models for predicting compound formation are only recently emerging [25,26].

The main experimental parameters include: (i) ion and evaporant species, (ii) the relative fluxes of the deposited material (e.g., B) and the ion species (e.g., N_2^+ plus N^+), and (iii) the substrate temperature. We have also found that other conditions such as background pressure and its makeup are also important affecting both the microstructure and properties of the resultant coating.

Spire's study of ion beam enhanced deposition of boron nitride (i-BN) systematically examined several properties of i-BN prepared over a range of boron to nitrogen arrival ratios (defined as R ratio) and for both low temperature and elevated substrate temperatures.

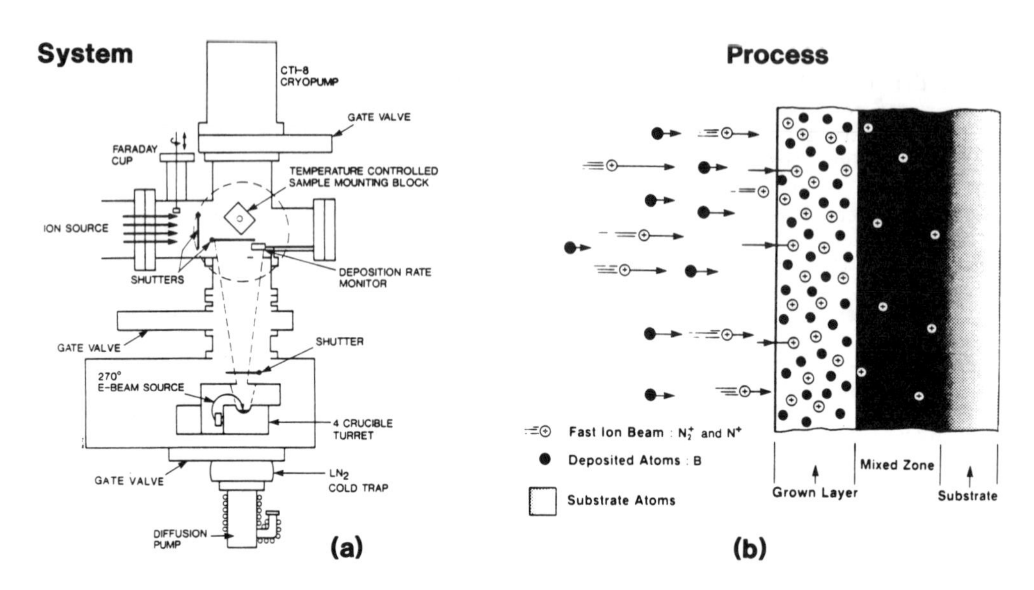

FIGURE 7. SCHEMATIC OF SPIRE'S ION BEAM ENHANCED (OR ASSISTED) DEPOSITION (IBED) SYSTEM (a) AND DEPICTION OF PROCESS (b) COMBINING ELECTRON BEAM EVAPORATION (BORON) WITH SIMULTANEOUS ION (NITROGEN) BOMBARDMENT.

C. APPLICATION TO NITRIDE FILMS

Previous studies [14,24] have shown the IBED process capable of producing stoichiometric nitride films at room temperature. Weissmantel demonstrated that γ-Si_3N_4 films formed when the nitrogen beam current passed a critical threshold with respect to the silicon evaporant. Cuomo et al. [24] also showed that AlN formed when the nitrogen to Al flux was above a critical value. In both of these cases any excess nitrogen appeared to be rejected.

C.1 PREVIOUS i-BN STUDIES

Weissmantel et al. [14] were some of the first to publish on the production of hard boron nitride films by ion beams (reactive ion plating). They found a quasiamorphous structure via TEM with IR spectra confirming B-N bonding states and x-ray data consistent with crystallites of cubic boron nitride. They postulated that in the wake of the violent collision cascade of the stopping ions local high-temperature, high-pressure areas could be produced as shown schematically in Figures 8 and 9 [27].

Shortly thereafter Satou and Fujimoto published a short note [28] claiming the formation of cubic boron nitride crystallites in an amorphous/hexagonal matrix. Their characterization consisted of electron micrography and electron diffraction. They evaporated boron and used a mass analyzed 40 keV N_2^+ beam (i.e., 20 keV N^+ions). This work was continued [18], using additional analysis of IR absorption, x-ray diffraction and electron microscopy. They give evidence that films having a B/N ratio larger than 0.9 have a cubic boron nitride structure along with the hexagonal phase.

Recently Sainty et al. [29] have published a comprehensive study using low energy (100 to 1500 eV) analyzed nitrogen beams on evaporated boron to produce boron nitride. Their films were characterized by EELS (electron energy loss spectroscopy), TEM,

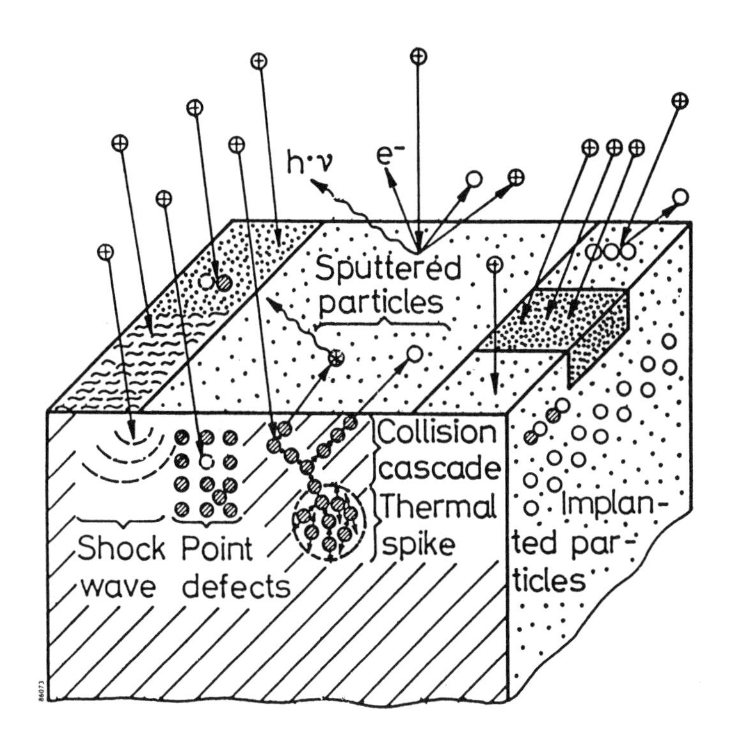

FIGURE 8. DEPICTION OF EVENTS INCLUDING DISPLACEMENT SPIKES OCCURRING DURING ION BOMBARDMENT OF A SURFACE. (C. Weissmantel)

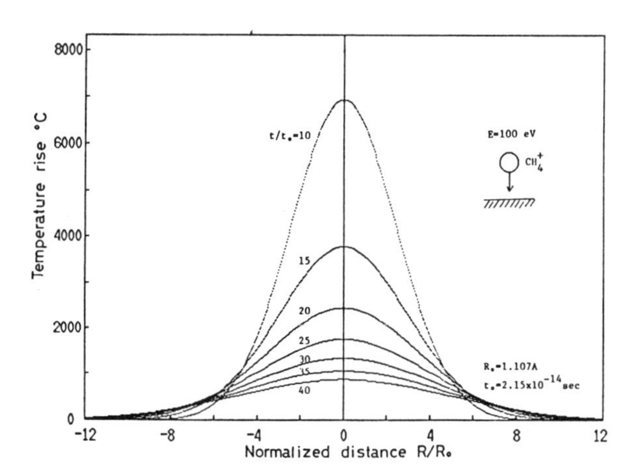

FIGURE 9. CALCULATED THERMAL SPIKE TEMPERATURE FOLLOWING ION BOMBARDMENT. (From Namba and Mori, Ref. 27)

ellipsometry, and microhardness. They found no evidence of the cubic phase, but only the hexagonal phase present, in an extremely fine tangled network of ribbons of random orientation.

C.2 SPIRE'S WORK

In the production of i-BN thin films by the IBED technique, the process parameters which are most relevant are the temperature of the substrate, the evaporation rate of the boron and the current density of the nitrogen ion beam at the substrate. Heating of the substrate due to the incidence of the ion beam alone typically results in a process temperature of 180° to 230°C for thin films (less than 1 μm) on non heat-sunk substrates. With the aid of a target stage heater, temperatures in excess of 400°C can be attained in the experimental chamber previously described. Evaporation rates for the boron are routinely in the range of 2 to 5 angstroms/sec for the films in this study and rate uniformity is generally kept within ±0.3 angstroms/sec of the desired programmed rate by the crystal monitor.

Changes in the nitrogen ion arrival rate are possible by varying the current density of the ion beam as measured by a probe near the substrate surface. The present measurement system does not allow for a mass analysis of the beam (i.e., N_2^+ to N^+ ratio) nor does it take into account the effect on the current reading by secondary electrons emitted from the probe surface under ion bombardment. Further difficulty in identifying the "true" nitrogen arrival rate is due to the unknown neutral flux produced by charge exchange neutralization.

Recently, attempts have been made to quantify some of these factors [26,30] using a Kaufman ion source of the type used in this study. A N_2^+:N^+ ratio of 89% N_2^+ to 11% N^+ has been determined as average for a beam energy of 500 to 1000 eV; with the neutralized fraction being 34% for beams of 500 eV in a working pressure of 2×10^{-4} torr. If a Faraday cup system equipped with a suppression bias is used to effectively eliminate the need to correct for a "secondary current," then a measured current density of 145 μ A/cm^2 with a boron evaporation rate of 2 angstroms/sec reduces to a B/N arrival rate of 1:1 for this experimental geometry.

C.2.1 SPIRE'S i-BN STUDY

Table III is a listing of selected i-BN depositions done for this study. In-house testing of Knoop microhardness as a function of load (grams) as well as ellipsometric determination of refractive index are routinely performed and are also listed here. Figure 10 illustrates a typical variation of microhardness with load as well as with incorporated B/N ratio of IBED i-BN thin films. Determination of the B/N ratio is done via nuclear reaction analysis using selected resonance reactions for boron and nitrogen and will be discussed in more detail later (see Section D.1). Other in-house analyses include friction and wear behavior (see Section D.5) and a more qualitative Moh's scratch hardness test (see Figure 11). It should be noted that this figure shows the general characteristics of the i-BN films; specifically, their ductile behavior (smooth, plastically deformed scratch edges) and their excellent adhesion to the substrate (no spalling or cracking along the scratch).

TABLE III. SELECTED IBED i-BN DEPOSITIONS.

Substrate	I ($\mu A/cm^2$)	E (eV)	Rate (Å/sec)	t (Å)	Microhardness KHN @ (Load)		Refractive Index (n)
Si (100)	155- 160	600	2.0	3500	1980	(5g)	-
Si (100) and 304 stainless steel	430	700	1.5-2.0	8786	2200	(5g)	2.49-2.7
Si (100)	300	1000	2.0	6000	2400	(5g)	2.1
Si(100 @ 400°C	450	550	1.5	2500	3600	(5g)	2.08
Si (100) @ 400°C	425	500	1.5-2.0	2500	1900	(5g)	1.96-2.05
Si (100) @ 400°C	1000	1000	2.0-2.5	6000	2000	(5g)	2.15-2.46
Si (100) @ 400°C	1000	500	2.0-2.5	12000	2500	(25g)	2.13-2.7
Si (100)	200	600	2.0	12000	1650	(5g)	2.65
304 stainless steel	300	600	3.0	11000	1740	(5g)	-

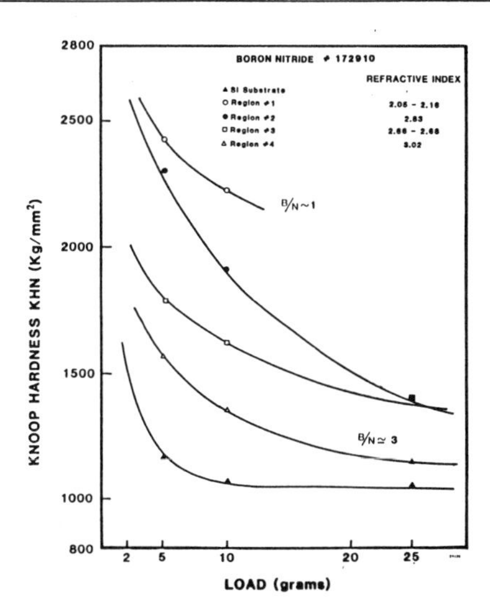

FIGURE 10. MICROHARDNESS VARIATION WITH B/N.

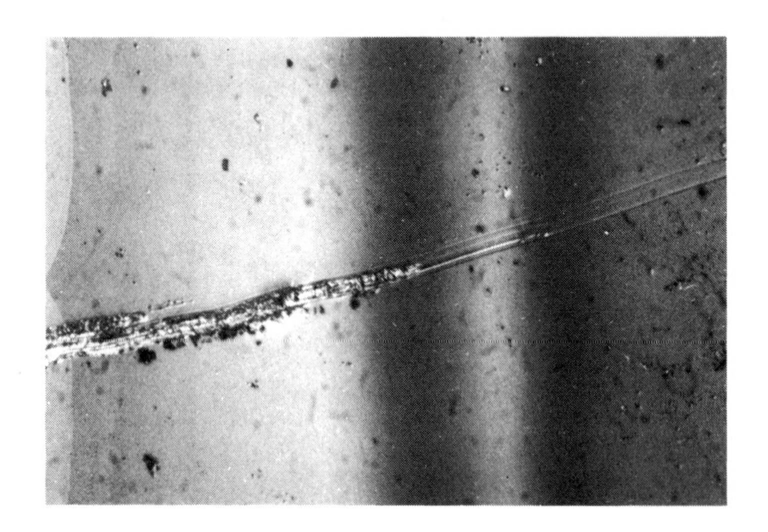

FIGURE 11. MOH'S #9 (Al_2O_3) SCRATCH ADHESION TEST ACROSS SILICON (left)/BN (right) BOUNDARY (10 g load).

As evidenced by the data thus far presented, IBED i-BN thin films generally exhibit a high microhardness (KHN = 2000-3000, i.e. 20-30 GPa) and good adhesion to the substrate. Also notable are the thermal and chemical stability of these films.

C.2.2 THERMAL STABILITY

The thermal stability of BN films has been studied by high temperature annealing. Figure 12 shows RBS spectra for a BN film on a Si substrate after (i) room temperature aging (two months) and (ii) after a subsequent high temperature (750°C, 2 hours) annealing treatment.

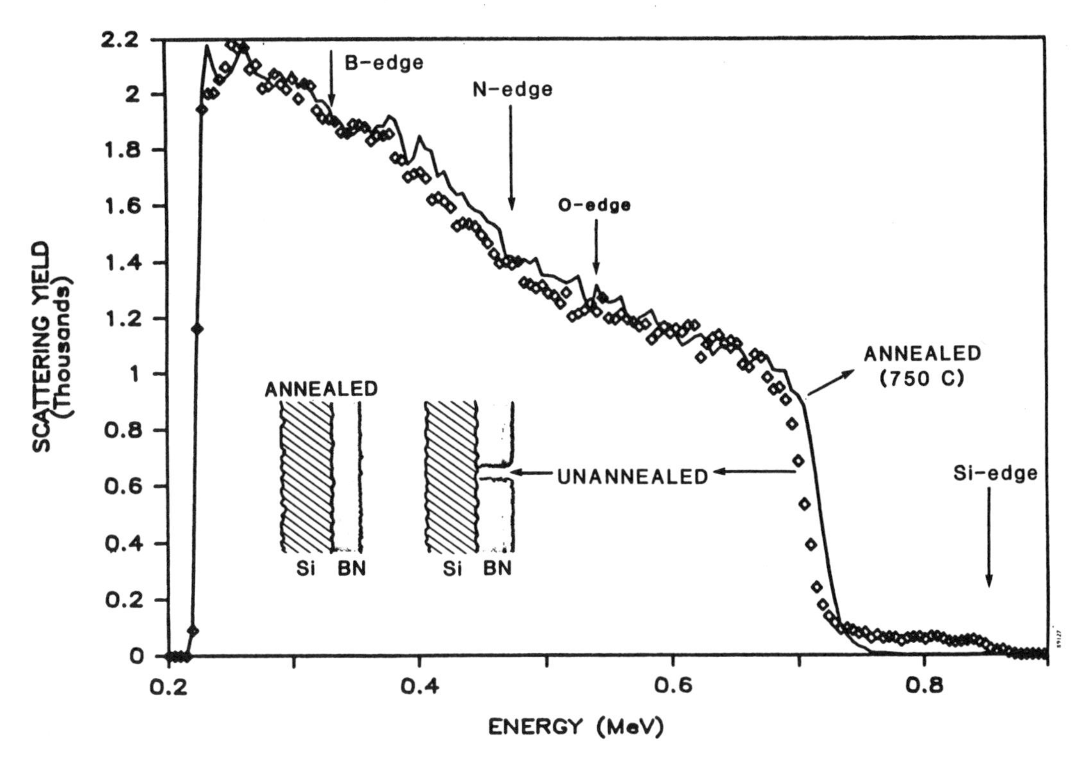

FIGURE 12. RBS SPECTRA OF UNANNEALED AND ANNEALED BN THIN FILMS (approximately 2000 angstroms) ON SILICON SUBSTRATE.

Although most of our IBED samples are free from any pinholes, in this particular sample (with a thickness of less than 2000 angstrom) RBS indicates the presence of pinholes in the unannealed condition. Upon annealing for two hours at 750°C in a nitrogen ambient, all of the pinholes filled up. On the other hand, RBS explicitly indicates that thin films become thinner as a result of annealing, indicating that during annealing some lateral diffusion occurs. However, no mechanical degradation has been observed for temperatures up to 900°C. RBS also shows little or no oxygen incorporation in this film either before or after the anneal, indicating the excellent resistance to oxidation which is characteristic of these i-BN films.

C.2.3 CHEMICAL STABILITY

The best BN coatings were exposed to concentrated nitric acid for periods of several hours to days. Although the substrate material was clearly attacked by the acid, there were no visible changes in the appearance of the films, and no changes of the mechanical properties could be discerned in hardness measurements. BN coatings that had oxygen contamination, on the other hand, degraded quickly in appearance and hardness after a few days' exposure to atmospheric humidity and were quickly destroyed by exposure to acid. These results indicate the necessity of a clean vacuum environment to ensure film stability.

D. ANALYSIS OF i-BN COATINGS

Strong activation involved in many ion beam assisted deposition processes does not only improve the quality and performance of stable compound films, e.g. titanium nitride, but also allows the formation of metastable structures with novel properties. Among these, diamond and diamond-like, as well as hard boron nitride films, are of special interest due to their many superior characteristics.

Deposition of compound films is generally complicated by the need to accurately control the composition. Ion beam processes, being normally far from the equilibrium conditions, require special attention in this respect. The increased chemical activity of ionized species also attracts many impurities, and contamination with carbon, oxygen, and hydrogen are common in many practical coatings.

In this study we have analyzed the boron and nitrogen concentrations in BN films produced by ion beam assisted deposition (i.e. i-BN films). It is assumed that the main properties of the films are determined by the boron to nitrogen ratio. That is why it is essential to know the effect of the process parameters on this ratio. In addition, impurities, especially hydrogen contamination, were examined. Hydrogen is known to possess the active role in hydrogenated i-C films occupying unfilled tetrahedral bondings [31]. In ion-plated hard BN films deposited using NH_3 gas, high concentrations of hydrogen have been observed [32]. Moreover, IR spectroscopy has revealed H-B bonding in these films. For this reason, the incorporation of hydrogen could be expected also in the case where hydrogen is available only as a contaminant in the process.

D.1 NUCLEAR REACTION ANALYSIS (NRA)

The concentrations of boron and nitrogen were measured [33] using the nuclear resonance resonance reactions $^{11}B(p,\gamma)^{12}C$ at E_p = 163 keV, and $^{15}N(p,\alpha\gamma)^{12}C$ at E_p = 429 keV, respectively. The width of the $^{15}N(p,\alpha\gamma)^{12}C$ resonance at E_p = 429 keV is extremely narrow, Γ = 120 eV, allowing precise depth profiling. In this case the depth resolution was determined by the energy resolution of the accelerator used, 400 eV (FWHM) at the energy involved. This corresponds to a depth resolution of 5 nm at the surface of the BN sample. The width of the $^{11}B(p,\gamma)^{12}C$ resonance at E_p = 163 keV is, on the contrary, very broad, Γ = 5.7 keV. In this case the depth resolution was determined by the width of the resonance alone. Consequently, the depth resolution of the boron profiling is only 40 nm at the surface of the BN sample. Because of the low energy of the $^{11}B(p,\gamma)^{12}C$ resonance the molecular beam H_2^+ at the acceleration voltage ≥ 326 keV was used. Both in the nitrogen and boron case the gamma-rays were detected using a NaI (Tl) detector 12.7 x 10.2 cm^2 in size. The measured gamma-ray yields were converted into the absolute concentrations with the help of TiN and TiB_2 calibration samples respectively.

Hydrogen analysis was performed with the forward recoil spectroscopy technique (FRES) [34] using the He^+ beam at an energy of 2 MeV. For the determination of the absolute hydrogen incorporation the FRES spectrum from Kapton®* was used together with the aid of computer program RUMP [35]. Complementary analyses of oxygen and carbon were carried out utilizing the deuterium reactions $^{12}C(d,p)^{13}C$, $^{16}O(d,p)^{17}O$, and $^{16}O(d,\alpha)^{14}N$ [36, 37] at the energy of E_d = 925 keV.

*Kapton is a registered trademark for a polyimide material manufactured by DuPont Electronics, Wilmington, DE.

The nitrogen, boron, and hydrogen analyses were performed on all samples, whereas the carbon and oxygen analyses were carried out only on selected samples.

D.1.1 RESULTS AND DISCUSSIONS

Shown in Figures 13 and 14 are examples of nitrogen and boron distributions in two different i-BNfilms. The sample #1103 can be found to be formed mainly of the main constituents boron and nitrogen with the average ratio [B]/[N] = 1.5. In the case of the sample #0626 the nitrogen distribution is very inhomogeneous and the sum of the boron and nitrogen concentrations deviates remarkedly from 100%. This indicates the presence of an extra constituent (i.e., carbon) which was verified later with deuterium measurements.

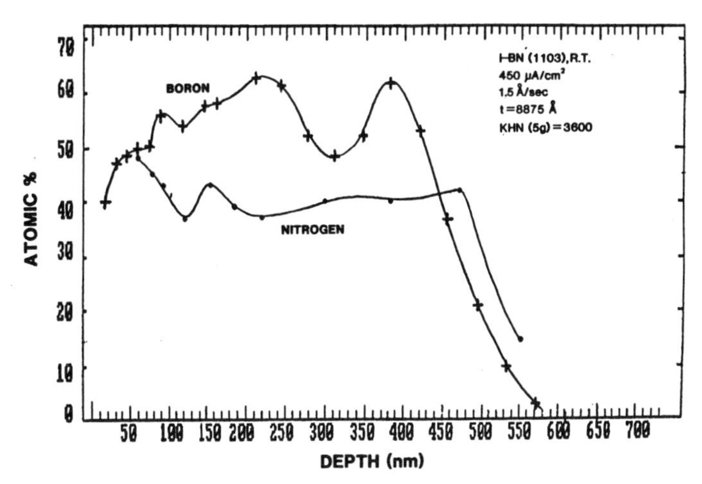

FIGURE 13. NITROGEN AND BORON DISTRIBUTIONS IN THE SAMPLE #1103.

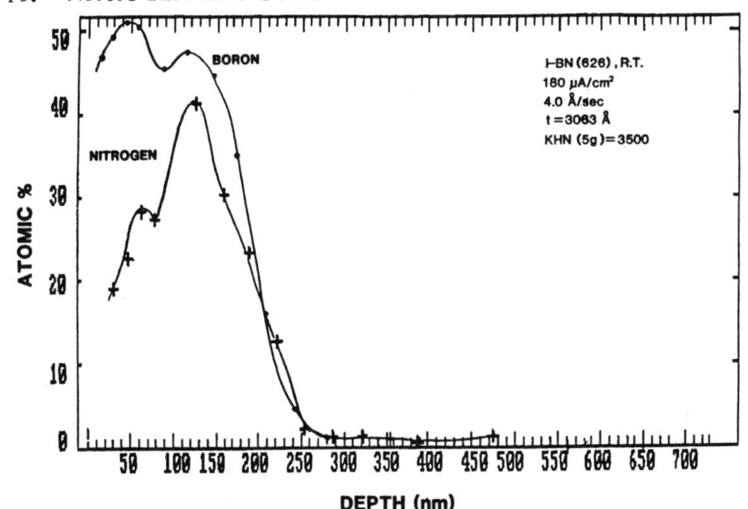

FIGURE 14. NITROGEN AND BORON DISTRIBUTIONS IN THE SAMPLE #0626.

The hydrogen contamination varied strongly. In most cases the contamination was small but detectable, e.g. 0.2 - 2.0 at. %. The surface contamination, however, was always very pronounced. In Figure 15 can be seen the FRES spectra from the samples #0626 and #0121. The hydrogen contamination of the former sample was the smallest of the samples examined. The incorporation of hydrogen in the sample #0121 was the highest observed, and the average concentration corresponds to 12 at.%. The high hydrogen concentrations ([H]$\geq$3.0 at. %) were measured in three samples, two of which were deposited at the elevated temperature (400°C).

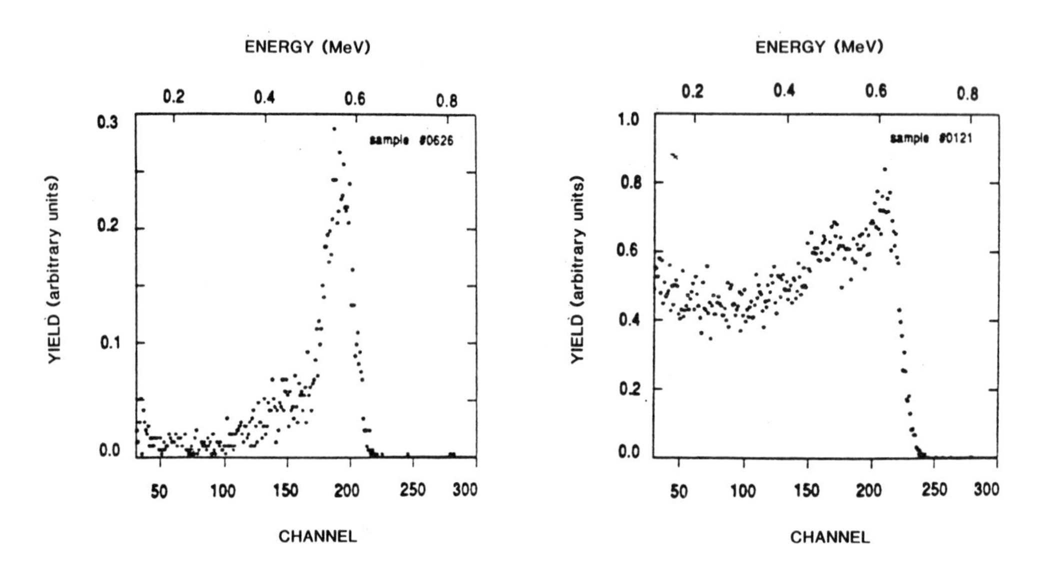

FIGURE 15. FRES SPECTRA OF THE SAMPLES #0626 and #0121.

Deuterium induced particle spectra were also taken from some selected samples. Shown in Figure 16 are the spectra from the samples #0626 and #0121. The signals from different elements are indicated in the figure. The most striking feature in the case of the sample #0626 is the strong signal from the $^{12}C(d,p)^{13}C$ reaction. Because the sum of boron and nitrogen in this sample was much less than 100% (Figure 14), it can be concluded on the basis of Figure 16 that carbon is the third main constituent of this film. This carbon contamination has been subsequently traced back to originating from the electron beam inadvertently striking the graphite crucible containing the boron evaporation charge.

The highest level of hydrogen contamination (i.e. 12 at.%) was seen in sample #0121 and is attributed to contamination from residual water vapor and hydrocarbons in the vacuum.

In both samples of Figure 16, detectable amounts of oxygen are visible. The oxygen signal was very small in all samples corresponding to only a few percent. However, in the case of the sample #0121 with the high hydrogen contamination the oxygen signal is slightly stronger indicating that at least part of the hydrogen originates in water vapor and that some water is incorporated in the film.

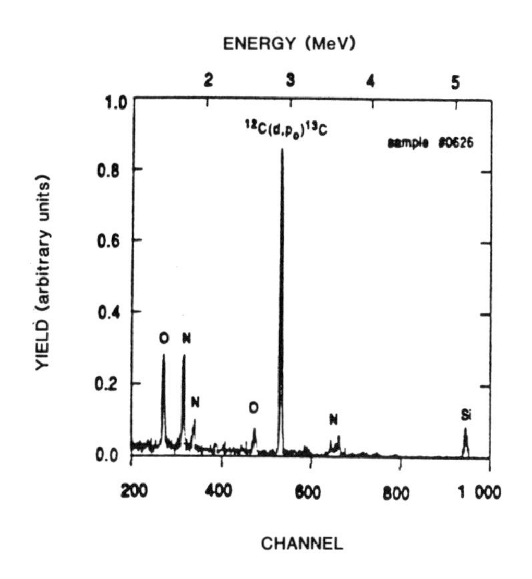

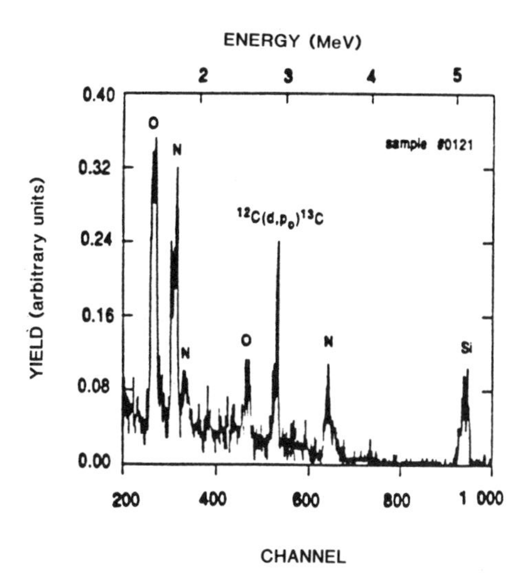

FIGURE 16. DEUTERIUM INDUCED PARTICLE SPECTRA OF THE SAMPLES #0626 AND #0121.

To relate the composition of the samples to the deposition conditions the [B]/[N] ratio was plotted as a function of the quantity R defined as:

$$R = \frac{\text{ion beam current density } (\mu A/cm^2)}{\text{arrival rate of evaporant } (\text{Å}/s)}$$

In Figure 17 the data for the depositions at low temperature as well as elevated temperatures are shown. In the low temperature case, only the samples with the low contamination level and reliable analysis were involved, whereas in the case of the elevated temperature, all three samples have been taken into account. The general tendency, as can also be expected, in both cases is the increasing [B]/[N] ratio with the decreasing R. In almost all cases the hyperstoichiometric boron concentration was observed. This is consistent with the observation of Weissmantel et al., who reported the boron concentrations from almost 100% to 50% in their i-BN films fabricated with the ion-plating technique [14]. Moreover, as can be seen in Figure 17 a higher temperature seems to favor the hyperstoichiometric boron concentrations.

The chemical composition of the i-BN films seems to have good correlation with the mechanical properties and this is discussed in Section D.5. It is interesting to note that the sample with the highest carbon contamination possesses the best tribological properties. This was thought to be due to the formation of boron carbide; however, ESCA analysis does not indicate B-C bonding in the form of B_4C.

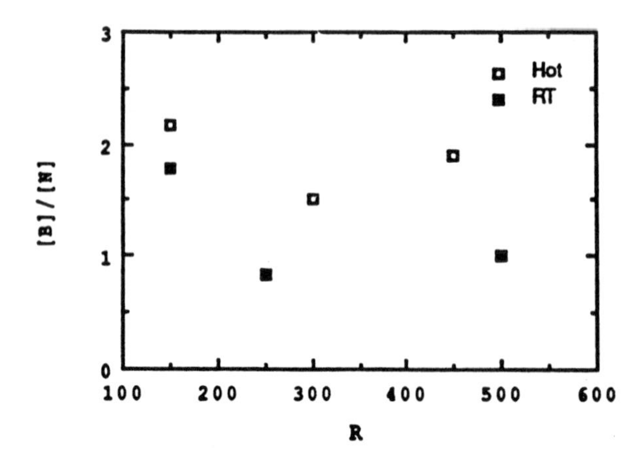

FIGURE 17. THE DEPENDENCE OF THE [B]/[N] RATIO ON THE DEPOSITION PARAMETERS.

D.1.2 CONCLUSIONS

A comprehensive chemical analysis based on the ion beam methods has been performed on several i-BN films. All films (except one) had the hyperstoichiometric boron concentration. The ratio [B]/[N] approached the theoretical value at the highest current densities of the nitrogen beam.

In three samples high hydrogen contamination was found although pure nitrogen was used in the ion source. This impurity level is related to the mechanical behavior of the films. In one case, high carbon contamination was observed and correlates with improved mechanical properties.

D.2 AUGER AND XPS ANALYSIS

Results from NRA of IBED i-BN films (Section D.1) indicate that many films are hyperstoichiometric (boron rich) as confirmed by Auger spectroscopy and XPS (ESCA) analysis.

An Auger spectrum of a cubic boron nitride reference sample (General Electric "BORAZON") appears in Figure 18(a). Figures 18(b) and 18(c) are spectra from an IBED i-BN film and show an apparent increase in boron enrichment with depth. An XPS depth profile of a similarly prepared BN film (Figure 19) also illustrates the hyperstoichiometric condition (B/N ratio) varying in the range of 2 to 3.

Included with the XPS survey is an interesting analysis which indicates that the relative amounts of boron nitride and elemental boron change several times throughout the depth profile. Figure 20 shows a three-dimensional montage of the boron (1s) binding energy peak as a function of depth in an IBED i-BN film. The high binding energy peak is due to boron in a B-N binding state, while the lower peak is attributable to elemental boron. The peak position varies depending upon the relative concentrations of BN and B present at that depth. It is not known at this time whether these variations are due to subtle changes in the growth dynamics of the BN system or whether they are the result of non-uniformities in the atomic arrival rate ratios of the various reactive species involved (i.e., B, N_2^+, N^+, N_2^0).

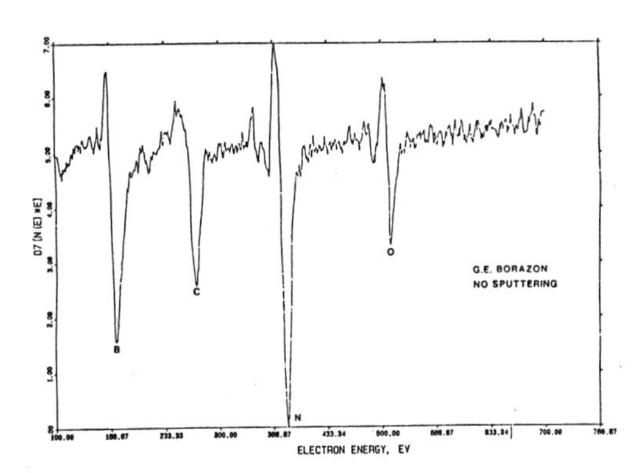

FIGURE 18(a). AES DATA FOR G.E. CBN REFERENCE SAMPLE.

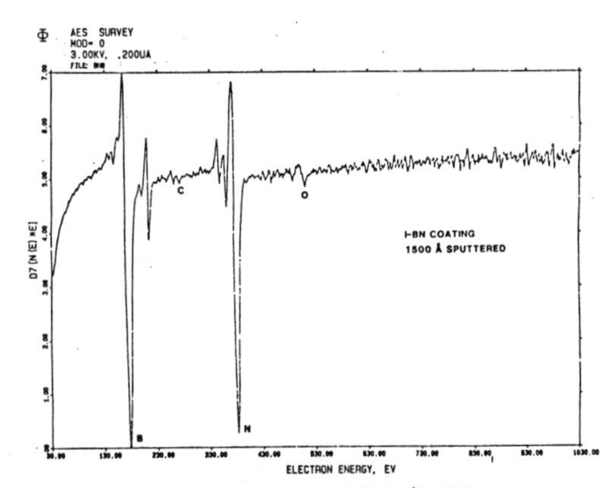

FIGURE 18(b). AES DATA FOR i-BN SAMPLE (1500 angstroms sputtered).

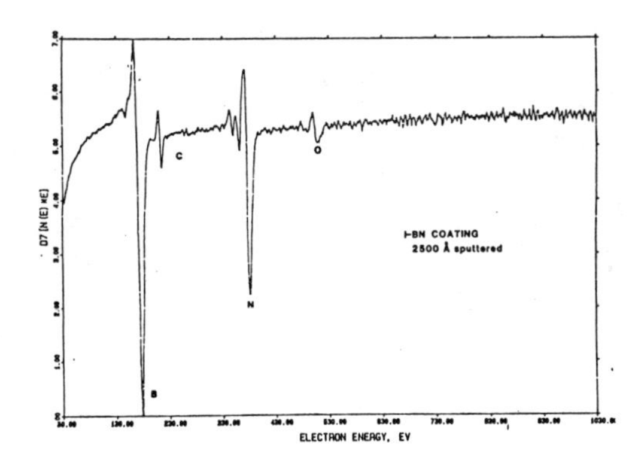

FIGURE 18(c). AES DATA FOR i-BN COATING (2500 angstroms sputtered).

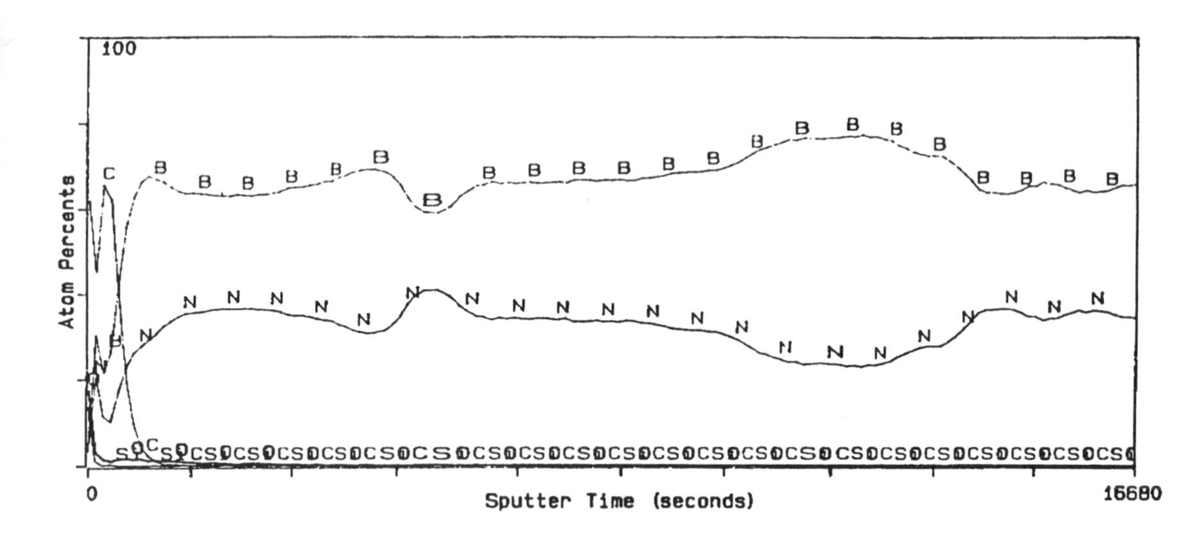

FIGURE 19. XPS DEPTH PROFILE OF IBED i-BN SAMPLE.

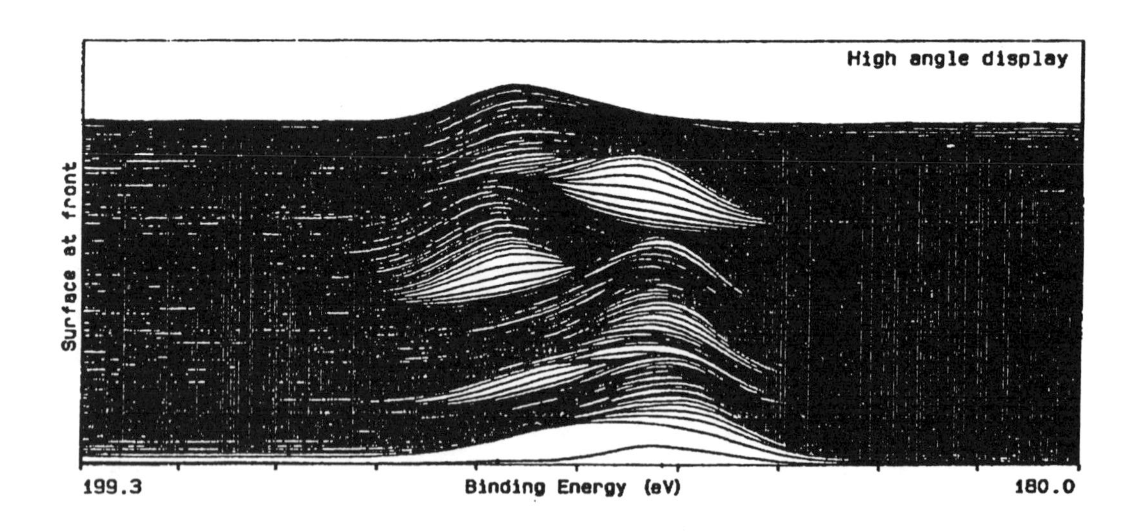

FIGURE 20. THREE-DIMENSIONAL MONTAGE OF BORON (1s) BINDING ENERGY PEAK AS A FUNCTION OF DEPTH IN IBED i-BN.

D.3 STRUCTURE AND CRYSTALLINITY

The boron nitride system has many similarities to the naturally occurring carbon system. Boron nitride exists in a soft, lubricious hexagonal phase (similar to graphite) as well as the super-hard cubic phase (similar to diamond). A weakly bonded, layered network of trigonally coordinated (sp^2) sites is found in the former while the latter consists of tetrahedrally coordinated (sp^3) sites which account for its many unique physical properties. Both systems can exist (less commonly) in a tetrahedrally coordinated, hexagonal close-packed structure known as wurtzite type.

Electron energy loss spectroscopy (EELS), and transmission electron microscopy (TEM) with selected area diffraction (SAD) analyses have been performed on selected i-BN samples in this study in an attempt to identify the crystalline nature of these films. There is some evidence to suggest that there may be differences in the nucleation process for crystallite growth between thin (approximately 450 angstrom) and thick (12,000 angstrom) films, since films of these thicknesses deposited under identical conditions are quite different in TEM. While a more detailed investigation into this hypothesis is being conducted, we present some preliminary results here.

A thin film of i-BN (~450 angstrom) was examined by EELS and TEM/SAD. Both techniques confirmed a featureless, amorphous matrix of BN. The EELS measurements in the core electron excitation region and the plasmon-loss region shown in Figure 21 agree well with those of a BN (hexagonal) standard Figure 22. Location of the B-K edge and N-K edge and the position of the main plasmon peak at 23.9 eV are indicative of a loosely packed hexagonal BN structure. The diffraction pattern (Figure 23) for this thin film indicates an amorphous structure and the TEM image was featureless and showed no crystallites.

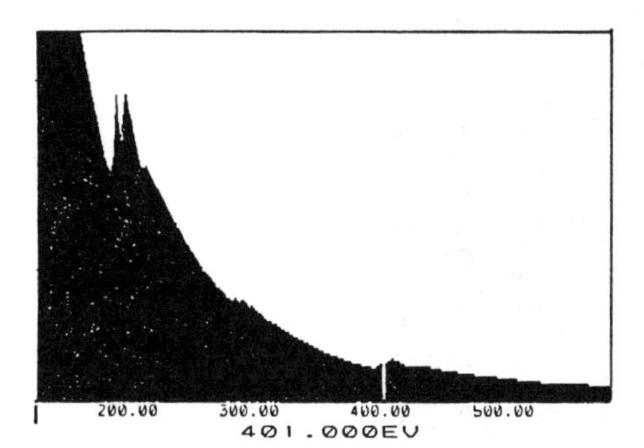

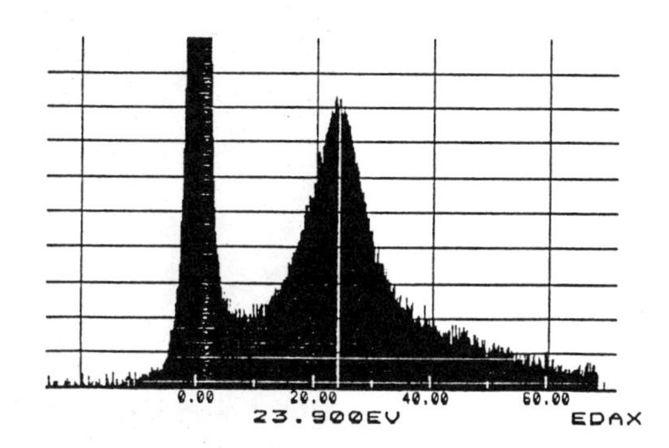

FIGURE 21. CORE ELECTRON EXCITATION REGION (a) AND PLASMON LOSS REGION (b) FOR IBED I-BN SAMPLE.

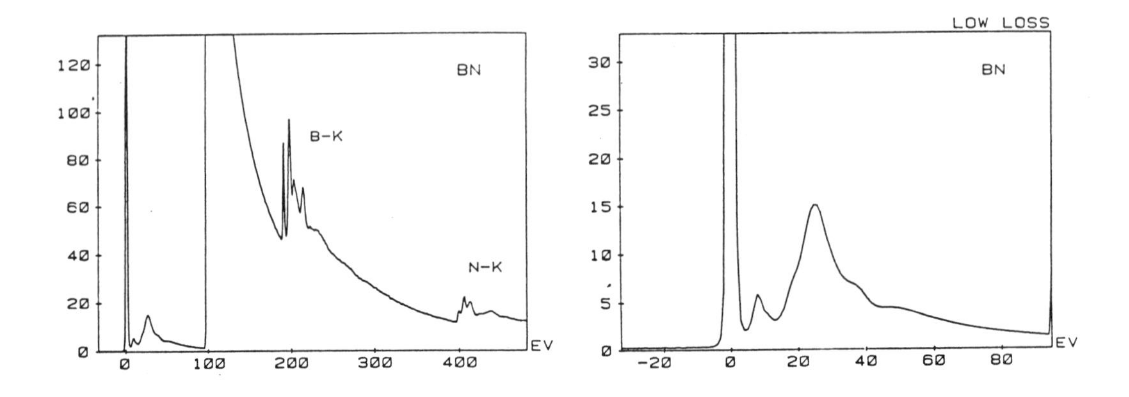

FIGURE 22. CORE ELECTRON EXCITATION REGION (a) AND PLASMON LOSS REGION (b) FOR HEXAGONAL BN STANDARD.

FIGURE 23. DIFFRACTION PATTERN SHOWING AMORPHOUS CONDITION OF 450 ANGSTROM i-BN FILM.

In contrast to the thin film characteristics, a thicker film (12,000 angstrom) was similarly analyzed by TEM/SAD and indicated a more structured system with crystallites of 200-500 angstrom cross section present. Figure 24 shows the TEM image and the associated SAD pattern. Analysis of the d-spacing ratios from this pattern is given in Table IV and clearly indicates the crystallites to be cubic BN.

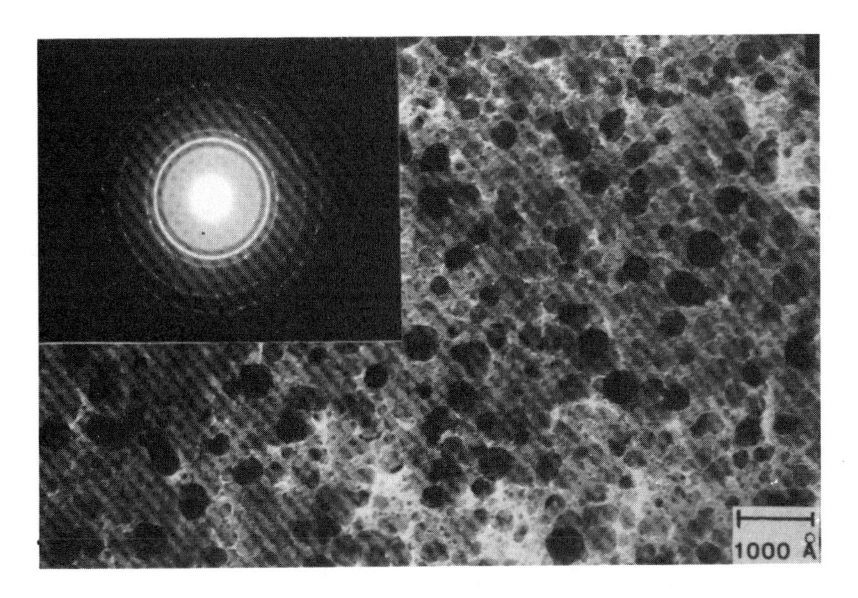

FIGURE 24. TEM IMAGE OF 12,000 ANGSTROMS i-BN FILM SHOWING CUBIC BN CRYSTALLITES. SAD pattern (inset).

TABLE IV. d-SPACING RATIOS* FOR THE BORON NITRIDE SYSTEM

IBED i-BN	Cubic BN	Hex.BN	Wurtzite BN
1.153	$\frac{(111)}{(200)}$=1.154	$\frac{(002)}{(100)}$=1.534	$\frac{(100)}{(002)}$=1.046
1.636	$\frac{(111)}{(220)}$=1.632	$\frac{(002)}{(101)}$=1.614	$\frac{(100)}{(101)}$=1.129
1.909	$\frac{(111)}{(311)}$=1.915	$\frac{(002)}{(102)}$=1.831	$\frac{(100)}{(102)}$=1.447

* Ratios for cubic BN derived from JCPDS card #35-1365. Ratios for hexagonal BN derived from JCPDS card #34-421. Ratios for wurtzite structure BN derived from JCPDS card #26-773.

Further work needs to be done in the investigation of how process parameters affect the nucleation and growth of second phase crystallites. The observed lack of precipitates for thin films (450 angstroms) vs. the results shown in Figure 24 makes it tempting to attribute the nucleation to be thickness or stress dependent.

D.4 INFRARED AND RAMAN SPECTRA ANALYSIS

IR transmission spectra for BN are prevalent in the literature (see Figure 25) for films which were deposited by a variety of methods. There is moderate agreement among investigators in the interpretation of these spectra with regard to the significance and location (wavenumber) of the major peaks, although the inconsistencies are enough to suggest that IR measurements alone are not sufficient to conclusively identify the crystal structure of a BN sample.

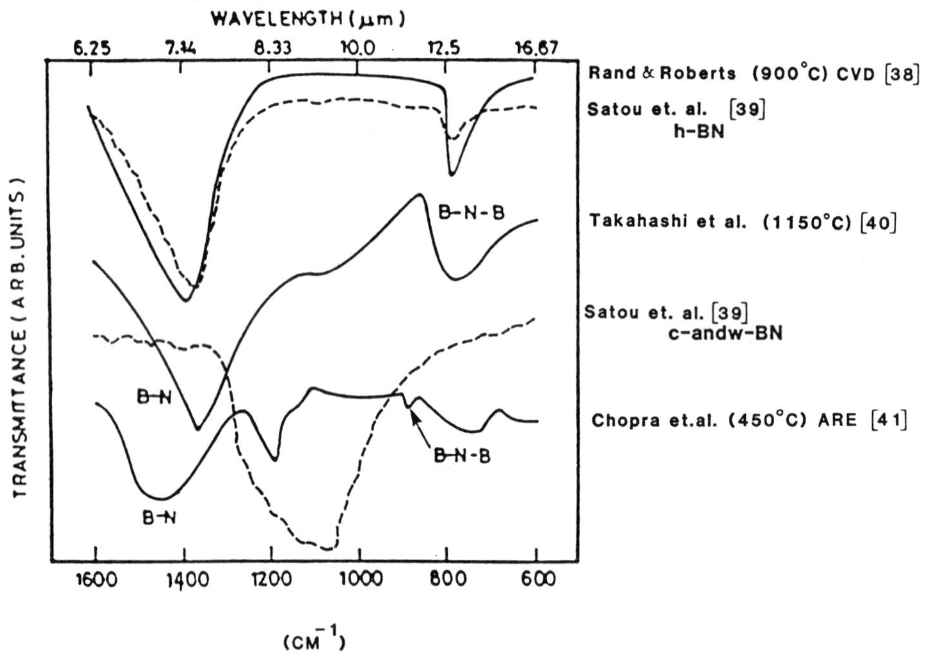

FIGURE 25. VARIOUS LITERATURE IR SPECTRA FOR BORON NITRIDE.

There are two peaks, however, which are generally attributed to be indicative of hexagonal phase BN. One rather broad peak is in the 1370 cm^{-1} to 1400 cm^{-1} range and is attributed to B-N bond stretching (in plane) while the other (usually smaller and somewhat more well defined) is located in the 750 cm^{-1} to 780 cm^{-1} range and is associated with B-N-B bending (out of plane). Occasionally seen are peaks at 2500 cm^{-1} (B-H bonds) and 3430 cm^{-1} (hydroxyl impurities or NH_2 groups) in films prepared by CVD type methods.

Among the smaller group of investigators who have claimed to identify cubic BN, there is more disagreement in the identification of peaks specific to the cubic phase.

Some researchers [42,43] attribute the peaks just mentioned (i.e., 1400 cm^{-1} and 800 cm^{-1}) to cubic phase, while others using reactive diode sputtering [44] attribute a peak that shifts from 810 cm^{-1} to 785 cm^{-1} with increasing bias to denote c-BN. Adding to the uncertainty are those findings [45,46] that claim a peak at 1050 cm^{-1} to 1100 cm^{-1} is characteristic of cubic BN.

Results of a study by Spire on selected IBED i-BN thin films are shown in Figure 26 and suggest a cubic phase (crystallites) in a hexagonal matrix due to peaks at 1100 cm^{-1} and 1400 cm^{-1}, 750 cm^{-1}, respectively. As mentioned in Section D.3, this has been confirmed by TEM/SAD analysis.

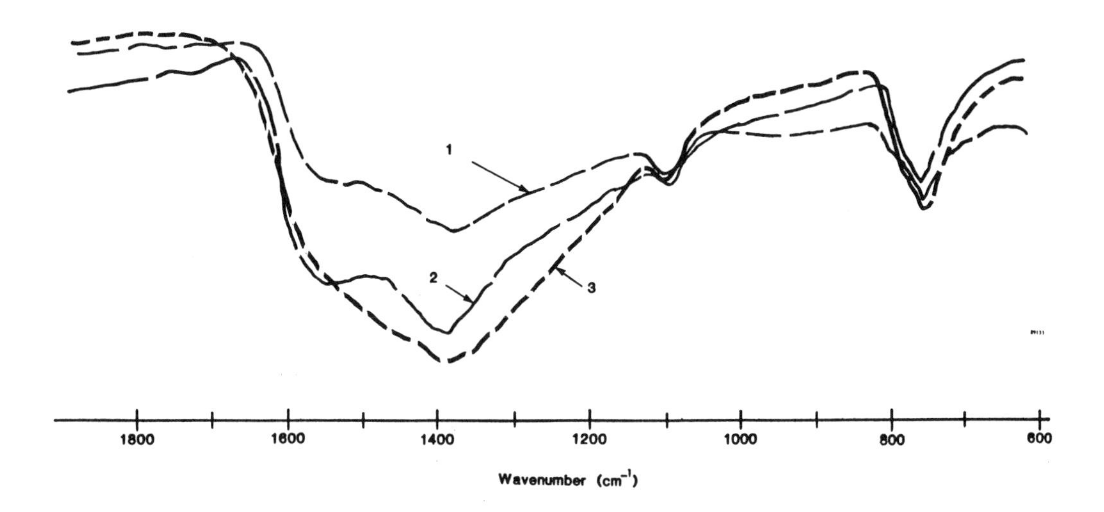

FIGURE 26. IR SPECTRA OF SELECTED IBED i-BN THIN FILMS.

In contrast to the conflicting interpretations which abound in IR studies of boron nitride, Raman spectroscopy appears to afford a clearer distinction between the structural phases (cubic vs. hexagonal) of this system.

Although the presence of peaks and their positions in the IR and Raman spectra of a material are often similar, they need not be so and indeed a substance may be inactive in infra-red and active in Raman. This difference is due to the fact that IR spectra are more dependent on molecular vibrations involving a change in dipole moment of the molecule, whereas Raman spectra are sensitive to changes in the polarizability of a molecule. As an example consider Cl_2, whose molecular vibrations, by symmetry, involve no net displacement of charge (dipole moment) and is therefore inactive in infra-red. However, the polarizability does change (with bond length) during vibration, and thus Cl_2 is active in Raman.

Figure 27(a) shows a Raman spectrum of a hexagonal-BN standard with characteristic peak at 1366 cm^{-1}. Figure 27(b) is the Raman of a cubic-BN standard showing the two characteristic peaks at 1055 cm^{-1} and 1306 cm^{-1}. The two spectra are clearly different. Figure 28 shows the Raman of an IBED i-BN film. A small peak at 1300 cm^{-1} is present while no peak is visible at the h-BN position of 1366 cm^{-1}. The spectrum suggests a cubic phase BN. The absence of the other cubic peak may be due to the fact that ratios of intensities of the two peaks (LO to TO modes) have been found to be as great as 10 in zincblende type crystal structures, [47] and at the very least have been found to be dependent on many factors [48].

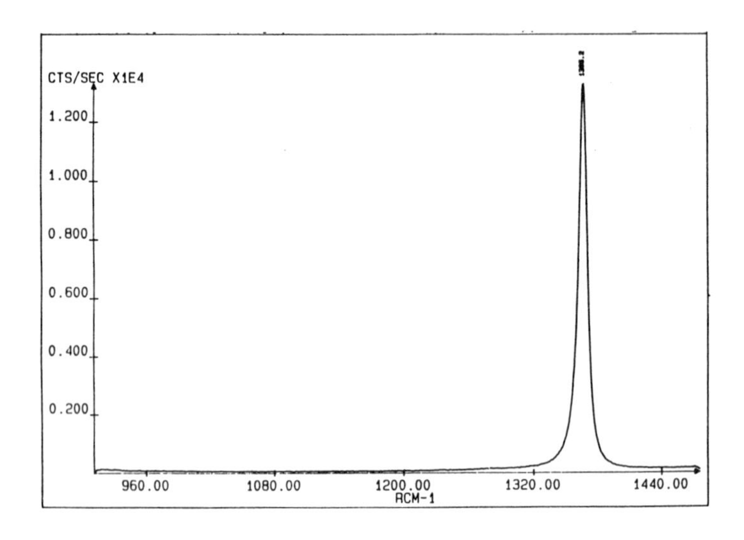

FIGURE 27(a). RAMAN SPECTRUM OF HEXAGONAL BN STANDARD.

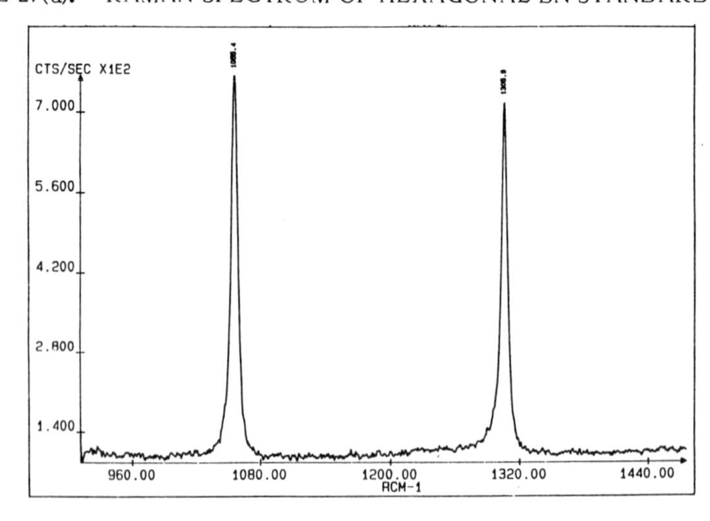

FIGURE 27(b). RAMAN SPECTRUM OF CUBIC BN STANDARD.

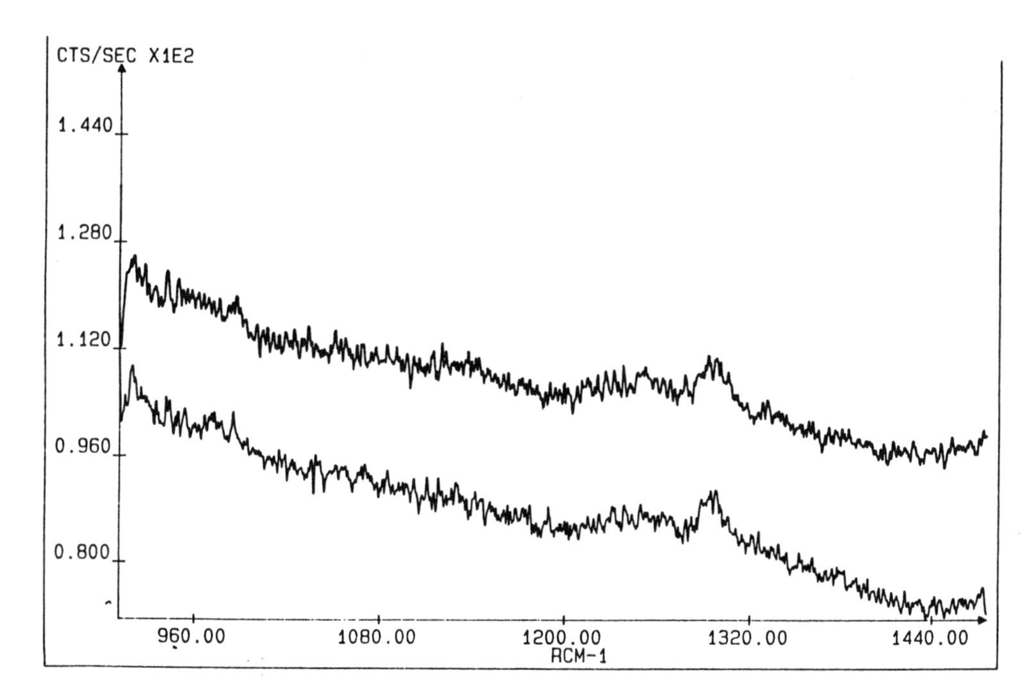

FIGURE 28. RAMAN SPECTRUM OF IBED i-BN.

D.5 FRICTION AND WEAR BEHAVIOR

Because of their high hardness and refractory nature, many nitride compounds are excellent candidates for coatings in a variety of high mechanical and/or thermal stress applications [49]. Other advantageous characteristics for a tribological coating would be excellent adhesion to its substrate as well as low friction and wear against a variety of counterface materials.

IBED i-BN films are being studied for their friction and wear behavior because the IBED process, as described earlier, characteristically produces highly adherent coatings and boron nitride as a material offers many unique properties especially in its cubic phase.

The friction and wear behavior of the coatings were investigated using an in-house built ball-on-disc apparatus. The test sequence involves a combination of adhesive and abrasive wear. Either continuous circular motion or reciprocating line motion of ball on coating is possible. Ball materials compared were silicon nitride and 440C stainless steel. All tests were run unlubricated in air, on coatings which were deposited on Si (100) wafers.

The load on the ball normal to the test sample can be varied by means of adding calibrated mass to the weight pan. During testing, the test stage is moved beneath the stationary loaded ball and the horizontal displacements of the ball/lever arm assembly are recorded on a strip chart recorder from the output of a load cell in the lever arm. Prior to the test, the recorder output is calibrated under a known load/displacement so that under the true test conditions the recorded signal yields the friction force of the ball/coating system. The coefficient of friction, μ , can now be calculated as the ratio of the normal force to the friction force.

The wear for the various systems was evaluated by measuring the depth profile of the wear scar seen on the coating using a Dektak profilometer and comparing it to the size of the wear flat on the ball surface under microscopic inspection. A new ball surface was used for each test. Estimates of coating volume loss were made from the wear scar profile and scar length (approximately 2 mm). By measuring the diameter of the contact area of the ball, the volume loss of the ball can be calculated in a manner similar to Hartley [50]. A dimensionless wear parameter, K, can be derived for the ball following Archard [51]:

$$K = 3H(\Delta V)/Ns$$

where H, Δ V, N and s are respectively the hardness of the ball (KHN = 700), the volume loss of the ball, the load, and the sliding distance. Figure 29 shows a typical wear track profile and associated ball wear flat.

The variation in coefficient of friction with sliding distance is summarized in Figures 30 and 31 for the Si_3N_4 ball and the 440C ball respectively. Additionally Figure 30 includes the boron to nitrogen ratio of the i-BN films determined by nuclear reaction resonance analysis as well as a qualitative mention of hydrogen content, [H], from forward recoil spectrometry (Section D.1). Figure 31 includes the wear parameter, K, for the 440C systems investigated, as well as estimates of coating volume loss.

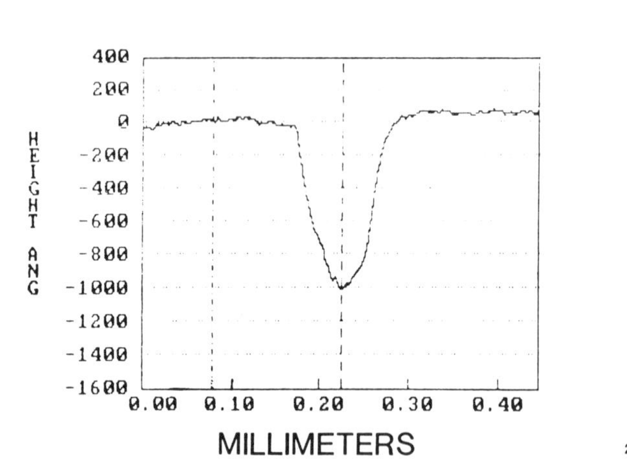

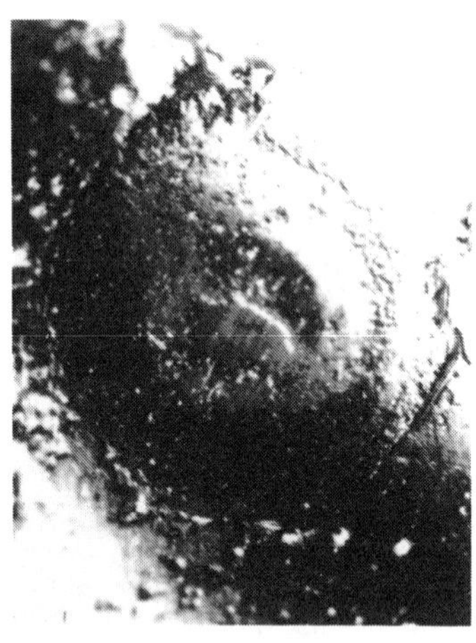

FIGURE 29. REPRESENTATIVE WEAR TRACK PROFILE AND ASSOCIATED BALL WEAR (400X) FOR 440C STAINLESS STEEL BALL ON i-BN SYSTEM.

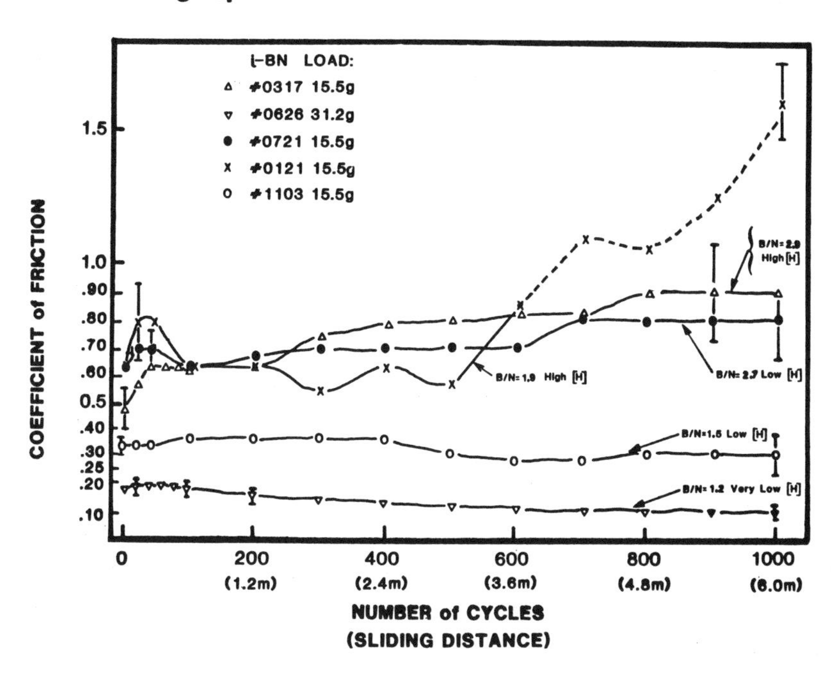

FIGURE 30. COEFFICIENT OF FRICTION VERSUS SLIDING DISTANCE. (Si_3N_4 ball, dry, on IBED nitride coatings.) From Ref. [37].

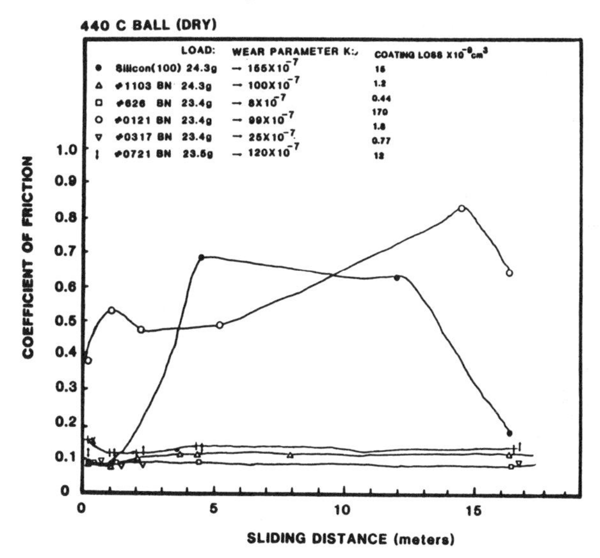

FIGURE 31. COEFFICIENT OF FRICTION VERSUS SLIDING DISTANCE. (440C ball, dry, on IBED nitride coatings.) From Ref. [37].

There appears to be a general increase in friction with increasing B/N ratio in the Si_3N_4 ball system while little difference is seen with the 440C system. Hydrogen content of the film seems to have an effect on the friction in both ball systems, i.e., friction increasing with [H]. Of particular interest is coating #0121 whose friction was among the highest in both systems. Its hydrogen concentration, [H] = 12%, was the highest measured of all the films. It should be noted that this coating exhibited brittle behavior and that it was worn through after about 600 cycles. Data beyond this point are influenced by accumulated debris along the wear track.

In the i-BN system, adhesive transfer of ball material to the coating occurred only with sample #0317. After the 440C ball test, several patches (< 1000 angstom thick) could be seen on the coating. A more continuous distribution of patches (approximately 1 micron thick) was present after the Si_3N_4 ball test Figure 32 shows a SEM micrograph of this wear track. Compare this figure to Figure 33, which shows an SEM of the wear track for sample #0626. Although the test for #0626 was run with twice the load and 20 times the number of cycles as sample #0317, essentially no wear was visible. Once again, this reinforces the conclusions from Figure 30 that a low B/N ratio and low [H] result in films with better mechanical properties in friction and wear. It should be noted that NRA of sample #0626 also indicated the presence of some carbon contaminations, as mentioned earlier, and this may also contribute to the low friction/wear behavior.

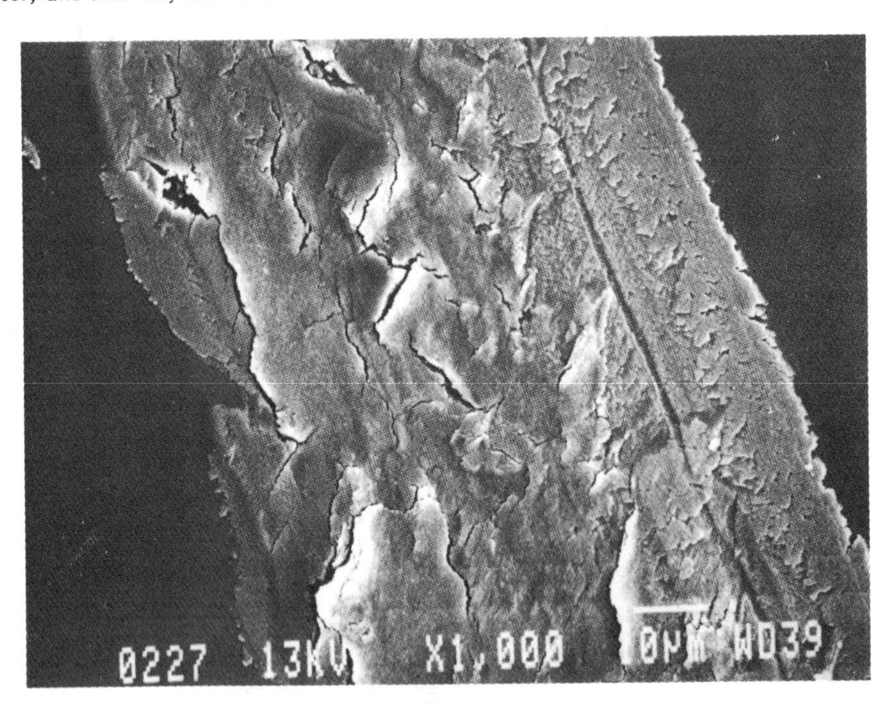

FIGURE 32. SEM MICROGRAPH SHOWING ADHESIVE TRANSFER TO COATING #0317 FOLLOWING Si_3N_4 BALL TEST.

FIGURE 33. SEM MICROGRAPH OF WEAR TRACK FOR SAMPLE #0626.

In summary, we can state that many of the i-BN coatings show low friction ($\mu \sim 0.1$) against 440C stainless steel and Si_3N_4 counterfaces. The coatings generally show a ductile behavior with very good adhesion to the Si(100) substrate. Friction seems to increase in the i-BN system with increasing B/N ratio and also when hydrogen content is higher. Hydrogen incorporation is also seen to have a deleterious effect on mechanical behavior.

E. POSSIBLE i-BN APPLICATIONS

The advent of relatively low temperature processes, such as the two described in this chapter, for the deposition of boron nitride thin films onto a variety of substrate materials, has sparked investigation into utilizing this unique material in an ever widening variety of applications of commercial interest.

E.1 OPTICAL

Spire IBED i-BN thin films have been investigated as possible optical baffle coatings for use in telescopes. Its high spectral emittance in the 8-12 micron window (Figure 34) coupled with its demonstrated resistance to high energy pulsed electron radiation (Figure 35) has made it an interesting candidate material.

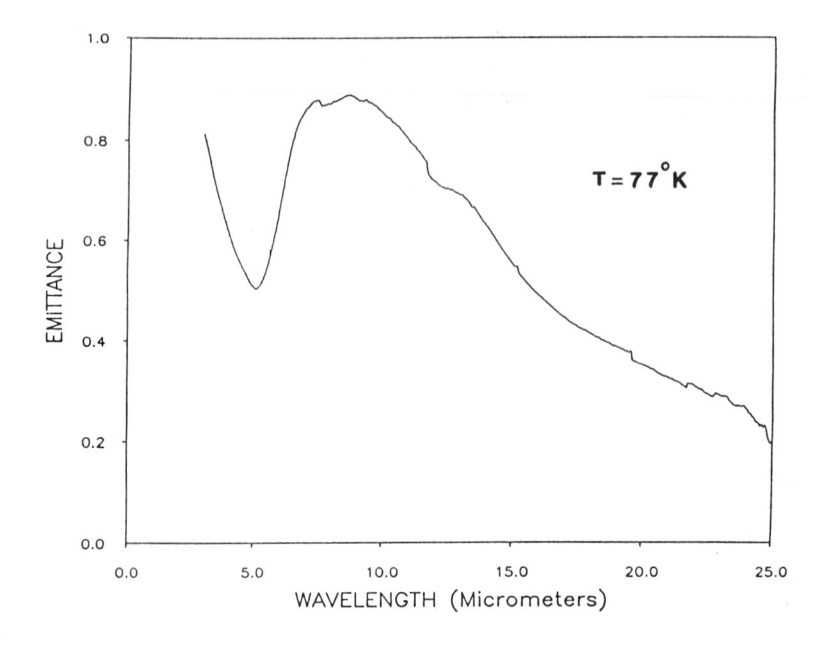

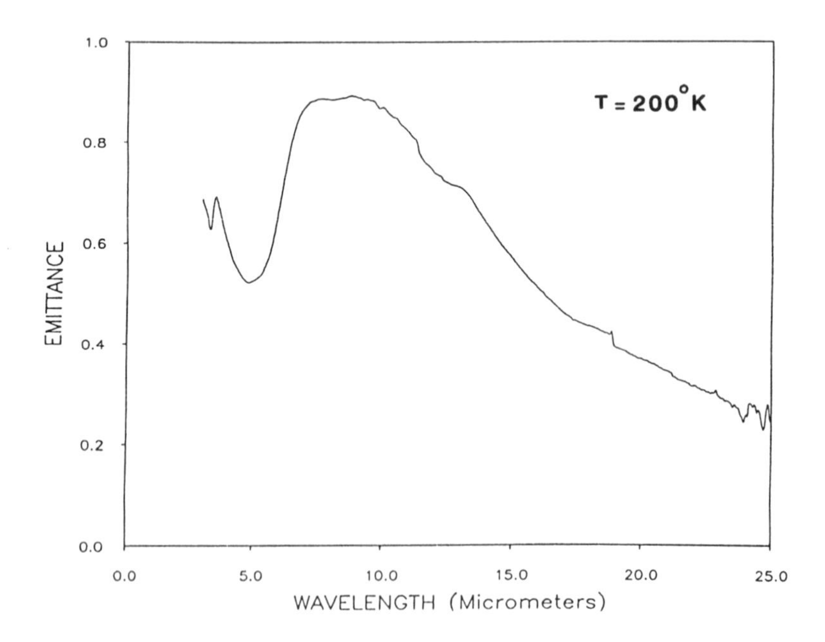

FIGURE 34. SPECTRAL EMITTANCE OF BORON NITRIDE ON ETCHED ALUMINUM AT 77°K AND 200°K.

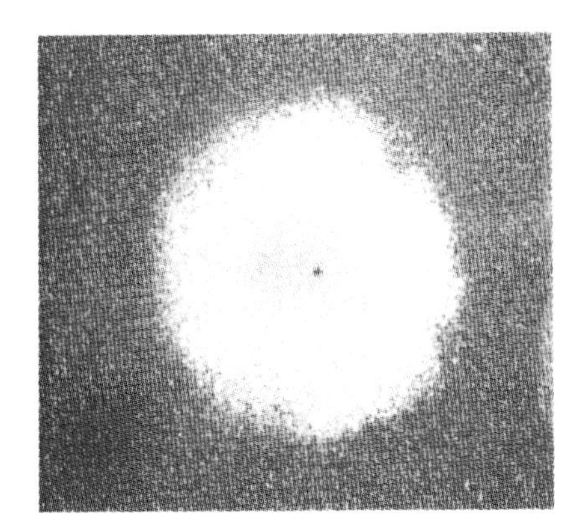

FIGURE 35. BLOW-OFF SCAR OF i-BN ON Al SUBSTRATE FOLLOWING PULSED ELECTRON BOMBARDMENT (3.58 cal/60 ns pulse). The outer edge of the scar where delamination was arrested received an energy flux of 0.6 cal/cm^2.

E.2 LOW-Z COATING

Research at Spire in conjunction with the DOE, MIT Fusion Center and Oak Ridge National Laboratories has been investigating using i-BN as a low-average Z ($\bar{Z}$=6), refractory coating for use in plasma confinement systems. Low Z materials are preferred for use in these plasma environments because radiation losses from them produce less contamination and stability problems for the plasma. Testing at ORNL in a neutral beam test apparatus delivered 1100 pulses of 200 W/cm^2 over two seconds every five seconds. The i-BN coating survived as well as a TiC coating which is widely used now in plasma systems and has the potential advantage of having a lower Z value.

Research sponsored by both government and private sector funding has generated much promise for the eventual commercialization of i-BN thin film technology into areas which have only recently been envisioned by today's technological advances.

G. ACKNOWLEDGEMENTS

The authors wish to gratefully acknowledge the contributions of the following individuals in providing many of the analytical results: J.P. Hirvonen, University of Helsinki (NRA, friction tests); T.C. Chou, Engelhard Corporation (XPS, RAMAN); Peter Kullen, Carnegie-Mellon (TEM); Don Potter, U. Conn. (TEM); Ray Egerton, S.U.N.Y. at Stonybrook (EELS); David Smith, Chomerics (IR).

The work presented here has been supported, in part, by the following Small Business Innovation Research (SBIR) programs:

National Science Foundation	Grant No. DMR-8212902
Department of Defense (Air Force)	Contract No. F33615-87-C-5203 Contract Monitor: Bob McConnell
Department of Energy:	Contract No. DE-AC02-87ER80449 Contract Monitor: Don Beard

G. REFERENCES

1. Wentorf, R.H., Jr.: J. Chem. Physics, 1957, 26, 956.

2. Aisenberg, S. and Chabot, R.: J. Appl. Phys., 1971, 42, 2953.

3. Weissmantel, C.; Reisse, G.; Erler, J.J.; Henny, F.; Bewilogua, K.; Ebersbach, U.; and Schrurer, C.: Thin Solid Films, 1979, 63, 315.

4. Berg, S. and Andersson, I.P.: Thin Solid Films, 1979, 58, 117.

5. Glashtold, S.A.; Kutol'in, S.A.; Shakarian, L.M.; Bielova, L.F.; and Komarova, G.M.: Elektron. Tekh., 1970, 12, 58.

6. Weissmantel, C.; Bewilogua, K.; Dietrich, D.; Erler, H.J.; Hinnerberg, H.J.; Klose, S.; Nowick, W.; and Reisse, G.: Thin Solid Films, 1982, 96, 31.

7. Shanfield, S. and Wolfson, R.: J. Vac. Sci. Technol. A, 1983, 1, 323.

8. Kaufman, H.R.: J. Vac. Sci. Technol. A, 1986, 4(3), 764.

9. Halverson, W. and Quinto, D.T.: J. Vac. Sci. Technol. A, 1985, 3, 2141.

10. Malter, L.: Phys. Rev., 1936, 49, 478.

11. Schmolla, W. and Hartnagel, H.L.: Solid-State Electron., 1983, 26, 931.

12. Harper, J.M.E.; Cuomo, J.J.; Gambino, R.J.; and Kaufman, H.R. in "Ion Bombardment Modification of Surfaces," ed. O. Aucielo and R. Kelly, Elsevier Press, New York (1984).

13. Pranevicius, L.: Thin Solid Films, 1979, 63, 77.

14. Weissmantel, C.; Reisse, G.; Erler, H.J.; Henny, F.; Bewilogue, K.; Ebersvack, V.; Schurer, C.: Thin Solid Films, 1979, 63, 315.

15. Colligon, J.S.; Hill, A.E.; Kheyrandis, H.: Vacuum, 1984, 34(10/11), 843.

16. Martin, P.J.; Netterfield, R.P.; Sainty, W.G.: J. Appl. Phys., 1984, 55, 235.

17. Takagi, T.: Thin Solid Films, 1981, 92, 1.

18. Satou, M.; Andoh, Y.; Ogatag, K.; Suzuki, Y.; Matsuda, K.; Fujimonto, F.: Jpn. J. of Appl. Phys., 1985, 24(6), 656.

19. Sato, T.; Ohata, K.; Asahi, N.; Ono, Y.; Oka, Y.; Hashimoto, I.: J. Vac. Sci. Tech., 1986, A4(3), 784.

20. McNeil, J.R.; Barron, A.C.; Wilson, J.R.; Herrmann, W.C.: Applied Optics, 1984, 23(4), 552.

21. Kant, R.A.; Sartwell, B.D.; Singer, I.L.; Vardiman, R.G.: See Ref. 6, 1985, 915.

22. Kennemore, C.M.; Gibson, U.J.: Appl. Optics, 1984, 23(20), 3608.

23. Martin, P.J.; Netterfield, R.P.; Sainty, W.G.: J. Appl. Phys., 1984, 55(1), 235.

24. Cuomo, J.J.; Harper, J.M.E.; Guarnieri, G.R.; Yee, D.J.; Attanasio, L.J. et al.: J. Vac. Sci. Technol., 1982, 20, 349.

25. Netterfield, R.P.; Muller, K.-H.; McKenzie, D.R.; Goonan, J.J.; Martin, P.J.: J. Appl. Phys., 1988, 63(3), 760.

26. Hubler, G.K.: J. Mat. Science & Engin., 1989 (Proceedings of the 1988 SM^{2}IB conf.) to be published.

27. Namba, Y. and Mori, T.: J. Vac. Sci. Technol. A, 1985, 3(2), 322.

28. Satou, M. and Fujimoto, F.: Jpn J. Appl. Phys., 1983, 22, L171.

29. Sainty, W.G.; Martin, P.J.; Netterfield, R.P.; McKenzie, D.R.; Cockayne, D.J.H.; Dwarte, D.M.: J. Appl. Phys., 1988, 64(8), 3980.

30. Van Vechten, D.; Hubler, G.K.; Donovan, E.P.: J. Vac. Sci. and Tech., 1988, A6, 1934.

31. Weissmantel, C.: in Thin Films From Free Atoms and Particles, ed. by K.J. Klabunde, 1985, Academic Press, NY, 153.

32. Rother, B.; Zscheile, D.; Weissmantel, C.; Heiser, C.; Holzhuter, g., Leonhardt, G.; Reich, P.: Thin Solid Films, 1986, 142, 83.

33. Hirvonen, J-P and Hirvonen, J.K.: Proceedings of Materials Research Society, 1989, 128 (to be published).

34. Feldman, L.C. and Mayer, J.W.: Fundamentals of Surface and Thin Film Analysis, 1986 (North Holland), 31.

35. Doolittle, L.R.: Nucl. Inst. Meth., 1985, B9, 344.

36. Amsel, G.; Nadai, J.P.; D'Artemare, E.; David, D.; Girard, E.; and Moulin, J.: Nucl. Inst. Meth., 1971, 92, 481.

37. Tetreault, T.G.; Hirvonen, J.K.; Parker, G.R.; and Hirvonen, J-P: Proceedings of Materials Research Society, 1989, 140 (to be published).

38. Rand, M.J. and Roberts, J.F.: J. Electrochem. Soc., 1968, 115(4), 423.

39. Satou, M.; Yamaguchi, K.; Andoh, Y.; Suzuki, Y.; Matsuda, K.; Fujimoto, F.: Nucl. Instr. and Meth., 1985, B7/8, 910.

40. Takahashi, T.; Itoh, H.; Takeuchi, A.: J. Cryst. Growth, 1979, 47, 245.

41. Chopra, K.L.; Agarwal, V.; Vankar, V.D.; Deshpandey, C.V.; Bunshah, R.F.: Thin Solid Films, 1985, 126, 307.

42. Lin, P.; Deshpandey, C.; Doerr, J.; Bunshah, R.F.; Chopra, K.L.; Vankar, V.: Thin Solid Films, 1987, 153, 487.

43. Voskoboynikov, V.V.; Gritsenko, V.A.; Effimov, V.M.; Lesnikovskaya, V.E.; Edelman, F.L.: Phys. Status Solidi (A), 1976, 34, 85.

44. Seidel, K.H.; Reichelt, K.; Schaal, W.; Dimigen, H.: Thin Solid Films, 1987, 151, 243.

45. Gielisse, P.J.; Mitra, S.S.; Plendl, J.N.; Griffis, R.D.; Mansur, L.C.; Marshall, R.; Pascoe, E.A.: Phys. Rev., 1967, 155(3), 1039.

46. Inagawa, K.; Wantabe, K.; Fanaka, I.; Saitoh, S.; Itoh, A.: Proc. 9th ISIAT Symp., Tokyo, 1985, 299.

47. Couture-Mathieu, L.; Mathieu, J-P: Comptes-Rendus Academie Des Sciences (Paris), 1953, 236, 371.

48. Ushioda, S.; Pinczuk, A.; Taylor, W.; Burstein, E.: in II-V Semiconducting Compounds, 1967, ed. D.G. Thomas, W.A. Benjamin, New York.

49. Holleck, H.: J. Vac. Sci. Technol., Nov/Dec 1986, A4(6), 2661-2669.

50. Hartley, N.E.W.: Tribology International, April 1975, 65.

51. Archard, J.F.: J. Appl. Physics, 1953, 24, 981.

Materials Science Forum Vols. 54 & 55 (1990) pp. 111-140

PREPARATION OF BORON NITRIDE THIN FILMS BY CHEMICAL VAPOR DEPOSITION

K. Nakamura
Department of Chemistry, College of Humanities and Sciences
Nihon University, Sakurajosui Setagaya-ku, Tokyo 156, Japan

ABSTRACT

Two methods are applied for the preparation of boron nitride films, which are molecular flow chemical vapor deposition (MFCVD) and metal organic chemical vapor deposition (MOCVD). The MFCVD which is chemical vapor deposition at a pressure in molecular flow region is superior in high controll efficiency of the deposition rate and composition of the product. The preparation conditions of boron nitride films in MFCVD are clarified. The MOCVD which is known as a preparation method of the compound semiconductors is also applied to the preparation of boron nitride films. In MOCVD, triethylboron and trimethoxyborane were used as boron source gases. The preparation conditions are clarified and the reaction mechanism is discussed. A "tympanic membrane" which bears tensile stress supported by a substrate ring is made by born nitride films with an intention to be applied to x-ray lithography mask.

I. INTRODUCTION

Crystal structures of boron nitride are very similar to those of carbon which have the structures of hexagonal and cubic symmetries. Recently, an increased attention is paid on the investigations of cubic boron nitride prepared at atmospheric pressure as the diamond films prepared by CVD. Boron nitride is transparent in the visible spectral region, highly resistant against chemical corrosion by acids and bases, refractory in vacuum up to 2000°C and inactive to many molten metals. Boron nitride is useful for electric insulator because of their high electric resistivity of the order of $10^{13}\Omega$-cm. Until now, however, these superior properties are not fully applied in industry. Investigations on new methods of preparation of high quality boron nitride should bring new possibilities for its applications.

Boron nitride films were first deposited using the reaction of trichloroboron with ammonia by Mayer et. al.(1). Befor this synthesis boron nitride had been prepared only in a powder form. After this first success, the deposition of boron nitride was carried out by using the reaction of a boron halide and ammonia (2,3).

The preparation of the films using the reaction of diborane with ammonia developed by Rand and Roberts(4) was epochal in contrast to the method using halogen compound sources. This method has been developed by the requirements of high purity boron nitride films to be used as electronic materials. Afterwards, Hirayama and Shono[5] demonstrated usefullness of boron nitride films deposited on silicon wafer as a diffusion source of boron. They fabricated silicon planer diodes and MIS memory diodes[6] by using this method. Hyder and Yep[7] measured electrical resistivity and dielectric constant of boron nitride films deposited by plasma assisted CVD to the same source gasses.

Several reports[8,9] on the boron nitride films in Bell Laboratory are interesting from the viewpoint of extending the possibilities of applications. In these repots, Murark et al[8] clarified the relation among composition, refractive index, stress, and deposition conditions of boron nitride films deposited by using diborane with ammonia in low pressure chemical vaper deposition (LPCVD). Adams and Capio[9] clarified the composition ratio of N, B and H in boron nitride films deposited by plasma CVD at the temperature range 250-600°C. Adams[10] reported also the preparation of boron nitride films deposited by pyrolysis of borazine at 300-650°C in 0.1 and 0.8Torr. The activation energy for deposition by this method using borazine was 9.6kcal/mol which was smaller than 20-25kcal/mol[9] obtained by using the reaction of diborane with ammonia. The purpose of these investigations at Bell Laboratory was to fabricate x-ray lithograph mask membrane, in particular, to deposit boron nitride film on silicon substrate with a tension stress about 1×10^9dyne/cm^2 [11].

On the other hand, application of boron nitride films as an insulator layer in III-V compound semiconductor divices was studied on the system of diborane with ammonia by Miyamoto et. al.[12] and Yamaguchi et. al.[13].

The boron nitride films deposited by using the boron halide source have been studied continuously[14,15]. The boron nitride film thinner than 1000Å deposited by using boron halide on quartz substrate by Boronian in 1972[16] have the refractive index of 1.9-2.0 and optical band gap energy 5.83eV. Sano et. al.[17] obtained boron nitride films with optical absorption edge of 5.8eV by CVD using a boron halide at 600-1100°C.

In 1980's, the studies of a plasma activated CVD[18] increased in order to lower the crystallization temperature and deposition temperature. Miyamoto et. al. [12] prepared insulating boron nitride thin films at the temperaturs below 300°C by the radio frequency (RF) glow discharge decomposition of a B_2H_6+NH_3+H_2 gas mixture. The optical and structural properties of the boron nitride films by plasma enhanced CVD were investigated as a function of the deposition temperature between 200 and 400°C, and NH_3 to B_2H_6 ratios, betwen 1 and 4 by Yuzuriha and Hess [19].

Sputtering is also used for the fabrication of boron nitride films. It was reported by Rother[20] that turbostratic structure boron nitride film was deposited by sputtering. Yasuda[21] also reported the deposition of amorphous boron nitride films by sputtering. Boron nitride films have been fabricated also by the pyrolysis of trichloroborazine[22], hexachloroborazine[23], borazine[10] and triethylaminborane[24] known as adduct compounds.

Boron nitride films have been prepared so far by CVD, plasma CVD, suputtering, etc. The preparation methods, however,have not been developed fully yet. The molecular flow chemical vapor deposition(MFCVD) and metal organic chemical vapor deposition(MOCVD) are developed by the auther as new methods for preparation of boron nitride films by using the source gasses of decaborane and triethylboron.

MFCVD is the method in which the deposition is carried out in the pressure region of molecular flow. This method bears high controll efficiencies of the deposition rate and the composition of the films, because the impingement frequency of the source gas onto the substrate surface is directly proportional to the pressure of the gas introduced. It is studied how the quality of the deposited film is affected by the substrate temperature and the source gas molar ratio. The condition is clarified in which stoichiometric boron nitride is obtained. It is shown that MFCVD is an effctive method for preparing boron nitride films. MOCVD have been known as the preparation method of compound semiconductors such as GaAs. When boron nitride is considered to apply as an insulator layar of III-V compound semiconductors, the consistency of apparatus should be an advantageous method for the sake of stability of the devices. The qualities of boron nitride films prepared by MOCVD are far better than those of the films prepared through the other metods. This method is inexpensive compared with the method using diborane. When one wishes to analyze the deposition mechanism of the film, it is important to consider the decomposition mechanism of organometalic compounds.

In order to clarify the decomposition mechanism, pyrolysis of triethylboron (TEB) is carried out in argon, hydrogen, and ammonia atmospheres. The gases generated by pyrolysis of TEB are analyzed by a mass spectrometer, IR spectrometer and gas chromatographer. The dissociation mechanism of three ethyl group bonded to boron atom in TEB molecule is disscussed from the results of the analyses. The deposition mechanism of boron nitride films is also discussed from the generated gases by pyrolysis of TEB in ammonia atmosphere.

High quality boron nitride films have been prepared by MOCVD using the reaction of triethyl boron with ammonia. It is shown how the deposition rate and composition of the products are affected by the deposition conditions.

Next, trimethoxyborane (TMOB) which has a chemical bond between boron and oxygen in the molecule is applied as a boron source instead of TEB. The difference in reactivity between TMOB and TEB will give useful information on the reaction mechanism in MOCVD. The reaction between trimethoxyborane (TMOB) with ammonia is also used in MOCVD method.

Finally, the fabrication of "tympanic membrane" is tried as one of the applications of boron nitride films. The "tympanic membrane" is held in tension like a skin of tympani by a supporting frame. In the fabrication of the membrane, it is an important problem how to apply tensile stress on the film. The tensile stress is shown to be controlled by thermal stress when the substrate material is chosen which has smaller thermal expansion coefficient than that of boron nitride. The stress control based on thermal hysteresis is invented to the solution of this problem.

II. PREPARATION OF BORON NITRIDE FILMS BY MFCVD

In general, chemical vapor deposition(CVD) have been investigated at atmospheric and moderately reduced pressures(>10^{-3}Torr). When the CVD is carried out in a pressure of molecular flow region (<10^{-3}Torr), the deposition of the elements or compound film is controlled by the impingement frequency of the molecules on the substrate, so that presise control of the composition and the deposition rate is expected on film preparation. As this method has advantages compared with the conventional CVD, we called this method as molecular flow chemical vapor deposition (MF-CVD)[(25)].

Recentry, a gas source molecular beam epitaxy (MBE) was proposed as a thin film growth technique which is close to our idea. However , up to the present, the gas source MBE is only used for the preparation of compound semiconductor single crystal films such as GaAs.

1. EXPERIMENTS

Boron nitride films were prepared by MF-CVD using the reaction of decaborane with ammonia. although diborane is mostly used as source material for boron nitride films prepared by CVD, diborane is a very dangerous gas which requires much care in handling. Pentaborane (B_5H_9;-46.5°C mp,60°C bp) and decaborane ($B_{10}H_{14}$;99.6°C mp,213°C bp) can be used as alternative boron sources. Decaborane is more safe and stable than pentaborane and diborane, and furthermore, decaborane is advantageous because high purity decaborane is easily obtained in the laboratory with a sublimation technique.

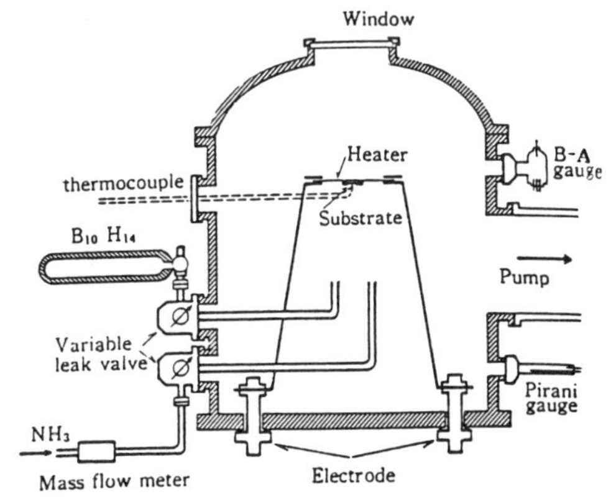

Fig.1 Schimatic diagram of the apparatus for deposition of boron nitride films by MFCVD

Figure 1 shows the schematic diagram of the apparatus used. Deposition of boron nitride films are carried out as follows. The vacuum chamber was evacuated by an oil diffusion pump and rotary pump system, the ultimate pressure being 2×10^{-7}Torr. Sapphire and silicon were used as substrates.

The substrates were fixed on the tungsten sheet heater (0.05mm thick) and were heated at the temperatures between 300 and 1200°C by resistant heating of the tungsten. A tantalum sheet, which was cut into a wedge shape in order to induce the temperature gradient on the substrate, was also used as substrate. By this method, we obtained samples grown at various temperatures between 300 and 800°C by one run. The temperature of the sapphire and silicon substrate were estimated from that of the tungsten heater, which was measured by an optical pyrometer and thermocouples. The relation between the temperature of the substrate and that of the heater was obtained in advance. In the case of tantalum substrates, the substrate temperature was measured at three points by thermocouples welded on the substrate surface.

Table 1 Experimental condition data of MFCVD

Decaborane	2×10^{-5} Torr
Ammonia	2×10^{-5}~8×10^{-4} Torr
$NH_3/B_{10}H_{14}$	1 ~ 40
Temperature	300° ~ 1150°C
Substrate	sapphire, Ta, Si

The decaborane was sublimated by heating at about 70°C and was introduced into the growth chamber at the pressure of 2×10^{-5}Torr through the variable leak valve, and then ammonia was introduced at the pressure between 2×10^{-5} and 8×10^{-4}Torr. The vacuum valve was throttled to maintain the pressure gradient between the growth chamber and the diffusion pump. The pressure was measured as a corresponding nitrogen pressure. The deposition time was varied from 30 to 300 min, depending on the gas pressure and the substrate temperature. The conditions of the film deposition are summarized in Table 1.

The crystal structure of the deposited films were studied with x-ray diffraction. Infrared transmission spectra of the films were mesured in the wavelength range of 2.5-50μm using an infrared spectrophotometer(DS-701G,Jasco). An electron probe x-ray microanalyzer(EPMA, EMX-SE, Shimadzu) was used to analyze the composition of the films. Measurements were carried out at 7kV accelleration voltage, 0.1μA sample current, and 100μm beam diameter by using the lead stearate as a spectral crystal. The film thickness was measured at the masked parts by the mycrostylus profilometer (DEKTAK II,Sloan). The deposition rate was calculated from the thickness.

2. RESULTS AND DISCUSSION

2.1. COMPOSITION

The composition of the films was determined from the peak area ratio of boron $K\alpha_2$ and nitro-

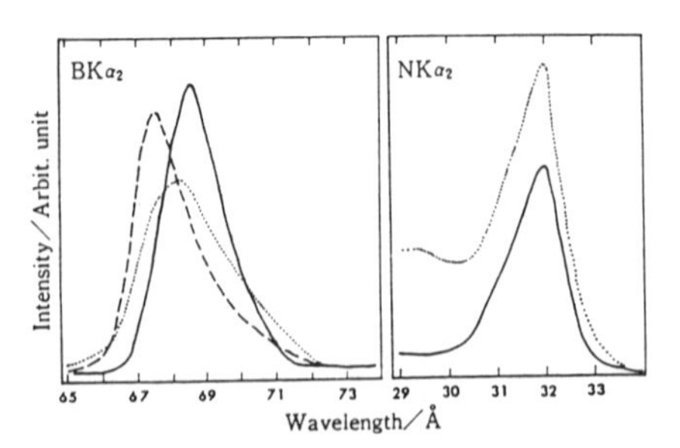

Fig.2 Typical peak profile of $Bk\alpha_2$ and $Nk\alpha_2$ on boron nitride analyzed by EPMA. Solid, dotted and broken lines correspond to B, $BN_{0.5}$ and BN, respectively

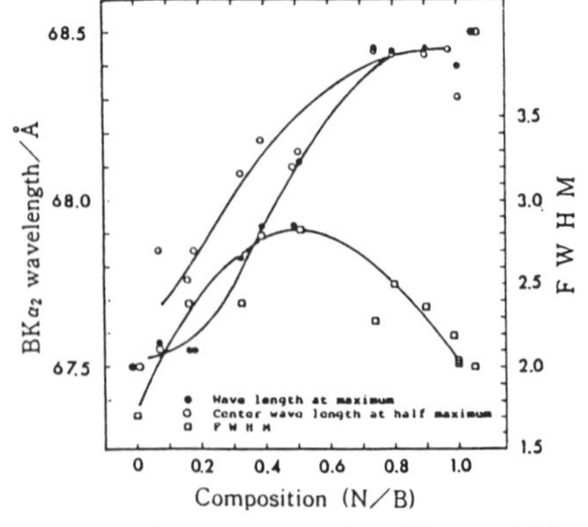

Fig.3 Dependence of composition of boron nitride film on the wavelength of $Bk\alpha_2$. Closed circles show the wavelength at maximum, open circle the center wavelength at half maximum and open squares the FWHM

gen $K\alpha_2$ analyzed by EPMA. Typical peak profiles of boron $K\alpha_2$ and nitrogen $K\alpha_2$ of boron and boron nitride films are shown in Fig.2. The chemical shift of $K\alpha_2$ peak of boron in boron nitride is estimated to be 1.0Å. The chemical shift of that of nitrogen is less than 0.2Å and is smaller than that of boron. In Figs.3 and 4, the composition dependence of the peak wavelengths, the center wavelengths, and the full widths at half-maximum (FWHM) of boron $K\alpha_2$ and nitrogen $K\alpha_2$, are shown. The shift of the peak wavelength of boron $K\alpha_2$ from 67.5Å to 68.5Å implies the decrease of the gap between the 1s and 2s energy levels in the boron atom. In order to understand this chemical shift, detailed knowledge of the electronic structure of boron and boron nitride is required. These electronic structure have been studied experimentally by ESCA(26,27), AES(28), x-ray emission measurements(29),and theoretically from the band structure calculation of boron(30,31) and boron nitride(32,33). By adding the energy of boron $K\alpha_2$ 183.7eV(67.5Å) in the boron films and the valence band energy of boron 8.43eV(27), we obtain 192eV for the boron 1s energy level in the boron film. We also obtain 194eV for the boron 1s and 400.5eV for nitrogen 1s by adding the energy of boron $K\alpha_2$ 181eV(68.5Å) and the nitrogen $K\alpha_2$ 385eV(32Å) in the boron nitride films from the EPMA experiment to valence band energy of boron nitride. The valence band of boron nitride is known to lie at 13eV(30).

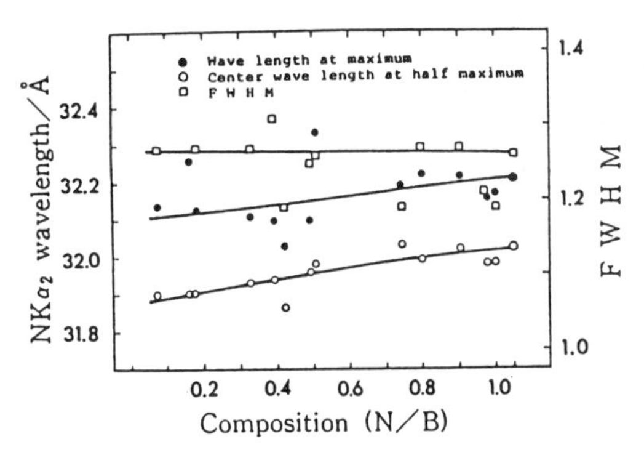

Fig.4 Effect of composition of boron nitride film on the wavelength of Nkα2; symbols correspond to those in Fig.4

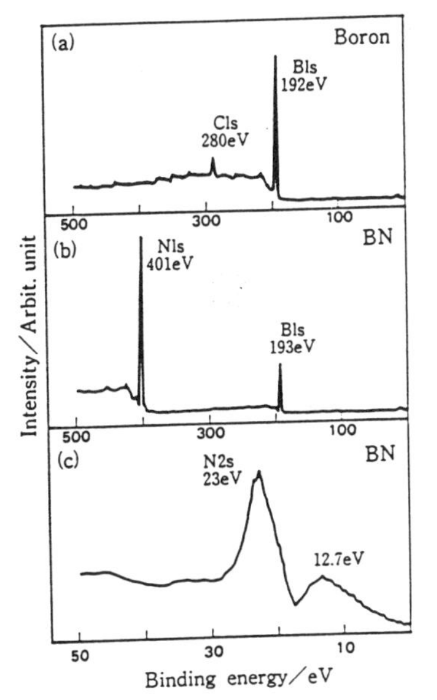

Fig.5 ESCA spectra for boron and boron nitride

The ESCA spectra for the boron film and the stoichiometric boron nitride film are shown in Fig.5. In the spectrum shown in part(a), the peaks at 192eV and 280eV correspond to boron 1s and carbon 1s states, respectively. This carbon peak is considered to be due to cabon impurities in the film or adsorbed carbon on the surface. In the spectrum shown in part (b), the peaks at 193eV and 401eV correspond to boron 1s and nitrogen 1s states, respectively. The spectrum shown in part (c) shows the valence band(0-50eV) of the boron nitride films. The peak at 23eV corresponds to nitrogen 2s states. The broad peak at 12.7eV marks the 2p electrons of boron and nitrogen. Since these values agree with the value (13eV) given above, the assignment of the x-ray emission seems resonable.

The chemical shift of boron $K\alpha_2$ is in agreement with the value reported by Grusserbaure(34), where the chemical shifts of boron $K\alpha_2$ for several boron compounds(BN, B_2O_3, B_4C, BP) were measured by EPMA. The shift is explained by the localization of electrons due to the difference in electro-negativities. Bonding state of boron and nitrogen in a nonstoichiometric region of films deposited is clarified from the chemical shifts.

It is found from Fig.3 that the chemical shift of boron $K\alpha_2$ changes continuously from pure boron to boron nitride in the nonstoichiometric composition region. However, above the composition N/B=0.75, the chemical shift of boron $K\alpha_2$ is almost the same value as that of stoichiometric boron nitride. It is considered that the electronic structure of boron nitride film with the composition N/B=0.75 is nearly the same structure as that of stoichiometric boron nitride. The FWHM of boron $K\alpha_2$ peak increases with an increase of the composition, N/B. It has a maximum value of

2.80Å at about N/B=0.5 and then decreases to 2.0Å at N/B=1.0. The increase of the FWHM is considered to be caused by superposition of the boron $K\alpha_2$ of boron on that of boron nitride. The dependences of the peak wavelength and the FWHM of the nitrogen $K\alpha_2$ on composition of the film are shown in Fig.4. These dependences are smaller than those of boron $K\alpha_2$. This fact suggests that all the nitrogen atoms included in the films are always bound to boron atoms. We can therefore conclude that the deposited films ar the mixture of boron and boron nitride in the nonstoichiometric composition region.

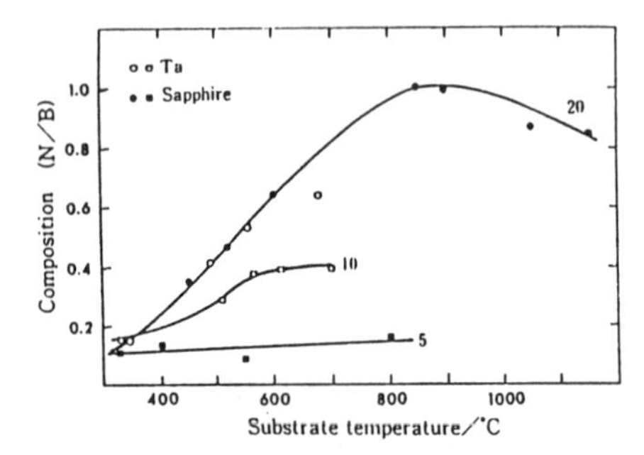

Fig.6 Composition of boron nitride as a function of substrate temperature

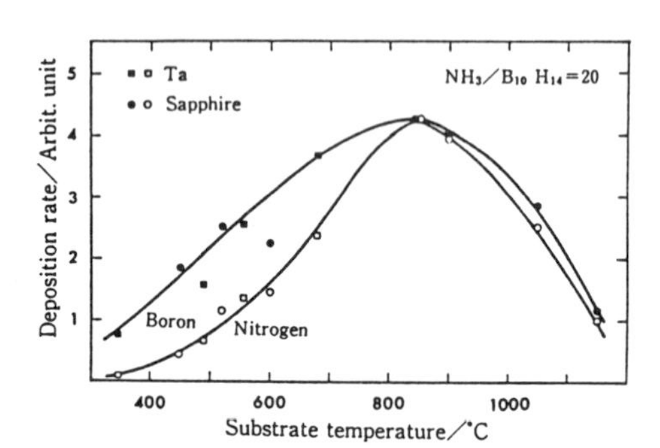

Fig.7 The deposition rate of boron and nitrogen as a function of substrate temperature

2.2. DEPOSITION RATE

The dependence of the composition of the films on the substrate temperature is shown in Fig.6 at thetemperatures between 300 to 1200°C. The composition of the film depends strongly on the substrate temperature when the pressure ratio of the reaction gases ($NH_3/B_{10}H_{14}$) is 20, whereas the dependence is weak when $NH_3/B_{10}H_{14}$=1. When $NH_3/B_{10}H_{14}$=20, the nitrogen content in the films increases with increasing substrate temperature in the temperature range 350 - 850°C, and then decreases gently. As the intensities of boron and nitrogen x-ray fluorescence are proportional to the amount of the boron and nitrogen per unit area, respectively, the deposition rates of boron and nitrogen are given by dividing the x-ray fluorescence intensities of boron and nitrogen by the deposition time. The temperature dependences of deposition rate of boron and nitrogen are shown in Fig.7. As mentioned befor, it is considered that nitrogen atom cannot stay on the substrate surface without bonding to boron atoms. The dependence of nitrogen deposition rate normalized by boron deposition rate on the substrate temperature is shown in Fig.8. The figure shows that the deposition rate is determined by the kinetic process in which the number of nitrogen atoms bonded to boron atoms increases with temperature between 350 and 850°C, but decreases with an increase of temperature above 850°C. The latter fact should be due to shortening of the residence time of nitrogen on the surface at high temperatures. The reaction of nitrogen with boron requires enough residence time of nitrogen on the deposition surface. The relation between the logarithm of nitrogen deposition rate and the reciprocal absolute temperature is shown in Fig.9. We obtain the activation energies for this reaction 6.3 and 0.4kcal/mol at temperatures above and below 850°C, respectively. The activation energy of 6.3kcal/mol is corresoonds to reaction of boron with nitrogen atoms. However, detailed information on this small activation energy is un-

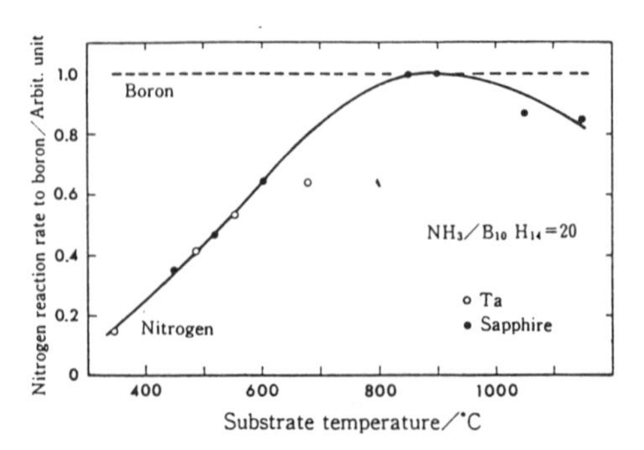

Fig.8 Normalized deposition rate of nitrogen by deposition rate of boron as a function of substrate temperature

known.

At the constant decaborane pressure of 2×10^{-5}Torr and the substrate temperatures of 850 and 550°C, the dependence of the film composition on the ammonia pressure at the pressure range of $2\times10^{-5} \sim 8\times10^{-4}$Torr is shown in Fig.10. It is found from the figure that , at 850°C, the composition increases with the increase of ammonia pressure up to $NH_3/B_{10}H_{14}$=20 and then saturates at N/B=1.0.

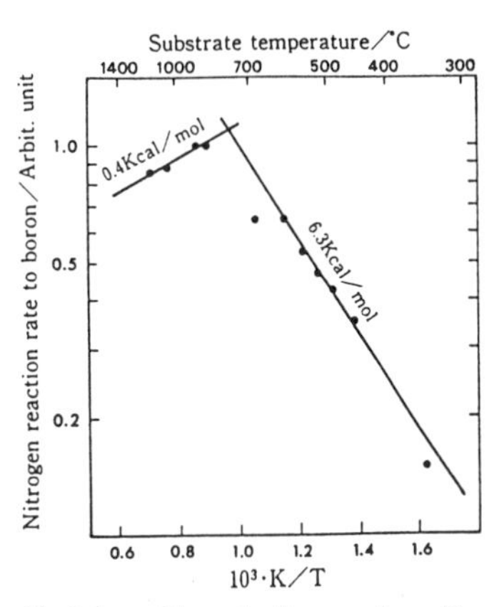

Fig.9 Logarithm of nitrogen deposition rate as a function of reciprocal absolute temperature

In general, the amount of gas molecules, S, impinging onto the substrate surface is given from the gas dynamics as

$$S=P(Mw/2RTg)=5.8\times10^{-2}P(Mw/Tg)^{1/2} \text{ (g/cm}^2) \quad (1)$$

where P is the partial pressure in Torr, Mw the molecular weight, R the gas constant, and Tg is the gas temperature which is considered to be room temperature(i.e., Tg=300K). By using the Avogadro number, Eq. (1) is rewritten to give the number of impingement molecules(ν), as

$$\nu=3.5\times10^{22}P/(Mw/T)^{1/2} \quad \text{(molecule/cm}^2\text{s)} \quad (2)$$

When P=2×10^{-5} Torr, Mw=122, and Tg=300K are substituted in Eq. (2), $\nu_{B_{10}H_{14}}=3.7\times10^{15}$ molecule/cm^2-s is obtained in the case of decaborane. In the case of ammonia, substituting the value of Mw=17 and Tg=300K in Eq. (2) yields $\nu_{NH_3}=4.9\times10^{20}P$ molecule/cm^2-s. The compositions of the films are shown in Fig.11 as a function of the partial pressure of ammonia together with the ν_{NH_3} at the constant $\nu_{B_{10}H_{14}}=3.7\times10^{15}$ molecule/cm^2-s and the substrate temperatures of 600 and 850°C. At the substrate temperature of 850°C, the composition of the films increases with increasing ammonia partial pressure,P_{NH_3}, and the stoichiometric composition obtained above $P_{NH_3}=4\times10^{-4}$ Torr, where $\nu_{NH_3}/\nu_{B_{10}H_{14}}$=54. When the substrate temperature is 600°C, the composition (N/B) saturates in the region $P_{NH_3}>4\times10^{-4}$Torr, but the saturated N/B composition is far less than unity. The stoichiometric boron nitride can not be obtained at this temperature. If the substrate temperatures is 850°C, on the other hand, the stoichiometric boron nitride films are obtained when the ratio of impingement frequencies of nitrogen and boron atoms, ν_N/ν_B, is larger than 5.4. If the substrate temperatures is 850°C, the compositions of the films are

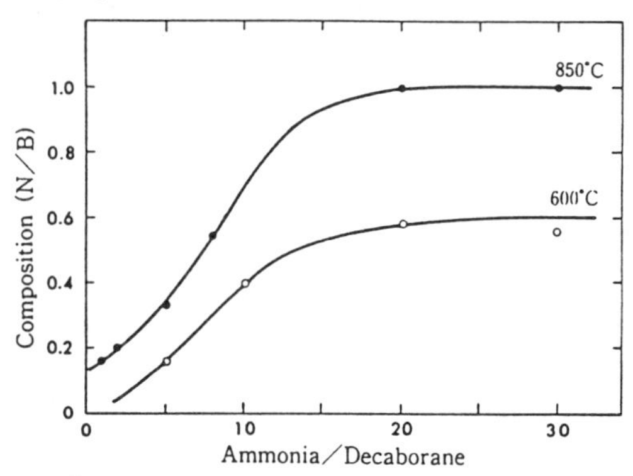

Fig.10 The effect of ammonia pressure on composition of boron nitride at the constant decaborane pressure 2x10⁵Torr and substrate temperature 600 and 800°C

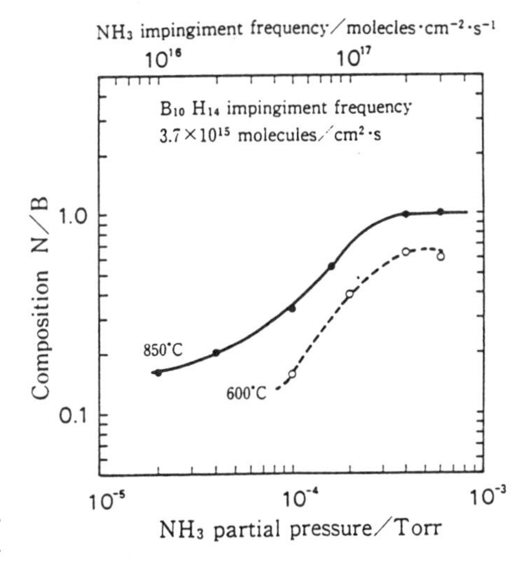

Fig.11 Composition of boron nitride as a function of ν_{NH_3} of the constant $\nu_{B_{10}H_{14}}$

determined by the impingement frequency ratio of decaborane and ammonia in the nonstoichiometric composition region.

2.3. PROPERTIES OF THE DEPOSITED FILMS

The color of the film changes continuously from black-brown to colorless with the increase of ammonia pressure. The x-ray diffraction analyses indicate that all the films obtained are amorphous. The results may be ascribed to the complex structure of decaborane.

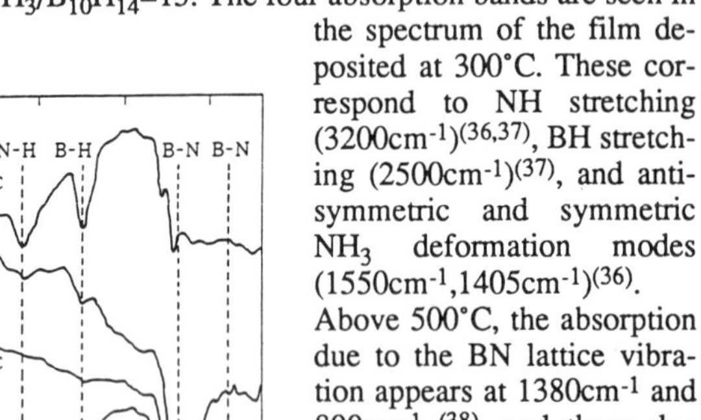

Fig.12 The adsorption coefficient of the stoichio-metric boron nitride film near absorption edge as a function of incident photone energy

The absorption coefficients of a colorless and transparent film deposited by MFCVD in the ultraviolet spectral region are shown in Fig.12 as a function of incident photon energy (hν). The absorption coefficient near the absorption edge for direct allowed transition is described as a function of photon energy hν as

$$\alpha \sim (h\nu - E_g)^{1/2} / h\nu \qquad (3)$$

where Eg is the optical band gap energy. The variation of $(\alpha h\nu)^2$ with incident photon energy is shown in Fig.13. This figure shows that the boron nitride obtained is a direct-transition type semiconductor. The energy gap, Eg, is estimated to be 5.90eV by the extrapolation of the linear part to the horizontal axis in the figure. This value is larger than the value of 5.80eV reported by Sano [17] and Zunger[35].

Figure 14 shows the infrared transmission spectra of the films deposited on silicon substrates at 300, 400, 500, and 700°C with the molar ratio of source gases $NH_3/B_{10}H_{14}$=15. The four absorption bands are seen in the spectrum of the film deposited at 300°C. These correspond to NH stretching (3200cm^{-1})[36,37], BH stretching (2500cm^{-1})[37], and anti-symmetric and symmetric NH_3 deformation modes (1550cm^{-1},1405cm^{-1})[36]. Above 500°C, the absorption due to the BN lattice vibration appears at 1380cm^{-1} and 800cm^{-1} [38], and those due to the BH and NH stretching vibrations disappear. Figure 15 shows the infrared transmission and reflection spectra of the film deposited at 1000°C. The characteristic absorption and reflection bands of boron nitride are found at 1380 and 800cm^{-1}, and 1565 and 800cm^{-1}, respectively.

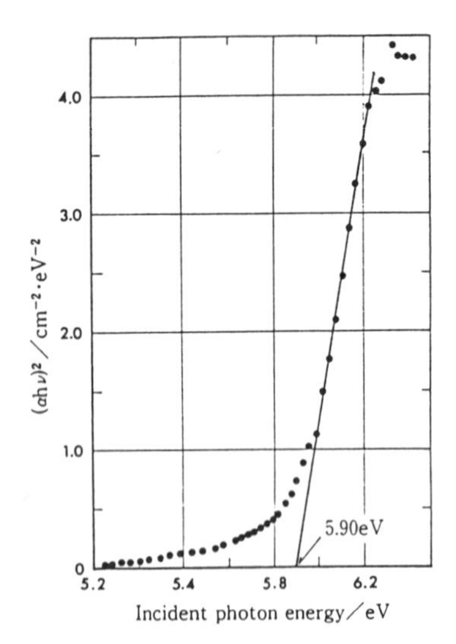

Fig.13 The absorption edge fitted to direct allowed transition

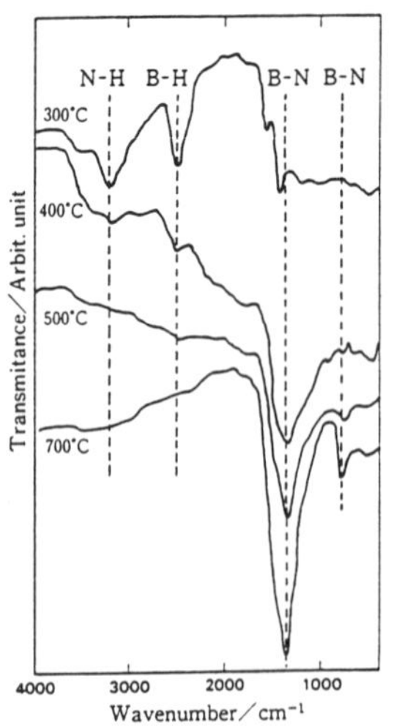

Fig.14 The effect of deposition temperature on the infrared transmission spectra

The following conclusion can be draw from the IR

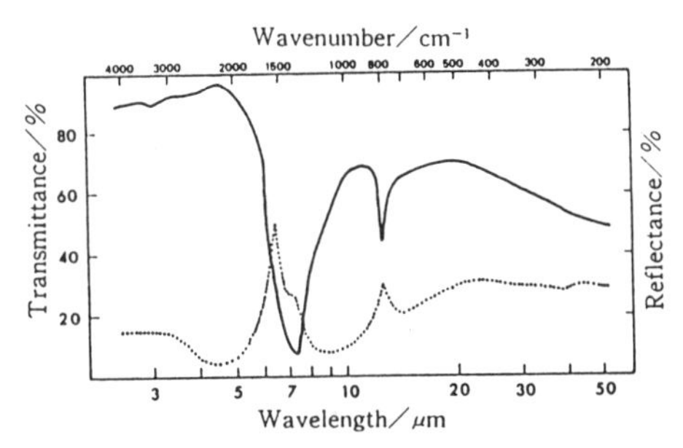

Fig.15 Infrared transmission and reflection spectra of deposited BN film at 1000°C, Solis and dotted lines represent the transmission and refrection spectra, respectively

spectra. The films deposited at 300°C by the reaction of decaborane with ammonia contain B-H and N-H bonds. B-N bonds are found to be formed even at a remarkably low temperature as 400°C. The film deposited at 500°C essentially consists of boron nitride. The temperature 700°C is enough to produce boron nitride because the spectrum of the film deposited at 700°C is almost the same as that of the film deposited at 1000°C.

III.A THE REACTION MECHANISM IN MOCVD

The growth of III-V compound semiconductors by MOCVD is put to practical use. However, its reaction mechanism has not been clarified. Especially, the pyrolysis mechanism of III group organo-metallic compounds is not known. The behavior of hydrocarbon has been a serious question in connection with the carbon impurities included in the films. In this chapter, part of the mechanism of MOCVD is clarified from the results of the studies on the deposition process of boron nitride films. We studied the pyrolysis of TEB and TMOB and showed that the behaviors of the pyrolyses of TEB and TMOB are different from each other. Trimethoxyborane is not an organometallic compound in a strict sense, TMOB was treated as an organo-metal. The deposition mechanism of boron nitride films is also discussed.

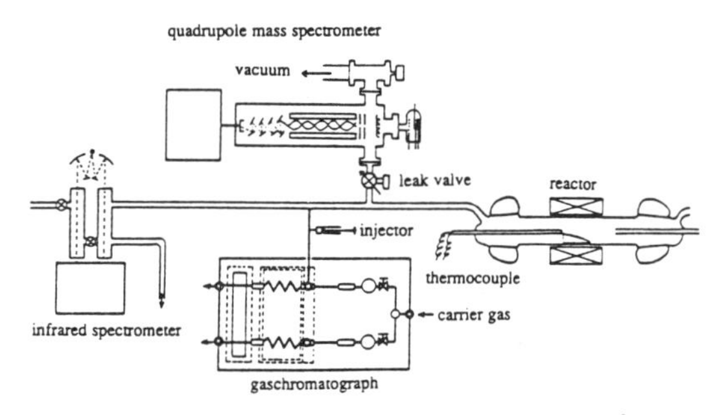

Fig.16 Schimatic diagram of the configuration of apparatus for analysis of the gasses pyrolyzed

1. EXPERIMENTS

Pyrolysis mechanism of TEB and TMOB can be made clear by analyzing the constituents of the exhausted gas. A schematic diagram of the apparatus used for the analysis of the exhaust gas is shown in Fig.16. It consists of a quadrupole mass spectrometer (ANELVA, NAG-531), a gas chromatographer (Yanagimoto, GCT-550T), and an infrared spectrometer (Simadzu, IR-400). Part of exhaust gas was introduced to the analyzer system.

The pyrolysis of TEB and TMOB was studied in the three atmospheres of argon, hydrogen and ammonia at the temperature range between room temperature and 1100°C with a step of 50 or 100°C. The mass analyses of exhaust gases were carried out by introducing the gases to the mass spectrometer at 1×10^{-6} ~ 5×10^{-6}Torr through a variable leak valve. The conditions of mass analysis are as follows: range of mass number m/e is 1-110, scanning time 100sec, ionization voltage 26.3eV. In the gas chromatograph analysis, a column of porous polymer beads (Porapuk waters) and that of Unibeads (Gaschrokogyo) were used in the experiments of TEB and TMOB, respectively. The column temperature was 50°C, and the carrier gas was helium whose flow rate was 20ml/ min .The sampling gas was 4ml. In the infrared analysis, a gas cell made of glassware with potassium bromide windows was used. Ammonia or mixed gas of TEB or TMOB and Ar, H_2 or NH_3 were sealed in the reference cell. The temperature of the saturators of TEB and TMOB were kept at 0°C by ice bath. Hydrogen or argon gases were used as a carrier gases of TEB and TMOB. Experi-

Table 2 Experimental condition of studing on pyrolysis mechanism of TEB

TEB	1.4 ml/min
NH_3	28 ml/min
H_2	0, 80 ml/min
Ar	0, 80 ml/min
Total	80, 120 ml/min
Temp.	20 ~ 1100°C

Table 3 Experimental condition of studing on pyrolysis mechanism of TMOB

TMOB	2.5, 5.0 ml/min
NH_3	0 ~ 50 ml/min
H_2	0 ~ 100 ml/min
Ar	0 ~ 100 ml/min
Temp.	20 ~ 1100°C

mental conditions for TEB and TMOB are summarized in Tables 2 and 3, respectively.

2. RESULTS AND DISCUSSION

Pyrolyses of TEB and TMOB in reducing(H_2) and inert(Ar) atmospheres were studied. Pyrolyses of them in ammonia was also studied. The pyrolyses of TEB and TMOB in these atmospheres were examined by analyzing the decomposed gases. The results of MASS, IR and GC due to pyrolysis of TEB in the three atmospheres are shown in Figs.17, 18 and 19, (A,B,C), respectively. These figures show the temperature dependence of the most significant peak in the obtained results. These values are not corrected for the absolute quantities, but are the bare results.

2.1. PYROLYSIS OF TRIETHYLBORON

(a) MASS, IR, AND GC MEASUREMENTS

In argon atmosphere, from a comparison among different peaks in Fig.17(A,B,C), it is proved that the mass m/e=41,69,98 and IR absorption at 2975 and 2890cm^{-1} are derived from TEB source. It is found from the figures that the TEB does not decompose until 300°C, and then decomposes rapidly with increasing temperature . Ethylene, a double bonded hydrocarbon, are found to be formed through the pyrolysis of TEB by the apearance of the peaks of m/e ratio 28, 26 and an IR absorption peak at 953cm^{-1}. This fact is confirmed by the obserbation of the spectrum of ethylene in a gas chromatography. The generation rate of ethylene increases steeply above 300°C, decreases gently above 500°C and then decreases steeply above 700°C. The species whose mass m/e=16 and

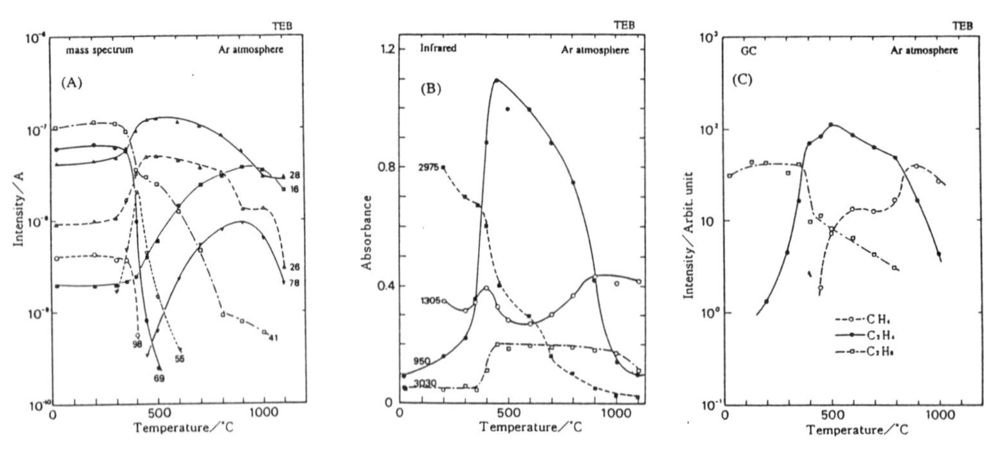

Fig.17 Relative intensity of the species in the pyrolitic gasses as a function of the pyrolysis temperature of TEB in argon atmosphere, (A) mass analysis, (B) gaschromatograph, (C) infrared

78 are identified to be methane and benzene, respectively, from the results of gas chromatography and IR spectra. The amounts of methane and benzene increase with increasing temperature between 300 and 900°C and depend on the pyrolysis rate of ethylene. The generation of ethane is not clearly observed in the mass spectra, but is identified from the results of IR measurements and gas chromatography, which shows that the ethane increases with increasing temperayure in the temperature region from 300° to 1000°C and decreases above 1000°C. It is proved from the above results that the ethyl groups in TEB are dissociated into ethylene in the pyrolysis of TEB in argon atmosphere.

The mass, IR absorption and GC due to pyrolysis of TEB in hydrogen atmosphere are shown in Fig.18(A.B.C). The intensities of mass m/e=41, 69, 98 derived from TEB itself decrease steeply above 300°C, similarly to those in argon atmosphere. The gases generated by pyrolysis in hydrogen atmosphere is mainly ethane (m/e=30, IR=3030cm^{-1}) while that in argon atmosphere was mainly ethylene. Many other hydrocarbons which are alkanes are also identified. The amount of methane increases with increasing pyrolysis temperature up to 350°C and is constant in the temperature range between 350°C and 700°C, and then increases again above 700°C. Benzene of mass m/e=78 was detected in argon atmosphere, but was not detected in hydrogen atmosphere.

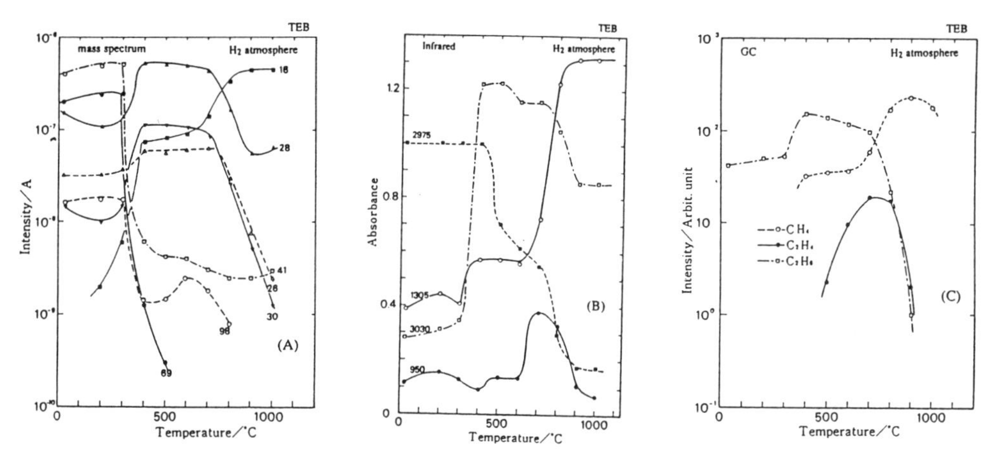

Fig.18 Relative intensity of the species in the pyrolitic gasses as a function of the pyrolysis temperature of TEB in hydrogen atmosphere, (A) mass analysis, (B) gaschromatograph, (C) infrared

In the pyrolysis of TEB in an ammonia atmosphere(Figs.19(A.B.C)), intensities of the molecles of m/e=41, 69, 98 derived from pyrolysis of TEB are small compared with those in argon and hydrogen atmosphere. Strong peaks of m/e= 56 and 85 were detected instead of 41,69 and 98. Those are considered to be the adducts bonded to ammonia. However, the temperature dependence of the intensities corresponding to m/e= 56 and 85 have maxima at the temperatures about 400°C and their temperature dependences are different from those of the molecules of masses 41 and 69 in the argon and hydrogen atmospheres. The gases generated by the increase of temperature above 300°C principally consist of ethane, ethylene and methane, as in the case of pyrolysis in argon and hydrogen atmospheres. Benzene of m/e=78 is generated in the ammonia atmosphere, which is different from that in the hydrogen atmosphere. The generation of benzene is in contrast to that of m/e=56 and 85. It is considered that the ethylene generated by pyrolysis of TEB changes to $C_2H_3NH_2$ (m/e=43) by the reaction with ammonia. The rate of production of benzene decreases with decreasing ethylene in the atmosphere. The absorption peaks at 1600 and 2500cm^{-1} in the IR spectrum weere not observed in the spectra obtained by pyrolysis in the argon and hydrogen atmospheres. The temperature dependence of this absorption coincidens with that of the peak of m/e=43 . The absorption

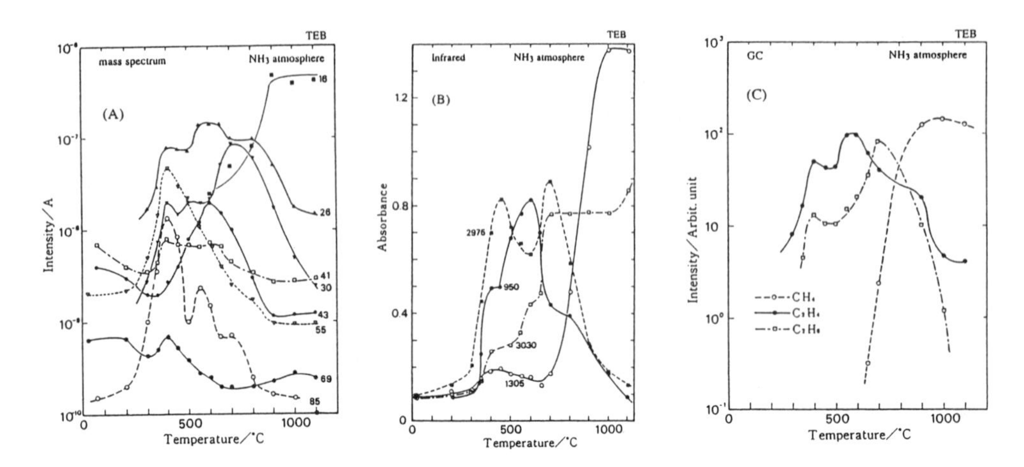

Fig.19 Relative intensity of the species in the pyrolitic gasses as a function of the pyrolysis temperature of TEB in ammonia atmosphere, (A) mass analysis, (B) gaschromatograph, (C) infrared

at 1600 and 2500cm^{-1} well correspond to in-plane stretching vibration of B-H bond. The absorption peaks at 1600cm^{-1} corresponds to the stretching vibration of NH_2. However, since the reference cell was filled with ammonia gas for absorption measurements, compound with B-H bond should be contained in the pyrolytic gasses.

The pyrolysis of TEB in the ammonia atmosphere presents an aspect between those in the argon and the hydrogen atmospheres. The formation of the adducts of ammonia with pyrolitic gases are confirmed.

(b) PYROLYSIS MECHANISM OF TEB

From the gas analysis of pyrolysis of TEB in the three atmospheres, it is clear that pyrolysis of TEB begins at about 250°C. Pyrolysis of TEB takes place through dissociation of the three ethyl groups bonded to boron atom. although methane was observed in the spectra of the pyrolysis in the three atmospheres, majority of the generated gas molecules through pyrolysis of TEB consists of mainly C_2H_n (n=4 and 6), i.e., ethylene and ethane, depending on the atmospheres. The generation of ethylene in argon atmosphere and the generation of ethane in hydrogen atmosphere are dominant. In ammonia atmosphere, the generation of ethane and that of ethylene are almost the same amount. However, the temperatures at which the generation rate of these gases have, maxima are different i.e., 550°C for ethane and 700°C for ethylene. It is difficult to explain the difference of the pyrolysis of TEB in argon and hydrogen atmospheres. However, it can be concluded from the analysis of generated gasses that the ethylene is generated by β-elimination in argon atmosphere and ethane is generated through hydrogen addition decomposition in hydrogen atmosphere. Some of the papers on MOCVD reaction mechanism describe that ethyl compounds pyrolyze through the generation of ethylene by the β-elimination. Recently, it is reported that ethane is produced at low temperatures through pyrolysis. The mechanism of β-elimination of ethylene from TEB is explained as follows. The boron atom in TEB molecule is bonded to three carbon atoms and the valence electrons of boron are put in an electron deficient state by one electron pair. Therefore, TEB is one of Lewis bases. The electron deficient boron, grabs hydrogen from the methyl group at the α-position of TEB. Double bond is thus formed between α and β-carbons, and finally the σ bond between β-carbon and boron is broken to release ethylene from TEB. In our experiments, the significant ethylene production in argon atmosphere is explained by the β-elimination, but the ethane generation in hydrogen atmosphere is not explicable by the similar mechanism. In the pyrolysis of TEB in hydrogen atmosphere, the boron atom in TEB molecule attract an atmospheric hydrogen molecule rather than an α-position hydrogen molecule. As the result , the bond between β-carbon and boron atoms breaks and an ethyl radical is eliminated. This ethyl radical combines often with an atmospheric hydrogen molecule to yield ethane molecule, and sometimes with another radical to convert to higher alkanes.

In ammonia atmosphere, an ammonia molecule combines with the boron atom in TEB molecule as an electron-donor and then ethyl radical eliminates. This ethyl radical converts to ethane as in the case of hydrogen atmosphere. However, it is not clear how ethylene is generated more abundantly than ethane at high temperatures. It is considered to be due to pyrolysis of ethane or predominate β-elimination at high temperatures.

(c) GROWTH MECHANISM OF BORON NITRIDE FILMS

In the pyrolysis of TEB in ammonia atmosphere, ethane and ethylene are produced through the reaction of TEB with ammonia. However, the deposition of boron nitride film cannot be explained

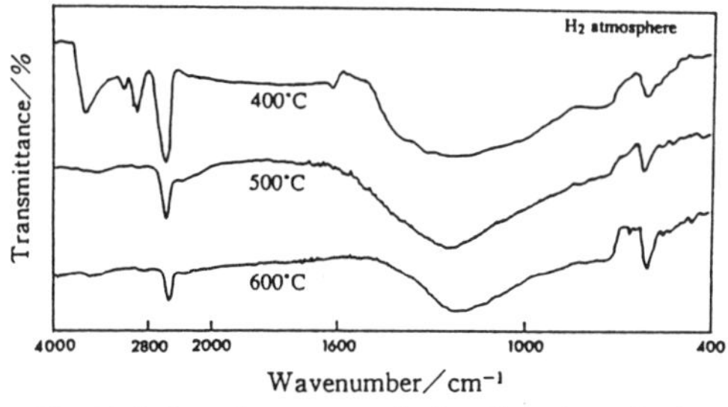

Fig.20 Infrared spectra of the film deposited by pyrolyzing of TEB in argon atmospher at the temperature of 400, 500 and 600 C

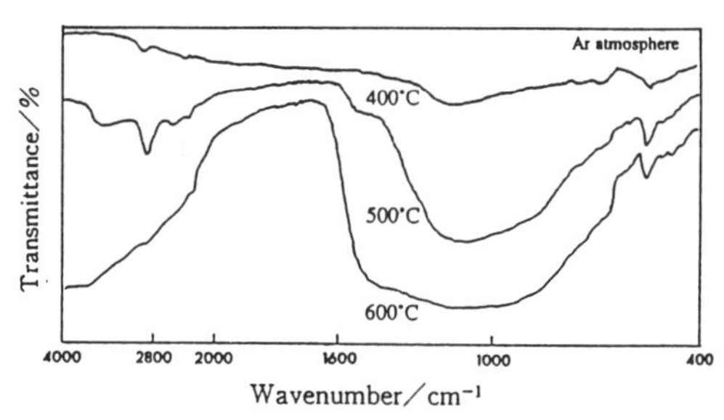

Fig.21 Infrared spectra of the film deposited by pyrolyzing of TEB in hydrogen atmospher at the temperature of 400, 500 and 600° C

by the pyrolysis mechanism only. The temperatures for the deposition of boron nitride films are much higher than that for the pyrolysis of TEB. The peaks corresponding to $(C_2H_5)_2BNH_2$ and $C_2H_5BHNH_2$ are found in the mass spectrum of TEB pyrolysis in ammonia atmosphere. It is inconceivable that these intermediate compounds are dominant species for deposition of boron nitride, because the intensities of these peakes have maxima at 400°C and becomes smaller with increa temperature, and finally the peaks of these intermediates cannot be observed at about 800°C. It is not clear whether the intermediate adducts are consumed by the deposition of the films or is decomposed at the temperature higher than 400°C.

Infrared spectra of the films deposited by the pyrolysis of TEB in argon and hydrogen atmospheres are shown in Figs.20 and 21, respectively. The deposition was carried out in the temperature range between 500 and 700°C. The absorption peaks at 1200 and 2500cm^{-1} correspond to B-C and B-H bonds, respectively, which suggests that the films deposited include H-B-C bonds. However, in the film deposited by pyrolysis of TEB below 700°C in ammonia atmosphere, an H-B bond dose not includs. This fact indicates that the pyrolysis of TEB is depressed by the introduction of ammonia. The depression is considered to be due to the formation of the adducts of TEB with ammonia. Because these adducts are volatile, the deposition does not occur on the substrate surface. These adduct compounds that dissociate into ethyl groups at the temperatures above 800°C deposit on the substrate to form a boron nitride fim. This fact suggests that the adduct compound contribute to deposit boron nitride.

2.2. PYROLYSIS OF TRIMETHOXYBORANE

(a) MASS MEASUREMENTS

The pyrolysis of trimethoxyborane was studied in three atmospheres of argon, hydrogen and ammonia in a similar method for TEB. The gasses generated by pyrolysis of TMOB were analyzed by mass spectrometry, gas chromatography and infrared absorption spectrometry. Emphasis of the study was put on the temperature dependence of gasses generated by the pyrolysis. The pyrolysis mechanism of TMOB is discussed primarily on the basis of the results of mass spectrometry, and is also explained from the results of gas chromatograph and IR spectra. The deposition mechanism of boron nitride films is also discussed from these results. The temperature dependences of gasses generated by the pyrolysis of TMOB in the atmospheres of argon, hydrogen and ammonia are shown in Figs.22,23 and 24(A,B), respectively.

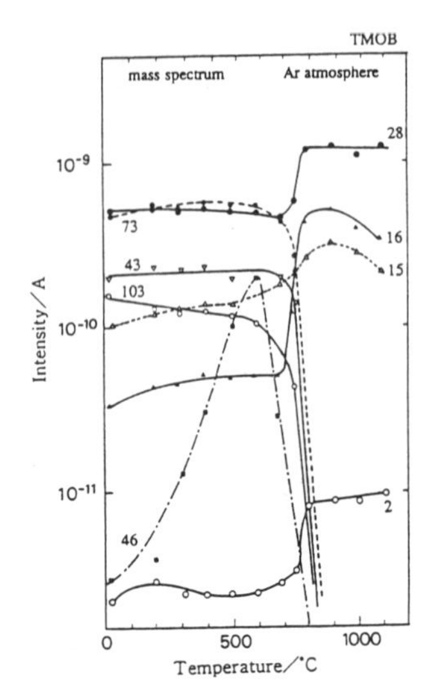

Fig.22 Relative intensity of the species in pyrolitic gasses as a function of the pyrolysis temperature of TMOB in the argon atmosphere

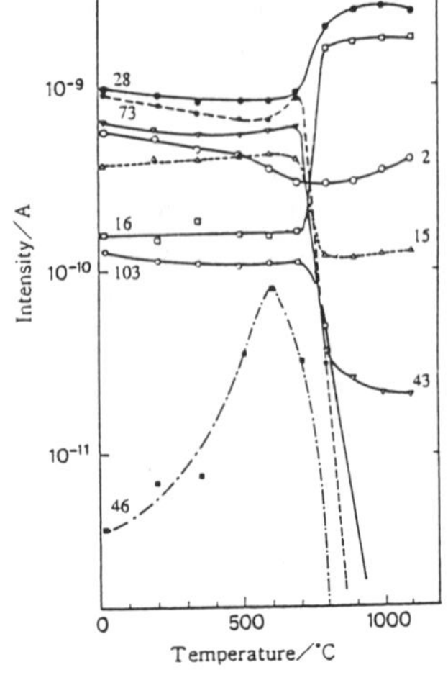

Fig.23 Relative intensity of the species in pyrolitic gasses as a function of the pyrolysis temperature of TMOB in the hydrogen atmosphere

In argon atmosphere

The mass peakes of TMOB are found at the mass numbers of 43, 73 and 103 as shown in Fig.22. The intensities of these peaks decrease steeply above 600°C. It is suggest that TMOB pyrolyzes at 600°C. The peak intensity of mass number 46 increases by the pyrolysis of TMOB. It is clear from the comparison with IR spectrum that the mass number 46 corresponds to dimethylether. Dimethylether decreases steeply above 600°C and TMOB also decreases steeply. Mass number 31 corresponds to methyl alcohol. They decrease above 700°C. In contrast to these, peak intensities of m/e=2, 15, 16, and 28 increase with temperatures above 600°C where TMOB pyrolizes. It was proved by comparing the results of IR spectra and gas chromatograph that the mass number 2, 1 and 28 correspond to hydrogen, methane and carbon monoxide, respectively. The peak intensity of mass number 18 corresponding to water is almost constant.

In hydrogen atmosphere

In Fig.23, the intensity of mass number 73, 103 and 43 derived from TEB itself decrease steeply at 700°C. Dimethyl ether(m/e=46) increases with increasing temperature and have a maximum at 600°C. Methyl alcohol (m/e=31,32) decreases with decreasing TMOB simultanously. Carbon monoxide (m/e=28) and methane (m/e=15) begins to increase at 700°C where the pyrolysis of TMOB begins.

In NH_3 + H_2 atmosphere

Figure.24(A) shows the temperature dependence of gases generated by the pyrolysis of TMOB in the atmosphere of hydrogen and ammonia The amount of trimethoxyborane (m/e=73, 104, and 43) decreases gently with increasing the temperature between 500 and 750°C, and decreases steeply above 750°C. This behavior differs from those in the hydrogen and argon atmospheres. The intensity of dimethylether (m/e=45) is smaller than that of alcohol. The generation of methyl alcohol(m/e=31 and 32) reaches a maximum around 600 and 700°C. The amount of ethane (m/e=30) increases gently up to 600°C, and decreases steeply until 1100°C. Carbon monoxide (m/e=28) increases at temperatures above 600°C. A slight decrease of ammonia (m/e=16,17) above 1000°C is considered to be due to a measurement error. Intensities of methane (m/e=15) and water (m/e=18) do not change with increasing of the pyrolysis temperature.

In NH_3 + Ar atmosphere

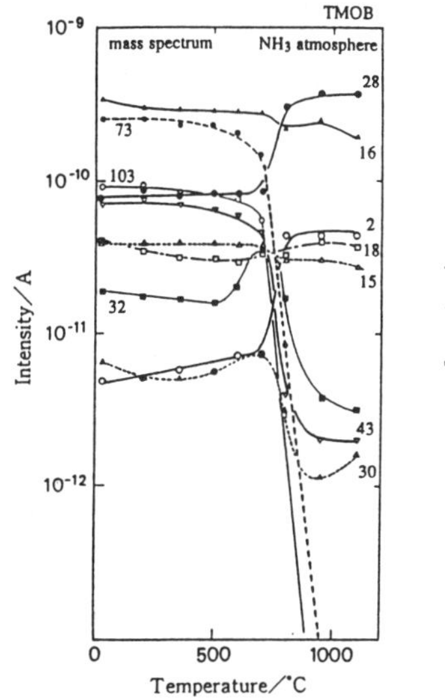

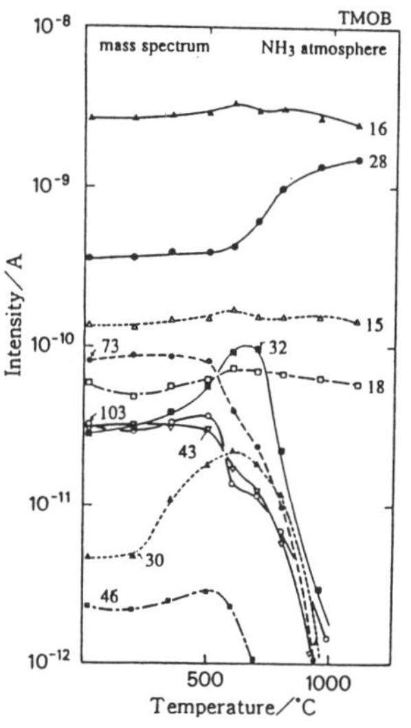

Fig.24 Relative intensity of the species in pyrolitic gasses as a function of the pyrolysis temperature of TMOB (A)in the hydrogen and ammonia and (B)in the argon and ammonia atmosphere

The amount of trimethoxyborane (m/e=73, 104, 43) decreases gently above 350°C and steeply decreases above 700°C as in the case of argon atmosphere (Fig.24(B)). The amount of ethylalcohol (m/e=31,32) begins to increase at 600°C and increases steeply above 700°C. Methane (m/e=15) and water (m/e=18) are almost constant. The peak of dimethylether (m/e=45 or 46) in the spectrum is small.

(Effect of NH_3/TMOB ratio)

The mass spectrum of the gases generated by changing the ratio of ammonia to TMOB (NH_3/TMOB= 0-3) at the constant temperature of 750°C is shown in Fig.25. The amount of ammonia (m/e=17) increases almost linearly with increasing ammonia. The intensity of trimetoxyborane and that of mass number 56 both decrease with increasing ammonia. Alcohol (m/e=32) and carbon monoxide (m/e=28) increase with increasing the ammonia. The amount of ethane (m/e=30) does not change by increasing ammonia. The peak of mass number 45 was not observed. The results for the ratio of ammonia to TMOB above 3 are not shown in here, but are similar to tha for the ratio 3.

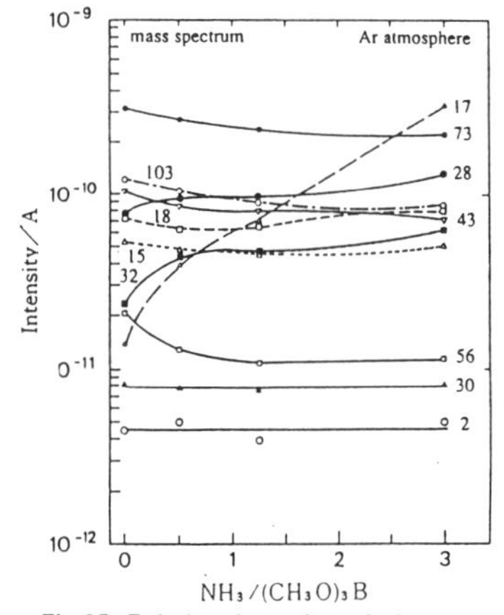

Fig.25 Relative intensity of the species in pyrolitic gasses as a function of the ratio of ammonia to TMOB

(b) MECHANISM OF PYROLYSIS OF TMOB

Pyrolysis of TMOB in ammonia atmosphere is different from those in argon and hydrogen atmospheres. The main gases generated by pyrolysis of TMOB are carbon monoxide, methane and hydrogen. The generation of these gases increases steeply at 600-700°C in all the atmospheres. No difference in the decomposition temperature is observed in the argon, hydrogen and ammonia atmospheres. However,the kinds of exhausted gases are quite different. The most predominantly generated gas is methanol in the ammonia atmosphere, whereas dimethylether is mainly produced in hydrogen and argon atmospheres.

This results is consistent with the observation that boron oxide films are deposited by pyrolysis of TMOB in argon and hydrogen atmosphere, whereas boron nitride films are deposited in ammonia atmosphere.

From the facts described above, the pyrolysis mechanism of TMOB is considered as follows. In argon and hydrogen atmospheres, methoxy radical dissociates from boron atom in TMOB, and it reacts with methyl group in TMOB, to form dimethylether. Anyway, all the bonds between oxygen and boron in TMOB are not broken by the dissociation. It is in agreement with the fact that the boron oxide film deposits by the pyrolysis of TMOB in argon and hydrogen atmospheres at temperatures above 600°C. Carbon monoxide, methane and hydrogen are generated by pyrolysis of dimethylether. From the fact that generation of carbon monoxide, methane and hydrogen at 700°C, it is considered that pyrolysis of TMOB proceeds rapidly although it is not clear whether the exhausted gases are due to pyrolysis of dimethylether or TMOB itself. In ammonia atmosphere, the starting temperature of pyrolysis of TMOB is influenced by the quantity of ammonia. The starting temperature of pyrolysis of TMOB in NH_3/TMOB=5 and 10 was 500°C and 350°C, respectively, and these temperatures are not affected by the kinds of carrier gases. It is remarkably different from the cases of argon and hydrogen atmospheres where pyrolysis of TMOB takes place above 700°C. In the ammonia atmosphere at the temperatures above 700°C, the amounts of carbon monoxide and hydrogen increase steeply with decreasing the amount the TMOB and methanol. The mechanism of pyrolysis of TMOB in the ammonia atmosphere is resolved next.

(c) GROWTH MECHANISM OF BORON NITRIDE FILMS

Deposition mechanism of boron nitride films by the reaction of TMOB with ammonia is closely related to the pyrolysis mechanism of TMOB. The methoxy group bonded to boron atom in TMOB isolates from the boron by breaking the bond between oxygen and boron but without breaking the carbon-oxygen bond. This fact does not contradict with the generation of methyl alcohol, carbon monoxide, and hydrogen, and the absence of methane. In the case of TMOB, on the other hand,the bond between boron and is broken by the nitrogen atom in ammonia which is co-ordinates to the boron atom. This explanation is consistent with the fact that the start temperature of pyrolysis TMOB is reduced with increasing ammonia quantities, and the deposition rate of boron nitride films increases with increasing ammonia, and only the stoichiometric boron nitride films deposit as described in the next chapter.

III.B. PREPARATION OF BORON NITRIDE FILMS BY MOCVD USING TEB

Recently in the study of preparations of compound semiconductors, CVD using organo-metallic compounds has been developed, and its achievements have become of major interest lately. This method called MOCVD was developed for the deposition of GaAs by Manasevit in 1968(39), and a number of related papers have been reported mainly concerning the growth of III-V compound semiconductors. Noting that boron nitride is also a III-V group compoumd, peparation of boron nitride films by MOCVD is described in this chapter.

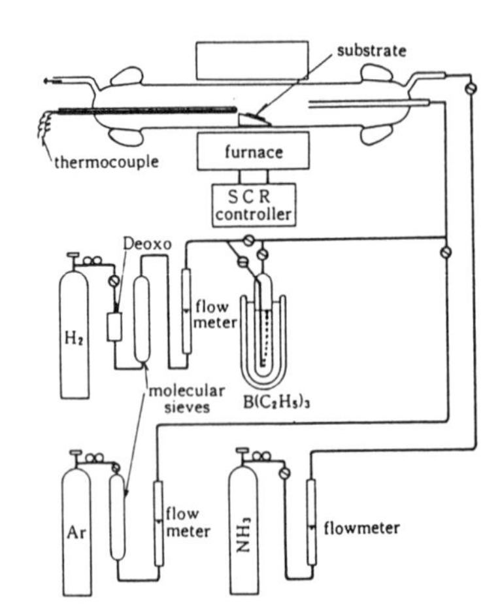

Fig.26 Schimatic diagram of the apparatus used for the deposition of boron nitride films by MOCVD.

1. EXPERIMENTS

Triethylboron(TEB,Arfa Products,98%), ammonia (99.9%), hydrogen (99.999%) and argon (99.995%) were used to grow boron nitride films by CVD. Sapphire ($12\times12\times0.6mm^3$) and silicon($12\times12\times0.25mm^3$) were used as substrates.

The schematic diagram of the apparatus for the deposition of boron nitride is shown in Fig.26. The fused quartz reaction tube has an inside diameter of 25mm and is 600mm long. Substrates were put on the substrate holder(sintered boron nitride) with the slope of 20 degrees from the horizon and heated by external heating furnace. The substrate temperature was measured by thermocouples attached at the rear of substrate holder.

Triethylboron was transported by bubbling hydrogen in TEB. The temperature of the TEB saturator was controlled between 0°C and 25°C. TEB is colorless clear liquid, and is moisture and air sensitive and pyrohoric. The vapor pressure at 0°C is 12.5Torr. The flow rate of TEB was determined from the loss of TEB in the saturator. The preparation conditions of boron nitride films are summarized in Table 4.

Table 4 Experimental conditions of MOCVD (TEB)

Growth temperature	750°C ~ 1200°C
$(C_2H_5)_3B$	0 ~ 2.8ml/min
NH_3	15 ~ 250ml/min
H_2	48 ~ 120ml/min
Total	60 ~ 300ml/min
Saturator temp.	0 ~ 25°C

2. RESULTS AND DISCUSSION

2.1. APPEARANCES OF THE FILMS

The changes of the color and the surface roughness of the films deposited at different flow rates of triethylboron and ammonia are shown in Fig.27. The stoichiometric boron nitride is colorless and transparent because of its wide energy band gap. However, some of the deposited boron nitride films are colored due to nonstoichiometry of the films. It was found that the color of the deposited films depends on the reactant molar ratio $NH_3/B(C_2H_5)_3$, and that the surfaces become rough with increasing flow rate. The black colored films with smooth surface were obtained when the films were deposited only with TEB or with TEB and a small amount of ammonia. X-ray diffraction analysis shows that the black colored films are amorphous. It is found from EPMA analysis that carbon is contained in the black film. At the deposition temperature of 1000°C, the colorless and transparent films were obtained when the flow rates of ammonia and TEB were in the ranges between 25 and 70 ml/min and 0.25 and 1.0ml/min, respectively. The color of the transparent films changed from colorless to dark brown by increasing the ratio of TEB to ammonia. Colorless films and brown films were found to be hexagonal boron nitride by x-ray diffraction analysis.

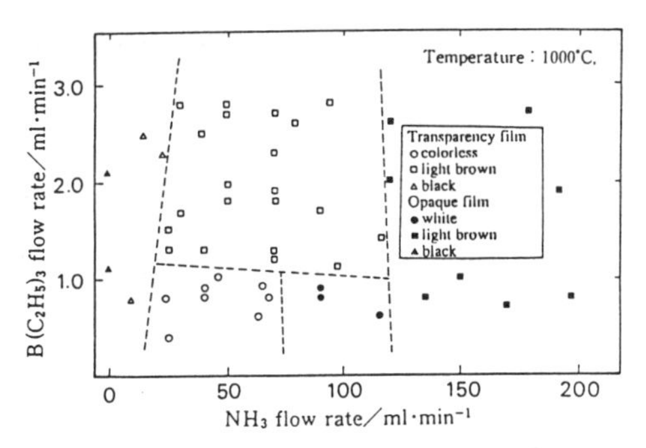

Fig.27 Relation between the flow rate of triethylboron and ammonia and the appearance of the films deposited at 1000 C

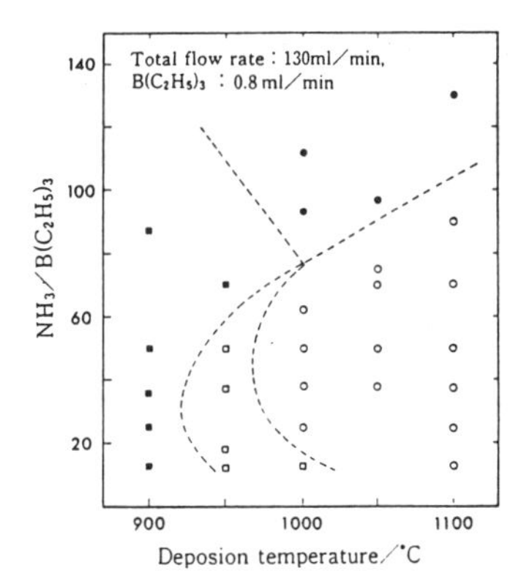

Fig.28 Relation between the NH_3/TEB and deposition temperature and the appearance of the films

Looking at Fig.27 along the line of TEB flow rates below 1.0 ml/min, the colorless films change from transparent to opaque at the ammonia flow rate 80ml/min. The brown-colored films deposited with the TEB flow rate above 1.0ml/min also change from transparent to opaque at ammonia flow rate 120ml/min. These results suggest that the deposition of opaque films is caused by gas phase reaction or uniform growth which is inhibited by numerous nucleation sites that increase with increasing source gas supply to substrate surface. This is consistent with the fact that the deposition rate increases with increasing flow rate.

Figure 28 shows how the appearance of the films deposited at constant TEB and total flow rate depends on temperature and TEB to ammonia ratio when deposition is carried out at constant TEB flow rate. The symbols in Fig.28 are the same as those in Fig.27. It can be seen that, at the low temperatures, a brown opaque film deposits but the film changes to a colorless transparent film with increasing temperature. When the flow rate of ammonia is more than enough, the opaque films deposit.

2.2. DEPOSITION RATE

When the total flow rate is increased by addition of argon, the deposition rate increases steeply with increasing total flow rate because the amount of reactant species arriving at the substrate surface increases with increasing total flow rate.

The deposition rate of the films on the sapphire substrate at the constant molar ratio $NH_3/B(C_2H_5)_3=30$ increases from 0.025μm/h at 750°C to 0.96μm/h at 1000°C and then decreases steeply to 0.235μm/h at 1100°C. At the temperatures below 1000°C, the deposition rate increases with temperature because of surface reactions. At the temperatures above 1000°C, on the other hand,the deposition rate decreases with temperature because of homogeneous reaction and the dissipation of the reactants at the reactor wall. White powder precipitated on the wall increases steeply above 1000°C.

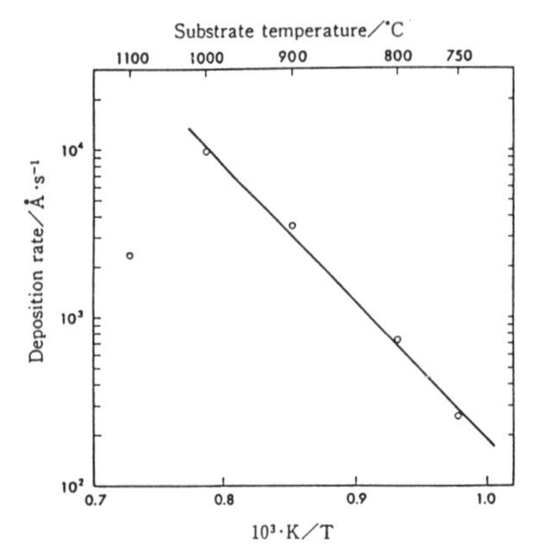

Fig.29 Logarithm of the deposition rate as a function of reciprocal absolute substrate temperature

Figure 29 shows the relation between the logarithm of the deposition rate and the reciprocal absolute temperature. An activation energy of 37.4kcal/mol was obtained for the reaction in the temperature range below 1000°C. This value is larger than that of 20.4-25.6[9] and 32.6Kcal/mol[13] obtained for the reaction of diborane with ammonia.

2.3 COMPOSITION OF THE FILMS

It is shown in Fig.30 that the composition of the films deposited at 1000°C changes from N/B=0.45 at $NH_3/B(C_2H_5)_3=5$ to 1.0 at 35, and then the composition remains constant for the reactant gas molar ratio $NH_3/B(C_2H_5)_3>35$. Stoichiometric boron nitride deposits over this point.

The extent of coloration of the films depends on the molar ratio of the reactants. It is found from the composition analysis that the coloration is caused by an excess of boron in the film.

Figure 31 shows the composition of the film as a function of the growth temperature at the molar ratios $NH_3/B(C_2H_5)_3=10$ and 30. In the case of the molar ratio equal to 10, the composition of the deposited films, B/N, increases from 0.46 at 800°C to 0.75 at 950°C and then decreases slightly to 0.73 at 1200°C. The composition is less than 1.0 even at 1000°C, which is believed to be due to the small ammonia ratio.

In the case of the molar ratio 30, the relation between the composition and the growth temperature have the same tendency as that for the ratio 10. However, the films deposited in the temperature range from 950°C to 1000°C have the stoichiometric composition.

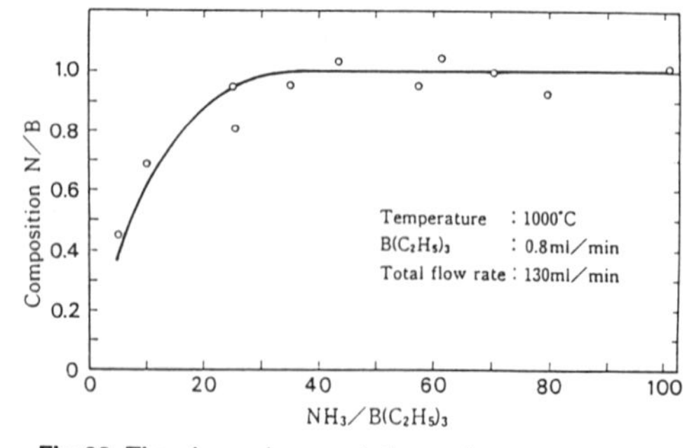

Fig.30 The dependence of the ratio of ammonia to TEB to the composition of the films deposited at the constant deposition temperature

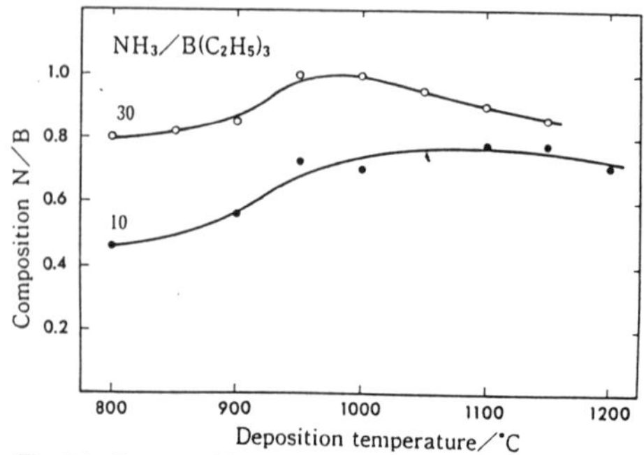

Fig.31 Composition of the boron nitride films deposited at the molar ratio 10 and 30 as a function of the deposition temperature

The increase of the composition with temperature below 950°C is considered to be due to thermally-activated reaction between boron and nitrogen. The decrease in composition above 1000°C can be explained by assuming that the deposition rate of boron atom decomposed from TEB exceeds the rate of combination of boron with nitrogen, and that the effective ammonia is decreased by decomposition at the reactor wall.

2.4. X-RAY DIFFRACTION OF THE FILMS

The films obtained were found to be hexagonal boron nitride by x-ray diffraction analysis. In most cases, only the (002) diffraction peak was observed, which indicates that the c-axis is preferentially oriented perpendicular to the substrate surface. However, for the opaque films with rough surfaces, diffraction peak other than the (002) peak were observed.

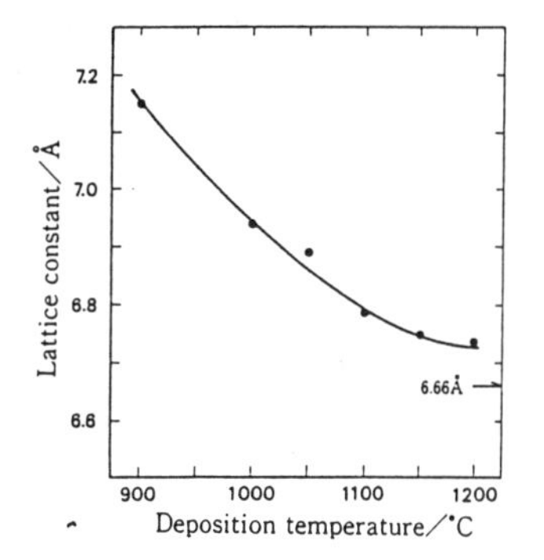

Fig.32 Latice constant C_0 of BN film deposited at various temperature

The diffraction pattern is similar to that of boron nitride powder. Since the film is deposited at high deposition rates, it is thought that the numerous nucleations occur on the substrate surface and the neuclei have different orientations from each other. The diffraction pattern of colorless transparent films deposited with low deposition rates and at high temperatures consists of only the diffraction from the (002) plane. The c-axis of crystal perpendicularly to the substrate plane when silicon wafer is used as a substrate. The lattice parameters along c-axis obtained from the x-ray diffraction measurements are shown in Fig.32 as a function of deposition temperature. The lattice parameter c_0 decreases with increasing the deposition temperature from 7.18Å at 1000°C to 6.73Å at 1300°C. However, the lattice parameter of the boron nitride film deposited at 1300°C remained larger than 6.66Å reported for bulk boron nitride. It is found that the higher deposition temperatures are required to obtaine the films with the value 6.66Å. The crystallite sizes calculated from FWHM of (002) diffraction peaks as a function of the deposition temperatures are shown in Fig.33. It is found that the size of the crystallite increases from 48Å to 153Å as the temperature is raised by only 250°C.

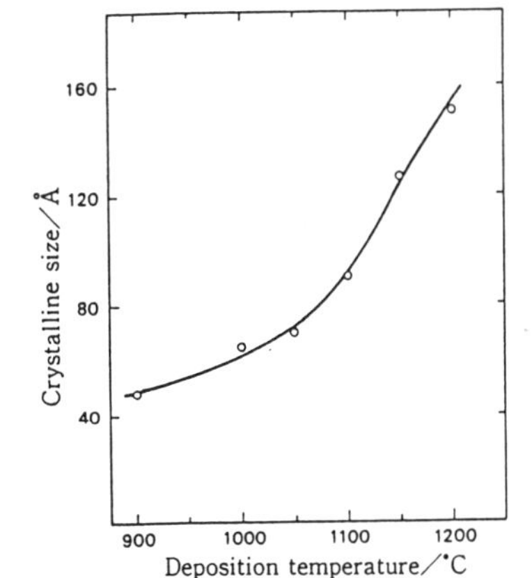

Fig.33 The crystalline size of the boron nitride films as a function of the deposition temperature

2.5. OPTICAL PROPERTIES OF THE FILMS

Figure 34 shows the transmission spectra in the wavelength range of 0.18-1.0μm for the films deposited by MOCVD on sapphire substrates at various compositions. The spectrua (a), (b), and (d), correspond to those of the films of the composition 0.4, 0.6, 0.8 and 1.0, respectively.

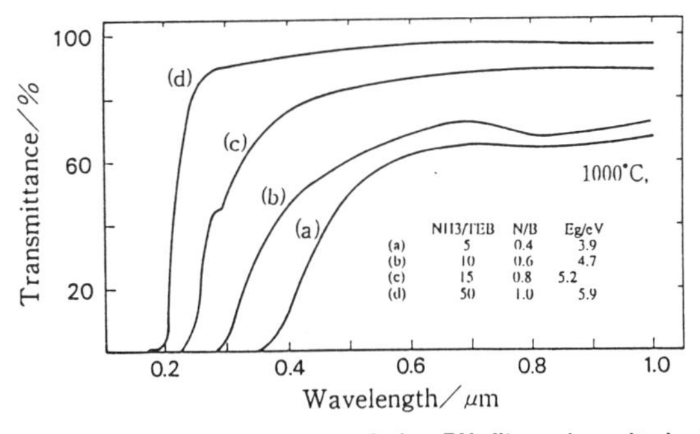

Fig.34 Transmission spectra of the BN films deposited at various condition (a) to (d) corresponding to the composition 0.4, 0.6, 0.8, and 1.0, respectively

The values of the square of absorp-

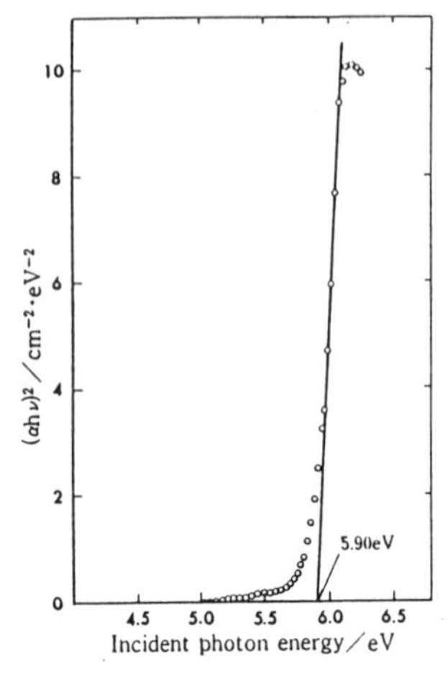

Fig.35 Absorption edge fitted to direct allowed transition

tion coefficients multiplied by photon energy, $(\alpha h\nu)^2$, of the films deposited at $NH_3/(C_2H_5)_3B$=20 and 1000°C are plotted in Fig.35 against incident photon energy. The figure shows that the values lie on a straight line. this suggests that boron nitride is a semiconductor having a direct energy band-gap. The energy gap, Eg, is estimated to be 5.90eV by extrapolation of the linear part to the horizontal axis in the figure. This value is in good agreement with the value obtained for the films deposited by MFCVD.

Infrared transmission and reflection spectra of the film of 5μm thick are shown in Fig.36. The figure shows that the reststrahlen band lies in the frequency range between 1600 and 1300cm^{-1}. This band corresponds to the in-plane vibration of boron nitride crystal structure. The absorption at 780cm^{-1} corresponds to the out of plane vibration. The oscillations in the spectra are due to interference effect at the air-film and film-substrate interfaces.

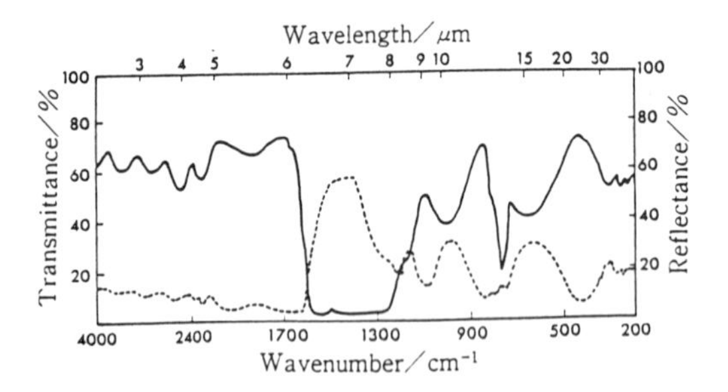

Fig.36 Infrared transmission and reflection spectra of boron nitride films deposited on Si

III.C. PREPARATIOM OF BORON NITRIDE FILMS BY MOCVD USING TMOB

High quality boron nitride films have been prepared by MOCVD using the reaction of triethylboron with ammonia. The effects of deposition conditions and the deposition rate on the composition of the films have been reported in the preceding sections. However, this method has a few weak points ,i.e., TEB is dangerous, poisonous and expensive.

Trimethoxyborane ($(CH_3O)_3B$,TMOB) is selected as an alternative source material to overcome these shortcomings. The physical properties of TMOB are similar to those of TEB, therefore, the same experimental apparatus as that for TEB can be used without alteration. TMOB is not ignite spontaneously in the air and is economical material, the price ratio of TEB/TMOB being about 30 at present.

Table 5 Experimental condition data of MOCVD (TMOB)

TMOB	3 ml/min
NH_3	3 ~ 270 ml/min
NH_3/TMOB	1 ~ 90
Total	250, 350 ml/min
Temperature	900 ~ 1300 °C
Substrate	Si (100)

1. EXPERIMENTS

TMOB(Arfa Products, 98% purity), ammonia(99.9%), hydrogen(99.99%) and nitrogen (99.9%) were used to grow boron nitride films. Silicon plates (16×16×0.35 mm^3) were used as substrates. TMOB was transported by bubbling with hydrogen or nitrogen. The temperature of the TMOB saturator was controlled at 0°C. TMOB is colorless clear liquid, and is moisture sensitive. The vapor pressure at 0°C is 37Torr. The flow rate of TMOB was controlled by changing the flow rate of carrier gas and was measured from the loss of TMOB in the saturator. The preparation condition of boron nitride films are summarized in Table 5. When the source gas ratio varied, the total flow rate was kept constant by adding argon gas in order to avoid influence of the flow rate on the deposition rate.

The analysis of the films deposited are carried out similarly as in the preceding chapters.

2. RESULTS AND DISCUSSION

2.1. DEPOSITION RATE

Figure 37 shows the deposition rate as a function of the temperature at the constant flow rates of TMOB=3.2ml/min, H_2=78ml/min, NH_3=251ml/min and total flow rate=357ml/min. The thickness of the film deposited at 900°C could not be obtained because it was too thin to be measured. Deposition rate increases from 64Å/min at 1000°C to 276Å/min at 1200°C with increasing temperature, and then decreases steeply above 1200°C. The increase of deposition rate with temperature can be attributed to the reaction rate determining step.

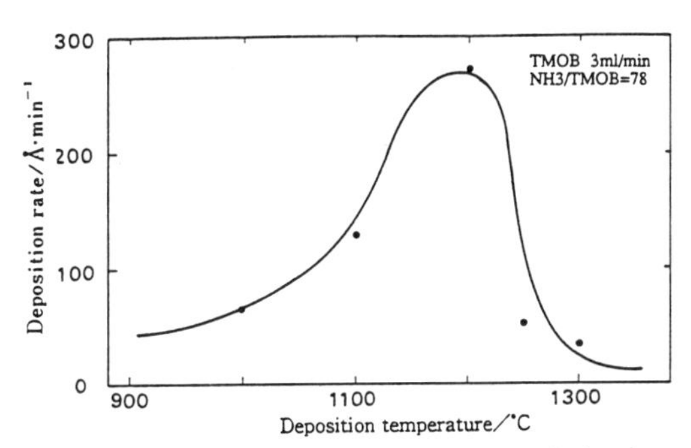

Fig.37 Deposition rate of the BN films deposited using TMOB as a function of the deposition temperature

The activation energy of this step is 27.1kcal/mol obtained from the results of Arrhenius plot. This value is smaller than 37.4kcal/mol in the case of TEB. The steep decrease of deposition rate above 1200°C is caused by the decrease of the supply gasses onto the substrate surface because of the inclease of deposition on the walls of the external furnace. This step corresponds to the diffusion rate determining step.

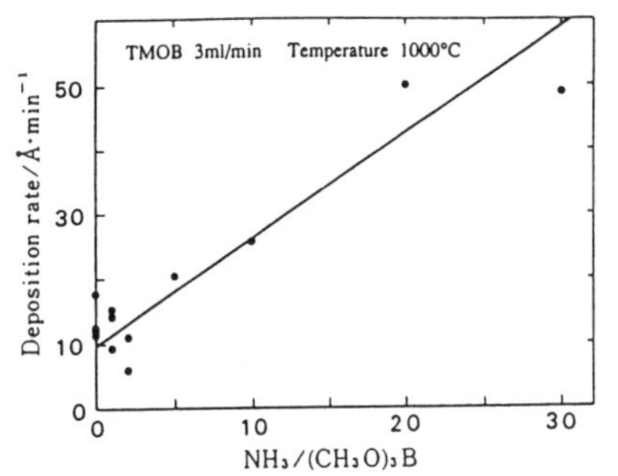

Fig.38 Deposition rate of the BN films deposited using TMOB as a function of the ratio of TMOB to ammonia

Figure 38 shows the deposition rate of boron nitride films as a function of the ratio of ammonia to TMOB at the constant flow rate of TMOB 3ml/min, totally 250ml/min and the temperature of 1000°C. The deposition rate increases with the ratio of ammonia to TMOB, but scatters at a small ratio. The deposition rate is about 13Å/min at zero ratio, i.e., by only the pyrolysis of TMOB. It decrease slightly at around the ratio of 2.0 and then increase linearly again up to about 50Å/min at the ratio of 30. The deposition rate increases to 160Å/min at the ratio of 80, which is not shown in the figure. The linear increase of the deposition rate with increasing the ratio of ammonia to TMOB shows that the ammonia-assisted decomposition of TMOB is the rate-determining step for the deposition rate.

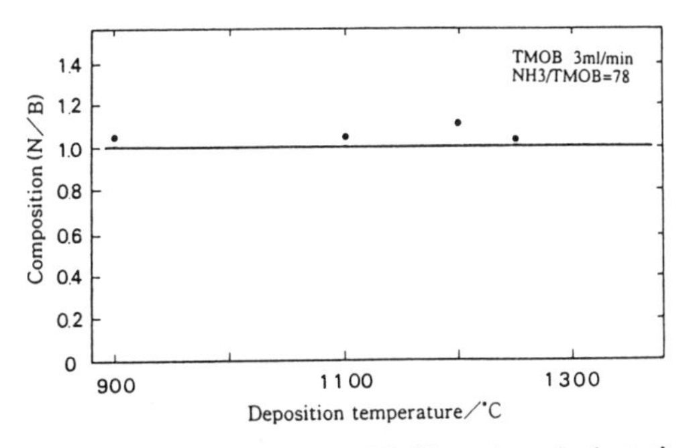

Fig.39 Composition of the BN films deposited at the constant source gasses as a function of the deposition temperature

2.2. COMPOSITION

The composition of deposited films as a function of the deposition temperature ranging from 900°C to 1300°C are shown in Fig.39. The ratio of ammonia to TMOB

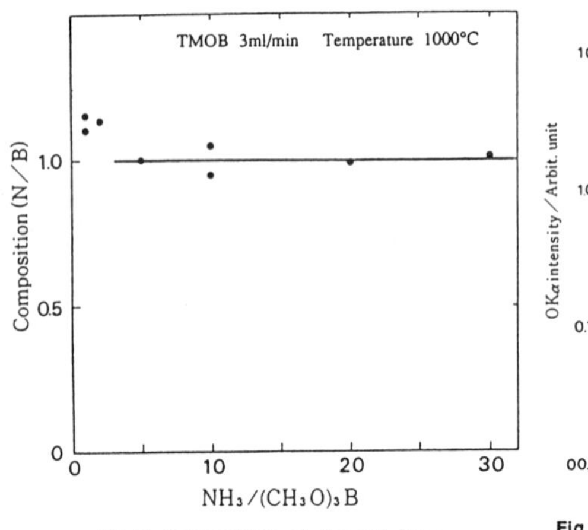

Fig.40 Composition of the BN films deposited at the constant deposition temperature 1000°C as a function of the ratio of TMOB to ammonia

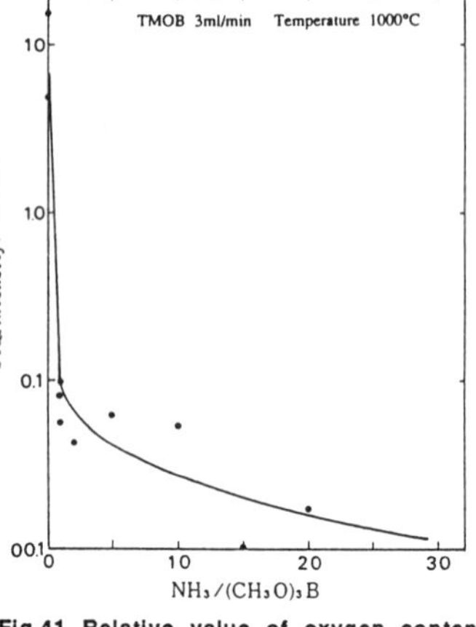

Fig.41 Relative value of oxygen content in BN films deposited by using TMOB as a function of the ratio of TMOB to ammonia

(NH_3/TMOB) is 78 and the total flow rate is 357ml/min. The scattered points from the line of unity in the figure may be caused by the inaccuracy of the analysis of the composition. It is shown that the composition dose not depend on the temperature and that all of the boron nitride films deposited have the stoichiometric composition. These facts involve clues to clarify the deposition mechanism.

The composition of the films is shown in Fig.40 as a function of the ratio of ammonia to TMOB in the rage from 0 to 30. The flow rate of TMOB was 3ml/min and the total flow rate250ml/min. The deposition temperature was kept at 1000°C. Figure 40 shows that stoichiometric boron nitride films can be obtained irrespectively of the ammonia to TMOB ratio with the exception of the ratio is so small. Carbon impurity was not obserbed within the detection limit of EPMA . However, oxygen was detected. Figure 41 show the oxygen content as a function of ammonia in the reaction gas. The relative oxygen content given in this figure was obtained by reducing the observed intensity by the boron $K\alpha_2$ intensity detected by the EPMA. It was found that the films deposited by the pyrolysis of TMOB consisted of boron oxide, and the oxygen content in the film decreases by adding ammonia. The ratio of oxygen to boron in the film deposited at NH_3/TMOB=30 was 0.001. This oxygen content will be disregarded as compared to stoichiometric boron nitride. The oxygen containing mechanism is described in a later section. A possible reason why the N/B composition ratio of the film exceed 1.0 at small ammonia/TMOB region (Fig.40) is considered to be due to this oxygen contents.

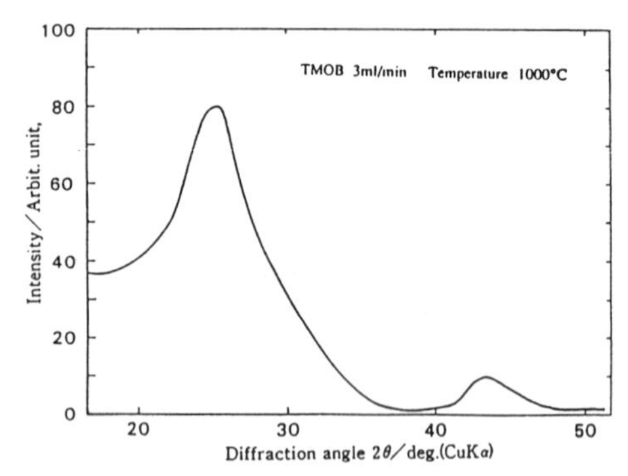

Fig.42 X-ray diffraction pattern of boron nitride film deposited by MOCVD using TMOB at 1000° C

2.3. PROPERTIES OF THE FILMS DEPOSITED BY MOCVD USING TMOB

Figure 42 shows the typical x-ray diffraction pattern of boron nitride films deposited on quartz substrate at 1000°C. The broad peakes indicate poor crystallinity of the films compared with the film made by MOCVD using TEB souce. One of the origins of difference in crystallinity is the difference in reaction temperatures. TMOB pyrolyzes at 700°C, and beings to deposit boron nitride at 900°C when mixed with ammonia. The reaction temperatures are generally lower in the case of TEB. The typical pyrolysis and deposition temperarures are 300°C and 700°C, respectively. Infrared transmission spectrum of the film deposited by MOCVD using TMOB is shown in Fig.43.

The figure shows a spectrum characteristic of boron nitride.

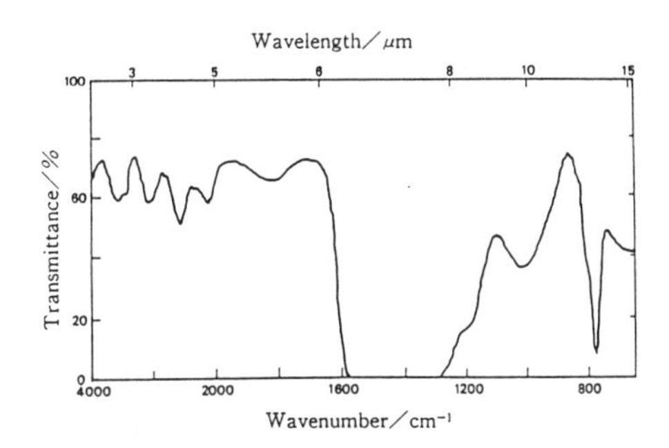

Fig.43 Infrared transmission spectra of boron nitride film deposited by MOCVD using TMOB

IV. BORON NITRIDE "TYMPANIC MEMBRANE"

Usually, thin films are deposited on substrates and used with the substrates. In a tympanic membrane, which we tried to make, the film is held in tension by a supporting frame without substrate. This is fabricated by etching only the central part of the substrate to make a frame-supported film looked like askin of tympani.

The x-ray lithography mask membrane which is attracted at present is considered as one of the applications of boron nitride tympanic membrane. Nowadays, semiconductor ICs called ultra LSI have been realized by very large integration. The enlargement of integration require the reduction of the line width in the circuit. Photo, electron beam, and x-ray lithographies are used at present as the drawing technique of the fine patterns of the circuit. The photo lithography is a most popular technique, and now almost all the commercial semiconductor chips are manufactured by use of this technique. The direct drawing technique using electron beam has considerable potential for a fine pattern drawing technique. However, the electron beam lithography will be used for a trial circuit because it is difficult to overcome the low through-put of this technique at present. An x-ray lithography has been noted during the last decade as a next generation lithography technique. However, the practical application of the x-ray lithgraphy is delayed by a lot of difficult problems to be solved. An x-ray lithography will become most suitable technique for DRAM(Dynamic Random Access Memory) above 256Mbits with quarter micro meter circuit line rule. There are many problems to be overcome in x-ray lithography to put it to practical use. One of the problems is about x-ray mask. The most important problem in the fabrication of the x-ray masks is about the material.

A lot of materials for x-ray mask membrane marerials have been developed so far. Among them are boron carbide (BC)(40), boron nitride (a-BN:H)(41,42), boron nitride doped with silicon (Si-BN:H)(43,44), silicon carbide (SiC)(45) and silicon nitride (SiN)(46,47,48) as inorganic compound materials and are beryllium and silicon as element materials. Boron nitride is a lowest Z material in these except for beryllium and boron carbide, and the thin membrane (<10μm) is transparent to x-ray and visible light.

In the case of boron nitride mask membrane, the problem has been pointed out on its stability (49,50). Boron nitride mask membranes so far have been fabricated to involve a large amount of hydrogen which realizes the tensile stress of 1×10^{9} dyne/cm^2 on the silicon substrate, and which is believed to be the cause for instability of the membrane. Studies to improve instability of the membrane have been made. The boron nitride series mask membrane have been deposited by CVD at low temperature and low pressure using ammonia and diborane as source gasses. A chemical formula for these membranes is written as a-BN:H as in the case of amorphous silicon a-Si:H. On the contraly, we can obtain pure boron nitride mask membrane is fabricated here through inexpensive method, i.e.,by the high temperature MOCVD using TEB as source gas. It is the purpose in this chaptor that the improvement of x-ray damage of these a-BN:H membrane and the solution of the stress control problem in later discribing.

1. STRESS IN BORON NITRIDE FILMS

The stress control in preparation of tympanic membrane is important for keeping the flat surface. The stress in the film deposited on a substrate is derived mainly from the difference of properties between the film and the substrate. The stress in the films is affected by many factors including preparation conditions. One of the factors is the difference in structure between film and the substrate, and some others are defects and imperfections of structure in the film. Since there are many factors affecting the stress, the control of the stress is difficult.

The structure of the film is affected by such as substrate material, composition, and deposition rate. Composition and the deposition rate of the films are affected by concentrations of gases and the deposition temperature.

The epitaxial growth method is emphasized on this point. However, since even the films grown by the epitaxial growth method generally include considerable amount of stress, preparation of films with controlled stress is an important problem. On the other hand, the magnitude of the stress required for the membrane has neither been manifested quantitatively. The required amount of stress is determined by a balance between the stress of the x-ray absorption pattern and that of the membrane substrate.

It is reported that a mask membrane having a stress about 1×10^9 dyne/cm^2 works well with carfully fabricated metal absorption pattern.

In the case of the high temperature deposition such as CVD, the stress of the film is presumed to be governed mainly by the thermal stress. Of course, the intrinsic stress generated during growth is also comprised.

If the thermal stress is much larger than the intrinsic stress, the film with tensile stress can be made by choosing the substrate material which has a smaller thermal expansion coefficient than that of the films. Boron nitride films deposited at high temperature have hexagonal crystal structure, in which physical properties are highly anisotropic. For example, thermal expansion coefficients are reported to be 1×10^{-6} deg^{-1} along a-axis and 40×10^{-6} deg^{-1} along c-axis [51]. Generally, boron nitride deposited on a crystalline substrate grows with c-axis oriented normal to the substrate surface as discribed in III.B.1.2.4 [52].

Thermal expansion coefficients of silicon which is the most popular substrate material are 2.3×10^{-6} deg^{-1} within (100)plane and 3.7×10^{-6} deg^{-1} within (111)plane, and are not small enough compared with boron nitride. Only quartz is a good canditate as a substrate material, because the thermal expansion coefficient of quartz is 5×10^{-7} deg^{-1}. The Vical glass which has a similar composition to quartz has the value of thermal expansion coefficient 8×10^{-7} deg^{-1}.

In the case of boron nitride films deposited on these low thermal expansion coefficient substrate at high temperature, thermal stress is generated in the film by cooling from deposition temperature to room temperature.

The thermal stress(σ_t) of the film deposited on a substrate is given by

$$\sigma_t = E_f(\alpha_f - \alpha_s)\,\Delta T/(1-\nu_f) \quad (4)$$

where E_f and ν_f are Young's modulus and Poisson's ratio of the film, respectively. α_f and α_s are thermal expansion coefficients of the film and the substrate, respectively, ΔT is the difference of deposition temperature and room temperature.

The thermal stress of boron nitride films are calculated to be $4.05\sim20.3\times10^8$ dyne/cm^2 for quartz and $-1.3\sim-17\times10^8$ dyne/cm^2 for silicon by using the values of $E_f/1-\nu_f=2.26\sim13\times10^{-11}$ dyne/cm^2 for BN[53], $\alpha_{BN}=1\sim1.55\times10^{-6}$ deg^{-1}, $\alpha_{Si}=3\times10^{-6}$ deg^{-1}, $\alpha_{quartz}=5\times10^{-7}$deg^{-1} and $\Delta T=800\sim1100$°C.

The results of this calculation indicate that the stress in boron nitride films deposited on quartz and silicon substrate is tensile and compressive, respectively. The value of the tensile stress in boron nitride film deposited on the quartz substrate is in the same oder of magnitude as the value for x-ray mask membrane reported. Boron nitride films on quartz substrate is thus expected to satisfy the requirements as an x-ray mask membrane.

Table 6 Experimental condition of preparation of BN films by MOCVD using TEB

TEB	5 ~ 20 ml/min
NH_3	50 ~ 400 ml/min
Total(+H_2)	500, 1000, 1400 ml/min
Temp.	770 ~ 940 °C
Dep.time	120 ~ 360 min

2. EXPERIMENTS

A reactor for the fabrication of boron ni-

tride tympanic membrare, two inches in diameter, is made up from a square shaped quartz tube ($5\times10\times40cm^3$, inside capacity). Halogen lamps were used for substrate heating. A gas flow system was similar to that used previously, except for the use of a mass flow controller. Quartz and silicon are used as the substrate. The purpose of this study is to search an adequate condition to get the tympanic membrane with proper tensile stress. It is studied how the stress of the film is affected by the kind of substrate, deposition temperature and gas flow rate. Growth condition of boron nitride film for the fabrication of the tympanic membrane studied are summarized in Table 6. Total stress of the films deposited on the silicon and the quartz substrates including the intrinsic and the thermal stresses were estimated by measuring curvature of the substrate with deposited film. The curvature was measured by a microstylus profilometer (DEKTAK II,Sloan).
The stress(σ) of the films was calculated using the following equation.

$$\sigma = E_s b^2 \delta / 3(1-\nu)\, dl^2 \qquad (5)$$

where b and d are the thicknesses of the substrate and the film, l is length measured of the substrate and δ is the magnitude of bending.

An x-ray diffraction of the film was measured and the relation of the lattice parameter and the stress of the films was also studied.

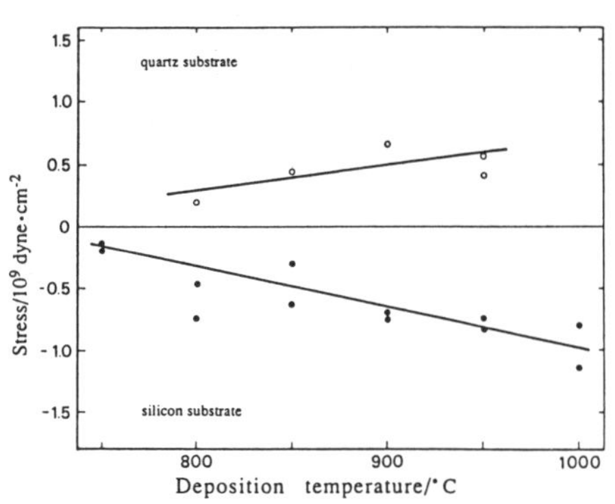

Fig.44 The stress of BN films deposited on the quartz and silicon substrate as a function of the deposition temperature

3. RESULTS AND DISCUSSION

The stresses in the films deposited on silicon and quartz substrates are shown in Fig.44 as a function of the deposition temperature. The magnitude of the stress in the films deposited on the both substrate increases linearly with increasing temperature. However, the stresses in the films deposited on the silicon substrate is compressive and that on the quartz substrates are compressive and tensile. This can be explained by the relative magnitudes of their thermal expansion coefficients i.e., 5×10^{-7} deg^{-1} for quartz, 3.5×10^{-6} deg^{-1} for silicon and about 1.5×10^{-6} deg^{-1} for boron nitride along the a-axis. However, the observed stress is larger than the value explained by thermal stress. This discrepancy suggests the existence of intrinsic stress in the films. The intrinsic stress in the deposited film increase slightly with increasing deposition temperature.

Although many origins are considered for the intrinsic stress in the films deposited, we present here some of the experimental results which would give insight in to the origin of the stresses. Figure 45 shows the dependence of the stress on the source gas ratio of ammonia to TEB. It is shown that the stress in the films decreases gently with increasing the flow rate ratio, and finally reaches a nearly constant value when the ratio is larger than twenty. In this molar flow rate ratio region, as mentioned in the previous chapter, the deposited product is the stoichiometric boron nitride. When the molar ratio changes from 15 to 20, the composition of the films shifts from boron excess to stoichiometric value. Therefore, it is understood that the stress is large when the composition of the films is boron rich, because thermal expansion coefficient of solid boron is larger than that of boron nitride.

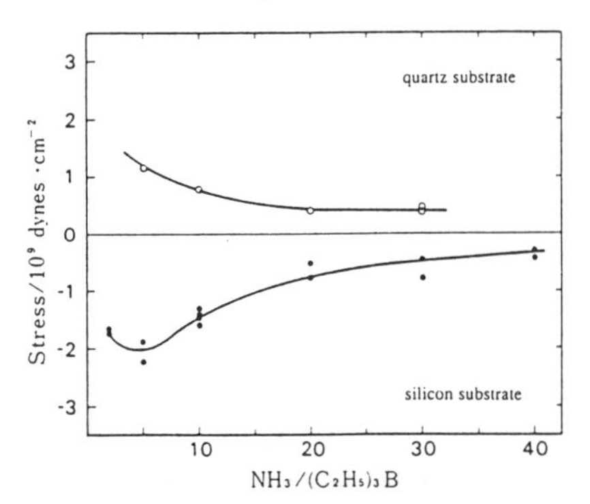

Fig.45 The stress of the boron nitride films deposited on quartz and silicon substrate as a function of the ratio of TEB to ammonia

The effect of thickness of the film is examined

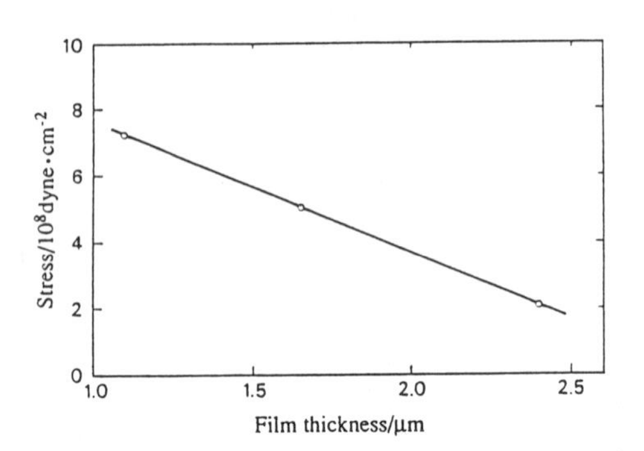

Fig.46 The stress of the boron nitride films as a function of the thickness of the BN films

next. Figure 46 shows the observed stress as a function of the thickness of film. The stress of the film decreases with increasing film thickness, because imperfection in the film is maximum at the interface between film and substrate, and so the total stress decreases gradually with film thickness. The relationships between stress and the deposition rate of the films deposited on silicon and quartz substrates are shown in Fig.47. On both substrates, the stress increases with increasing the deposition rate. A simple explanation for this phenomenon is that the increase of the deposition rate increases the imperfections in the film, and brings a tensile stress as a result of a shrinking force arising in the film.

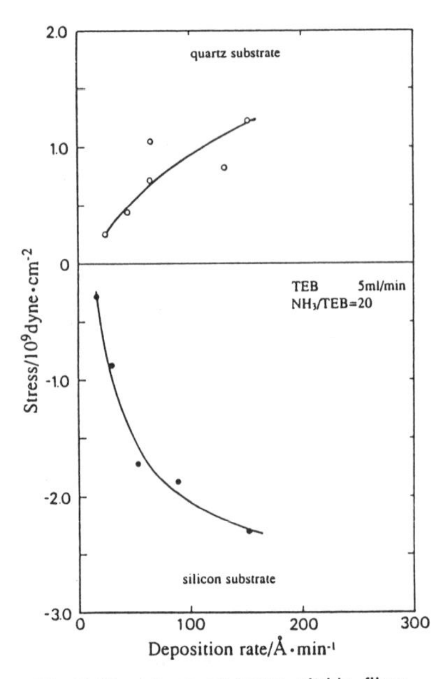

Fig.47 The stress of boron nitride films deposited on quartz and silicon substrate as a function of deposition rate by changing deposition temperature

An x-ray diffraction of boron nitride film was analyzed to examine the generation mechanism of the stress from a structural point of view. On the silicon substrate, only the diffraction peak around 25° corresponding to (002) plane of boron nitride crystal was observed. On the quartz substrate, however, the diffraction peak around 43° corresponding to (100) and (110) planes was also observed. These facts suggest that boron nitride film deposited on quartz substrate is grown in random direction. The films deposited are aggregation of microcrystals. These findings are consistent with the different structural properties of the two substrates, namely, the silicon substrate is a single crystal wafer with (100) surface but the quartz substrate is amorphous. It should be noted that the orientation of the crystallites in the films have serious effect on the stress of the film because of large anisotropic thermal expansion coefficient of boron nitride.

The interlayer spacing along the c-axis of the boron nitride films deposited on the silicon and quartz substrates are shown in Fig.48. The interlayer spaceing of the film deposited on the quartz substrate decreases with increasing deposition temperature, whereas that of the film deposited on the silicon substrate increases with increasing temperature. It is known that boron nitride crystal grows with c-axis perpendicular to the silicon wafer substrate surface. Because the in-plane thermal stress of the film deposited on silicon is a compression one, the interlayer spacing along the c-axis is lengthened, and its effect is enhanced as the deposition temperature is higher. When the crystal is grown so that the c-axis lies parallel to the substrate surface, the crystal suffers strong pulling force along c-axis from the substrate due to very large thermal expantsion coefficient of boron nitride along the c-axis. In the case of quartz substrate, where the crystallites are grown in random directions, the boron nitride film experiences combined effects of shortening c_0 due to in-plane tensile stress and that of lengthening due to out-of-plane tensile stress. The fact that the observed c_0 of the film grown on the quartz substrate is larger than that on the silicon substrate and also the value 6.66Å for bulk boron nitride suggests that the effect of out-of-plane tensile stress is significant. The temperature dependence of c_0 should be governed mainly by the in-plane tensile stress.

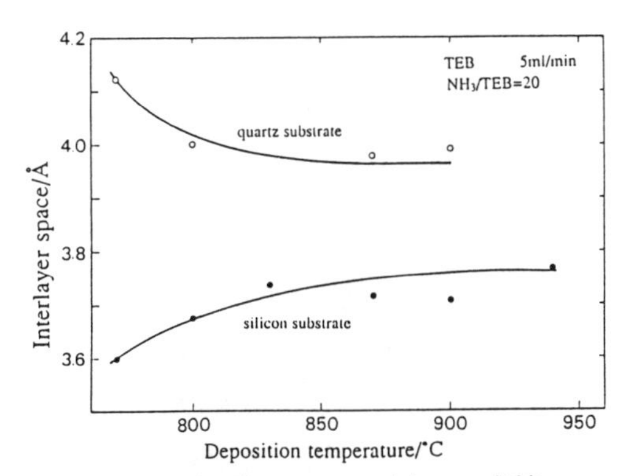

Fig.48 Interlayer space of boron nitride crystal deposited on quartz and silicon substrate as a function of deposition temperature

4. FABRICATION OF BORON NITRIDE TYMPANIC MEMBRANE.

The tympanic membrane is fabricated by etching the central part of the substrate. Boron nitride is inactive to etching reagents such as acids and bases. The processes are illustrated in Fig.49. The etching reagents used are fluoric acid (50%) and a mixture of fluoric acid and nitric acid for the quartz and silicon substrate, respectively. The tympanic membrane fabricated on quartz and silicon substrates are shown in Figs.50 and 51, respectively. The figures show that the films deposited on quartz and silicon substrate are flat and wrinkly, respectively. It is ascertained from these figures that the film have tensile and compressive stresses when the quartz and silicon is used as substrates, respectively.

The x-ray diffraction patterns of boron nitride films before and after the etching of the substrates are shown in Figs.52 and 53, respectively. The interlayer spacing along the c-axis, and the crystal size of the boron nitride films obtained from Figs.52 and 53 are summarized in Table 7. The interlayer distance along the c-axis is 4.134Å for the film deposited on quartz substrate and 3.559Å on silicon substrate. The interlayre spasing decreases from 4.134Å to 3.619Å on the quartz substrate and increases from 3.559Å to 3.735Å on the silicon substrate by etching the substrate. The stress of the boron nitride films deposited on the substrate is relaxated by removal of the substrate, resulting in the decrease or increase of interlayer distance. The FWHM of the diffraction peaks also decreases from 8.2° to 4.0° for the films on the quartz substrate and increases from 4.8° to 6.0° for the films on the silicon substrate. The cause of the large change of FWHM is considered that the microcrystals in the film are rearranged by the relaxation of the film stress.

The wrinkled membrane with compressive stress using the silicon substrate and the flat surface membrane with tensile stress using quartz substrate were obtained. This tympanic membrane was achived by using quartz substrate, and the stress was tensile with about 10^9dyne/cm^2. The tympanic membrane should be able to use as x-ray mask membrane. It was proved to be promising as x-ray mask membrane that a fine metal pattern could be fabricated on boron nitride film.

1 Substrate, BN deposited; 2 Wax coating; 3 Teflon frame; 4 Wax off; 5 Substrate etching by HF; 6 Substrate etched; 7 Teflon frame off; 8 Wax off

Fig.49 Boron nitride tympanic membrane fabrication process

Table 7 Results of x-ray diffraction of BN film on substrate or on substrate etching

Substrate	etching	2θ/deg	d/Å	FWHM/deg
quartz	before	21.5	4.134	8.2
	after	24.6	3.619	4.0
silicon	before	25.0	3.559	4.8
	after	23.8	3.735	6.0

V. CONCLUSION

The boron nitride films have been prepared by two new methods of MFCVD and MOCVD. MFCVD using the reaction of decaborane with ammonia is one

Fig.50 Boron nitride tympanic membrane fabricated on quartz substrate

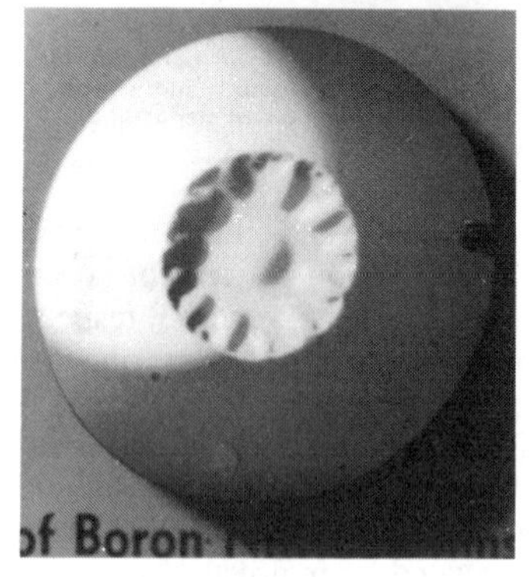

Fig.51 Boron nitride tympanic membrane fabricated on silicon substrate

of the preparation methods to boron nitride films with good controllabllity of the deposition rate and composition. MOCVD was also used to prepare boron nitride films. The triethylboron and trimethylborane were used as the boron source gases. The good quality boron nitride films were obtained by use of both sources. The reaction mechanisms are different from each other. In the case of the TEB source, the synthesis of boron nitride films are proceeds by the reaction between ammonia and boron which is deposited by the pyrolysis of TEB. On the other hand , in the case of TMOB, TMOB decomposes by the reaction with ammonia. The nonstoichiometric boron nitride films were deposited by using the TEB. However, in the case of TMOB, only stoichiometric films were obtained. The boron nitride films deposited by using TMOB is contaminated with oxygen

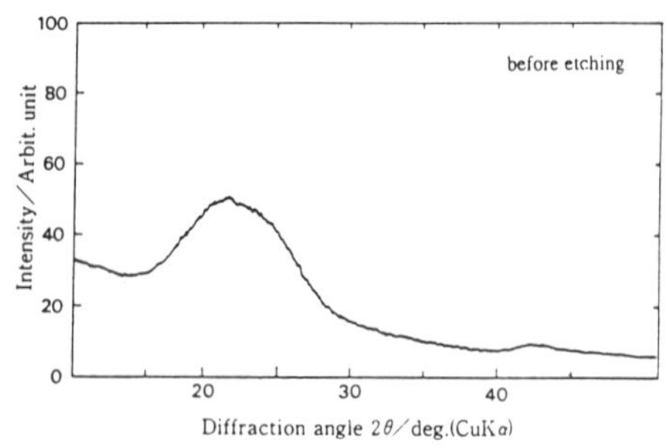

Fig.52 X-ray diffraction pattern of boron nitride film deposited on quartz substrate

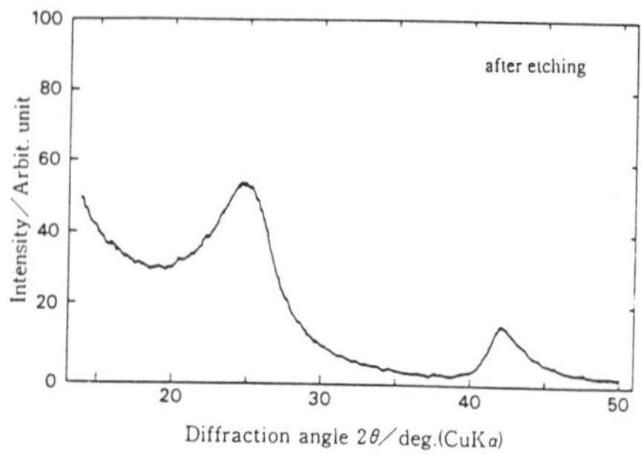

Fig.53 X-ray diffraction pattern of boron nitride tympanic membrane fabricated on quartz substrate

when ammonia concentration is low. The crystal of boron nitride films deposited by MFCVD and MOCVD are amorphous and aggregate of microcrystals with hexagonal structure, respectively. The films show the typical absorption of boron nitride at 1380 and 800cm^{-1}. The optical energy gap was estimated to be 5.9eV from the absorption spectra in the near ultraviolet spectrum. Tympanic membrane was fabricated successfully using boron nitride film deposited on quartz substrate. The membrane obtained is expected to be useful for an x-ray lithography mask membranes.

Acknowledgement
The author wishes to thank Dr. S. Yoshida and Dr. S. Miyajima for their helpful suggestions and a critical reading of the manuscript.

REFERENCE

(1) F.Meyer and R. Zappner, Ber. Dt. Chem. Ges., **54** 560 (1921)
(2) I.E.Campbell, C.F.Powell, D.H.Nowicki and B.W.Gonser : J. Electrochem. Soc.,**96**, 318 (1948)
(3) V.T.Renner: Z. Anorg. Allegem. Chem., **298**, 22 (1959)
(4) M.J.Rand and J.F.Roberts: J. Electrochem. Soc., **115**, 423 (1968)
(5) M.Hirayama and K.Shono: J. Electrochem. Soc., **122**, 1671 (1975)
(6) K.Shono, T.Kim and C.Kim: J. Electrochem. Soc., **127**, 1546 (1980), Japan J. Appl. Phys., **20**, 1901 (1981)
(7) S.D.Hyger and T.O.Yep: J. Electrochem. Soc., **123**,1721 (1976)
(8) S.P.Murarka, C.C.Chang, D.N.K.Wang, T.E.Smith: J. Electrochem. Soc, **126** 1952 (1979)
(9) A.C.Adams and C.D.Chapio: J. Electrochem. Soc., **127**, 399 (1980)
(10) A.C. Adams: J. Electrochem. Soc., **128**, 1378 (1981)
(11) R.Rudy, D.Baldwin and M.Karnezos: J. Vac. Sci. Technol., **B5** 272 (1987)
(12) H.Miyamoto, M.Hirose and Y.Osaka: Jpn. J. Appl. Phys., **22**, L216 (1983)
(13) E.Yamaguchi and M.Minakata: J. Appl. Phys., **55**, 3098 (1984)
(14) S.Motojima, Y.Tamura and K.Sugiyama: Thin Solid Films, **88**, 269 (1982)
(15) T.Matsuda, N.Uno, H.Nakae and T.Hirai: J. Mater. Sci., **21**, 649 (1986)
(16) W.Boronian: Mater. Res. Bull., **7**, 119 (1972)
(17) M.Sano and M.Aoki: Thin Solid Films, **83**, 247 (1981)
(18) W.Schmolla and H.L.Hartnagel, J.Phys.D.**15** L95 (1982)
(19) T.H.Yuzuriha and D.W.Hess, Thin Solid Films, **140** 199 (1986)
(20) B.Rother and C.Weissmantel: Phys. Stat. Sol.,(a) **87** **k**119 (1985)
(21) K.Yasuda, A.Yoshida, M.Takeda, H.Masuda, and I.Akasaki: Phys. Stat. Sol.,(a) **90** (1985)
(22) J.J.Gebhardt: Proc. 4th Int. Conf. on CVD, Boston, MA, Electrochem. Soc.,Princeton, NJ, p460 (1973)
(23) G.Constant and R.Feurer: J. Less-Common Met., **82** 113 (1981)
(24) W.Schmolla and H.L.Hartnagel: Solid State Electron, **26** 931 (1983)
(25) K.Nakamura: J. Electrochem. Soc., **132**, 1757 (1985)
(26) "ESCA Spectrometer System 5950 Manual," Section IV. Hewlett-Packard Co (1972)
(27) D.J.Joyner and D.M.Hercules: J. Chem. Phys., **72**, 1095 (1980)
(28) H.E.Bishop and J.C.Riviere: Appl. Phys. Lett., **16**, 21 (1970)
(29) V.AFomichev, T.M.Zmkina and I.I.Lyakhovskaya: Sov. Phys. Solid State, **12**, 123 (1970)
(30) G.Bambakidis and R.P.Wager: J. Phys. Chem. Solid, 42, 1023 (1981)
(31) F.Perrot: Phys. Rev. B, **23**, 2004 (1981)
(32) M.S.Nakhmanson and V.P.Simirnov: Sov. Phys. Solid State, **13**, 2763 (1972)
(33) F.C.Brown, P.Z.Bachrach and M.Skibowski: Phys. Rev. B, **13**, 2663 (1976)
(34) M.Grasserbauer: Microchim. Acta, **1**, 145 (1975)
(35) A.Zunger: Phys. Rev. B, **13**, 5560 (1976)
(36) J.Williams, R.L.Williams and J.C.Wright: J. Chem. Soc., 5816 (1963)
(37) W.C.Price, R.D.B.Fraser, T.S.Robinson and H.C.Longuet-Higgins: Discuss. Farady Soc., **9**, 131 (1950)
(38) R.Geick and C.H.Perry: Phys. Rev., **146**, 543 (1966)
(39) H.M.Manasevit: Appl. Phys. Lett., **12**, 156 (1968)
(40) A.P Neukermas, Proc. SPIE, **471** 96 (1984)

(41) D.L.Bros, Proc. SPIE, **333** 111 (1982)
(42) D.Maydan, G.A.Coquin, H.J.Levinstein, A.K.Shinha and D.N.K.Wang: J. Vac. Sci. Technol., **16** 1959 (1979)
(43) A.C.Adams, C.D.Capio, H.J.Levinsein, A.K.Sinha and D.N.Wang, US Patent, 4171489 (1979)
(44) M.P.Leoselter, H.J.Levinstein and D.Maydan: US Patent, 4253029 (1981)
(45) P.Blais, T.O'Keeffe, D.Tremere, M. Cresswell, SEMICON/West 1982 Tech. Prog. Proc., May 26-28, San Mateo, CA.
(46) K.Suzuki, J. Matsui, T. Kadota, T. Ono, Japan J. Apply Phys., **17** 1447 (1978)
(47) Y.Saito, H. Yoshida, I. Watanabe, S. Nakamura, J. VacSci. Technol., **B2** 63 (1984)
(48) M.Sekimoto, H. Yoshihara, T. Ohkubo, J. Vac. Sci. Technol., **21** 1017 (1982)
(49) W.A.Johnson, R.A.Levy, D.J.Resnik, T.E.Sanders, A.W.Yanof, H.Betz, H.Huber and H.Oertel: J. Vac. Sci. Technol., **B5**, 257 (1987)
(50) P.L.King, L.Pan, P.Pianetta, A.Shimkunas, P.Mauger and D.Seligson: J. Vac. Sci. Technol., **B6**, 162 (1988)
(51) Pyrolytic BN Catalog, Union Carbide Corporation
(52) K.Nakamura: J. Electrochem. Soc., 133 1120 (1986)
(53) D.S.Williams: J.Appl. Phys., **57** 2340 (1985)

Materials Science Forum Vols. 54 & 55 (1990) pp. 141-152

CHEMICAL VAPOR DEPOSITION OF TURBOSTRATIC AND HEXAGONAL BORON NITRIDE

K. Sugiyama and H. Itoh
Synthetic Crystal Research Laboratory
Facutly of Engineering, Nagoya University
Furo-cho, Chikusa-ku, Nagoya 464-01, Japan

ABSTRACT

Turbostratic or hexagonal boron nitride was chemically vapor deposited at relatively low temperatures on to various substrates such as iron group metals, graphite and silicon nitride in forms of thick films, free-standing shapes and infiltrated coatings. Chemical vapor deposition in the BCl_3-NH_3-H_2-Ar reactant system under normal and reduced pressures and plasma-enhanced chemical vapor deposition in the BCl_3-N_2-H_2-Ar reactant system were investigated in a low temperature range of 500 to 1300°C as well as pulse chemical vapor infiltration in the B_2H_6-NH_3-N_2-Ar reactant system. The structure (turbostratic or hexagonal), texture and the properties (thermal, optical and electrical) of the vapor deposited boron nitride were examined and evaluated.

1. INTRODUCTION

Hexagonal boron nitride (hBN) has unique chemical and electrical properties in a wide temperature range from room temperature to 2500°C [1,2], which give rise to the possibilities of technical applications to corrosion-resistant and insulating films, coatings or free-standing shapes [3-5]. Typical examples for such practical uses of "pyrolytic boron nitride" are heating crucibles, microwave guide tubes, high-temperature insulators, etc.

Since pure hBN has less sinterability, it is difficult to fabricate a bulky hBN compact by normal pressure sintering and even by hot-pressing unless appropriate additives are mixed to pure hBN powder [1]. On the other hand, Chemical Vapor Deposition (CVD) of hBN is a convenient procedure to prepare pure and dense films or shapes, because hBN can be synthesized and consolidated simultaneously in the vapor deposition process without sintering.

On analogy to the structural transformation of carbon, turbostratic boron nitride (tBN) is known to be formed on the intermediate stage of the crystallization process of amorphous boron nitride (aBN) [6,7]. tBN has a diffused structure of boron and nitrogen networks which have three dimensional disorder and have lattice parameters somewhat longer than those of hBN. It is interesting also to examine the stability and properties of vapor deposited tBN.

Several papers have been reported on CVD of boron nitride (hBN or tBN) films [8-10], free-standing bodies [11-15] and infiltrated compacts [16-19]. However, few papers on low temperature deposition of tBN or hBN films and coatings have been published except those on thin film growth [20-26] and properties [27-34] of aBN or hBN for electronic devices.

A basic research on the preparation of tBN and hBN thick films, shapes and infiltrated coatings has been developed in our laboratories in the past decade [35-38]. Various types of vapor deposition processes from normal pressure CVD to the pulse Chemical Vapor Infiltration (CVI) are described in the present paper. The structure and properties of the tBN or hBN synthesized at relatively low temperatures below 1300°C are evaluated.

2. PREPARATION OF BORON NITRIDE THICK FILM ON IRON GROUP METALS BY NORMAL PRESSURE CVD

2.1 Deposition Conditions of Thick tBN or hBN Film on Iron and Nickel Substrates

A tBN or hBN film up to about 100 μm thick was prepared on carbon steel and nickel substrates at deposition temperatures below 1200° C and at reactant gas pressure of 1 atm [35-37]. Figure 1 shows the schematic diagram of the reactor. In the center of a vertical quartz reactor tube (inside diameter: 25 mm) a carbon steel or nickel substrate was placed on a quartz pedestal. The substrate is heated by an r.f. induction coil. A mixed gas of NH_3 and H_2 was introduced to upper side of the reactor, while another mixed gas BCl_3 and Ar was introduced by a thin quartz pipe down to the reaction zone which is 2 cm apart above the substrate surface. Separation of the two paths for feeding NH_3 and BCl_3 followed by mixing the two streams near the substrate is important to avoid the formation of an adduct compound $H_3N:BCl_3$, which disturbs the direct surface reaction of the two species on the substrate.

The effects of HN_3 and BCl_3 flow rates on the film growth on to carbon steel substrate are shown in figure 2, where the deposition temperature is 1150°C and the flow rate of both H_2 and Ar are 30 ml/min. A uniform crystalline and semicrystalline

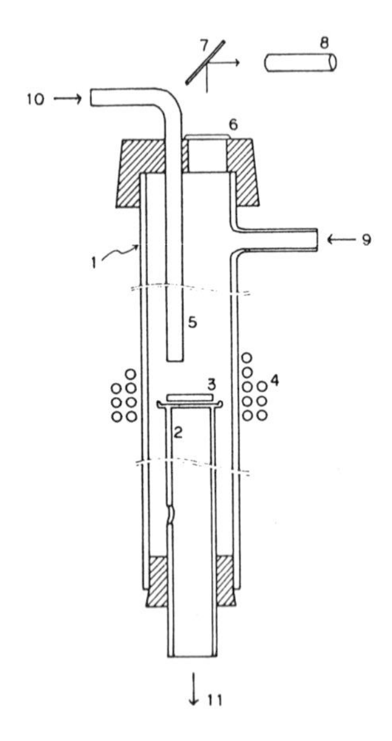

Fig. 1 Schematic diagram of the reactor; (1) quartz reactor tube, (2) quartz pedestal, (3) substrate, (4) r.f. induction coil, (5) quartz pipe, (6) observation window, (7) mirror, (8) pyrometer, (9) NH_3 + H_2, (10) BCl_3 + Ar, (11) outlet.

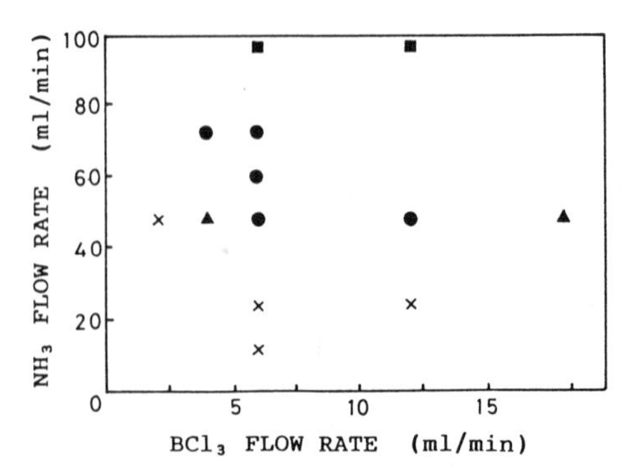

Fig. 2 Effects of NH_3 + BCl_3 flow rates on film growth; temperature: 1150 C, X :no deposit, ▲: semi-crystalline BN film, ●:crystalline BN film, ■: overgrowth of BN.

boron nitride was deposited at a narrow flow rate range of BCl_3 (4.2-12 ml/min) and NH_3 (48-72 ml/min). For the desired BN formation, a relatively higher concentration of NH_3 is required compared with that of BCl_3, because a part of NH_3 is consumed by the reaction with by-produced HCl to form NH_4Cl. The stagnation of the NH_4Cl vapor in the exhaust side prevents the steady state formation reaction of boron nitride, which results in the decrease in the growth rate of the film after the reaction time of about 90 min.

Figure 3 shows X-ray diffraction patterns which indicate the temperature dependence of the crystallinity of the products, where the reactant gas flow rates were kept constant at the above optimum deposition conditions. A broad diffraction peak, which appears at an angle lower than (002) line of JCPDS hBN, is confirmed in the patterns at the deposition temperature below 1100°C. This phase is considered to be tBN which has a lattice constant c ~ 3.5 A with the c axis elongated a little. Another sharp diffraction line of the (002) plane is found at a temperature above 1100° C at the same angle as that of hBN (c = 3.33 A), while the broad line at lower angle tends to disappear instead. No diffraction peak except (002) and (004) planes can be confirmed, so that the deposited film might have a structure with a highly preferred orientation to the c axis of hBN, or have a disordered structure of t'BN as Matsuda et al. pointed out [13].

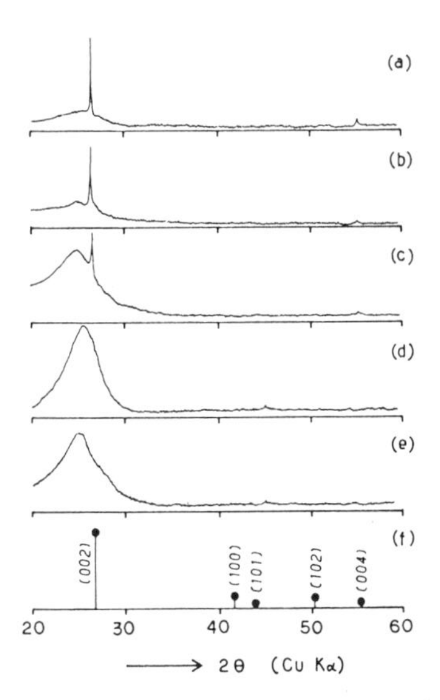

Fig. 3 X-ray diffraction patterns of vapor-deposited films; deposition temperature: (a) 1200°C, (b) 1150°C, (c) 1100°C, (d) 1050°C, (e) 1000°C, (f) JCPDS hBN.

Figure 4 shows a typical hBN (or t'BN) film of about 80μm thick which was obtained under the growth conditions of the optimum gas flow rates, growth temperature of 1200° C and reaction time of 120 min. This hBN film was easily separated from the carbon steel substrate due to a thermal shock during cooling of the specimen. The free-standing hBN thick film or plate appears to be translucent, but has a conical microstructure with some large nodular deposits. The film thickness increased linearly after 20 min with increasing the reaction time up to 90 min, where the growth rate is approximately 70 μm/h. A catalytic action of boron diffused iron seemed to play an important role in the formation of the first hBN layer at the initial stage of deposition. Similar catalytic effect was confirmed in case of using nickel substrate. Therefore, hBN film was successively prepared on boron diffused iron group metals at the deposition temperatures a slightly below the eutectic temperature of the M-B system (M = Fe or Ni). Analogous BN films were prepared on copper substrate at temperatures as low as 600-700°C [10].

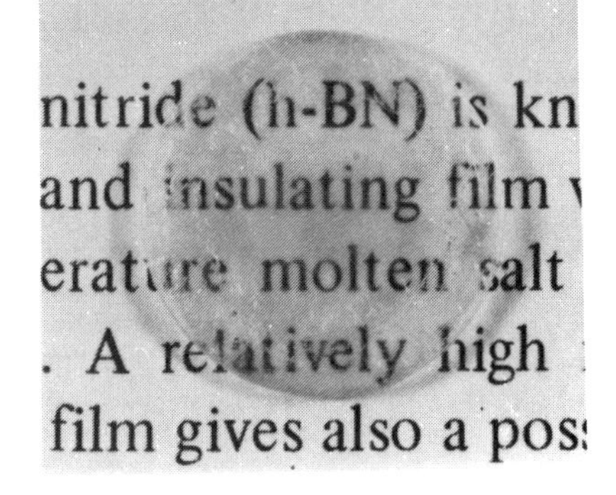

Fig. 4 Typical hBN film obtained at the optimum growth conditions.

2.2 Properties of hBN Thick Film

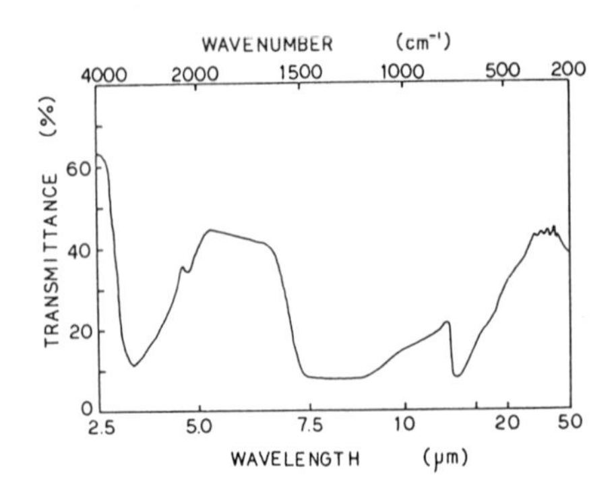

Fig. 5 IR transmission spectrum of the hBN film.

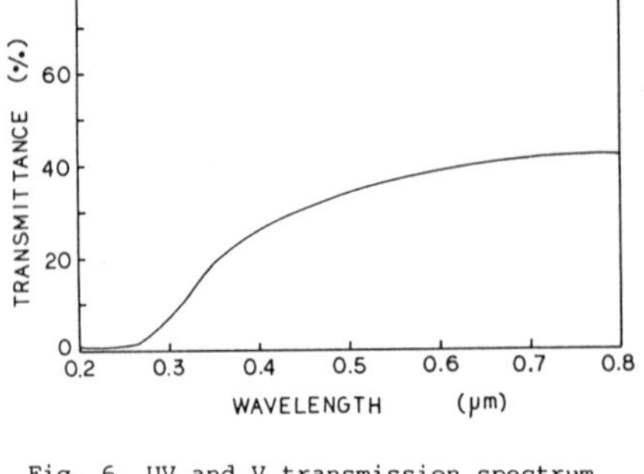

Fig. 6 UV and V transmission spectrum of the hBN film.

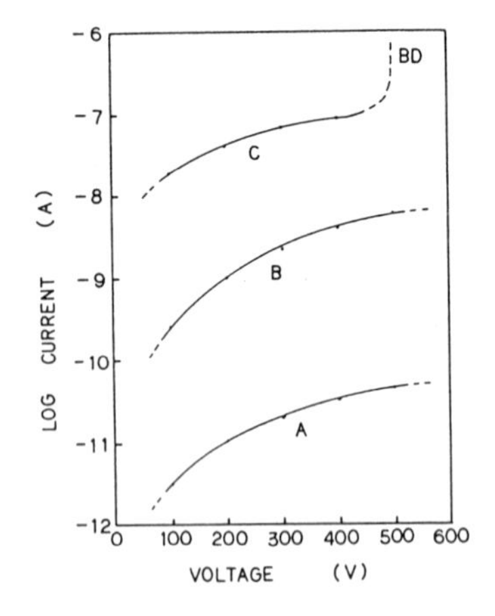

Fig. 7 Current-voltage characteristics of the hBN film; (A) room temperature (22°C), (B) 100°C, (C) 200°C, BD: breakdown region.

Optical properties of the as-grown hBN (or t'BN) films of 55 μm thick was examined. Figure 5 shows the IR transmission spectrum of the film. Some broad absorption peaks can be confirmed corresponding to the vibration of the B-N stretching (~1400 cm^{-1} and B-N-B bending (~800 cm^{-1}). Other peaks due to the B-H stretching (~2500 cm^{-1}) and O-H stretching (~3400 cm^{-1}) were found in the low wave length region. A small amount of water vapor will be adsorbed on the surface or intercalated into the as-grown hBN film [16,18,20,22,25]. A transmission spectrum of the same film in the UV and V regions is shown in figure 6. The transmittance of 20-40 % was found in the V region, which is represented by relatively good transparency of the film as shown in figure 4. The transmittance decreases gradually in the UV region and especially in the wave length less than 0.25 μm, the transmittance being only 1 %, which is similar to the optical behavior of the pyrolytic BN films [27,29]. The refractive indices of the BN film were 1.650-1.651, which is nearly consistent with the data by Murarka et al [24].

The current-voltage characteristics of the hBN film was measured in the temperature range from room temperature (22°C) to 200°C. Figure 7 shows a plot of Log I vs. the applied dc voltage (V) at three different temperatures. The ohmic behavior of the I-V characteristics can be seen in the low voltage region. A dielectric breakdown occurred at 500 V (~6.2×10^4 V/cm) at the temperature of 200°C, which is due to the leakage, probably through a grain boundary. From Arrhenius plot of the dc conductivity of the hBN film, the conductivity was found to be 4×10^{-13}-1.6×10^{-9} Sm^{-1} and the activation energy for conduction was 99.5 kJ/mol.

3. CVD COATING OF BORON NITRIDE ON GRAPHITE MANDREL

3.1 Deposition Conditions of tBN on Graphite Substrate

CVD of thick tBN coatings at relatively low temperatures on graphite mandrel was carried out under low pressures. A graphite mandrel (13.4 mm in diameter, 30 mm in length) was suspended via a tungsten wire and heated by r.f. induction coil in a vertical quartz reactor, into which the two streams of BCl_3 + Ar and NH_3 + H_2 were introduced by a concentric tube from the bottom. As suggested from the data in 2.1, boron diffused layer into the substrate seems to act as a catalyst for the formation of BN. Prior to the BN coating, boron was precoated on graphite substrate under the following conditions in order to induce the formation of boron nitride; total pressure: 160 Torr, temperature: 1200°C, reaction time: 20 min, flow rates: BCl_3 6 ml/min, H_2 80 ml/min, Ar: 30 ml/min. The optimum flow rates for the subsequent BN coating were: BCl_3: 6 ml/min, NH_3: 80 ml/min, H_2: 35 ml/min, Ar: 30 ml/min.

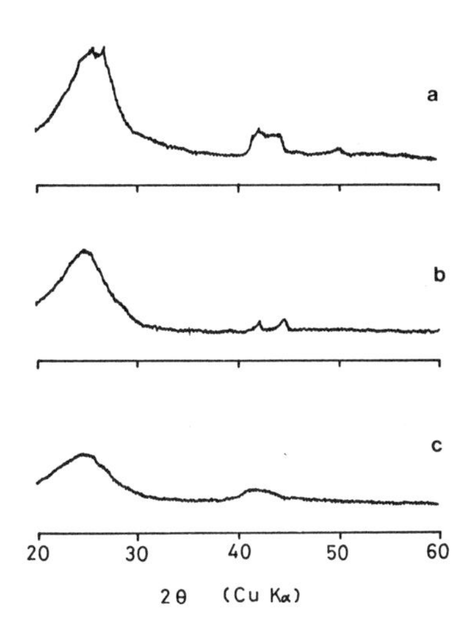

Fig. 8 X-ray diffraction patterns of vapor-deposited tBN film; deposition temperature: (a) 1300°C, (b) 1200°C, (c) 1000°C.

Figure 8 shows the dependence of deposition temperature on the X-ray diffraction patterns of the specimen at the total pressure of 160 Torr. Broad lines corresponding to tBN are observed in the temperatures as low as 1000°C, the crystallinity increasing slightly with deposition temperature up to 1300°C. No hBN deposited in all experiments below 1300°C. The formation behavior is similar to the results by Matsuda et al. [13].

Figure 9 shows the relation between the weight of deposit and the total pressure at a constant deposition temperature, time and gas flow rates. A good linear relationship can be seen in the pressure range of 5 to 260 Torr. The waste gases are removed so fast in the low pressure range that the deposition rate would increased with decreasing the total pressure, since the deposition of boron nitride is kinetically limited by the desorption of waste gases from the substrate, as described in 2.1. The crystallinity of the deposits was found to increase with a decrease in the total pressure [13], which would be attributed to the decrease in supersaturation of the reactant gas.

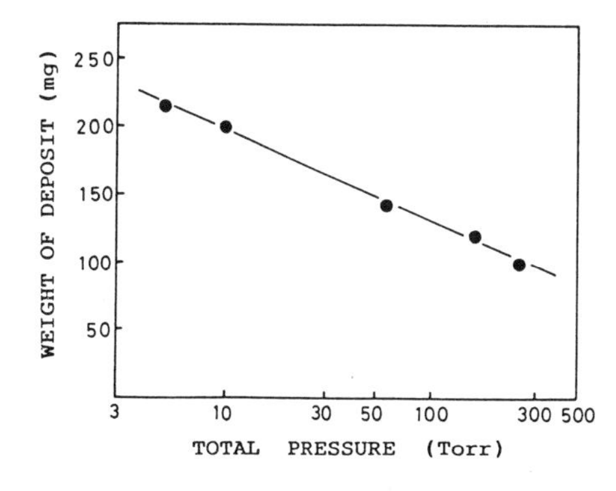

Fig. 9 Relationship between weight of deposit and total pressure; deposition temperature: 1200° C, reaction time: 60 min.

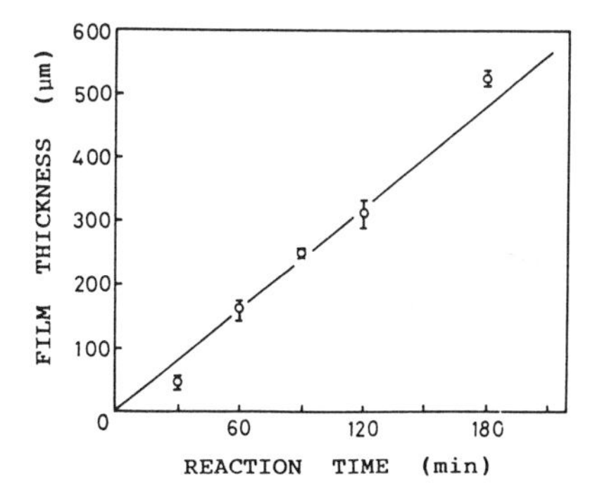

Fig.10 Relationship between film thickness of tBN and reaction time; deposition temperature: 1200° C, total pressure: 160 Torr.

3.2 Texture of the tBN Coatings

Figure 10 shows the relationship between thickness of the grown tBN film and reaction time, where the pressure and deposition temperature are 160 Torr and 1200° C, respectively. The growth rate is found to be 150 μm/h and constant even at the reaction time longer than 160 min, when a steady state deposition reaction is maintained. Cone like deposits of tBN grains which are composed of smaller grains apparently grew with increasing the reaction time. By the precoating of boron, an interlayer with several μm thick was observed between the tBN film and the graphite substrate, as shown in figure 11. This layer could not be identified by X-ray diffraction, but amorphous boron or amorphous B-C-N phase would be formed. This interlayer adheres tightly to the graphite substrate, so that the deposited tBN might be used as tBN coated graphite shapes. In order to separate the tBN film from the mandrel, an oxidation treatment was carried out at 600°C for several hours. The free-standing tBN crucible separated from the mandrel was as translucent as the boron nitride film observed in figure 4.

Fig.11 tBN thick coating on boron precoated graphite mandrel; deposition temperature: 1200° C, total pressure: 160 Torr.

4. PLASMA ENHANCED CVD OF BORON NITRIDE FILMS WITH HIGH PREFERRED ORIENTATION

4.1 Growth Conditions of BN Film in a Plasma Atmosphere

Thick boron nitride films were prepared by plasma enhanced CVD at temperatures below 800°C using a reactant gas mixture of the BCl_3-N_2-H_2-Ar system, where the formation of BN would proceed under non-equilibrium conditions by the following reaction.

$$BCl_3 + 1/2N_2 + 3/2H_2 \longrightarrow BN + 3HCl \quad (1)$$

or

$$BCl_3 + 2N_2 + 6H_2 \longrightarrow BN + 3NH_4Cl \quad (2)$$

In a horizontal quartz reactor tube (inside diameter: 25 mm), graphite substrate was mounted on a graphite susceptor which was grounded by tungsten wire. The plasma discharge was generated by an induction coil (frequency: 13.56 MHz), which was coupled with a matching box and an r.f. oscillator. In order to obtain tBN film highly oriented to the c axis, AlN was undercoated by the following plasma CVD [39] prior to the BN coating.

$$AlBr_3 + 1/2N_2 + 3/2H_2 \longrightarrow AlN + 3HBr \quad (3)$$

Figure 12 shows the pressure dependence on the deposition rate of boron nitride, where the deposition temperature and reaction time was kept constant at 610°C and 20 min, respectively. A maximum growth rate was observed at 23 Torr. In the pressure range lower than 20 Torr, the net growth rate to the substrate decreased due to a high linear velocity of the reactant gas. This tendency is contrary to that in the low pressure CVD as shown in figure 9. On the other hand, the amount of activated species such as NH_x^+, NH_x, etc. would decrease in the higher pressure range with the result of decrease in the growth rate. Cone-like deposits with the grain diameters of 5 to 10 μm were

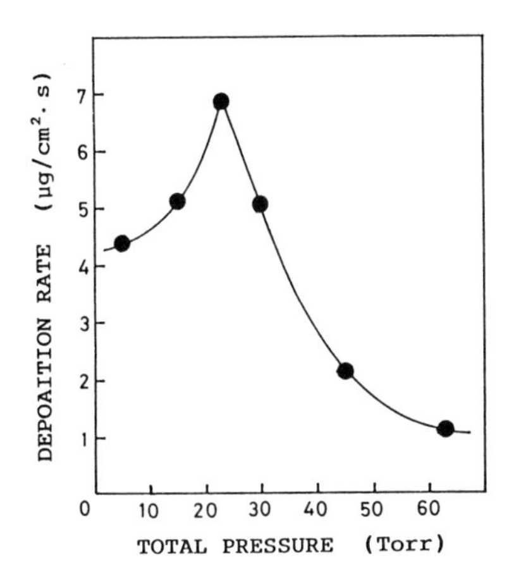

Fig.12 Relationship between the deposition rate of BN and the total pressure; substrate temperature: 610° C, reaction time: 20 min.

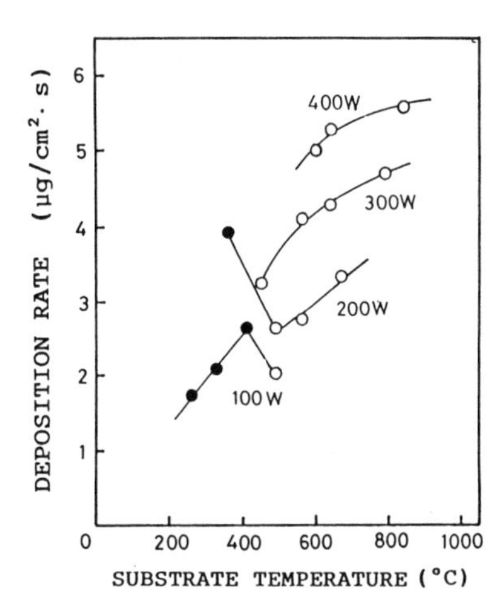

Fig.13 Variation of the deposition rate with substrate temperature at various r.f. power; total pressure: 7 Torr, reaction time: 30 min.

aggregated on a large area of the substrate surface. The weight of deposit increased linearly with the reaction time as well as in figure 10, although the growth rate was lower (about 30 μm/h).

Figure 13 shows the variation of the deposition rate with the substrate temperature at the r.f. powers of 100 to 400 W. The deposits obtained at temperatures below 450°C are found to contain an appreciable amount of NH_4Cl by X-ray diffraction, which disappears by the heat-treatment of as-grown films. Arrhenius plot of the deposition rate vs. substrate temperature above 450°C gave an activation energy for BN formation reaction of 3.3-8.5 kJ/mol, which is smaller than that for the kinetically controlled formation reaction of boron nitride [14,15]. The deposition of BN in a plasma atmosphere would be controlled by diffusion of the activated species to the substrate surface.

4.2 Evaluation of the Grown BN Film

IR absorption spectrum of the grown BN film by KBr tablet method showed that an absorption peak due to O-H stretching (3400 cm^{-1}) was observed besides those due to the B-N stretching and B-N-B bending, as in the case of normal pressure CVD of boron nitride in 2.2. It was confirmed by X-ray diffraction that the deposits are mostly tBN, but they contains a small amount of hBN (or t'BN). By SEM observation of the film surface, the grain size of cone-like deposits decreases with increasing the deposition temperature, which is contrary to the variation of film texture observed in the normal pressure CVD.

Oxidation resistance was examined by treating the BN coated graphite in air from room temperature to 1000°C. The specimen initiated to be oxidized gradually from 800°C, whereas graphite was oxidized quickly at 700°C. In a reducing atmosphere of $Ar-H_2$ at 4 Torr, the plasma etching rate of the film obtained at 400 W was 0.054 $mg/cm^2 \cdot h$, which is 100 times as high as that of graphite.

BN films with high preferred orientation to the c axis were grown on the AlN precoated substrate. The AlN film was grown at the pressure of 10 Torr and the temperature of 700°C (plasma power: 300 W). Figure 14 shows microhardness of the BN film as a function of the degree of preferred orientation of AlN which is defined as the ratio of X-ray diffraction intensities (the peak-height

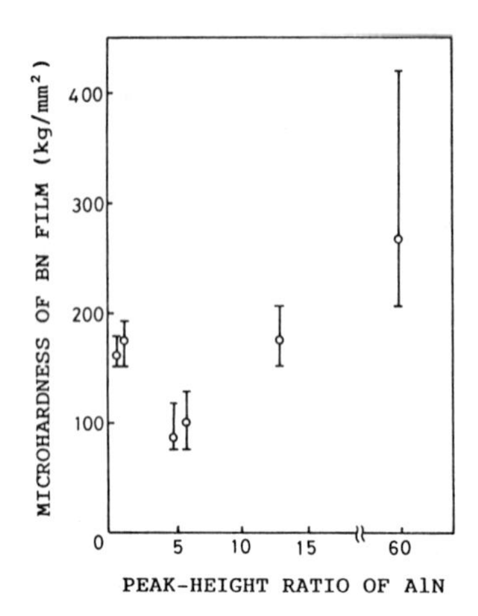

Fig.14 Deposition rate and microhardness of the BN film vs. the degree of preferred orientation of AlN; reaction time: 20 min.

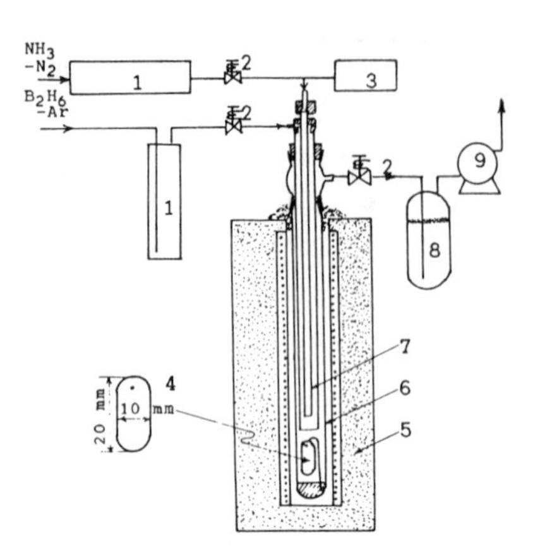

Fig.15 Apparatus for pulse CVI of BN on to Si_3N_4 powder preform; (1) reservoir, (2) magnetic valve, (3) pressure gauge, (4) powder preform, (5) furnace, (6) vessel, (7) concentric tube, (8) filter, (9) vacuum pump.

ratio of AlN, R = (002)/(100)). It is interesting that the microhardness increases with the increase in the degree of preferred orientation of AlN and attains 400 kg/mm^2 at R = 60, although the deposition rate is almost constant.

5. CHEMICAL VAPOR INFILTRATION OF BORON NITRIDE INTO Si_3N_4 PREFORM

A composite compact of the BN-Si_3N_4 system is expected as a machinable and high strength composite material [40]. We have developed the pulse CVI method, where the by-produced or waste gases are removed during the evacuation step, and the reactant fresh gases can be filled instantaneously into deep levels of the preforms in each pulse. Therefore, the pulse CVI is expected to prepare a dense infiltrated compact and to reduce the operation time markedly [41,42]. Figure 15 shows the experimental apparatus used for the pulse CVI of BN to Si_3N_4 powder preform [38]. The source gas streams of NH_3-N_2 and B_2H_6-Ar were stored in the separate reservoirs, from which gas streams were allowed to flow into the reaction vessel via a concentric tube in 10 msec to attain 690 Torr by opening of entrance magnetic valves. The source gas composition was adjusted to 0.5% B_2H_6, 16% NH_3, 8% Ar and 75.6% N_2. Si_3N_4 powder with the average particle size of 5 μm, was pressed under 100 MPa into a disc (1.4 mm thick) as shown in figure 15, using starch solution as a binder. The pressed disc was heated in a nitrogen flow at 1000°C for 6 h. The density, porosity and carbon content of the heat-treated compact were 1.75 g/cm^3, 45% and 0.4 wt%, respectively. The path to gas outlet was set in the clearance between the outer face of the concentric tube and the inner wall of the vessel in order to avoid mixing with fresh source gas streams. It took 5 sec to evacuate the vessel down to the pressure below 1.5 Torr.

From the relationship between the weight increase per pulse and holding time, the deposition rate was so high that the reaction finished within 0.3 sec at the treatment temperature of 1200°C. However, it took 5 sec to finish the deposition of boron nitride at 1000°C. Figure 16 shows the relationship between the weight increase of infiltrated BN per pulse and the number of pulses, in which the thickness of the surface film was measured from the SEM image of the cross-section, followed by multiplication by the density of BN, and then the weight of infiltrated BN was estimated by subtraction of the weight of surface film from the total weight increase. It can be seen that the surface film growth is dominant over 1000 pulses at 1200°C with a hold time of 1 sec; however, BN infiltration continues over 3000 pulses at 1000°C with the same hold time. Even if the deposition temperature is low, the surface film grows from the source gas over the surface when the hold time is as long as 5 sec.

Figure 17 shows the relationship between the three-point flexural strength of the infiltrated compact at room temperature and the number of pulses. The curves shows that thickening of the surface film contributes to a small extent to the increase in flexural strength; therefore, the strength of the specimen obtained at 1200°C is nearly saturated above 2000 pulses. On the other hand, the strength of the specimen obtained at 1000°C increases with the number of pulses at least up to 3000 pulses. Thus, the pulse CVI is found to be an effective infiltration technique of boron nitride into a powder preform.

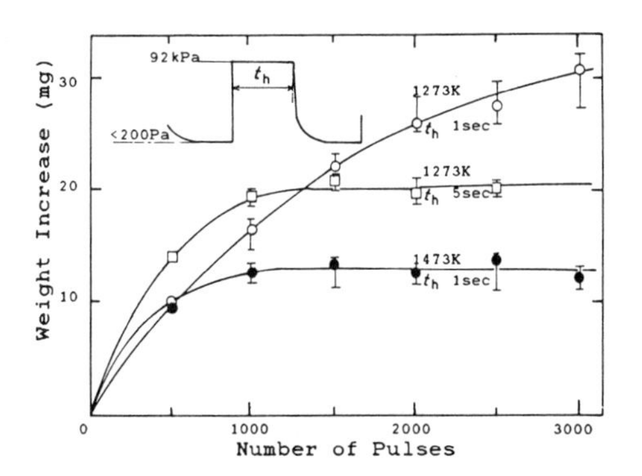

Fig.16 Relationship between weight of infiltrated BN and number of pulses.

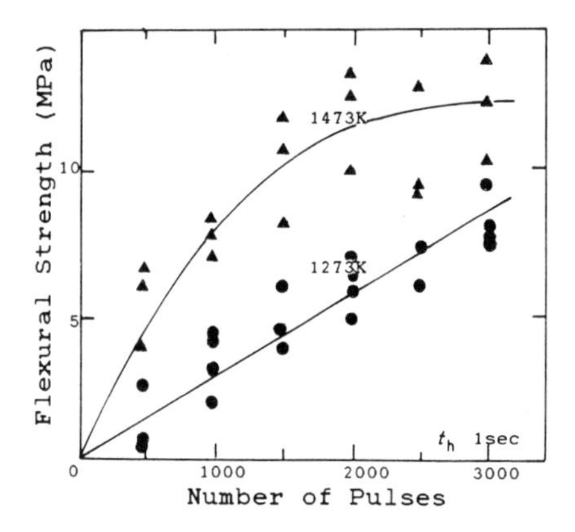

Fig.17 Relationship between three-point flexural strength and number of pulses; hold time: 1 sec.

6. CONCLUSIONS

Chemical vapor deposition and infiltration of boron nitride on to various substrates or into preforms were investigated with an emphasis on low temperature deposition of tBN or hBN below 1300°C.

1) Boron nitride (tBN and hBN) thick films were prepared on carbon steel and nickel substrates by normal pressure CVD from the reactant system, BCl_3-NH_3-H_2-Ar. A translucent hBN (or t'BN) film with high preferred orientation to the c axis was grown at 1200°C with the aid of a catalytic action of boron diffused iron-group metals.

2) Tight coatings of tBN were obtained on graphite mandrel at 1000-1300°C by low pressure CVD in the reactant system, BCl_3-NH_3-H_2-Ar. Boron precoating on graphite mandrel was effective to induce the formation of boron nitride and to form an adherent interlayer between the graphite substrate and tBN layer. The tBN coating would be used not only as tBN coated graphite shapes, but also as free-standing shapes by oxidation treatment of the coated graphite mandrel at 600°C for several hours.

3) Thick tBN films were prepared in a lower temperature range (500-800°C) by the plasma enhanced CVD in the BCl_3-N_2-H_2-Ar reactant system. tBN films which have high preferred orientation to the c axis were grown on a highly oriented AlN precoated film.

4) A composite compact of the Si_3N_4-BN system was fabricated by the pulse CVI of the NH_3-N_2-B_2H_6-Ar reactant system into Si_3N_4 powder preform. Boron nitride was infiltrated effectively at 1000°C into a deep level of the porous Si_3N_4 preform.

REFERENCES

1) Ingles, T.A. and Popper, P.: "Special Ceramics" (Academic Press, London, 1960) p.144
2) Economy, J. and Lin, R.: in "Boron and Refractory Borides" (Springer-Verlarg, Berlin, 1977) p.552
3) Powell, C., Oxley, J. and Blocher, J. Jr.: "Vapor Deposition" (Wiley, New York, 1962) p.663
4) Belforti, D., Blum, S. and Bovarnick, B.: Nature, 1961, 4779, 901
5) Basche, M. and Schiff, D.: Mater. Design Eng., 1964 (2), 78
6) Wells, A.F.: "Structural Inorganic Chemistry" (Clarendon Press, Oxford, 1975) p.847
7) Thomas, J., Weston, N.E. and O'Connor, T.E.: J. Am. Chem. Soc., 1963, 84, 4619
8) Gafri, O., Grill, A., Itzhak, D., Inspector, A. and Avni, R.: Thin Solid Films, 1980, 72, 523
9) Sano, M. and Aoki, M.: Thin Solid Films: 1981, 83, 247
10) Motojima, S., Tamura, Y. and Sugiyama, K.: Thin Solid Films, 1982, 88, 269
11) Clerc, G. and Gerlach, P.: in "Proc. 5th Intern. Conf. C.V.D.", (Electrochem. Soc., New Jersey, 1975) p.777
12) Malé, G. and Salanoubat, D.: in "Proc. 7th Intern. Conf. C.V.D.", (Electrochem. Soc., New Jersey, 1979) p.391
13) Matsuda, T., Uno, N., Nakae, H. and Hirai, T.: J. Mater. Sci., 1986, 21, 649
14) Matsuda, T., Nakae, H. and Hirai, T.: J. Mater. Sci., 1988, 23, 509
15) Tanji, H., Monden, K. and Ide, M.: in "Proc. 10th Intern. Conf. C.V.D." (Electrochem. Soc., New Jersey, 1987) p.562
16) Gebhardt, J.J.: in "Proc. 4th Intern. Conf. C.V.D.", (Electrochem. Soc., New Jersey, 1973) p.460
17) Hannache, H., Quenisset, J.M., Naslain, R. and Heraud, L: J. Mater. Sci., 1984, 19, 202
18) Singh, R.N.: in "Proc. 10th Intern. Conf. C.V.D.", (Electrochem. Soc., New Jersey, 1978) p.543
19) Singh, R.N. and Brun, M.K.: Advanced Ceram. Mater., 1988, 3, 235
20) Rand, M.J. and Roberts, J.F.: J. Electrochem. Soc., 1968, 115, 423
21) Pierson, H.O.: J. Composite Mater., 1975, 9, 228
22) Hirayama, M. and Shono, K.: J. Electrochem. Soc., 1975, 122, 1671
23) Hyder, S.B. and Yep, T.O.: J. Electrochem. Soc., 1976, 123, 1721
24) Murarka, S.P., Chang, C.C., Wang, D.N.K. and Smith, T.E.: J. Electrochem. Soc., 1979, 126, 1951
25) Adams, A.C. and Capio, C.D.: J. Electrochem. Soc., 1980, 127, 399
26) Matsumoto, O., Sasaki, M., Suzuki, H., Seshimo, H. and Uyama, H.: in "Proc. 10th Intern. Conf. C.V.D.", (Electrochem. Soc., New Jersey, 1987) p.552
27) Baronian, W.: Mater. Res. Bull., 1972, 7, 119
28) Stach, J. and Turley, A.: J. Electrochem. Soc., 1974, 121, 722
29) Zunger, A., Katzir, A. and Halperin, A.: Phys. Rev. B, 1976, 13, 5560
30) Kimura, T., Yamamoto, K. and Yugo, S.: Japn. J. Appl. Phys., 1978, 17, 1871
31) Hoffman, D.M., Doll, G.L. and Eklund, P.C.: Phys. Rev. B, 1984, 30, 6051
32) Kim, C., Sohn, B.K. and Shono, K.: J. Electrochem. Soc., 1984, 131, 1384
33) Adams, A.C.: J. Electrochem. Soc.: 1981, 128, 1378
34) Miyamoto, H., Hirose, M. and Osaka, Y.: Japn. J. Apll. Phys., 1983, 22, L216
35) Takahashi, T., Itoh, H. and Takeuchi, A.: J. Crystal Growth, 1979, 47, 245

36) Takahashi, T., Itoh, H. and Ohtake, A.: Yogyo-kyokai-Shi, 1981, 89, 63
37) Takahashi, T., Itoh, H. and Kuroda, M.: J. Crystal Growth, 1981, 53, 418
38) Sugiyama, K. and Ohsawa, Y.: J. Mater. Sci. Lett., 1988, 7, 1221
39) Itoh, H., Kato, M. and Sugiyama, K.: Thin Solid Films, 1987, 146, 255
40) Besman, T.M.: J. Am. Ceram. Soc., 1986, 69, 69
41) Sugiyama, K. and Nakamura, T.: J. Mater. Sci. Lett., 1987, 6 331
42) Sugiyama, K. and Yamamoto, E.: in "Proc. 10th Intern. Conf. C.V.D.", (Electrochem. Soc., New Jersey, 1987) p.1041

Materials Science Forum Vols. 54 & 55 (1990) pp. 153-164

GROWTH OF BN THIN FILMS BY PULSED LASER EVAPORATION

P.T. Murray
Research Institute, University of Dayton, Dayton, OH 45469, USA

ABSTRACT

Thin films of BN have been grown by pulsed laser evaporation. The films were analyzed by in-situ Auger and x-ray photoelectron spectroscopy as well as by ex-situ grazing incidence x-ray diffraction and Raman spectroscopy. The results indicate that films grown at room temperature, in high vacuum, were N-deficient. Time-of-flight analysis indicated that pulsed laser evaporation of a BN target resulted in the ejection of primarily atomic species, with an effective translational temperature of 2600K. The film results are consistent with thermodynamic calculations, which indicate that elemental B formation, with concomitant N_2 desorption, is energetically favored over stoichiometric BN formation. Film growth was also carried out in an NH_3 atmosphere, with substrate temperatures from 1000°C to room temperature. It was found that stoichiometric BN films can be grown at substrate temperatures as low as 500°C.

1. INTRODUCTION

Solid lubricant films are required for a variety of high precision, space-borne applications such as satellite mechanism bearings, gears, and splines. Although MoS_2 films are currently the material of choice, there still exists a need for materials with superior thermal stability and whose tribological properties are not degraded by air exposure. BN is one such candidate material. Thin films of BN have been grown by a number of techniques [1-5] including chemical vapor deposition (CVD). Murarka and coworkers [6] found that a substrate temperature of 800°C was required to grow stoichiometric BN films by CVD; at lower temperatures, the films were N-deficient. Adams and coworkers [7] investigated the properties of films grown, by CVD, between 250 and 600°C. Their results indicated that the films thus grown had a composition of B_2NH_x. Yamaguchi and Minakata [8] used a two-furnace technique and found the films grown above 500°C to be stoichiometric. BN films have also been grown by sputtering [9,10]. A reactive sputtering technique was carried out by Noreika and Francombe [11]. Wiggins and coworkers [12] grew BN films by rf sputtering. It was found that such films could be grown without heating the substrate, but the films were not insulating and contained pinholes. The growth of BN has also been carried out by activated reactive evaporation [13], electron beam evaporation [14], ion beam deposition [15-20], and by a pulse plasma method [21]. Arya and D'Amico [22] have published a review of the various growth techniques.

Despite these advances, there still exists a need for improved deposition processes, particularly those which allow growth of stoichiometric BN films at lower substrate temperatures. The purpose of the work presented here was to determine the feasibility of growing stoichiometric BN films by pulsed laser evaporation (PLE). PLE is an emerging film growth technique which possesses a number of advantages over more conventional deposition processes. Among these advantages are the potential for congruent target evaporation, the capability of growing high purity films (since the target becomes its own crucible), and the relative ease with which even refractory materials can be evaporated.

2 EXPERIMENTAL

The apparatus used for BN film growth and analysis has been described previously [23]. The system consists of a deposition chamber which is directly connected to a Perkin-Elmer (Phi) Model 550 surface analysis system. The use of a specimen introduction port and transfer arm allowed film growth in the deposition chamber and subsequent surface analysis of the films without air exposure. The films were analyzed by using Auger electron spectroscopy (AES) and x-ray photoelectron spectroscopy (XPS). The frequency-doubled output (λ = 0.53 μm) of a Q-switched (15 ns pulse duration) Nd:YAG laser was used to evaporate the (hexagonal) BN target. The laser was focused to a 0.9 mm spot at the target and was scanned across the target to produce a uniform film. The target-substrate distance was approximately 3 cm. Films were grown on Si(100) substrates, which were resistively heated during film growth. The substrate temperature was determined by use of a calibrated infrared pyrometer. A number of films were grown under high vacuum conditions, in which case the base pressure of the deposition chamber was 2×10^{-6} Pa. The pressure increased to 2×10^{-5} Pa during film growth. A number of films were also grown with the deposition chamber back-filled with NH_3. The NH_3 was obtained from Matheson and was used without further purification. Film growth in this case was carried out with an NH_3 pressure of 0.65 Pa.

Auger spectra were recorded at 5 keV incident beam energy, with a current density of 2.5×10^{-3} A cm^{-2}. A 6 eV (peak-to-peak) modulation was used to obtain the derivative spectra. XPS spectra were acquired by using an Al Kα x-ray source operating at 600 W and by utilizing a 25 eV pass energy for electron energy analysis.

The films were also analyzed by grazing incidence x-ray diffraction. These measurements were carried out on a commercial Rigaku D/max-1B diffractometer which is equipped with a thin film attachment. The spectra were recorded using Cu Kα radiation, with the x-ray source operating at 1000 W.

Investigations were carried out on the nature of the species evaporated from the BN target. Such information was obtained by performing time-of-flight (TOF) measurements on the positive ions ablated from the target by laser irradiation. The TOF measurements were carried out in two modes, each of which provided a different type of information. The first entailed recording TOF mass spectra of the laser-evaporated ions, using a reflectron [24] spectrometer. Such a system is energy compensating (to second order) and allows for good mass resolution of ions with broad kinetic energy distributions, as might be anticipated for the case of laser-evaporated ions. These measurements were carried out by irradiating the BN target with the pulsed Nd:YAG laser and allowing the nascent ions to drift for several microseconds before a potential was applied to one of the acceleration plates. The net result of this procedure was to accelerate all (positive) ions along the axis of the drift tube to the same nominal kinetic energy. Ions of different mass, then, had different velocities, and appeared at the ion detector at different times.

The second mode entailed recording velocity distributions. These measurements involved grounding the acceleration plates used for the previous measurements and allowing the laser-evaporated ions to drift to an ion detector situated 15 cm from the target. Under these conditions, the time of arrival of an ion at the detector was determined by the velocity with which the ion was evaporated from the target. In both types of measurements, the analog output of the appropriate detector was recorded by using a home-built transient digitizer which had a dwell time of 12.5 ns per channel. The mass spectra and velocity distributions were acquired by averaging 50 laser shots.

3 RESULTS AND DISCUSSION

3.1 HIGH VACUUM FILM GROWTH

Presented in figure 1a is an Auger spectrum of the BN target used in this work; this spectrum was recorded with no precleaning (such as ion sputtering) of the specimen. There are B and N peaks present in the spectrum, as well as smaller peaks due to C and O. The latter were present only in the surface of the target material. The N:B peak-to-peak height ratio, as well as the shapes of the B and N peaks, are in excellent agreement with the spectrum of bulk, stoichiometric BN reported previously by Hanke and Muller [25].

Shown in figure 1b is an Auger spectrum of a film grown in high vacuum, on a Si substrate at room temperature. The N:B peak-to-peak height ratio of this film is substantially smaller than that of stoichiometric BN. Moreover, the B peak shape is considerably different than that shown in figure 1a. This peak shape is, in fact, indicative of metallic B. These two observations indicate that PLE of BN, under high vacuum conditions, produces films which are primarily metallic B with very little incorporated N.

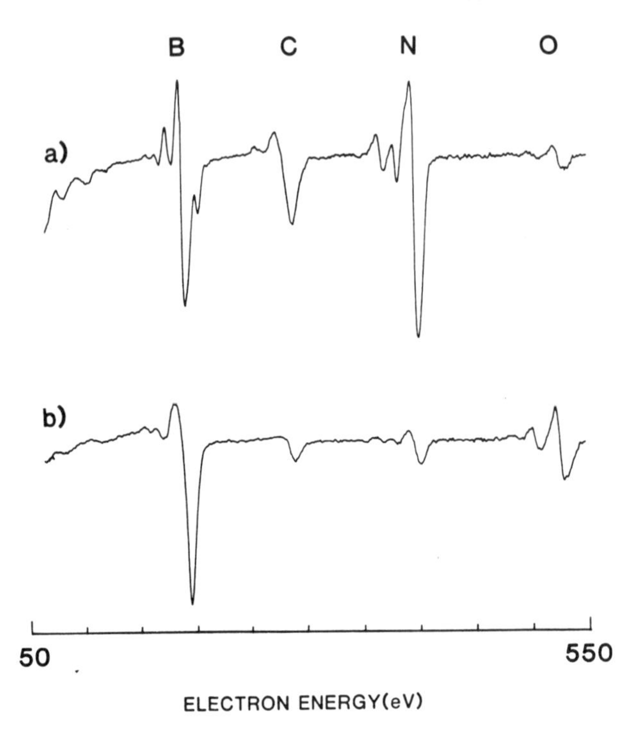

Figure 1. Auger electron spectra of (a) BN target and (b) film grown in high vacuum.

3.2 FILM GROWN IN NH_3

Film growth was also carried out in 0.65 Pa of NH_3. Shown in figures 2a, b, and c are Auger spectra of the films grown in NH_3 at substrate temperatures of 400, 500, and 1,000°C, respectively. The N:B peak-to-peak height ratio in all three spectra, as well as the N and B peak shapes, are very similar to that of the stoichiometric target. Shown in table 1 are the calculated atomic ratios obtained from the Auger data. The N/B ratios in all films grown in NH_3 are consistently near unity, with the film grown at 500°C exhibiting a somewhat larger value of 1.1. It can be concluded that growth of films in NH_3 does indeed restore the N stoichiometry, even at substrate temperatures as low as 400°C. The film grown at 1000°C contained the smallest impurity level of all samples examined in this work, with a C/B ratio of 8%. The detection of C on the surface of a target abraded in vacuum, as well as the consistent detection of C on all films grown, suggests that the BN target was the primary source of C in this work. The

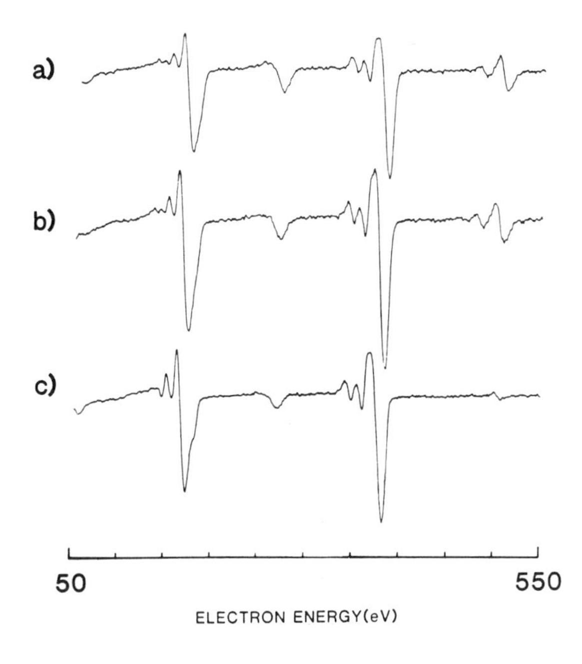

Figure 2. Auger electron spectra of films grown in 0.65 Pa of NH_3 at substrate temperatures of (a) 400°C, (b) 500°C, and (c) 1000°C.

film grown at 1000°C exhibited an O/B ratio of 2%, that grown at 400°C had a ratio of 9%, while the O/B ratio of the target was 3%. Thus, the films grown at lower substrate temperatures contained more O than did the target. O_2 and H_2O were listed by the supplier as the major contaminants in the NH_3; these are the most likely source of O in the low temperature films. The Auger results suggest that stoichiometric films of BN can be grown at substrate temperatures as low as 400°C.

Table 1

Atomic Ratios from Auger Data

Sample	N/B	C/B	O/B
BN target	1.00	0.31	0.03
High vacuum film*	0.19	0.11	0.13
400°C film	1.04	0.19	0.09
500°C film	1.11	0.11	0.07
1000°C film	1.05	0.08	0.02

*Change in B peak shape

The results of XPS analysis are compiled in table 2. Shown are the binding energies of the B(1s) and N(1s) photoelectron peaks for a film grown, in NH_3, at 500°C. The observed binding energies are in excellent agreement with data published previously by Hamrin and coworkers [26] on bulk, stoichiometric BN. The results of XPS indicate that stoichiometric films of BN can be grown at substrate temperatures as low as 500°C.

Table 2
Summary of XPS Results

	Binding Energy	
	B(1s)	N(1s)
500°C film	190.7	398.4
Bulk BN [26]	190.6	398.3

Shown in figures 3a and b are Raman spectra of the (hexagonal) BN target and that of a film grown at 500°C in NH_3, respectively. Both spectra exhibit peaks at 1366 cm^{-1}, although the peak appearing in the film spectrum is somewhat broader. This peak is a vibration assigned to the B-N stretching mode in hexagonal BN. The similarity of the two spectra suggests the film to be hexagonal BN.

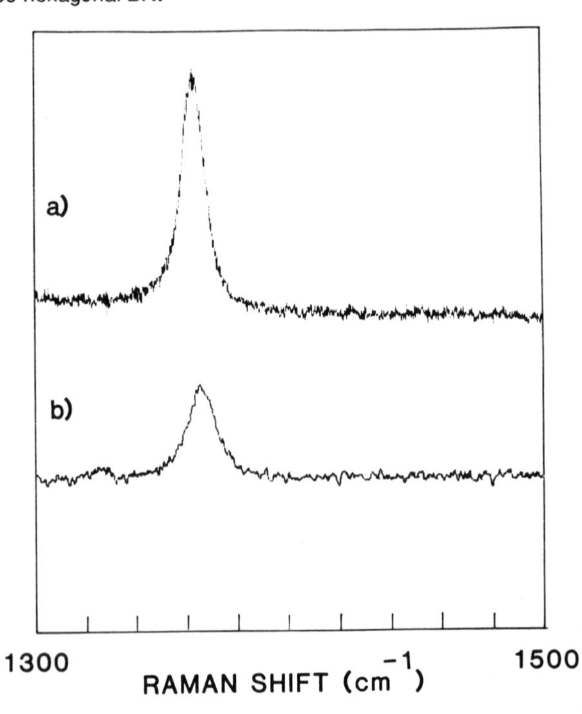

Figure 3. Raman spectra of (a) BN target, (b) film grown in 0.65 Pa of NH_3 at 500°C.

Presented in table 3 are the results of film and target analysis by graxing incidence x-ray diffraction. Data from the powder diffraction file [27] are also included for comparison. The film grown at 500°C exhibited an intense peak at 26.65 degrees, corresponding to a lattice parameter of 0.334 nm. A low intensity peak, with a corresponding lattice parameter of 0.164 nm was also observed. These lattice parameters are in excellent agreement with the data in the powder diffraction file [27] as well as those observed from a sample of the powdered BN target. These results suggest that the films grown at substrate temperatures as low as 500°C were hexagonal BN.

Table 3
X-ray Diffraction Data Summary

Observed				Powder File		
PLE Film		BN Target		Hexagonal BN		
d(nm)	I/I_o	d(nm)	I/I_o	hkl	I/I_o	d(nm)
0.334	100	0.334	100	(002)	100	0.33281
		0.217	9	(100)	15	0.21693
0.207	3	0.206	5	(101)	6	0.20619
0.182	1	0.181	6	(102)	9	0.18176
0.164	2	0.166	4	(004)	6	0.16636
0.117	1	0.117	2	(112)	5	0.11720
0.119	4					
0.260	3					

3.3 DYNAMICS OF BN EVAPORATION

Films grown in high vacuum, as discussed in 3.1, were found to be N-deficient. The dynamics of pulsed laser evaporation of BN was investigated, in order to better understand the N-deficient nature of these films. This was carried out by recording TOF distributions of the laser-evaporated ions. Shown in figure 4 is a TOF mass spectrum of the positive ions evaporated from the BN target; the laser power density at the target was identical to that used during film growth. There are peaks in the spectrum due to $^{10}B^+$ and $^{11}B^+$, as well as a smaller peak due to N^+. There is also an intense peak due to laser evaporated H^+. The nature of the broad peak appearing between 6 and 12 μs is not well established but is most likely due to some dynamical process such as charge exchange or collision-induced dissociation within the evaporant plume. There were also peaks observed at longer flight times (not shown) corresponding to various $BxNy^+$ cluster species. The intensity of these species was substantially less than that of the atomic species shown in figure 4. Of course, the relative intensities of the species seen in figure 4 are, most likely, not representative of the total relative intensities of the (more abundant) neutral species. Differences in ionization efficiency as well as neutralization probability make a direct comparison difficult. Nevertheless, this spectrum does indicate that the most intense ionized species evaporated from the BN target were B^+ and H^+.

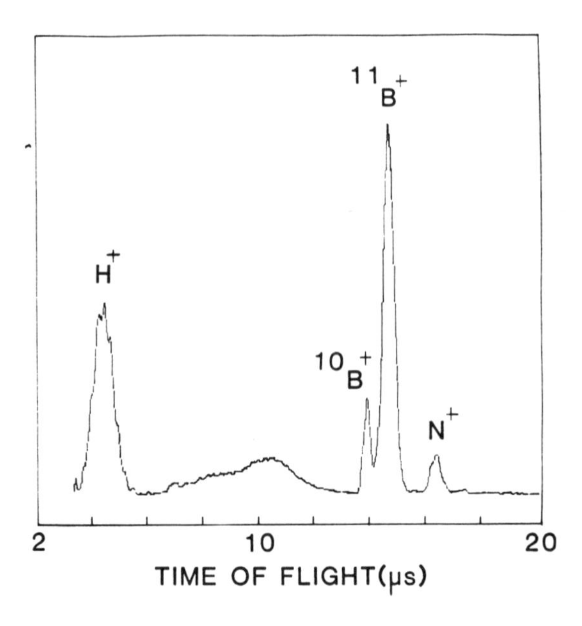

Figure 4. TOF mass spectrum of the positive ions ejected from the BN target by laser evaporation.

Presented in figure 5 is a TOF velocity distribution of the positive ions. There are two clearly separated maxima in this distribution. The dashed lines in figure 5 are the TOF distributions predicted by a Maxwellian distribution for H+ and B+, with both species at an effective translational temperature of 2600K. The fit to the experimental TOF data is excellent. It is interesting to note that a value of 2600K has been reported [28] as the decomposition temperature of BN. This close agreement may be fortuitous, since the evaporation conditions (laser power density at the target) were above the evaporation threshold.

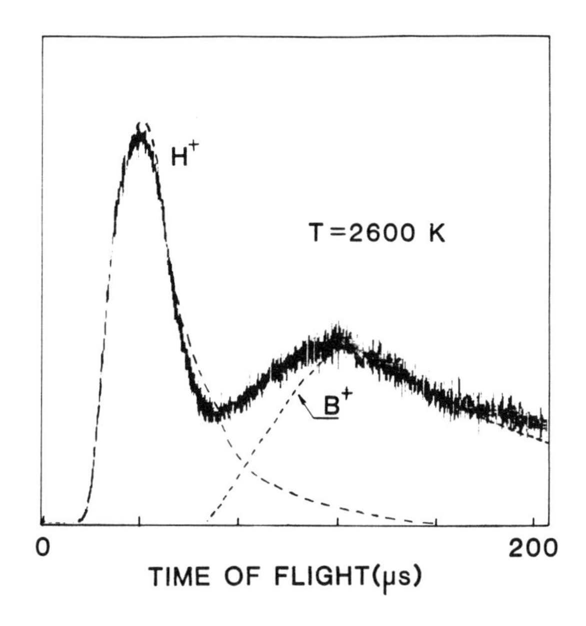

Figure 5. TOF velocity distribution of the positive ions ejected from the BN target by laser evaporation.

The TOF results indicate that PLE of BN, under the conditions used in this study, produces primarily atomic species with an effective temperature of 2600K. These results allow thermochemical calculations to be carried out on the two reactions

$$B(g) + 2N \rightarrow BN(s) + N(g) \qquad (1)$$

$$B(s) + N_2(g) \qquad (2)$$

with the reactants at 2600K and the products at 298K. The calculations indicate reaction (2) to be more exothermic by 231 kJ/mol. Thus, the N-deficient nature of the films can be attributed to the greater strength of the N-N bond compared to that of B-N. In order to confirm this assessment, residual gas analysis (by conventional quadrupole mass spectrometry) was carried out during BN irradiation. An intense peak due to molecular N_2 was observed during PLE of BN, thus confirming the implications of the calculations. Whether N atom recombination occurs primarily within the evaporant plume or on the surface of the growing films has yet to be determined.

SUMMARY

PLE of BN under high vacuum conditions produces films which are N-deficient. This can be understood in terms of the greater bond strength of N_2 versus BN. PLE of BN in an NH_3 atmosphere results in growth of oriented, stoichiometric films of BN. This can be accomplished at substrate temperatures as low as 500°C. These results indicate that PLE is a feasible method of growing stoichiometric films of BN for tribology or other applications.

ACKNOWLEDGEMENTS

The excellent technical support of Dave Dempsey is gratefully acknowledged. Thanks to Neil McDevitt for the Raman spectrum. This work was supported by the Materials Laboratory, Wright Research and Development Center, Wright-Patterson Air Force Base, OH.

REFERENCES

1. Rand, M. J. and J. Roberts, J. Electrochem. Soc., 1968, 115, 423.
2. Hirayama, M. and K. Shohno, J. Electrochem. Soc., 1975, 122, 1671.
3. Sano, M. and M. Aoki, Thin Solid Films, 1981, 83, 247.
4. Motojima, S., Y. Tamura, and K. Sugiyama, Thin Solid Films, 1982, 88, 269.
5. Takahashi, T., H. Itoh, and A. Takeuchi, J. Cryst. Growth, 1979, 47, 245.
6. Murarka, S. P., C. C. Chang, D. N. K. Wang, and T. E. Smith, J. Electrochem. Soc., 1979, 126, 1951.
7. Adams, A. C. and C. D. Capio, J. Electrochem. Soc., 1980, 127, 399.
8. Yamaguchi, E. and M. Minakata, J. Appl. Phys., 1984, 55, 3098.
9. Gashtold, V. N. , S. A. Kutol'n, L. M. Shakarian, L. F. Bielova, and G. M. Komarova, Elektron. Tekh., 1970, 12, 58.
10. Davidse, P. D. and L. I. Maissel, J. Appl. Phys., 1966, 37, 574.
11. Noreika, A. J. and M. H. Francombe, J. Vac. Sci. Technol., 1969, 6, 722.
12. Wiggins, M. D., C. R. Aita, and F. S. Hickernell, J. Vac. Sci. Technol., 1984, A2, 322.
13. Chopra, K. L., V. Agarwal, V. D. Vanker, C. V. Deshpandey, and R. F. Bunshah, Thin Solid Films, 1985, 126, 307 .
14. Lee, E. H. and H. Poppa, J. Vac. Sci. Technol., 1977, 14, 223.
15. Weissmantel, C., K. Bewilogua, D. Dietrich, H. J. Erler, H. J. Hinnerberg, S. Klose, W. Norwich, and G. Reise, Thin Solid Films, 1980, 72, 19.
16. Weissmantel, C., K. Bewilogua, K. Breuer, D. Dietrich, U. Ebersach, H. J. Erler, B. Rau, and G. Reisse, Thin Solid Films, 1982, 96, 31.
17. Shanfield , S. and R. Wolfson, J. Vac. Sci. Technol., 1983, A1, 323.
18. Miyoshi, K., D. H. Buckley, and T. Spalvins, J. Vac. Sci. Technol., 1985, A3, 2340.
19. Halverson , W. and D. T. Quinto, J. Vac. Sci. Technol., 1985, A3, 2141.
20. Guzman, L., F. Marchetti, L. Calliari, I. Scotoni, and F. Ferran, Thin Solid Films, 1984, 117, L63.
21. Sokolowski, M., J. Cryst. Growth, 1979, 46, 136.

22. Arya , S. P. S. and A. D'Amico, Thin Solid Films, 1988, 157, 267.

23. Donley, M. S., N. T. McDevitt, T. W. Haas, P. T. Murray, and J. T. Grant, Thin Solid Films, 1989, 168, 335.

24. Mamyrin, B. A., V. I. Krataev, D. V. Smith, and V. A. Zagulin, Zh. Eksp. Teor. Fiz., 1973, 64, 8; Sov. Phys.: JETP, 1973, 37, 45.

25. Hanke , G. and K. Muller, J. Vac. Sci. Technol., 1984, A2, 964.

26. Hamrin, K., G. Johansson, V. Gelius, C. Nordling, and K. Siegbahn, Phys. Scr., 1970, 1, 277.

27. Powder Diffraction File: Set 34, International Center for Diffraction Data, Swarthmore, PA, 1987, No. 34-421.

28. Chase, M. W., C. A. Davies, J. R. Downey, D. J. Frurip, R. A. McDonald, and A. N. Syverud, JANAF Thermochemical Tables, J. Phys. Chem. Ref. Data, 1985, 14, 1.

Materials Science Forum Vols. 54 & 55 (1990) pp. 165-192

HIGH DENSITY PHASES OF BN

P.K. Lam (a), R.M. Wentzcovitch (b) and M.L. Cohen (b)

(a) Dept. of Physics and Astronomy
University of Hawaii at Manoa
Honolulu, HI 96822, USA
(b) Department of Physics, University of California, Berkeley, CA 94720, USA
and
Material and Chemical Sciences Division
Lawrence Berkeley Laboratory, Berkeley, CA 94720, USA

ABSTRACT

An *ab initio* pseudopotential total energy method is used to investigate various electronic and structural properties of boron nitride, BN. Emphasis is placed on zincblende (cubic) BN but some results for the wurtzite phase are also presented for comparison. In particular, we examine the energetics along two transformation paths which connect the low density layered phases to the high density zincblende and wurtzite phases as well as the electronic band structure, charge density distribution, equilibrium structural properties, and possible ultra-high pressure structural phases of BN.

I. INTRODUCTION

There is considerable interest in the study of high density zincblende or wurtzite phases of boron nitride, BN. From a technological point of view, these materials have very useful physical properties such as extreme hardness, high melting points, and high thermal conductivity[1]. Theoretically, BN and other compounds formed by atoms from the first row of the Periodic Table behave differently than their counterparts in the later rows. The unusual properties are generally attributed to the very attractive nature of the electron-ion potentials of these atoms. The exact relationship between these properties and the potentials is not clearly understood yet.

BN exhibits solid phases[2] similar to the phases of carbon but different from those of the other group-IV or III-V solids. It is usually synthesized in the low density hexagonal phases with two- or three-layered stacking sequences (see figure 1). The two-layered form[3] has an AaAa... stacking sequence which we will refer to as hexagonal BN or *hBN*. This form differs slightly from hexagonal graphite which has an ABAB... stacking sequence. The three-layered form is similar to rhombohedral graphite; both have an ABC... stacking sequence. We

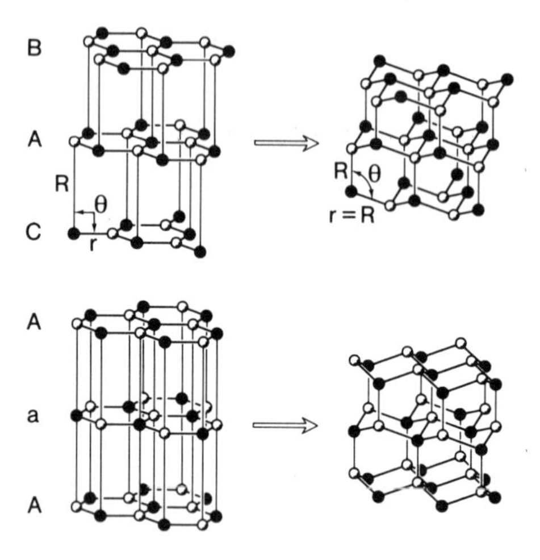

Fig. 1 Structural relationship between the low density layered phases and the high density four-fold coordinated structures; (top) rhombohedral BN to zincblende transition; (bottom) hexagonal BN to wurtzite transition.

will refer to this form as rhombohedral BN or *rBN*. The denser forms of BN, zincblende (*zBN*) and wurtzite (*wBN*), are analogous to cubic and hexagonal diamond. These forms can be produced directly, without catalysts, from *hBN* and *rBN* under static[4,5,6] and dynamic compression[7,8,9]. The products of the compression, zincblende or wurtzite, depend on the relative amount of *hBN* and *rBN* in the initial sample and the temperature and pressure at which the compression is performed[5,7]. Besides structural similarities, BN and C also manifest similarities in their electronic properties. Band structure calculations have indicated[10] that the fundamental gap of zincblende BN increases with pressure, a behavior exhibited also by cubic diamond[11] but not by other semiconductors with zincblende and diamond structures[12]. Because of their similarities, we will often draw comparison between BN and C.

In this paper, we present results of first-principles calculations on various properties of the dense phases of BN. Focus will be placed on zincblende BN, but some results for wurtzite BN are also included for comparison. There are previous calculations of the electronic band structure of *zBN*[13-23]. However, the results vary greatly among calculations. One possible reason for the large discrepancies may be the inaccurate treatment of the highly localized $2p$ states of B and N. In our calculation, we use two different bases to represent the wavefunctions, plane wave and localized orbitals. The results obtained with the two approaches are very similar, indicating good representation of the wavefunctions.

The paper is organized as follows. We present the calculational method in section II, an investigation of the energetics for the formation of the dense phases (*zBN* and *wBN*) from the layered phases (*rBN* and *hBN*) in section III, the electronic band structure of *zBN* and *wBN* and their behavior under compression in section IV, the electronic charge distribution of *zBN* in section V, the structural properties of *zBN* in section VI, and possible structural phases under ultra-high pressure in section VII.

II. CALCULATIONAL METHOD

The calculational method employed in our study is based on the pseudopotential-total-energy approach[24] within the local density approximation (LDA)[25]. The crystal total energy can be expressed as a sum of terms:

$$E_{tot} = E_{kin} + E_{e-ion} + E_{e-e} + E_{ion-ion}, \tag{1}$$

where E_{kin} is the kinetic energy of the valence electrons, E_{e-ion} is the interaction energy between the valence electrons and the ion cores (which is represented by pseudopotentials), E_{e-e} is the electron-electron interaction energy (Hartree and exchange correlation within LDA), and $E_{ion-ion}$ is the ion-ion Coulomb interaction energy (Ewald energy). The Wigner interpolation formula[26] is used for the exchange and correlation functional. The electron-ion interactions are represented by nonlocal ionic pseudopotentials with s, p, and d symmetries and are generated using the method proposed by Hamann, Schlüter and Chiang[27]. The pseudopotentials can reproduce the all-electron atomic eigenvalues and excitation energies to within 1.5 mRy over a wide range of excited configurations (excitation energies up to 1.5 Ry were tested). The one-electron crystal wavefunctions satisfy the Kohn-Sham equations[25],

$$\left\{ \frac{-\hbar^2}{2m}\nabla^2 + V_{ion}(r) + V_H[\rho(r)] + \mu_{xc}[\rho(r)] \right\} \psi_i(r) = \varepsilon_i \ \psi_i(r), \tag{2}$$

where V_{ion} is the sum of the ionic pseudopotentials from all the sites, V_H and μ_{xc} are the Hartree and exchange-correlation potentials which depend on the local density $\rho(r)$. The density is related back to the one-electron wavefunctions by

$$\rho(r) = e \sum_{i,(\varepsilon_i < E_f)} |\psi_i(r)|^2, \tag{3}$$

where E_f is the Fermi level. Thus, the crystal electronic wavefunctions must be calculated self-consistently.

Two different methods are used to represent the wavefunctions, plane waves and localized orbitals. The plane waves are generally more suitable for systems with high symmetries while the localized orbitals can be used for systems with low symmetries. When possible, both methods are used to check the stability of the results with respect to basis functions. In the plane wave expansion, the wavefunctions contain plane waves up to a kinetic energy cutoff, E_{pw}, of 60 Ry. This large E_{pw} for BN is required because of the very attractive feature of the N's p-potential and hence the wavefunction is very localized. The corresponding number of plane waves is approximately 950 per molecule. To diagonalize the large Hamiltonian matrix, the residual minimization method is used[28]. The localized orbital basis is used when calculating the energetics for the transitions from the layered to the dense phases. The wavefunctions are expanded in Bloch sums of Gaussian orbitals[29] of the form:

$$f_l(\vec{r})=e^{-ar^2}r^l Y_{lm}(\theta,\phi). \tag{4}$$

The crystal potential is expanded in plane wave up to 64 Ry and is iterated to full self-consistency using the method of Chan *et al.* [30]. The localized basis set consists of sixteen orbitals per atom, four orbital symmetries (s, p_x, p_y, and p_z) and four decay constants (a) for each orbital symmetry. For simplicity, the decay constants for each type of atom are chosen to follow a geometric progression, i.e., $a_2 = \mathrm{r}\, a_1$, etc., therefore the decay constants are fixed once the smallest and the largest constants are determined. The optimal decay constants are obtained by minimizing the crystal total energy with respect to the smallest and the largest decay constants. This procedure are performed at the two end points of the low density to high density transition path. The sets of decay constants turn out to be very similar. For N, one set of decay constants is satisfactory along the path, while for B the decay constants for the layered phases are found to be slightly smaller than those for the denser phases. The smallest and the largest decay constants for N are 0.23 and 5.8 in units of of inverse Bohr radius. The corresponding decay constants for B in the dense phases are 0.17 and 5.3, and those in the layered phases are 0.15 and 5.0. For the intermediary structures, the decay constants for B are linearly interpolated from those at the two extremes of the transition as functions of the interlayer distance. Since we are comparing energies of the layered phases with those of the dense phases where the decay constants are different, we need to have an idea of how sensitive is the energy to the decay constants. We therefore calculate the energy of the layered phases using the decay constants of the dense phases. We find that the energy changes by only 0.03 eV per molecule.

The charge density is obtained by summing over all the occupied states. For the summation, a uniform grid of k-points in the Brillouin zone is used. For calculating the cohesive energy and bulk modulus of *zBN*, 10 k-points in the irreducible Brillouin zone (IBZ) are sufficient. For the *rBN*$\rightarrow$*zBN* transition, 25 k-points in the IBZ are used. For the *hBN*$\rightarrow$*wBN* transition, the IBZ is slightly different and 30 k-points are used. For the ultra-high pressure phases, the number of k-points in the IBZ changes with each structure, but they are chosen to ensure the uncertainty in the total energy is less than 2 mRy/pair. Because of the existence of a Fermi surface, a much larger number of k-points is used for the metallic phases. For the β-Sn and rocksalt structures we use 100 and 60 k-points, respectively, although expansions up to 280 k-points have also been tested. Finally, the total energies are computed for various volumes and are fitted to the Murnaghan[31] and the and Birch equations of state[32] to obtain the bulk modulus and its pressure derivative.

III. TRANSITION FROM THE LAYERED PHASES TO THE FOUR–FOLD PHASES

Zincblende and wurtzite BN can be formed by applying high pressure and high temperature to layered BN[4-9]. The synthesis procedures for these dense phases are relatively well established. Their production depends on the starting materials and experimental conditions involved, such as pressure, temperature, and presence of catalysts. Pressure-temperature phase diagrams for BN are also available. However, the individual role played by these factors at the microscopic level is not yet well understood. A knowledge of the energetics for the transition would be valuable. We report here a theoretical investigation of the transition energetics. Because of the large unit cell involved, we use the localized basis formalism[29,30] as described in section II. We obtain the structural energies of BN along specific transition paths. Similar studies exist for the graphite to diamond transitions[33-36]. Comparison will be made between carbon and BN on different aspects of the transition, such as the barrier heights and the formation of bonds linking the original graphitic layers.

a) DESCRIPTION OF THE STRUCTURAL TRANSITIONS

The most common form of layered BN at ambient pressure and temperature is the hexagonal phase (*hBN*). The rhombohedral layered form (*rBN*) can also be formed at ambient pressure[37]. While both dense phases, wurtzite (*wBN*) and zincblende (*zBN*), can be synthesized from *hBN*, the formation of *wBN* has been shown to require a lower temperature than the formation of *zBN* [5,8] mainly because of the closer structural relation between *wBN* and *hBN* than between *zBN* and *hBN*. This situation is similar to the syntheses of hexagonal and cubic diamond from (bernal) graphite[33]. However, under shock compression only *wBN* is produced from *hBN* and only *zBN* is produced *rBN* [9] which further indicates that the initial stacking sequence plays an important role in the transformations. This evidence also supports models which suggest that the transformations are proceeded by diffusionless mechanisms with well defined epitaxial relationships between the parent and the newly formed phases[8]. Under shock compression, same conditions of pressure and temperature are required in both transitions which indicates that the energetics are very similar.

Although there are many similarities between the *hBN* $\rightarrow$ *wBN* transition and the graphite $\rightarrow$ hexagonal diamond transition, their microscopic mechanisms appear to be different. This is essentially caused by the difference in the stacking sequence of the initial layered phase of BN (AaAa...) and that of (bernal) graphite (ABAB...). As pointed out by Bundy and Kasper[38], in order to achieve the observed epitaxial relationship between the parent and the daughter phases, the graphite layers must deform into the "boat" instead of the "chair" configuration. A distortion of this kind, combined with intra-layer bond stretching and inter-layer bond-compression, results in a hexagonal diamond phase whose c-axis is perpendicular to the graphitic c-axis. The possibility of a similar mechanism, "boat" rather than "chair" deformation, be operative in the layered BN to *wBN* transition was raised[5,39]. However, Riter[39] argued that such mechanism is more likely if the layered structure were "boron shadowing boron and nitrogen shadowing nitrogen" in adjacent layers. This possibility is not addressed in the present study because a comparative calculation of this process would require much larger unit cells and prohibitive computational costs.

b) RESULTS

The geometrical relationships between the low density layered structures and the high density four-fold coordinated structures are shown in figure 1. In these transformations, the symmetries of the original phases remain unchanged along the transformation paths. The rhombohedral to zincblende ($rBN \rightarrow zBN$) transition has a C_{3v} symmetry while the hexagonal to wurtzite ($hBN \rightarrow wBN$) transition has a D_{6h}^4 symmetry. The transitions can be characterized by three internal structural parameters, R, r, and Θ, where R is the inter-layer bond length, r is the intra-layer bond length, and Θ is the buckling angle. The transformations can be described by decreasing the inter-layer bond lengths, buckling the honeycomb layers into chair configurations, and simultaneously stretching the intra-layer bond lengths. In the hexagonal layered limit, the three structural variables assume the values: buckling angle $\Theta = 90^o$, intra-layer bond length $r = 1.45$ Å (comparing with 1.42 Å in C), and inter-layer bond length $R = 3.34$ Å (3.35 Å in C). In the dense limit they are: $\Theta = 109.47^o$, $r = R = 1.57$ Å (1.54 Å in C) [40].

Following this convention, the internal structural energy per molecule becomes a function of the three selected variables, $E(R, r, \Theta)$. Once one of them is chosen to be the independent variable, the other two variables are obtained by minimizing the total energy with respect to them. A convenient independent variable is the inter-layer bond length R, which in practice can be modified by application of stress along the soft [0001] "graphitic" direction. It should be noted that a different independent variable could have been chosen, giving rise to a different transformation path. For example, the cell volume is the appropriate independent variable for studying the effect of hydrostatic pressures.

Our convention gives rise to three functions of R: $E_{min}(R)$, $r_{min}(R)$, and $\Theta_{min}(R)$. These functions correspond, respectively, to the minimum value of the energy at a given R, the value of r at which this minimum occurs, and the value of Θ at the same minimum. They are shown in figure 2. The two curves in figure 2-a correspond to the two transitions studied: the full line corresponds to the $rBN \rightarrow zBN$ transition, in which the structural energy minimizations are performed as mentioned above, while the dashed line corresponds to $hBN \rightarrow wBN$ transition, in which the same values of the parameters R, r, and Θ obtained in the $rBN \rightarrow zBN$ transition are used.

There are close similarities between the present transitions and the equivalent ones in C[35,36]. A common feature of the transitions in both the BN and the C systems is the sudden change in the intra-layer bond lengths $r(R)$ and buckling angle $\Theta(R)$. As the layers approach each other the buckling angle Θ is initially very insensitive to the inter-layer distance R but changes suddenly at a faster rate when R decreases to below a critical value, around 2.6 Å (2.9 Å in C). This happens at the point where the interaction between layers starts to increase and the energy barrier starts to rise. The behavior of the intra-layer bond length $r(R)$ is quite similar, but the sudden change in behavior does not occur until a later stage, closer to the top of the barrier, which corresponds to an inter-layer bond length of approximately 2.2 Å (2.2 Å in C). Beyond this point the structures collapse into the dense forms.

The behavior of the energy barrier, buckling angle, and intra-layer bond length can be rationalized in the following way. The electron-ion interaction energy favors bringing the two layers closer together because the electrons from one layer can experience the attractive Coulomb potential from the positive ions of the other layer. Although there is also Coulomb repulsion between electrons from both layers, the attractive potentials win because they are arose from point ions (actually ions of the size of core states). The initial rise in energy when the layers are brought together is caused by the increase in quantum mechanical kinetic energy resulting from overlapping electronic charges from adjacent layers. Initially the electron-ion interaction energy is negligible compared to the kinetic energy because the positive ions are almost fully screened when the layers are far apart. There is also an increase in Coulomb repulsion between the inter-layer charge density and the in-plane σ bond charge. As an initial relief of this Coulomb repulsion, the buckling angle Θ increases (without increasing the intra-layer bond length). As the transition approaches the top of the energy barrier, it

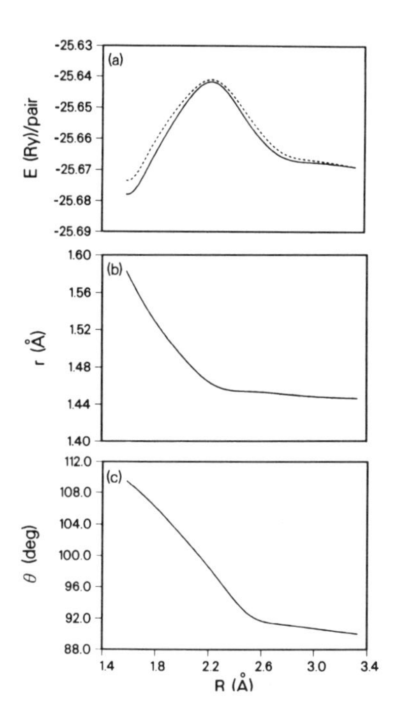

Fig. 2 - a) Total energy per BN molecule for the *rBN*→ *zBN* (full line) and *hBN*→ *wBN* (dashed line) transitions, b) intra-lyer bond length, and c) buckling angle as functions of the inter-layer bond length. The left side corresponds to the dense limit and the right corresponds to the layered limit.

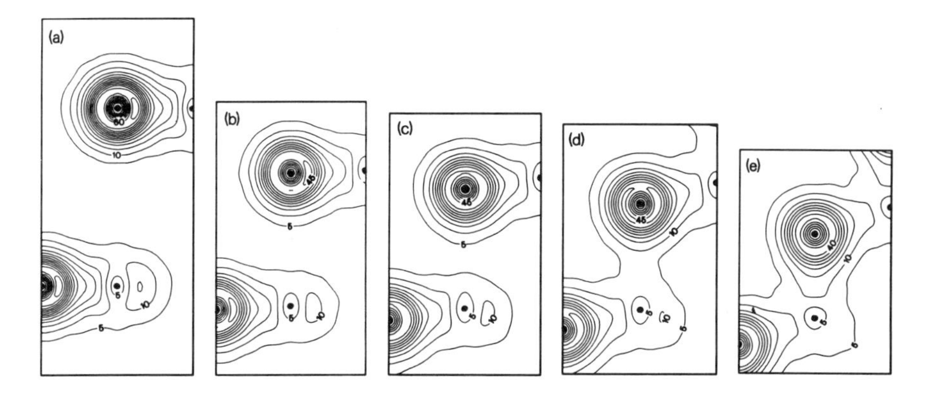

Fig. 3 - Charge density in the (1$\bar{1}$00) plane for various points along the path in the *rBN*→ *zBN* transition. They correspond to inter-layer bond (R) lengths of a) 3.34 Å, b) 2.50 Å, c) 2.25 Å, d) 2.00 Å, and e) 1.56 Å. Unit is in e/cell.

signifies that the electron-ion interaction energy in the inter-layer bond is beginning to overtake the kinetic energy. At $R \approx 2.2$ Å, the electrons from one layer can now experience the bare Coulomb attraction from the positive ions in the other layer. At this point, the electrons from the in-plane σ bond flow to the inter-layer bond, hence the in-plane bond weakens and $r(R)$ increases. The increase in the inter-layer bond charge causes Θ to increase further. The above scenario is supported by our charge density analysis. The corresponding charge density contours at various stages along the *rBN*$\rightarrow$ *zBN* transition path are shown in figure 3. The $(1\bar{1}00)$ plane, which is perpendicular to the hexagonal plane, is shown. This plane contains the in-plane σ bonds (derived from sp^2 orbitals), the π bonds in the graphitic limit, and as well as the sp^3 bonds in the dense limit. Note that we have expressed the charge density in electrons per unit cell, i.e., the absolute unit is obtained by dividing the value on the figure by the corresponding unit cell volume. The inter-layer charge density increases when the layers approach each other. Figures 3-a through 3-c correspond to the right hand side of the barrier in figure 2-a. Up to the top of the barrier, which corresponds to figure 3-c, the integrity of the σ bond is preserved. The formation of bonds linking the layers happens in the succeeding figures 3-d and 3-e, which correspond to the left hand side of the barrier in figure 2-a. In this region of the transition, the layered structures collapse into the dense phases. The lowering of the total energy in this region is followed by a corresponding rearrangement of the sp^2 bonds into sp^3. Although we have not calculated the corresponding charge densities in the *hBN*$\rightarrow$ *wBN* transition, the bond formation is expected to be very similar because the same situation was found in the graphite $\rightarrow$ diamond transition[36].

Quantitatively, the BN system requires smaller activation energies than the graphite $\rightarrow$ diamond transitions. The barrier heights for the transitions in BN are approximately 0.39 eV/pair while they are 0.66 eV/pair in the C system. The smaller barrier is consistent with the fact that BN is less covalent and more ionic than C. As we have suggested, the barrier height is mainly caused by the kinetic energy increase of the overlapping charges in the inter-layer region and the Coulomb repulsion between the in-plane σ bond charge and the inter-layer charge. Both the kinetic energy and the Coulomb repulsion are weaker in BN because the bond charge is smaller in BN compared to C. The buckling angle also increases earlier in C (at $R = 2.9$ Å) than in BN (at $R = 2.6$ Å) for the same reason of the larger bond charge in C.

There are other quantitative features of our calculation which need to be examined more closely. Firstly, according to our calculated energies, *zBN* is stabler than *rBN* by about 0.12 eV/pair in contrast with carbon where the layered phase is stabler. This result does not include the zero-point vibrational energy. For carbon, where the phonon spectra of both diamond and graphite are available[41,42], the zero-point energies are 0.36 eV/pair and 0.33 eV/pair, respectively[43]. The entire phonon spectrum is not available for the layered BN nor the four-fold BN. Some information on the phonon modes of *zBN* and *hBN* are available. The phonon modes of *wBN* and *rBN* are expected to be close to those of *zBN* and *hBN*, respectively. The LO and TO modes of *zBN* are reported to be 1305 cm^{-1} and 1054 cm^{-1}[44] while the LO and TO modes of *hBN* are 1610 cm^{-1} and 1367 cm^{-1}[45]. The higher optical frequencies for *hBN* is related to the shorter in-plane BN bond compared to the bond in *zBN*. However, the weaker inter-planar interaction will produce some low frequency modes in the layered BN. The overall zero-point energies of the two structures will be nearly equal. In view of the fact that there are many similarities between C and BN, the difference in zero-point energies between *zBN* and *rBN* is also expected to be only a few hundredths of an eV per pair. We should mention that there is a report of the specific heat of *hBN* at low temperatures[46]. The specific heat follows a T^2 law which is derivable from a two dimensional (2D) Debye model. The fitted Debye temperature is 598 K for *hBN*[46], in contrast to 1700 K for *zBN*[47]. If we use the value of 598 K for the Debye temperature of *hBN* and estimate the zero-point energy using a 2-D Debye model for the phonon density of states, we would get a very small value. A close inspection of the experimental data indicates that the T^2

dependence of the specific heat is only valid up to ≈ 60 K; the specific heat varies more closely to a linear T-dependence for temperature up to 300 K (the linear T-dependence is consistent with a weak inter-planar interaction which gives rise to an almost 1-D spectrum at low frequencies). A "Debye temperature" derived from such a low temperature data will not be reliable for estimation of the zero-point energy of the entire spectrum. Furthermore, Ref. 46 also quoted a Debye temperature of 767 K for graphite; in this case the zero-point energy is known to be much larger than that estimated by this low temperature value of the Debye temperature. We believe that the inclusion of the zero-point energies will not alter the order of stabilities of *zBN* and *rBN* although it will reduce the energy difference from 0.12 eV/pair to ≈ 0.09 eV/pair (using data for carbon).

Secondly, the local density approximation (LDA) is likely to underestimate the cohesion in the layered phases because the charge density in the inter-planar region is very small; LDA is more appropriate for uniform and high charge densities. The fact that the calculated energies for *hBN* and *rBN* are nearly equal indicates that the LDA is inadequate in describing the inter-layer bonding. In the case of carbon, the calculated LDA cohesive energies for graphite and diamond are 15.73 eV/pair and 15.74 eV/pair, respectively[43]. With the inclusion of the zero-point energies, the calculated cohesive energies are 15.40 eV/pair and 15.38 eV/pair[43]. Within the numerical accuracy, these two calculated energies are practically the same. The experimental cohesive energies for graphite and diamond are 14.76 eV/pair and 14.70 eV/pair, respectively. Hence, the calculated energy difference between graphite and diamond (with inclusion of zero-point energy) underestimates the experimental value by ≈ 0.04 eV/pair. Assuming similar error in BN, *zBN* is still stabler than *rBN* by ≈ 0.05 eV/pair. If our estimates are indeed correct, then the *rBN* to *zBN* transition should be exothermic. No accurate thermal data for BN has been reported. An indirect evidence that *zBN* may be stabler than *rBN* or *hBN* is from the very early report of cubic BN by Wentorf[4]. Wentorf reported that the temperature required to transform *zBN* back to *hBN* is higher than the temperature needed to transform *hBN* to *zBN* indicating that the barrier height is higher for the *zBN* to *hBN* transition or equivalently *zBN* is lower in energy than *hBN*. However, a calculation of the heat of formation by Sirota and Kofman[48], basing on measured specific heats, gives a positive heat of formation (510 cal/mole or 0.02 eV/pair) for the *hBN* to *zBN* transition, i.e, *hBN* is stabler. More accurate thermodynamic data are needed to clarify the relative stability issue.

Although the above corrections for the calculated energies are only rough estimates, our conclusion that *zBN* is stabler than the layered phases is consistent with general chemical trends. The low coordination (layered) structure is usually preferred by materials which are capable of forming strong covalent bond. For example, carbon forms layered structures while silicon does not. This trend can be rationalized as follows. The cohesive energy depends on both the strength and the number of bonds. However, strong bonds are usually associate with short bond distances which can accommodate only a few nearest neighbors because of Pauli exclusion and electrostatic repulsion between bond charges. That is, there is a trade off between the strength and the number of bonds. Even in the case of carbon, graphite is only slightly stabler than diamond, it is conceivable that the order of stability between the layered and four-fold phases would be reversed for the less covalent material such as BN.

Let us use our calculated total energies to interpret some of the experimental results for the transitions under static and shock compression. First of all, our calculated barrier for the *hBN*→ *wBN* transition is nearly the same as that for the *rBN*→ *zBN* transition. This result is consistent with the shock compression result that same condition of pressure and temperature are required in both transitions. In the static compression experiment, both *wBN* and *zBN* are produced from *hBN*. Although we did not investigate the *hBN*→ *zBN* transition, we can infer about its energetics from the shock compression experiment. Only *wBN* is produced from *hBN* and only *zBN* is produced from *rBN* under shock compression. This can be attributed to the fact that the *hBN*→ *zBN* and the *rBN*→ *wBN* transitions would have required a rearrangement of the BN layers. It is reasonable to expect that the activation barrier is higher and the reaction time is longer for these transitions. Hence, neither the $hBN \rightarrow zBN$ nor the $rBN \rightarrow wBN$ transition is observed in shock compression because there is not enough time for the kinetics. Combining this information with our calculated result that zBN is stabler than wBN, see figure 2, we can understand the temperature dependence of the static compression results. At low temperature or short reaction time, the product is wBN[5] rather than zBN because there is not enough kinetic energy to overcome the higher $hBN \rightarrow zBN$ barrier. However, at high temperature the $hBN \rightarrow zBN$ barrier can be overcome and the product is mainly zBN[5] because it is stabler. The calculated cohesive energies of zBN and wBN are 13.5 eV and 13.4 eV, respectively, using the localized orbital basis. Similar results are obtained using the plane wave basis[49].

Finally, our study reveals that, along the $rBN \rightarrow zBN$ or the $hBN \rightarrow wBN$ transition path, the structures remain as insulators. Initially, the LDA gap is approximately 4.6 eV in the rhombohedral structure. It decreases to approximately 3.5 eV near the top of the barrier and increases again to 4.5 eV in the zincblende structure.

IV. ELECTRONIC BAND STRUCTURE

a) ZINCBLENDE BN

The important features of the zBN band structure are summarized in table I, where the results of previous calculations and available experimental data are reported in chronological order. Unfortunately there is only limited experimental data. In 1962, Philipp and Taft[50] identified structures in the region of 9 to 10 eV and a prominent peak at 14.5 eV in the reflectivity data. In 1968, a K-band X-ray emission spectroscopy study[51] reported upper bound values of 6 eV for the indirect gap and 22 eV for the full valence band width. However, a later UV absorption experiment[52] gave a lower bound of 6.4 eV for the indirect gap. On the theoretical side, although several calculations[13-23] have been reported, the calculated results differ considerably. The calculated indirect gap ranges from 3.0 to 10.5 eV and the symmetries assigned to the indirect and direct gap transitions differ among calculations (see table I). This large discrepancy may be caused by inaccurate treatments of the highly localized 2p states of B and N.

Our calculated band structure is shown in figure 4. The full valence band width is 20.3 eV, which is in good agreement with the value of ≈22 eV reported by soft X-ray spectroscopy. Our calculated indirect gap ($\Gamma_{15}^{v} \rightarrow X_{1}^{c}$) is 4.2 eV while the measured gap is ≈ 6 eV. It is well known that the LDA underestimates the gap by 30 to 50%. This deficiency can be remedied by performing an elaborate calculation of the electron self-energy. Such calculation has been performed for diamond[53], in which case the conduction band moves up by about 1.5 eV to 2 eV depending where the k-point is located in the Brillouin zone. We simulate this correction in BN by a rigid shift of the conduction bands by 1.8 eV, which would bring the calculated gap closer to the experimental value. This rigid shift would also bring the optical transitions at the L point to a closer agreement with the reflectivity data. The calculated transitions $L_{3}^{v} \rightarrow L_{1}^{c}$ and $L_{3}^{v} \rightarrow L_{3}^{c}$ are 12.1 and 12.6 eV, respectively. If the shift of 1.8 eV is assumed, these transitions become 13.9 and 14.4 eV which are close to the value of 14.5 eV reported for the strong peak in the reflectivity spectrum[50]. The majority of the calculations gives Γ_{15}^{c} as the lowest conduction state at Γ, however two EPM calculations[19,21] reported Γ_{1}^{c} as the lowest conduction state. Generally, as the ionicity increases, the energy of Γ_{1}^{c} is lowered with respect to the Γ_{15}^{c}. For example, BeO, has Γ_{1}^{c} below Γ_{15}^{c}[12] while diamond, on the other hand, has Γ_{1}^{c} above Γ_{15}^{c}[29]. *A priori*, both orderings seem possible in BN. According to our calculation, Γ_{15}^{c} is the lowest conduction state at Γ.

Table I -Summary of important features of the band structure of *zBN*. The values interpolated from published figures are denoted by asterisks (*).

	Lower valence bandwidth (eV)	Upper valence bandwidth (eV)	Full valence bandwidth (eV)	Minimum Gaps (eV) indirect	Minimum Gaps (eV) direct
			Theoretical		
a	5.6	3.6	18.0	10.5 ($\Gamma^v_{15} \rightarrow X^c_3$)	14.3 ($\Gamma^v_{15} \rightarrow \Gamma^c_{15}$)
b	7.0	5.1	12.6	-	-
c	5.0	12.7*	23.2*	3.0 ($\Gamma^v_{15} \rightarrow X^c_1$)	7.6 ($\Gamma^v_{15} \rightarrow \Gamma^c_{15}$)
d	4.5	8.0	17.8	7.2 ($\Gamma^v_{15} \rightarrow X^c_1$)	8.9 ($\Gamma^v_{15} \rightarrow \Gamma^c_{15}$)
e	6.8*	16.5	27.5	7.6 ($\Gamma^v_{15} \rightarrow X^c_1$) *or* ($\Gamma^v_{15} \rightarrow L^c_1$)	8.4 ($\Gamma^v_{15} \rightarrow \Gamma^c_1$)
f	5.1	9.0	19.1	8.7 ($\Gamma^v_{15} \rightarrow \Delta_{min}$)	10.8 ($\Gamma^v_{15} \rightarrow \Gamma^c_{15}$)
g	6.5*	11.3*	20.4	4.3 ($\Gamma^v_{15} \rightarrow X^c_1$) *or* ($\Gamma^v_{15} \rightarrow K^c_1$)	9.2 ($X^v_5 \rightarrow X^c_1$)
h	7.3	14.0	27.4	8.0 ($\Gamma^v_{15} \rightarrow X^c_1$) *or* ($\Gamma^v_{15} \rightarrow L^c_1$)	14.3 ($\Gamma^v_{15} \rightarrow \Gamma^c_1$)
i	9.3*	15.0*	30.0*	11.3* ($\Gamma^v_{15} \rightarrow X^c_1$)	14.6 ($\Gamma^v_{15} \rightarrow \Gamma^c_{15}$)
j	5.5	9.7	19.9	7.0 ($\Gamma^v_{15} \rightarrow X^c_1$)	9.9 ($\Gamma^v_{15} \rightarrow \Gamma^c_{15}$)
k	5.9	10.8	20.3	4.2 ($\Gamma^v_{15} \rightarrow X^c_1$)	8.6 ($\Gamma^v_{15} \rightarrow \Gamma^c_{15}$)
			Experimental		
l	-	-	<22.0	< 6.0	-
m	-	-	-	≈ 6.4	-

[a] OPW-Pseudopotential (Reference 14)

[b] Linear Combination of Bonding Orbitals (Reference 15)

[c] OPW (Reference 16)

[d] APW (Reference 17)

[e] Nonlocal EPM (Reference 19)

[f] *Ab Initio* calculation (Reference 13)

[g] Plane-wave-gaussian mixed basis (Reference 20)

[h] Nonlocal EPM (Reference 21)

[i] HF exact exchange (Reference 22)

[j] Semi-*Ab Initio* approach (Reference 23)

[k] Present calculation and Reference 10

[l] Soft X-ray spectroscopy (Reference 51)

[m] UV absorption (Reference 52)

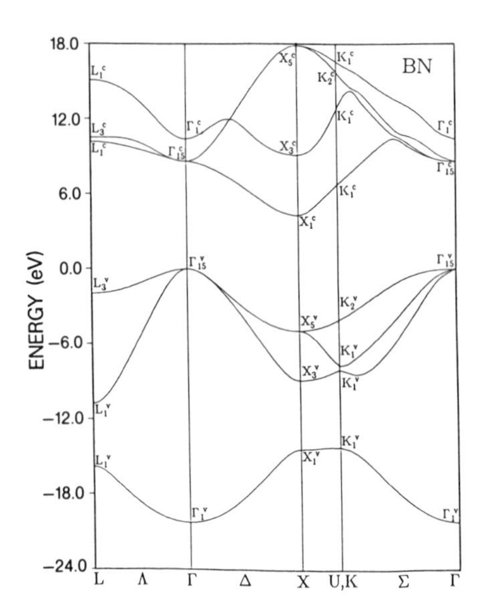

Fig. 4 - Electronic band structure of zincblende BN at the calculated equilibrium volume. Energies are measured from the top of the valence band.

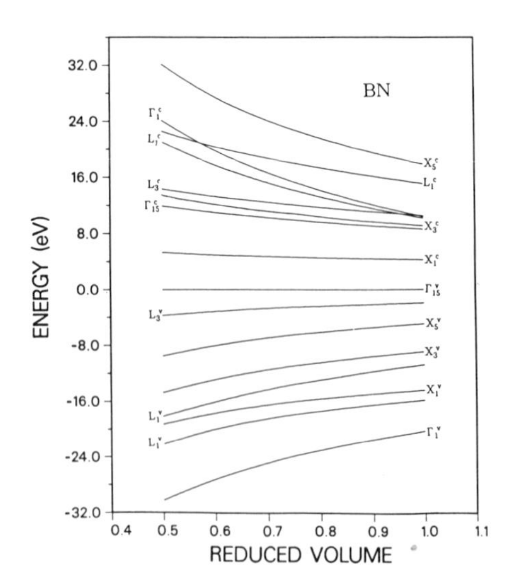

Fig. 5 - Volume dependence of the valence and conduction band energies for zincblende BN. The volumes are normalized by the calculated equilibrium volume.

It is informative to compare the band structure of BN with that of diamond[29,51]. The crystal electron-ion potential for BN can be expressed in terms of a symmetric part, $V_s = (V_N + V_B)/2$, and an anti-symmetric part, $V_a = (V_N - V_B)/2$. If the symmetric part is similar to the potential for C, then the differences in the band structures of BN and C are caused mainly by the anti-symmetric part. One evidence that the V_s for BN is similar to the C-potential is that both BN and C have similar full valence bandwidths; the LDA valence bandwidth is 20.3 eV in BN and 21.7 eV in C. The self-energy correction changes the full valence bandwidth to 22.7 eV in C. The experimental valence bandwidth is 22 eV in BN and 24.2 eV in C, that is, the calculated bandwidths are off by about 2 eV for BN and C. The degree of ionicity (V_a) can be inferred from the gap at X, $X_1^v \rightarrow X_3^v$ is 5.5 eV, where the corresponding states in diamond are degenerate. The gap form X_5^v to X_1^c, on the other hand, is almost unaffected by V_a; it is 9.2 eV in BN and 10.8 eV in C. The direct gap at Γ from Γ_{15}^v to Γ_{15}^c is 8.6 eV in BN and 5.5 eV in C.

The behavior of the electronic states under pressure is examined by calculating the energy eigenvalues at several high symmetry points in the Brillouin zone for different volumes ranging from $V_o/2$ to V_o, where V_o is the calculated equilibrium volume. The results are shown in figure 5. The change in eigenvalues are measured with respect to Γ_{15}^v. The full valence band width increases by 50% when the volume is reduced by 50%. The calculated fundamental indirect gap ($\Gamma_{15}^v \rightarrow X_1^c$) increases from 4.2 to 5.3 eV. This increase of the band gap with compression is different from that observed in other semiconductors, but similar to that found in diamond[2,11]. One possible origin for the increase in the band gaps under pressure for C and BN is the relatively large energy separation of the d excited atomic orbitals from the Fermi level[11]. When the volume changes from V_o to $V_o/2$, the degree of ionicity decreases; the separation of the levels, X_1^v and X_3^v, decreases from 5.5 to 4.6 eV. For the conduction states, Γ_1^c, L_1^c and X_5^c are the most sensitive to the compression. This behavior is in qualitative agreement with the calculation of Zunger and Freeman[13].

b) WURTZITE BN

There are very few calculations on the electronic band structure of *wBN*. We found one previous band structure calculation on this structure by Park et al.[54]. Our results and theirs are very similar. Our calculated LDA gap ($\Gamma \rightarrow K$) is 4.9 eV, same as that obtained by Park et al. For comparison, the calculated LDA band gap for hexagonal diamond is also from Γ to K but the gap is only 3.3 eV[55]. The details of the band structure of wurtzite BN are shown in figure 6 and the behavior of the energy levels under compression in figure 7. Unlike *zBN*, the gap in *wBN* decreases under compression. This behavior also occurs in carbon where the gap increases with pressure in cubic diamond but decreases in hexagonal diamond[55]. Fahy and Louie proposed an explanation for the behavior of the gap in cubic and hexagonal diamond[55]. They argued, basing on a tight-binding picture of $s-$ and $p-$ orbitals, that there are conditions for the conduction band state to achieve the lowest energy. The conditions are: (1) the "atomic condition" - the conduction state should have mainly p-character because the bonding-antibonding splitting is much smaller for the p-state than for the s-state; (2) the "interstitial condition" - the conduction state should contain linear combination of anti-bonding orbitals that are in phase so that the kinetic energy in the interstitial site will be small. They showed that it is possible to construct a state satisfying both conditions for the hexagonal diamond structure but not for the cubic diamond structure. The frustration of the atomic and interstitial conditions in the cubic diamond structure leads to a larger gap (the calculated LDA gaps are 4.3 eV for cubic diamond and 3.3 eV for hexagonal diamond). They further argued that as the structures are compressed, the interactions of the anti-bonding orbitals increase and the effect of frustration increases the difference in the conduction band minimum between cubic and hexagonal diamond. There is a difficulty in extending the above argument to BN

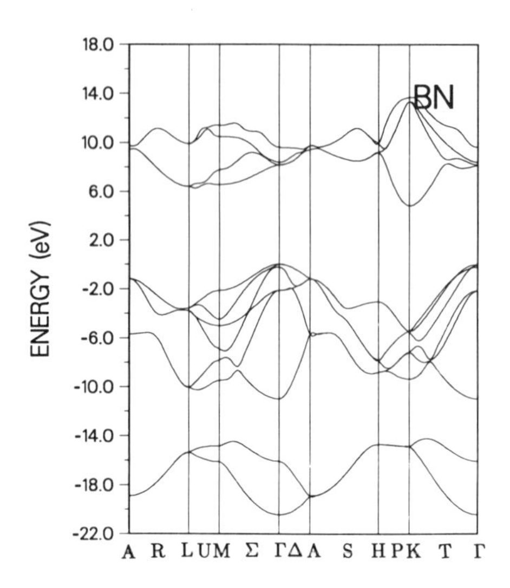

Fig. 6 - Electronic band structure of wurtzite BN at the experimental equilibrium volume. Energies are measured from the top of the valence band.

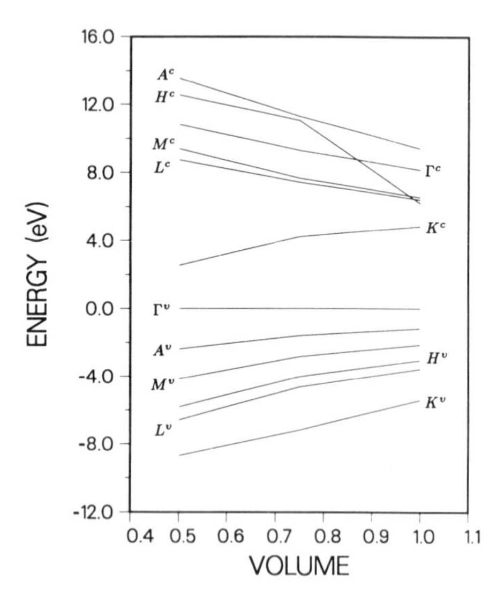

Fig. 7 - Volume dependence of the valence and conduction band energies for wurtzite BN. The volumes are normalized by the calculated equilibrium volume.

because the calculated band gap for *zBN* is 4.2 eV which is smaller than that for *wBN*, 4.9 eV, implying that the "frustration" in the cubic phase did not cause the gap to be larger. At present, we do not have a consistent explanation. Further analysis of this behavior is necessary.

V. ELECTRONIC CHARGE DISTRIBUTION IN ZINCBLENDE BN

The valence charge density contours of *zBN* at the equilibrium volume, V_o = 79.78 $(a.u.)^3$, and at $0.45V_o$ are shown in figure 8. Three planes are shown, (i) and (ii) are the (110) planes containing the B-N bond, (iii) and (iv) are the (100) planes containing N, and (v) and (vii) are the (100) planes containing B. The charge density of BN shows features of a typical III-V compounds. The charge density localizes near the N atom and levels off toward the B atom. The variation of the charge density can be seen more easily in figure 9 which shows a linear plot of the charge density along the B-N direction. The contributions to the charge density from the first and the upper three valence bands are shown separately in figure 9. An angular momentum decomposition[56,57] of the states in the valence bands shows that the first band is predominantly N *s*-state while the upper three bands are predominantly N *p*-states with some admixtures of B *p*-state. At the equilibrium volume, the density at the peak is approximately 42 e/unit-cell. For comparison, in BeO, its isoelectronic II-VI[58], the maximum value is about 60 e/unit-cell while in diamond, its isoelectronic group-IV, the maximum value is 23 e/unit-cell (however, in diamond, there are two local maxima along the bond). The charge density is about 12 e/unit-cell in the "bonding" region, mid-way between N and B, as compared to the average density of 8 e/unit-cell. At compressed volume, $0.45V_o$, the charge density becomes more uniform as indicated by more evenly spaced contour lines. The charge density in the "bonding" region and near the B atom has increased by more than a factor of two (in absolute unit of $e/(a.u.)^3$) while the maximum charge density near the N atom remains almost as the same. The behavior can be explained as follows. The increase in the charge density is caused mainly by the reduction in volume. However, the N atom cannot accommodate additional density, so some of the charge flow to the bonding region and the B atom. This increase in the charge density causes a large increase in the electron kinetic energy. The B-N distance is quite short in the zincblende structure, the kinetic energy can be lowered by moving the B away from the N, that is transforming into another structure; see section VII for more discussion on pressure-induced structural transition.

VI. STRUCTURAL PROPERTIES OF ZINCBLENDE BN

In this section, we will examine the bulk modulus, the equation of state, and the equilibrium volume of *zBN*. Despite its unusual physical properties and its potential commercial importance, there had been no experimental determination of the bulk modulus *Bo* and its pressure derivative *Bo'* until recently. A previously reported value of 465 GPa[2] basing on an empirical relation of elastic constants does not seem reasonable because this value is larger than the accepted value for diamond (442 GPa)[2]. Generally, the bulk modulus decreases with ionicity[59,60]. Here we report the results obtained from our pseudopotential total energy calculations. To check the accuracy of our calculated results, both the plane wave and the localized orbital bases are used for comparison. The values of *Bo* and *Bo'* are obtained by fitting the calculated total energies to the Murnaghan[31] and the Birch[32] equations of

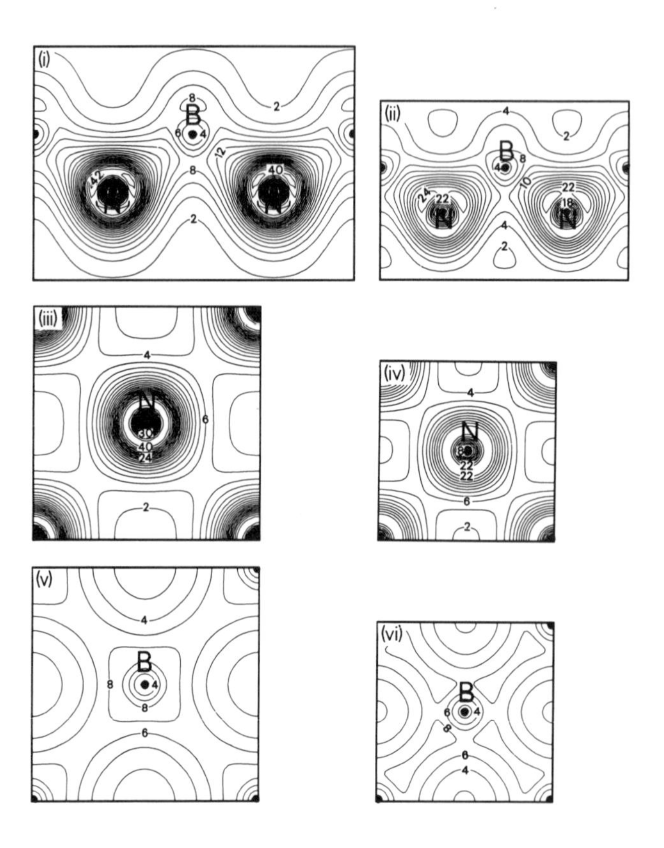

Fig. 8 - Total valence charge density contours of *zBN* at equilibrium volume, V_o and at 0.45 V_o. The contours are shown in the (110) plane at V_o (i) and at 0.45 V_o (ii); the (100) plane containing N at V_o (iii) and at 0.45 V_o (i); the (100) plane containing B at V_o (v) and at 0.45 V_o (vi). Units are in e/unit-cell and contour intervals are in steps of 2 units.

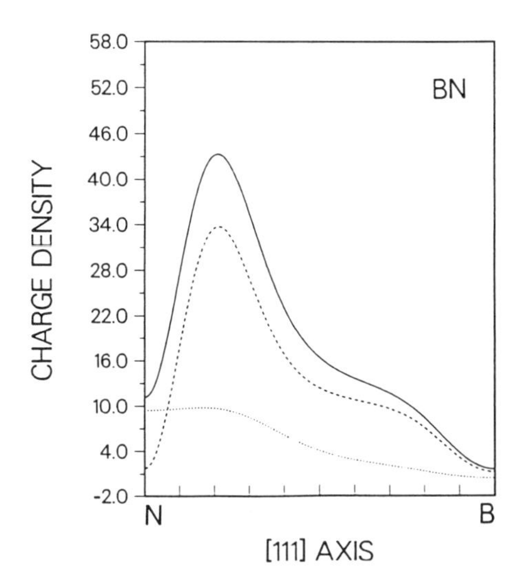

Fig. 9 - Total valence charge density along the bond of BN; ---- contribution from the upper three valence bands; .... contribution from the first band.

state. These results are also compared with a recent experimental bulk modulus measurement of *zBN* under static compression[61]. There is excellent agreement between theory and experiment. Other properties such as the equilibrium lattice constant, cohesive energy, and frequency of the transverse optical vibrational mode at $\vec{q} = 0$, TO(Γ), are also presented.

The total energies are calculated for twelve volumes between 50 and 90 (a.u.)^3. In order to compare the results with the experimental data at 300 K, we use the Debye model to calculate the Helmholtz free energy, F = E -TS, for the phonon degree of freedom. The use of the Debye model for calculating the vibrational energy is purely empirical and is only an estimate. We use the experimental values of Θ_D = 1700 K for the Debye temperature and γ_o = 1.87 for the Grüneisen parameter[62]. These values are referenced to the equilibrium volume at ambient conditions. We assume that γ/V is constant because of the empirical success of this approximation[63]. At 300 K, the dominant correction from the phonon is the zero-point energy. The effect of compression through γ is to increase the vibrational energy with decreasing volume, so the equilibrium volume is larger than that obtained from the static lattice energies. The total energy as a function of volume is shown in figure 10. The filled circles represent the static lattice energies, the open circles represent the energies with phonon free energy included, and the solid curves are fits of the energies using a third-order polynomial in the Eulerian finite strain parameter, $f = [(V_0/V)^{2/3}-1]/2$. The fitting function used is identical to the Birch equation of states for the pressure-volume relation. The arrows indicate the experimental and theoretical values for the volume at zero pressure. A plot of the pressure vs volume is shown in figure 11.

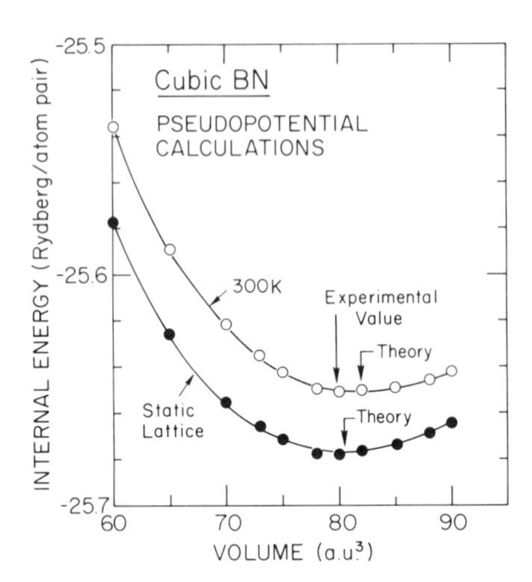

Fig. 10 - The calculated internal (total) energy of cubic BN (*zBN*). The filled circles denote the static lattice results. The open circles denote the energies with the phonon free energy, approximated by the Debye model, included. The two curves show the Eulerian finite strain fits to the calculated energies. The arrows indicate the experimental and theoretical values for the equilibrium volume at zero pressure.

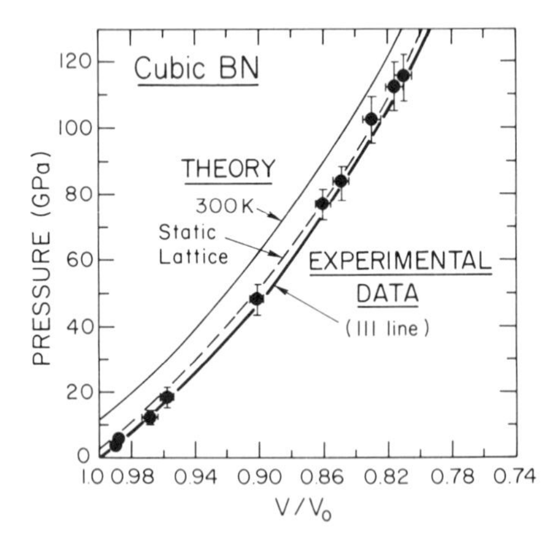

Fig. 11 - Experimental pressure-volume data for cubic BN at 300 K (circles with error bars, ref. 61) are compared with theoretical isotherms for the static lattice (dashed curve) and at 300 K with phonon free energy included (thin solid line). The thick solid line is the Birch-Murnaghan isotherm fit to the experimental data. The volume is monitored by the (111) x-ray diffraction line.

The computed equilibrium lattice constant, a, bulk modulus, B_o, cohesive energy, E_{coh}, and frequency of the transverse optical vibrational mode at $\vec{q} = 0$, TO(Γ), are given in table II. To test the dependence of a and B_o on the exchange and correlation functional, we also use the Hedin and Lundqvist formulation[64]. We find that the bulk modulus increases by about 3% while the lattice constant decreases by about 1%. Furthermore, the values of a and B_o are found to be rather insensitive to different samplings of $\vec{k}$ points in the Brillouin zone integrations. The lattice constant of BN is similar to that of diamond but the bulk modulus is smaller than that of diamond. It is interesting to compare our results for B_o with the value obtained using an empirical relation[59,60], $B_o = (1971 - 220\lambda)\, d^{-3.5}$ where d is the nearest neighbor distance in Å and λ is an empirical parameter which accounts for the effect of ionicity; $\lambda = 0, 1, 2$ for group-IV, III-V, and II-VI semiconductors, respectively. This relation gives $B_o = 364$ GPa for BN, in very good agreement with our results.

The cohesive energy is obtained by subtracting the total ground state energy of the solid from those of the isolated atoms. A correction from the phonon free energy is estimated by using the Debye model. As mentioned earlier, the major contribution from the phonon is the

Table II - Lattice constants, bulk moduli, cohesive energies and phonon frequencies of the TO(Γ) modes for BN in the zincblende structure.

		a(Å)	B_o(GPa)	E_{coh}(eV)	f_{TO}(THz)
BN	Exp.	3.615[a]	369[b]	≈13.2[c]	31.6[d] 32.0[e]
	Calc.	3.606[f] 3.618[g] 3.625[h]	367[f] 370[g] 368[h] 364[i]	13.3[f] 13.5[g] 13.5[h] 12.8[j]	30.1[f]

[a] Reference 2

[b] Reference 61

[c] estimated by comparing to the cohesive energy of layered BN which was obtained using References 66 and 67 (see text)

[d] Raman spectroscopy (Reference 44)

[e] IR absorption (Reference 69)

[f] Present calculation with plane wave basis (fit to the Murnaghan equation)

[g] Present calculation with localized orbital basis (fit to the Murnaghan equation)

[h] Present calculation with localized basis basis (fit to the Birch equation)

[i] Empirical relation $1751\, d^{-3.5}$ (Reference 59 and 60)

[j] Numerical-basis set LCAO (Reference 13)

zero-point energy. For BN this correction is approximately 0.33 eV/pair. The temperature dependent part of the phonon energy and the entropy (-TS) are very small at room temperature, ≈ 0.01 eV. A correction is made to take into account the spin polarization energy of the atoms. Atomic calculations using spin polarized and spin unpolarized density functionals[65] are performed. Spin-polarization lowers the atomic total energy by 2.5 and 0.9 eV for N and B respectively. Unfortunately there is no experimental value for the cohesive energy of *zBN*. However, according to reference 48, the experimental value for the cohesive energy for the zincblende phase should differ from that of *hBN* by only ≈ 0.02 eV/pair (510 cal/mole). So we use the experimental cohesive energy of *hBN* for comparison. For the hexagonal phase the experimental value can be obtained from the heat of formation[66] and the heat of atomization of the elements[67].

The frequency of the transverse optical phonon mode at the Brillouin zone center, Γ, is calculated using the frozen-phonon approximation[11,68]. This phonon mode is simulated by displacing one ion with respect to the other along the (111) direction. The change in energy, ΔE, as a function of the displacement, u, can be related to the phonon frequency by assuming simple harmonic motion, i.e., $\Delta E = (1/2) M_\mu \omega^2 u^2$ where M_μ is the reduced mass. To eliminate the odd power terms in $\Delta E(u)$, we plot the quantity $[\Delta E(u)+\Delta E(-u)]/2$ versus u^2. The slope at $u^2=0$ gives $(1/2) M_\mu \omega^2$. The calculated frequency is within ≈ 3% of the experimental values[44,69].

VII. ULTRA–HIGH PRESSURE PHASES OF BN

In this section, we explore some possible structural phase transitions of *zBN* under very high pressures. Phillips and Van Vechten developed an ionicity scale, basing on a dielectric theory[70], to predict the first pressure induced structural phase transitions. Generally, there are two likely structural candidates into which the family of diamond and zincblende semiconductors transform under isotropic pressure, the β-Sn and the rocksalt structures. Empirically, the β-Sn structure is usually preferred by low ionicity zincblende compounds while the rocksalt structure is preferred by high ionicity zincblende compounds. This trend is also the conclusion of a recent quantum mechanical calculation[71] which studied the behavior of a hypothetical AB compound in the same row of the Periodic Table as Ge. Starting with the same potential for both atoms, a localized Gaussian potential is added to one and subtracted from the other potential to simulate the effect of ionicity. However, in that study the symmetrical part of the potential was not varied (that is the degree of covalency was kept fixed) therefore it may not be applicable to elements in other rows of the Periodic Table, in particular the elements with very attractive pseudopotential such as those in the early rows. For example, an important exception to the above trends is carbon. Calculations indicate that its first high pressure phase is the BC8 phase[72,73]. Furthermore the cubic structure (which is the group-IV equivalence of the rocksalt structure) has a lower energy than the β-Sn structure even though the ionicity is zero. It has been suggested that the lack of p-states in the core and the large separation in energy of the d-orbitals from the Fermi level in C contribute to this behavior[73]. In fact the β-Sn phase of carbon has recently been shown to be unstable[55].

Although there are various structural candidates for the high pressure phase, we restrict our investigation to simple structures such as the β-Sn and rocksalt. We have calculated the total energies E_{tot} for the zincblende, β-Sn, and rocksalt structures at several volumes. The energies are fitted to the Murnaghan equation of state[31] and are shown in figure 12. The calculated values for the transition pressures and volumes are summarized in table III. In addition to the rocksalt and β-Sn phases, we have also examined the wurtzite phase (see insert

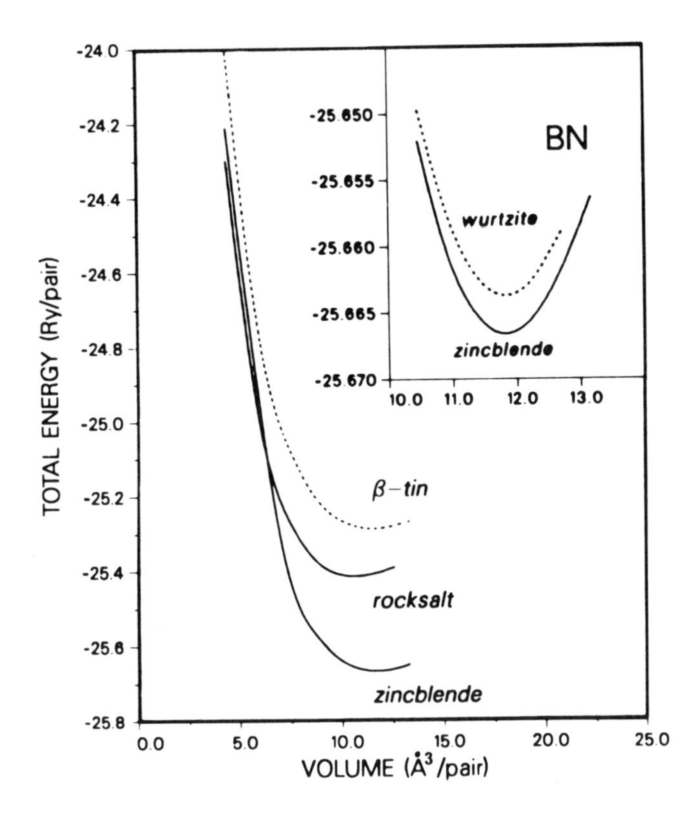

Fig. 12 - Calculated total energies of BN as a function of primitive cell volume in the zincblende, wurtzite, rocksalt, and β-Sn structures.

Table III - Transition pressures and transition volumes for the zincblende (ZB) to rocksalt (RS) structures of BN. The volumes are given in fraction of the equilibrium volume of *zBN* at zero pressure.

	P_t(Mbar)	V_t(ZB)	V_t(RS)
BN	11	0.45	0.42

of figure 12). The energy of the wurtzite structure lies entirely above that of zincblende for all volumes considered, therefore there is no pressure-induced transition from zincblende to wurtzite. As mentioned earlier in section III, the wurtzite structure is only slightly higher in energy than the zincblende phase ($\approx$ 0.04 eV/pair using the plane wave basis and $\approx$ 0.1 eV/pair using the localized orbitals basis). The rocksalt structure is definitely lower in energy than the β-Sn structure. Let us examine the β-Sn structure more closely. The β-Sn structure can, in principle, be obtained from the zincblende structure continuously by changing the c/a ratio of the tetragonal cell; in the zincblende structure c/a is equal to $\sqrt{2}$ while in the β-Sn it is typically 0.55. However for BN it is not possible to find a value of c/a at which this structure is stable or metastable[49]. The total energy curve of the β-Sn in figure 12 is for an assumed c/a ratio of 0.58 (this is the ratio obtained for BP and BAs[49]). To see whether a local energy minimum for the β-Sn structure would develop at compressed volume, we calculate the total energy for a wide range of c/a ratios at a volume equal to $0.45V_o$, where V_o is the equilibrium volume in the zincblende structure, see figure 13. The E_{tot} versus c/a ratio curve has no local minimum although there is some structure around c/a$\approx$0.55[74]. Similar behavior has also been found in diamond[55]. As shown previously for the case of silicon[75], the local minimum in the total energy curve corresponds to a local minimum in the Ewald energy around c/a=0.55. The Ewald energy is the electrostatic energy of a periodic array of point positive ions embedded in a compensating negative charge of uniform density. The Ewald energy is only one component of the total energy, see equation 1. If the other components are relatively insensitive to the c/a ratio then the total energy will follow the behavior of the Ewald energy. In Si where the electron charge density is relatively uniform, the Ewald energy is a good representation of the electrostatic energy and the behavior of the total energy follows that of the Ewald energy. In BN, the electron charge distribution is far from uniform. Figure 14 shows the charge density of BN in the β-Sn structure at $0.45V_o$ and c/a=0.58. Even at such highly compressed volumes where the electronic band structure is metallic, the charge density is still highly concentrated around the N atoms. The charge density contours for the rocksalt structure at the equilibrium volume of the zincblende structure, V_o, and at the transition volume, 0.42 V_o are shown in figure 15. At the transition volume, the charge density shows features typical of a rocksalt structure, where the large anions touch each other in a fcc lattice and the small cations fit in between.

A comparison between the phase diagrams of BN (figure 12) and C[73] reveals some clues concerning the relative stability of β-Sn and rocksalt. In carbon, the minimum energy of the β-Sn structure is 5.64 eV/pair above that of the diamond structure and the simple cubic is 5.32 eV/pair above that of the diamond[73]. In BN, the minimum energy of the β-Sn structure is 5.2 eV/pair above that of the zincblende structure but the rocksalt structure is only 3.4 eV/pair above that of the zincblende. It is interesting to note that the β-Sn energy with respect to the zincblende is almost the same in BN as in C; ionicity seems to have no effect. However, the rocksalt energy is relatively lower in BN than in C, indicating that ionicity plays a role as expected from conventional models. The difference in the rocksalt energies in BN and C accounts for the large difference in the predicted zincblende to rocksalt transition pressures. In BN, the calculated transition pressure is 11 Mbar while it is 23 Mbar in C. In view of the above comparison between BN and C, one may be tempted to hypothesize that the energy of the β-Sn with respect to that of the zincblende depends mainly on the symmetric part of potential and not the ionicity. However, this is not true for other III-V compounds, such as AlP and GaAs[76], where the ionicity also increases the β-Sn energy with respect to the zincblende. Let us focus on homopolar materials first. There seems to be a general trend that as the electron-ion potential increases the energies of both β-Sn and rocksalt increases with respect to that of zincblende. For example, E(β-Sn) - E(zb) = 0.50 eV/pair, 0.54 eV/pair, and 5.64eV/pair for Ge, Si, and C, respectively and E(rs) - E(zb) = 0.62 eV/pair, 0.70 eV/pair, and 3.52 eV/pair for Ge, Si, and C, respectively[73,75]. Initially, the rocksalt energy seems to increase fast than that of the β-Sn, but as the electron-ion potential gets very strong, the β-Sn

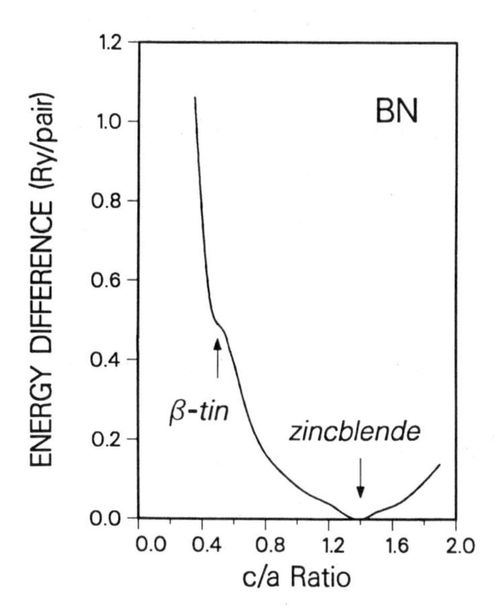

Fig. 13 - Calculated total energies as a function of the c/a ratio for BN assuming a primitive cell volume equal to $0.45V_o$.

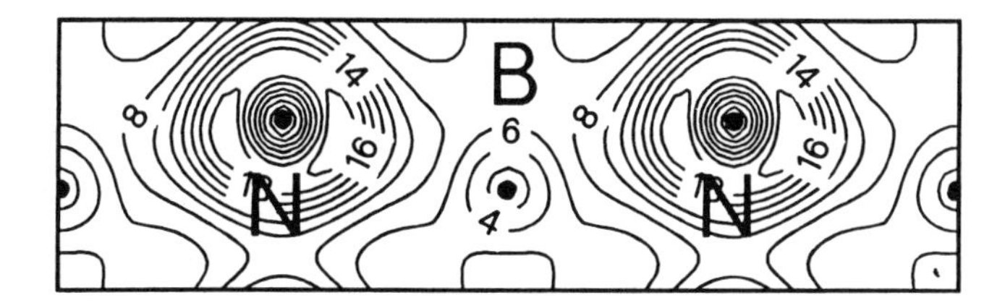

Fig. 14 - Valence charge density distribution for BN in the β-Sn structure assuming primitive cell volume equal to $0.45V_o$ and c/a ratio equal 0.58.

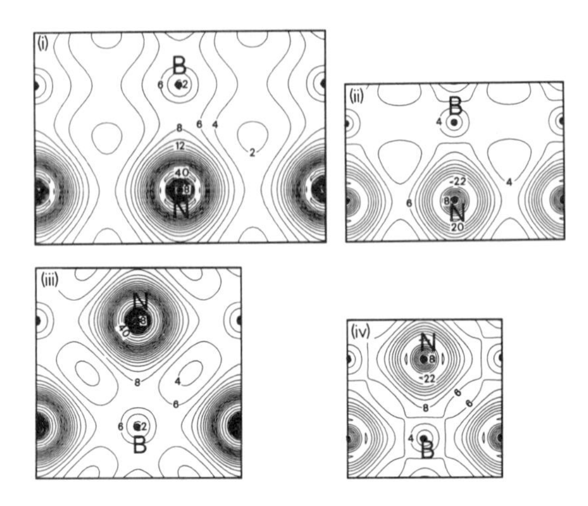

Fig. 15 - Charge density contours for BN in the rocksalt structure at the equilibrium volume of the zincblende, V_o, and at the transition volume, 0.42 V_o. Contours are shown in the (110) plane at V_o (i) and at 0.42 V_o (ii); (100) plane at V_o (iii) and at 0.42 V_o (iv). Units are in e/unit-cell and contours are in steps of 2 units.

energy increases faster than the rocksalt. This behavior could be caused by the fact that the β-Sn structure has a large d-character in its valence states and the energy level of the atomic d-state increases from Si to C. In the reverse direction, the energy of the β-Sn lowers toward that of the zincblende as the electron-ion potential gets weaker, such as from C to Sn. As mentioned in section VI, the electron-ion potential of a heteropolar (AB) compound can be written as a sum of two terms, a symmetric part, $V_s = (V_A + V_B)/2$ and an anti-symmetric part, $V_a = (V_A - V_B)/2$. The anti-symmetric potential indicates the degree of ionicity. We can investigate the effect of ionicity by comparing the III-V compounds with their isoelectronic group-IV, providing that the symmetric potential of the III-V is very similar to that of the group-IV. Generally the β-Sn energy increases as ionicity increases. For example, E(β-Sn) - E(zb) = 0.75 eV/pair for GaAs[76] as compared to 0.54 eV/pair for Ge[75]. An explanation for this is that the energy of the β-Sn is lower when the charge density is more uniform (the Ewald term), hence when ionicity introduces more inhomogeneity in the charge density, the β-Sn energy goes up. But, as indicated earlier this trend is not obeyed when comparing BN with C. Perhaps, it is because the charge density in C is already very non-uniform. Generally, the rocksalt energy decreases as ionicity increases. For example, E(rs) - E(zb) = 0.48 eV/pair for AlP[76] as compared to 0.70 eV/pair for Si[75] and 3.4 eV/pair for BN as compared to 5.32 eV/pair for C[73]. However, there seems to be exception to this trend also. For example, E(rs) - E(zb) = 0.68 eV/pair for GaAs[75] as compared to 0.62 eV/pair for Ge[75]. It is clear that it is not easy to establish an empirical rule governing the relative stability of structures. Some general trends are emerging but there are also exceptions. First-principles calculations, which give accurate microscopic description of structural properties, have helped to improve our understanding. However, much more work still needs to be done.

VIII. SUMMARY

Various electronic properties and structural transformations of BN are analyzed by calculating the structural energy using a first-principles total energy method. We investigate two transformation paths which bring the layered phases (*hBN* and *rBN*) to the four-folded phases (*wBN* and *zBN*) without altering the structural symmetry. An activation barrier of $\approx$ 0.39 eV/pair is predicted for the both transformations. The general conclusions concerning the transformations drawn from the calculated energetics agree well with experimental observations. However, much more work, both experimental and theoretical, is needed to clarify some issues. For example, one conclusion from the our calculated result is the possibility of the zincblende phase of BN being stabler than the hexagonal phase. More accurate thermodynamical data and detailed information of the phonon modes are needed. We also present results for the electronic band structure, the structural properties, and the electronic charge density distribution of the zincblende phase. Good agreement with available experimental data is obtained. Finally, some possible structural transformations of BN under very high pressure are also examined. We find that the rocksalt structure is preferred over the β-Sn.

ACKNOWLEDGEMENTS

PKL would like to acknowledge the San Diego Supercomputer Center for providing computing resources. RMW and MLC were supported by National Science Foundation Grant No. DMR8818404 and by the Director, Office of Energy Research, Office of Basic Energy Sciences, Materials Sciences Division of the U.S. Department of Energy under contract No. DE-AC03-76SF00098.

REFERENCES

(1) Golikova, O. A.: Phys. Status Solid(a); 1979, 51, 11

(2) Bornstëin, Landolt: "Numerical Data and Functional Relationships in Science and Technology" - Crystal and Solid State Physics, vol. III (Springer, Berlin, 1972).

(3) Pease, R. S.: Acta Cryst., 1952, 5, ~~536~~ 356

(4) Wentorf, R. H., Jr.: J. Chem. Phys., 1957, 26, 956

(5) Bundy, F. P. and R. H. Wentorf, Jr.: J. Chem. Phys., 1963, 38, 1144

(6) Wakatsuki, M., K. Ichinose, and T. Aoki: Mat. Res. Bull., 1972, 7, 999

(7) Dulin, I. N., L. V. Al'tshuler, V. Ya Vashchenko, and V. N. Zubarev: Sov. Phys. Sol. State, 1969, 11, 1016

(8) Soma, T., A. Sawaoka, and S. Saito: Mat. Res. Bull., 1974, 7, 755

(9) Sato, T., T. Ishii, and N. Setaka: J. Amer. Cer. Soc., 1982, C-162
(10) Wentzcovitch, R. M., K. J. Chang, and M. L. Cohen: Phys. Rev. B, 1986, 34, 1071
(11) Fahy, S., K. J. Chang, S. G. Louie, and M. L. Cohen: Phys. Rev. B, 1987, 35, 5856
(12) Chang, K. J., S. Froyen, and M. L. Cohen: Solid Sate Commun., 1984, 50, 105
(13) Zunger, A. and A. J. Freeman: Phys. Rev. B, 1978, 17, 2030
(14) Kleinman, L. and J. C. Phillips: Phys. Rev., 1960, 117, 460
(15) Stocker, D.: Proc. R. Soc. Lond., 1962, 270, 397
(16) Bassani, F. and M. Yoshimine: Phys. Rev., 1963, 130, 20
(17) Wiff, D. R. and R. Keown: J. Chem. Phys., 1967, 47, 3133
(18) Phillips, J. C.: J. Chem. Phys., 1968, 48, 5740
(19) Hemstreet, L. A. and C. Y. Fong: Phys. Rev. B, 1972, 6, 1464
(20) Hwang, H. C. and J. Henkel: Phys. Rev. B, 1978, 17, 4100
(21) Tsay, Y. F., A. Vaidynanathan, and S. S. Mitra: Phys. Rev. B, 1979, 19, 5422
(22) Dovesi, R., C. Pisani, C. Roetti, and P. Dellarole: Phys. Rev. B, 1981, 24, 4170
(23) Huang, M. Z. and W. Y. Ching: J. Phys. Chem. Solids, 1985, 46(8), 977
(24) Cohen, M. L.: Phys. Scr. T, 1982, 1, 5
(25) Hohenberg, D. and W. Kohn: Phys. Rev., 1964, 136, B864; W. Kohn and L. J. Sham, Phys. Rev., 1965, 140, A1133
(26) Wigner, E.: Trans. Faraday Soc., 1938 34, 678
(27) Hamann, D. R., M. Schlüter, and C. Chiang: Phys. Rev. Lett., 1979, 43, 1494
(28) Bendt, P. and A. Zunger: Solar Energy Research Institute Tech. Rep. TP-212-1698.
(29) Chelikowsky, J. R. and S. G. Louie: Phys. Rev. B, 1984, 29, 3470
(30) Chan, C. T., D. Vanderbilt, and S. G. Louie: Phys. Rev. B, 1986, 33, 2455; *ibid*, 1986, 34, 8791
(31) Murnaghan, F. D.: Proc. Nat. Acad. Sci. U.S.A., 1944, 30, 244
(32) Birch, F.: J. Geophys. Res., 1978, 83, 1257
(33) Bundy, F. P., H. T. Hall, H. M. Strong, and R. H. Wentorf, Jr.: Nature (London), 1955, 176, 51; for a review on high pressure properties of graphite see Clarke, R. and C. Uher, Adv. in Phys.: 1984, 33, 469
(34) Kertesz M. and R. Hoffman: J. Solid State Chem., 1984, 54, 313
(35) Fahy, S., S. G. Louie, and M. L. Cohen: Phys. Rev. B., 1986, 34, 1191
(36) Fahy, S., S. G. Louie, and M. L. Cohen: Phys. Rev. B, 1987, 35, 7623
(37) Ishii, T., T. Sato, Y. Sekikawa, and M. Iwata: J. Cryst. Growth, 1981, 52, 285
(38) Bundy F. P. and J. S. Kasper: J. Chem. Phys., 1967, 46, 3437
(39) Riter, J. R. Jr.: J. Chem. Phys., 1973, 59, 1538
(40) These values differ slightly in the wurtzite structure.
(41) Warren, J. L., J. L. Yarnell, G. Dolling, and R. A. Cowley: Phys. Rev., 1967, 158, 805
(42) Nicklow, R. M., N. Wakabayashi, and H. G. Smith: Phys. Rev. B, 1972, 5, 4951
(43) Yin M. T. and M. L. Cohen: Phys. Rev. B, 1984, 29, 6996
(44) Sanjurjo, J. A., E. Lopez-Cruz, P. Vogl, and M. Cardona: Phys. Rev. B, 1983, 28, 4579
(45) Geick, R., C. H. Perry, and G. Rupprecht: Phys. Rev., 1966, 146, 543

(46) Dworkin, A. S., D. J. Sasmor, and E. R. Van Artsdalen: Phys. Rev., 1954, 22, 837
(47) Sirota, N. N. and N. A. Kofman: Sov. Phys. Dokl., 1976, 21, 516
(48) Sirota, N. N. and N. A. Kofman: Sov. Phys. Dokl, 1979, 24(12), 1001
(49) Wentzcovitch, R. M., M. L. Cohen, and P. K. Lam: Phys. Rev. B, 1987, 38, 6058
(50) Philipp H. R. and E. A. Taft: Phys. Rev., 1962, 127, 159
(51) Fomichev, V. A. and M. A. Rumsh: J. Phys. Chem. Solids, 1968, 29, 1015
(52) Chrenko, R. M.: Solid State Commun., 1974, 14, 511
(53) Hybertsen, M. S. and S. G. Louie: Phys. Rev. Lett., 1985, 55, 1418
(54) Park, K. T., K. Terakura, and N. Hamada: J. Phys. C, 1987, 20, 1241
(55) Fahy S. and S. G. Louie: Phys. Rev. B, 1987, 36, 3373
(56) Lam, P. K. and M. L. Cohen: Phys. Rev. B, 1983, 27, 5986
(57) Wentzcovitch, R. M., S. L. Richardson, and M. L. Cohen: Phys. Lett. A, 1986, 114, 203
(58) Chang K. J. and M. L. Cohen: Solid Sate Commun., 1984, 50, 486
(59) Cohen, M. L.: Phys. Rev. B, 1985, 32, 7988
(60) Lam, P. K., M. L. Cohen, and G. Martinez: Phys. Rev. B, 1987, 35, 9190
(61) Knittle, E., R. M. Wentzcovitch, R. Jeanloz, and M. L. Cohen: Nature (in press)
(62) Devries, R. C.:*GE Report# 72CRD* 178, 1978
(63) McQueen, R. G., S. P. Marsh, J. W. Taylor, J. N. Fritz, and W. J. Carter: in *High–VelocityImpactPhenomena*, 1970, R. Kinslow, ed., Academic Press, Orlando, Fl., 294-419
(64) Hedin L. and B. I. Lundqvist: J. Phys. C, 1971, 4, 2064
(65) Louie, S. G., S. Froyen, and M. L. Cohen: Phys. Rev. B, 1982, 26, 1738
(66) Wagman, D. D., W. H. Evans, I. Halow, V. B. Parker, S. M. Baley, and R. H. Shumm: " National Bureau of Standards"- Technical Note 270, United States Government Printing Office, Washington D. C. 176 (1968).
(67) Kittel, C.: "Introduction to Solid State Physics", 5th over ed., John Wiley and Sons, New York, 74 (1976).
(68) Lam, P. K. and M. L. Cohen: Phys. Rev. B., 1982, 25, 6139
(69) Gielisse, P. J., S. S. Mitra, J. N. Pendl, R. D. Griffis, L. C. Mansur, R. Marshall, and Pescoe: Phys. Rev. B, 1967, 155(3), 1039
(70) Phillips, J. C.: "Bonds and Bands in Semiconductors", Academic Press, New York (1973); Van Vechten J. A. and J. C. Phillips, Phys. Rev. B, 1970, 2160
(71) Chelikowsky, J. R.: Phys. Rev. Lett., 1986, 56, 961
(72) Biswas, R., R. Martin, R. J. Needs, and O. H. Nielsen: Phys. Rev. B, 1984, 30, 3210
(73) Yin M. T. and M. L. Cohen: Phys. Rev. Lett., 1983, 50, 2006
(74) The existence of a local minimum in the total energy also depends on the number of k-points used to perform the summations over the Brillouin Zone. For sets containing more than 60 k-points, the local minimum disappears.
(75) Yin, M. T. and M. L. Cohen: Phys. Rev. B, 1982, 26, 5668
(76) Froyen, S. and M. L. Cohen: Phys. Rev. B, 1983, 28, 3258

Materials Science Forum Vols. 54 & 55 (1990) pp. 193-206

THE MICROSTRUCTURE OF BORON NITRIDE THIN FILMS

D.R. McKenzie (a), W.G. Sainty (b) and D. Green (a)
(a) School of Physics, University of Sydney, Sydney NSW 2006, Australia
(b) C.S.I.R.O., Division of Applied Physics, NSW 2070, Australia

ABSTRACT

Boron nitride has three principal crystallographic forms: hexagonal, diamond cubic and wurtzite. The determination of the actual structure occurring in a particular thin film specimen is discussed. Diffraction methods must be used with caution owing to a coincidence in the positions and intensities of the principal maxima of the diffraction patterns. The determination of the Radial Density Function (RDF) allows a resolution of the ambiguity. Methods based on electron band structure as probed by electron energy loss and optical spectroscopies are also of value. Methods based on lattice properties such as infrared and Raman spectroscopies are useful to distinguish the hexagonal from the other forms.

INTRODUCTION

Boron nitride, BN, in the crystalline state has two principal forms [1]: hexagonal, which we shall term h-BN, and cubic, c-BN. Boron nitride also crystallises in the wurtzite structure [2]. Hexagonal boron nitride is prepared as a greasy white solid with a density of 2.25×10^3 kg m^{-3} by heating elemental boron in an ammonia atmosphere. The structure of the three principal forms of BN are shown in figure 1.

The structure of h-BN is related to that of graphite in that it consists of sheets of atoms in three-fold coordination. The bonding orbitals of both atoms are sp^2 hybrids, which may be thought of as arising from the resonance transfer of an electron from the N atom to the B atom, giving both atoms the electronic configuration of carbon in graphite. The stacking of the layers in h-BN, however, is different to that in ordinary graphite, with unlike atoms in adjoining layers directly overlying each other. The physical properties of h-BN and graphite, however, are not similar since h-BN is non-absorbing at visible wavelengths and is electrically insulating, whereas graphite is almost metallic in properties. The differences arise from differences in the electronic band structures. In h-BN no overlap occurs between π bands [3] in contrast to the situation in graphite [4]. In fact h-BN has a direct gap of 5.8ev which makes it a good insulator.

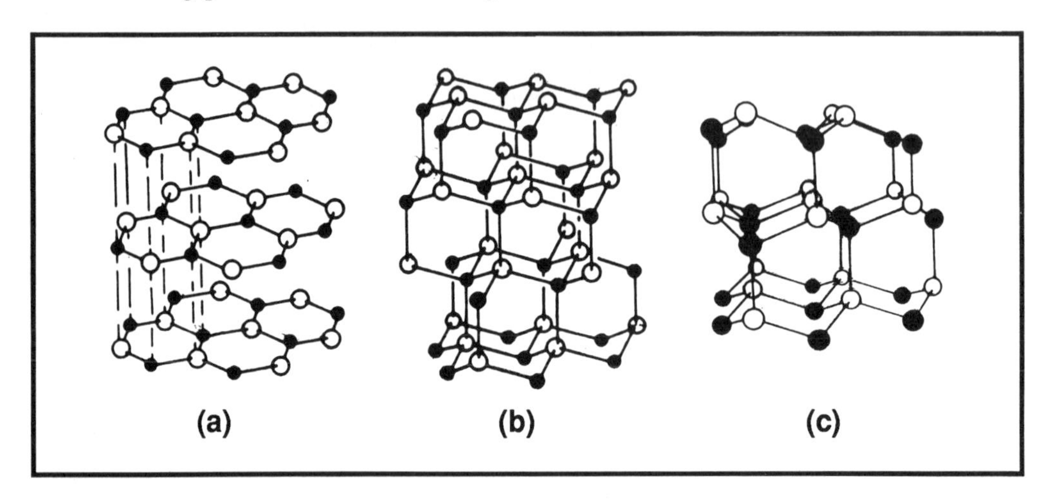

FIGURE 1 The structures of boron nitride, (a) hexagonal, (b) cubic, (c) wurtzite.

Cubic boron nitride may be prepared from the hexagonal form by heating h-BN with catalysts at high pressures [5]. This form has zinc blende or diamond strucutre (see figure 1), has a hardness comparable to diamond and a considerably higher density than h-BN, of 3.48×10^3 Kg m^{-3}. The bonding orbitals in c-BN are the tetrahedrally directed sp^3 hydrids as in diamond.

When deposited as a thin film, boron nitride has the potential to form material with the h-BN or c-BN structures or possibly the wurtzite structure; or alternatively an amorphous structure having the corresponding local structure of these materials, that is, coplanar sp^2 or tetrahedral sp^3 bonding. The techniques for determining the atomic arrangement applying to a particular specimen may be classified according to whether they are direct, that is, they attempt to measure atom positions directly, or indirect, where they use a property of the structure to identify its presence. The direct methods we shall discuss are essentially diffraction methods, while indirect methods rely on either a lattice or electronic property.

DIFFRACTION METHODS FOR MICROSTRUCTURE DETERMINATION

Thin film specimens rarely consist of parallel-sided single crystals; usually they are a collection of small crystals with a distribution of sizes and orientations. Alternatively, they show an amorphous microstructure without true crystalline regions. Such polycrystalline or amorphous structures are conveniently described by a reduced density function (RDF), which describes the deviation from the average density with distance from the centre of an atom.

An average RDF, G(r), is readily obtained by Fourier sine transformation of the measured intensity I(s) obtained by neutron, x-ray or electron diffraction.

$$G(r) = 8\pi \int_0^{s_{max}} \frac{I(s)-N\langle f^2\rangle}{N\langle f\rangle^2}\, s\, e^{-\beta s^2} \sin(2\pi sr)\, ds \qquad ...(1)$$

where f is the atomic form factor, N is a scale factor, s is the scattering variable($s = \frac{2 \sin\theta}{\lambda}$), β is an artificial temperature factor, and $\langle\,\rangle$ represents an average with respect to elemental composition. Measurements are made over a range of s out to some maximum value, s_{max}.

Structural information sufficient to distinguish c-BN from h-BN is obtained by measurement of G(r) alone, without evaluation of the partial RDFs for each atom type.

Electron diffraction is especially convenient for measurements of thin films consisting of light elements such as boron nitride and the reader is referred to a recent paper [6] for a discussion of current practice in electron diffraction from polycrystalline and amorphous thin films. The electron diffraction pattern may now conveniently be acquired in a transmission electron microscope fitted with post specimen scan coils and an electron energy loss spectrometer. The method is superior to the still commonly used photographic recording of electron diffraction patterns in the accuracy to which the variable s is determined. Low accuracy in s leads to special difficulties in the use of the raw electron diffraction intensity, I(s), in the identification of the phases of BN. Like graphite, h-BN often shows a tendency to disorder in the stacking of the layers, so that diffraction from crystal planes of the type (hkl) with $l > 0$ are often weak or absent. Such structures are referred to as 'turbostratic'. This results in a marked tendency of h-BN to mimic the diffraction pattern of c-BN. The difficulty is illustrated in figure 2 in which the diffraction intensity has been calculated for specimens consisting of randomly oriented crystallites for both h-BN and c-BN. In the case of h-BN, single layers containing 72 atoms have been used and for c-BN roughly cubic crystallites containing 216 atoms have been used. The principal maxima of both diffraction patterns are quite similar in shape and differ only slightly in position. Further confusion arises if there is any degree of preferred orientation. If the position of the first diffraction ring only is used for identification, then as can be seen from figure 2(a), there is a precise overlap and it is impossible to distinguish h-BN from c-BN. Further out in the diffraction pattern (see figure 2(b)), none of the major features differ by more than 3% in position.

The situation is much improved if the diffraction intensity is transformed to an RDF: figure 3 shows the G(r) functions resulting from the complete sets of data of figure 2 out to $s_{max} = 5.0\ A^{-1}$.

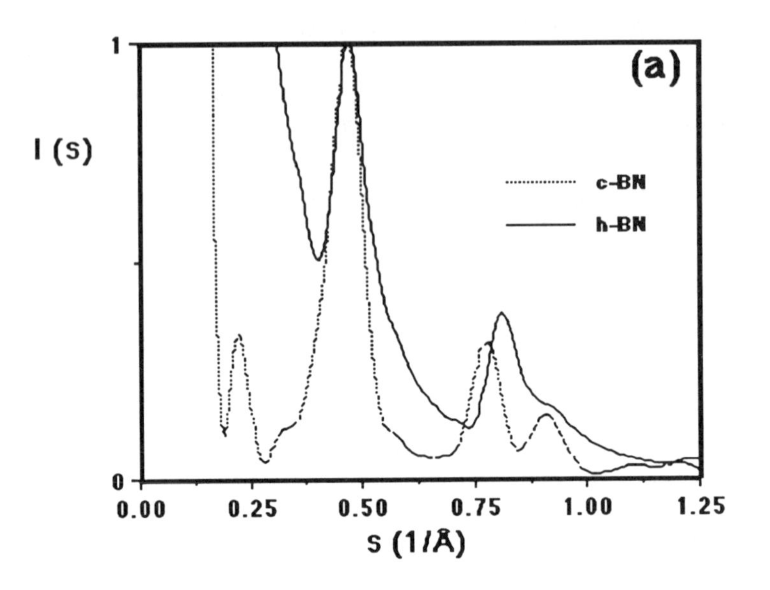

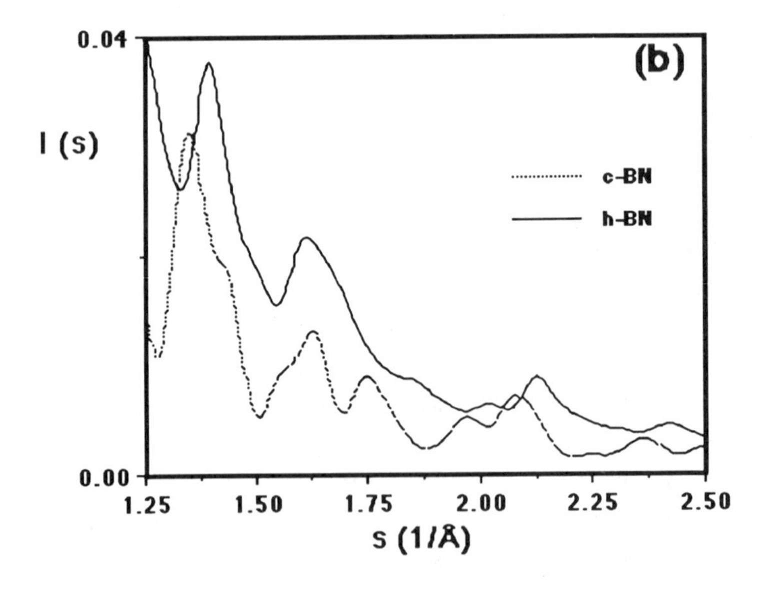

FIGURE 2 The calculated diffraction intensity for a single layer of the h-BN structure containing 72 atoms compared with that of an approximately cubical volume containing 216 atoms of the c-BN structure. The plots (a) and (b) are different sections of the same graph.

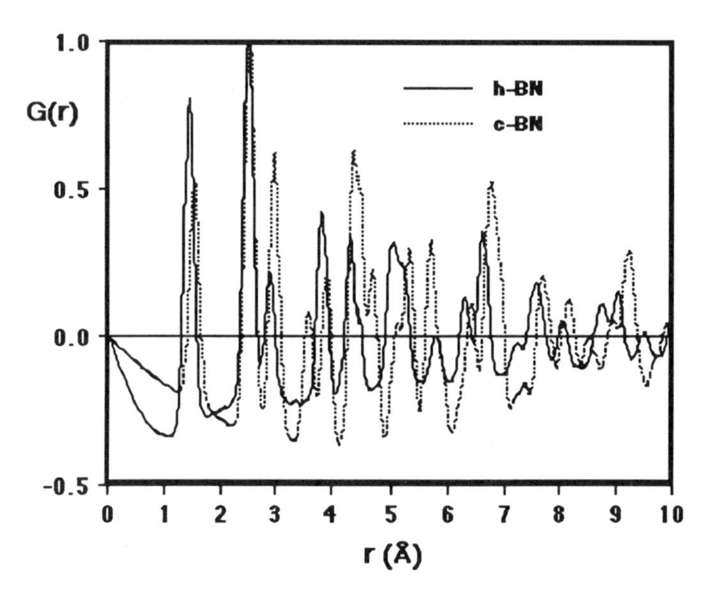

FIGURE 3 The G(r) functions resulting from Fourier transformation of the intensities of figure 2 according to equation 1. Small angle scattering has been sensibly removed.

While reasonably close coincidences in position occur for the first three peaks, the relative intensity of the second and third peaks is very different in the two structures. Even more useful for distinguishing h-BN from c-BN is the G(r) function beyond the third peak. c-BN shows two doublet peaks while h-BN shows only two single peaks, the first of which does not coincide with any c-BN peak.

CASE STUDY 1 : ION BEAM SYNTHESISED BN FILMS

Boron nitride of stoichiometric composition may be produced in thin film form by evaporation of boron in vacuum and bombardment of the growing film with nitrogen ions [7].

The films have a density of $(2.27 \pm 0.05) \times 10^3$ Kg m^{-3} and a Vickers hardness of greater than 1900 Hv (10g loading). The G(r) function for a typical film is shown in figure 4 and is compared to the theoretical result for a single layer of h-BN. The excellent fit is convincing evidence that the observed structure is indeed h-BN and that the structure is turbostratic.

Additional information was obtained by high resolution imaging of the films using a 300kV electron microscope. The micrograph of figure 5 shows that the film consists of ribbon-like structures containing h-BN layers. The correlation of atom positions between layers must be poor given the large number of regions of small radius of curvature which will cause displacement of one layer relative to its neighbours. In addition, edge dislocations are common and will also lead to errors

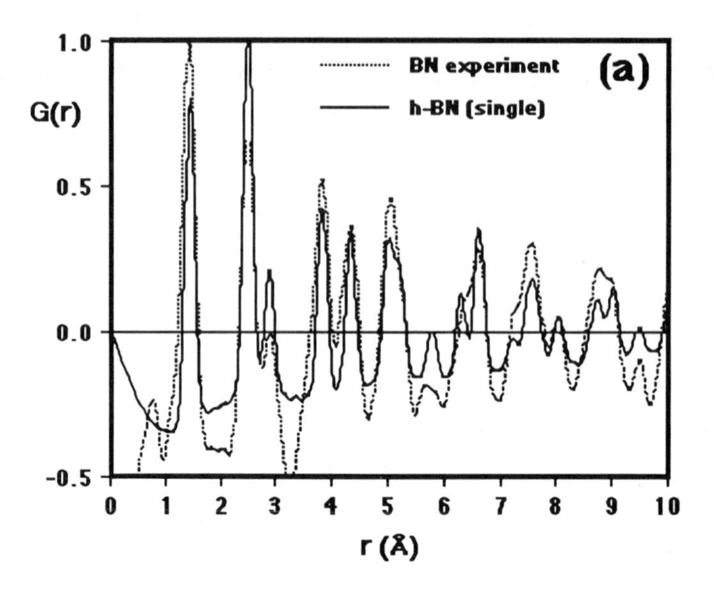

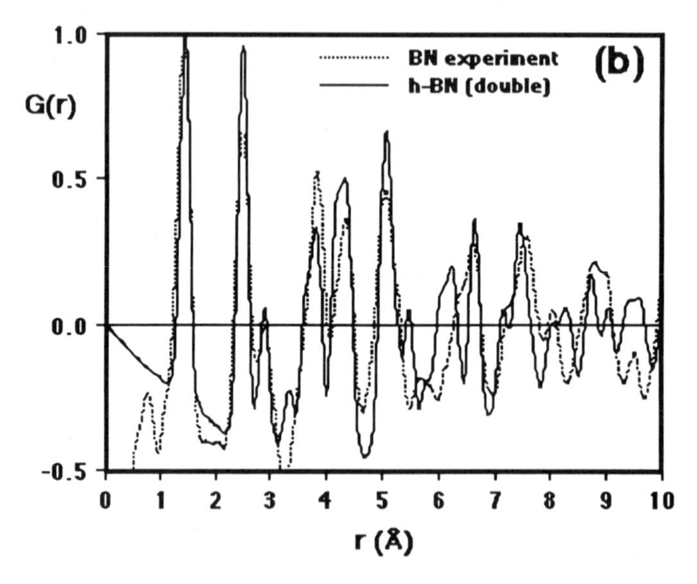

FIGURE 4 (a) shows the experimental G(r) for a BN film prepared by ion beam synthesis[7] compared with the G(r) calculated (see figure 3) for a single layer of h-BN containing 72 atoms.

(b) shows the same experimental G(r) compared with the G(r) calculated for an hexagonal prism consisting of two layers of the h-BN containing 144 atoms.

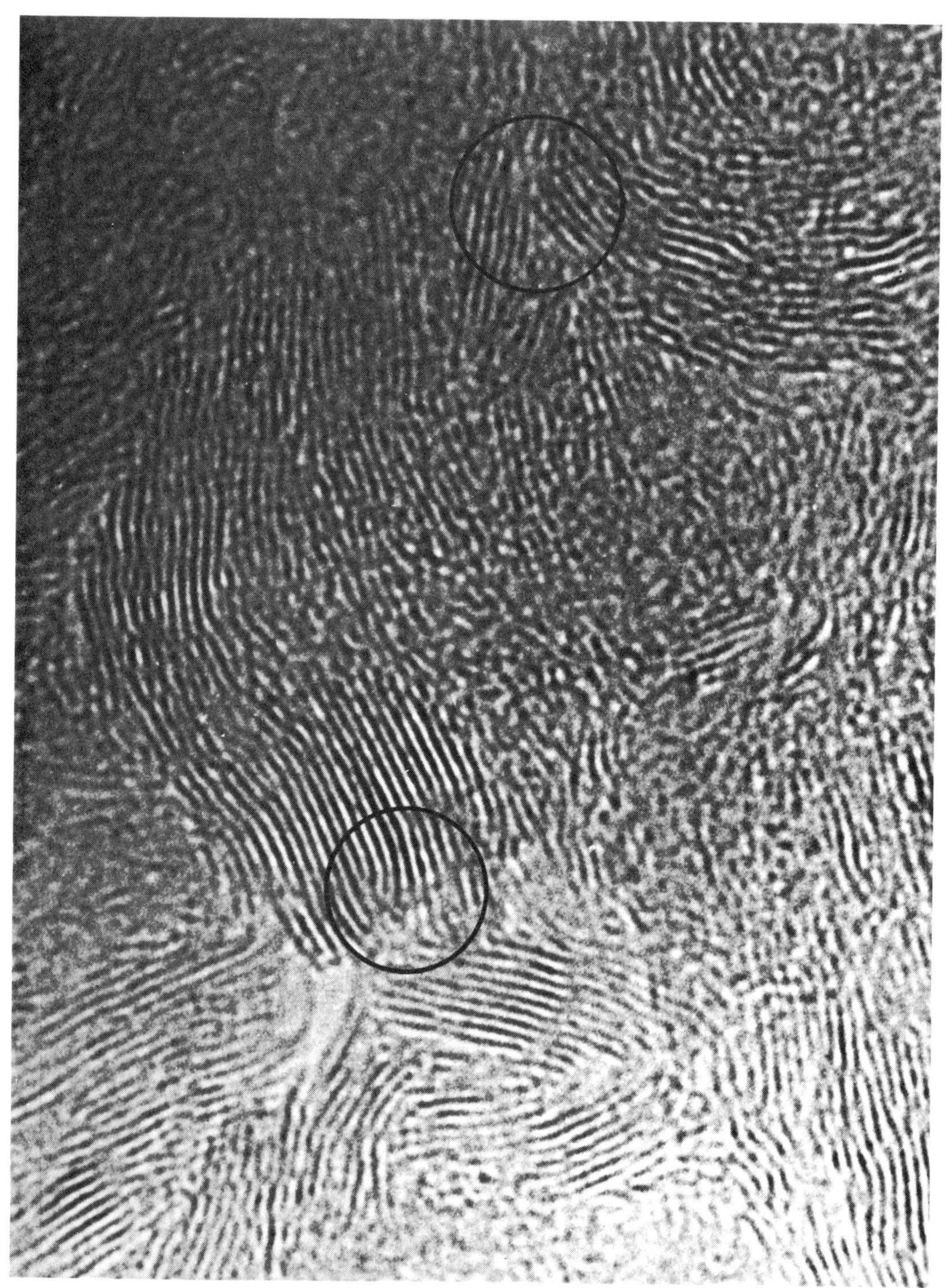

FIGURE 5 A high resolution micrograph of a BN film prepared by ion beam synthesis shows well developed parallel fringes corresponding to the 3.3Å spacing of h-BN layers. Strong curvature and edge dislocations (circled) are indicative of turbostratic structure.

LATTICE PROPERTY MEASUREMENTS FOR MICROSTRUCTURE DETERMINATION

The lattice properties of materials can be used as 'fingerprints' of the microstructure in order to provide supporting evidence for the identification of phases. The most convenient of these techniques is the measurement of the frequencies of particular modes of vibration of the lattice at the centre of the Brillouin Zone as determined by infrared and Raman spectroscopies. Infrared reflectance measurements [8] show that c-BN has a zone-centre transverse optical mode frequency of $1065cm^{-1}$ and a zone-centre longitudinal optical mode frequency of $1340cm^{-1}$. The infrared transmission spectrum of thin films of c-BN will therefore show a single absorption maximum at $1065cm^{-1}$ if single phonon processes are dominant.

Hexagonal boron nitride has two strong infrared-active zone-centre vibrational modes, [9] the in-plane transverse optical mode at $1367cm^{-1}$ and the out-of-plane transverse optical mode at $783cm^{-1}$. The mode at $1367cm^{-1}$ is also Raman active. The infrared transmission spectrum of h-BN will therefore show two absorption maxima at $1367cm^{-1}$ and $783cm^{-1}$. On the basis of the two quite distinct infrared absorption spectra for c-BN and h-BN, useful microstructural conclusions can be drawn. For example, let us look at a case study.

CASE STUDY 2 : PLASMA CVD PREPARED BN

Chayahara et al [10] have reported on the properties of BN films grown by ECR plasma CVD of mixtures of diborone, B_2H_6 and nitrogen, N_2. RF plasma excitation was used to introduce a negative bias of the substrate with respect to the plasma, causing ion bombardment of the growing film. When the magnitude of the negative bias was 200V or greater, the infrared spectrum developed an absorption peak at approximately $1070cm^{-1}$ characteristic of c-BN, whereas at smaller bias values the principal absorptions occurred at $1370cm^{-1}$ and $800cm^{-1}$, approximately the characteristic frequences for h-BN. These results imply that the growth of h-BN in the deposited film is inhibited by selective sputtering of the hexagonal phase under the ion impact induced by the self bias.

The three forms of BN have different IR spectra. Figure 6 shows the results obtained for the wurtzite form of BN produced by laser pulse vapour deposition [2] together with spectra for c-BN and h-BN produced by activated reactive evaporation at different RF substrate biases [11].

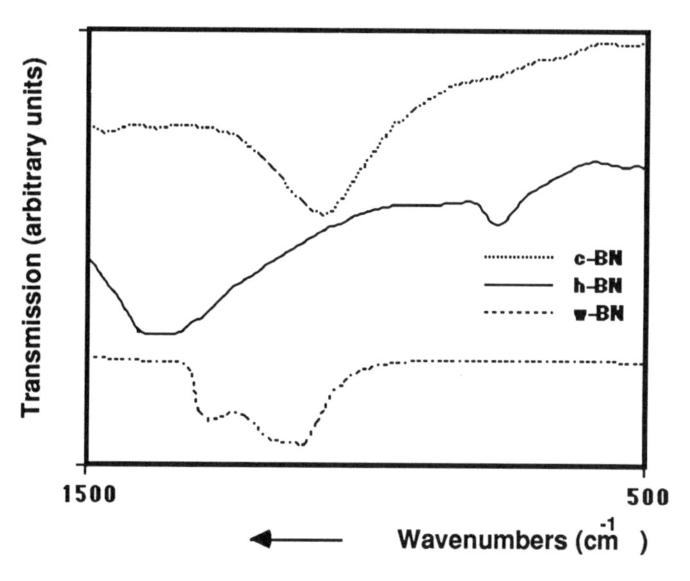

FIGURE 6 Parts of the infrared absorption spectra for the three forms of BN.

METHODS BASED ON ELECTRONIC STATES

The electronic energy levels of a solid are determined both by the nature of the atoms and the structure of the solid. In the case of h-BN and c-BN, the distribution of these levels is sufficiently different to enable a good discrimination to be obtained by methods which probe the electronic energy level structure. We shall discuss two methods, the first in which the probe is a photon, and the second in which it is an electron.

The absorption of light by a material is described by the (complex) dielectric response function $\varepsilon = \varepsilon_1 + i\,\varepsilon_2$, or equally well, by the complex refractive index $\mathbf{n} = n + ik$. These response functions may be determined over the visible and UV regions by photometry or ellipsometry. The absorption of light in the visible and UV by boron nitride causes electronic transitions from occupied to unoccupied energy levels. The lowest energy transitions in h-BN are those in which the electron moves from the π band to the unoccupied π^* band [3]. This gives rise to the maxima in ε_2 at 5.2eV and 6.2eV. Since the lowest observed transition in c-BN is at 6.4eV [12] and the onset of strong absorption does not occur until 8eV [13], the $\pi \rightarrow \pi^*$ transition in h-BN is a good indicator of the presence of h-BN.

For thin film specimens, the measurement of ε is conveniently done by means of electron energy loss spectroscopy. The energy loss spectrum is, under appropriate conditions, proportional to $-\operatorname{Im}(1/\varepsilon) = \varepsilon_2/|\varepsilon|$. The complex dielectric function ε may be determined with reasonable precision over a wide energy range by means of an algorithm [14] based on the Kramers-Krönig relation.

However, since a local maximum in ε_2 will lead to a local maximum in $\varepsilon_2/|\varepsilon|$ provided that $|\varepsilon|$ is slowly varying over the same region, it is sufficient to examine the energy loss function directly for the presence of a peak corresponding to the $\pi \rightarrow \pi^*$ transition. Figure 7 shows the results obtained for the h-BN film of Case Study 1 which exhibits a feature corresponding to this transition. The spectrum for c-BN over the same region is also shown for comparison and does not show a peak in this region. Both spectra were obtained using Gatan EELS spectrometer fitted to an electron microscope operating at 100kV.

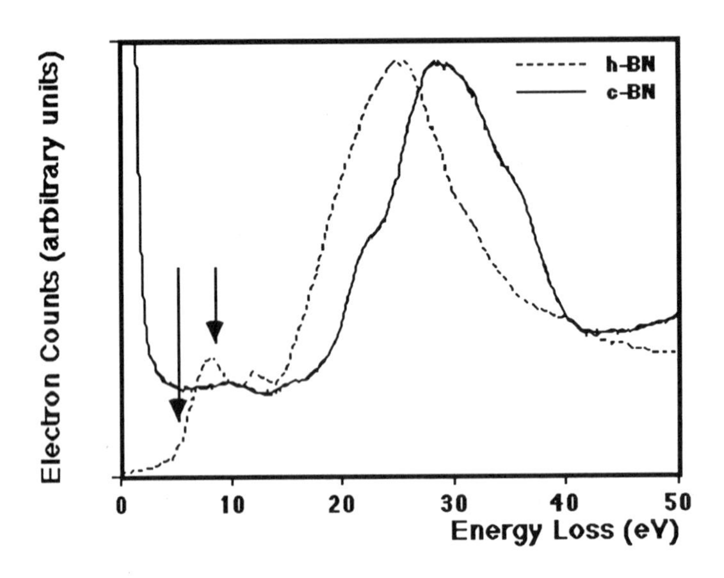

FIGURE 7 Electron energy loss spectra in the plasmon region of a h-BN film and c-BN crystals are compared The onset of absorption is shown by an arrow in each case, for h-BN at 6 eV and for c-BN at 9 eV.

The energy loss spectrum contains additional information of value in characterising a thin film. The dominant maximum in this spectrum is due to the excitation known as a 'plasmon', a longitudinal oscillation of the valence electrons as a whole against the atom cores. The peak occurs when $|\varepsilon|$ passes through a minimum, and under the approximation of free electron model [15] this frequency is

$$\omega_p = \sqrt{\frac{Ne^2}{\varepsilon_0 m}} \quad ...(2)$$

where N is the number density of valence electrons in the specimen. For both h-BN and c-BN there are four such electrons per atom, so that a measurement of ω_p gives an estimate of the density of the specimen. Since the densities of h-BN and c-BN are in the ratio 1.55, a good discrimination between them should be possible on the basis of the plasmon frequency. The peak plasmon energy positions in figure 7 are at 25.5eV and 28.5eV, respectively, so that a clear distinction is indeed possible. In addition to the different plasmon peak positions, the onset of absorption at low energies can be distinguished as shown in the figure.

The energy loss spectrum at higher energy losses, in the vicinity of the boron and nitrogen K-edge features also contains useful microstructural information. The boron edges for a h-BN film prepared by ion beam synthesis are shown in figure 8.

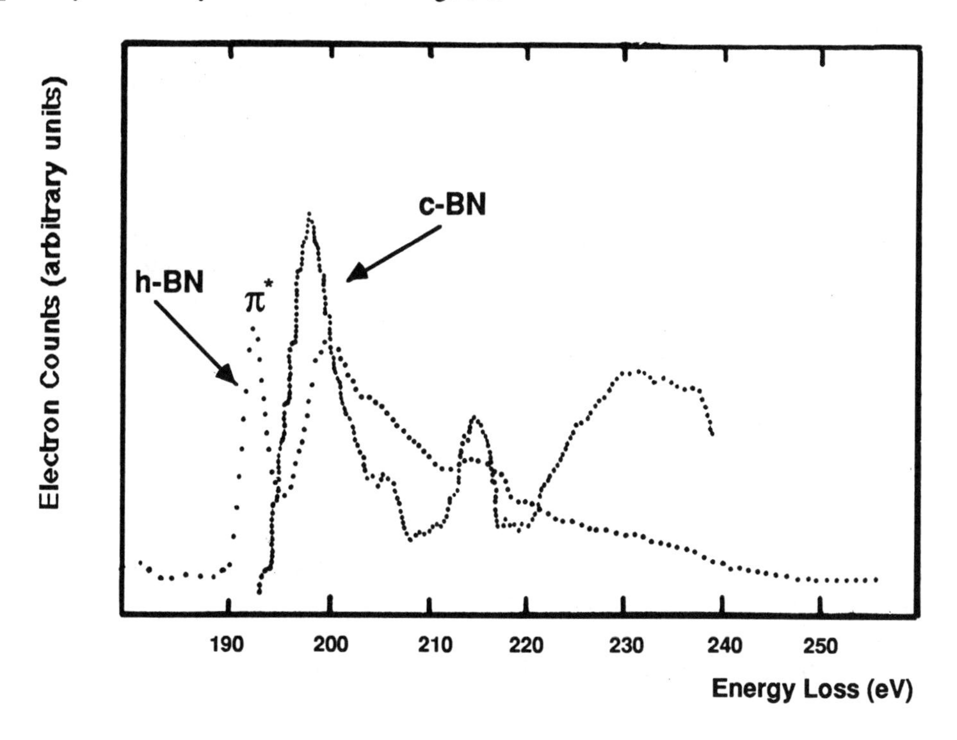

FIGURE 8 Electron energy loss spectrum in the vicinity of the K-edge of boron. The characteristic signature the h-BN structure is shown.

The graph, in fact, shows the detailed variation of the cross section for a transition of a K-shell electron to an unoccupied level and reflects the density of electronic states in the unoccupied bands. One such band in h-BN is highly characteristic, namely the relatively narrow band labelled π^* representing the unoccupied anti-bonding level of the π electronic states for atoms in sp^2 hybridization. The fact that both B and N (see figure 9) atoms show such a feature confirms the view that both are in sp^2 hybridization as the analogy with graphite suggests. c-BN, on the other hand, shows no trace of the π^* feature on the B (see figure 8) or N (not shown) edges . The π^* edge feature is therefore an excellent indicator of the presence of a h-BN phase.

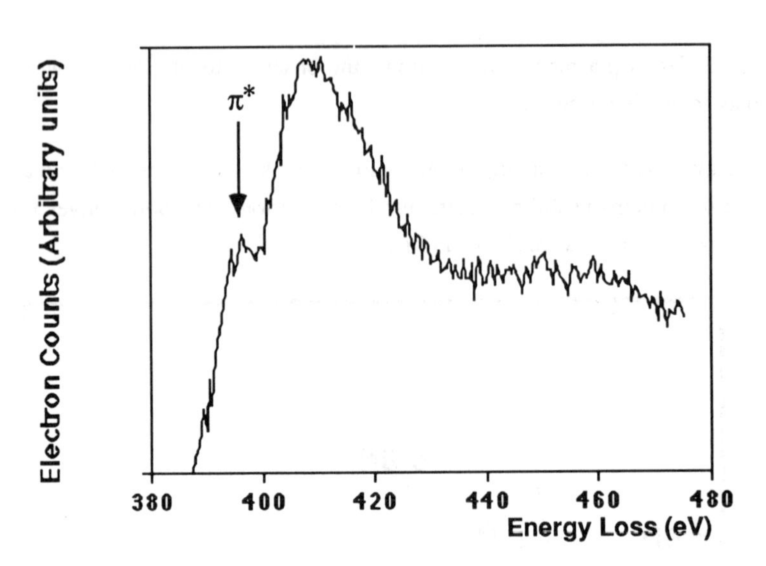

FIGURE 9 Electron energy loss spectrum in the vicinity of the nitrogen K-edge for ion beam synthesised h-BN.

CONCLUSIONS

We have seen that the diffraction pattern alone without further analysis does not provide sufficient evidence to determine the microstructure of a boron nitride thin film. Other methods should also be used to confirm any hypothesis.

If a definitive statement is required concerning the presence of cubic boron nitride then we recommend that all of the techniques discussed here be used.

REFERENCES

1) Wyckoff, R.W.G.: Crystal Structures Volume 1. 2nd Edition., John Wiley, New York, p. 184 (1963).
2) Kessler, G., Bauer, H.D., Pompe, W. and Scheibe, H.J.: Thin Solid Films, 147, L45 (1987).
3) Zunger, A., Katzier, A. and Halperin, A.: Phys. Rev. B13, 5560 (1976).
4) Buchner, U.: Phys. Stat. Sol. (b) 81, 227 (1977).
5) Wentorf, R.H.: J. Chem. Phys. 34, 809 (1961).
6) Cockayne, D.J.H. and McKenzie, D.R.: Acta Cryst. A44, 870 (1988).
7) Sainty, W.G., Martin, P.J., Netterfield, R.P., McKenzie, D.R., Cockayne, D.J.H. and Dwarte, D.M.: J. App. Phys., 64, 3980 (1988).

8) Gielisse, P.J., Mitra, S.S., Plendl, J.N., Griffis, R.D., Mansur, L.C., Marshall, R. and Pascoe, E.A.: Phys. Rev. B155, 1039 (1967).
9) Geick, R., Perry, C.H. and Rupprecht, G.: Phys. Rev. 146, 543 (1966).
10) Charahara, A., Yokoyama, H., Imura, T., Osaka Y. and Fujisawa M.: Appl. Surf. Sci. 33/34, 561 (1988).
11) Inagawa, K., Watanabe, K., Ohsone, H., Saitoh, K. and Itoh, A.: J.Vac. Sci. and Tech., A5 (4) 2696.
12) Chrenko, R.M.: Solid State Commun. 14, 511 (1974).
13) Tsay, Y.F., Vaidyanathan, A. and Mitra, S.S.: Phys. Rev. B19, 5422 (1979).
14) White, S.B. and McKenzie, D.R.: J. Electron Spect. Relat. Phenom., 43, 53 (1987).
15) Egerton, R.F.: Electron Energy Loss Spectroscopy in the Electron Microscope, Plenum Press, New York, (1986), p. 159.

Materials Science Forum Vols. 54 & 55 (1990) pp. 207-228

LOW ENERGY AUGER TRANSITIONS OF h-BN AND c-BN: AN EXAMPLE OF THE INFLUENCE OF CRYSTALLOGRAPHIC STRUCTURE ON THE AUGER DECAY

G. Hanke (a), M. Kramer (b)* and K. Müller (b)

(a) FAG Kugelfischer KGaA, Dept. of Development and Construction
Georg-Schäfer-Strasse 1, D-8720 Schweinfurt, FRG
(b) Institut für Angewandte Physik, Lehrstuhl für Festkörperphysik
of The University of Erlangen Nürnberg, Staudtstr. 7, D-8520 Erlangen, FRG

ABSTRACT

Auger spectra obtained from cubic and hexagonal BN are presented and compared with those observed for boron, boron carbide and oxide. Characteristic changes in these low energy spectra depend on the additional decay channels supplied by the bonding partners for the recombination of the boron core hole. Supported by SXS-data the main features can be understood in terms of a selfconvolution model of the p-like DOS. Deviations from this model can be discussed by partial localization in the bonding resulting in different final states. Especially the existance of an extra peak at the low energy side of the B-KVV transitions of the nitrides and the oxide can be explained by this model. Based on ELS-measurements a high energy peak in the Auger multiplet of h-BN may be the direct nonradiative recombination of a core exciton. The comparison of the new Auger data for c-BN with those observed from h-BN reveals differences which are not expected by regarding their quite similar local DOS from SXS spectra. The differences are attributed to the different sp^3 and sp^2 hybridiziation leading to distinct localizations.

---

*) Present address: Gossen GmbH, Nägelsbach Str. 25, 8520 Erlangen, FRG

1. INTRODUCTION

The Auger electrons from core -valence-valence (CVV) transitions carry a lot of informations about the chemical enviroment of their source atom [1]. This fact was demonstrated in a systematic manner by the comparison of low energy Auger spectra of the nitrides and oxides of the light elements lithium, beryllium and boron [2]. As a general feature the CVV-transitions of the electropositive partner in these binary compounds appear to be split into several lines. The chemical environment of an atom can not only be changed by different binding partners but also by different structural arrangments of the surrounding atoms. So the CVV-transitions of carbon in graphite, diamond and carbides show distinguishing features being characteristical for the respective structure [3].

Boron nitride is the simplest III-V compound. Like carbon it exists in a hexagonal (h-BN) and a cubic (c-BN) modification, which are isoelectronic and isostructural to graphite and diamond. It is interesting to get Auger spectra from these two boron nitride structures, because their comparison can be helpful for a better understanding of the influence of crystallographic structures on the electronic states of an atom.

Only Auger transitions of h-BN are published [2,4,5]. Hitherto no Auger data from c-BN are presented in literature because of the the unsolved problems with charging while recording electron spectra from this highly insulating and chemically very stable solid.

The aim of this article is to present the first Auger spectrum from c-BN. The preparational techniques for obtaining these experimental data are described. For a better understanding of the observed features in the Auger transitions of c-BN and h-BN their lineshapes will be compared with the spectra of pure boron, boron carbide and boron oxide. Models interpreting the features of the Auger lines are supported by measured ELS spectra and published SXS data.

2. EXPERIMENTAL DETAILS

The Auger transitions were excited by an electron beam of I_p = 10 µA at a primary energy of E_p = 2.5 keV. The spectra were obtained with a 150^0 hemispherical energy analyser (VG ESCA 3) run in the $\Delta E/E$ = const. mode except for the ELS measurements where the analyser was used in the ΔE = const. mode with the advantage of a better resolution for electrons with kinetic energies above 200 eV. Because of the energy dependence of the effective cross section for core excitation the ELS spectra of boron were induced by an electron beam of I_p = 5 µA at an energy of E_p = 950 eV.

The signals S(E) were recorded using lock-in technique in the first derivative mode with a modulation voltage of 0.1 V_{pp}. After integration of the dS(E)/dE spectra the data were corrected by the analyser transmission function, which is in the $\Delta E/E$ mode proportional to E. A background substraction was performed by a n-th order polynomial function approximation [1]. By deconvolution with the destribution of the elastically reflected electrons at the same energy the broadening of the Auger line by

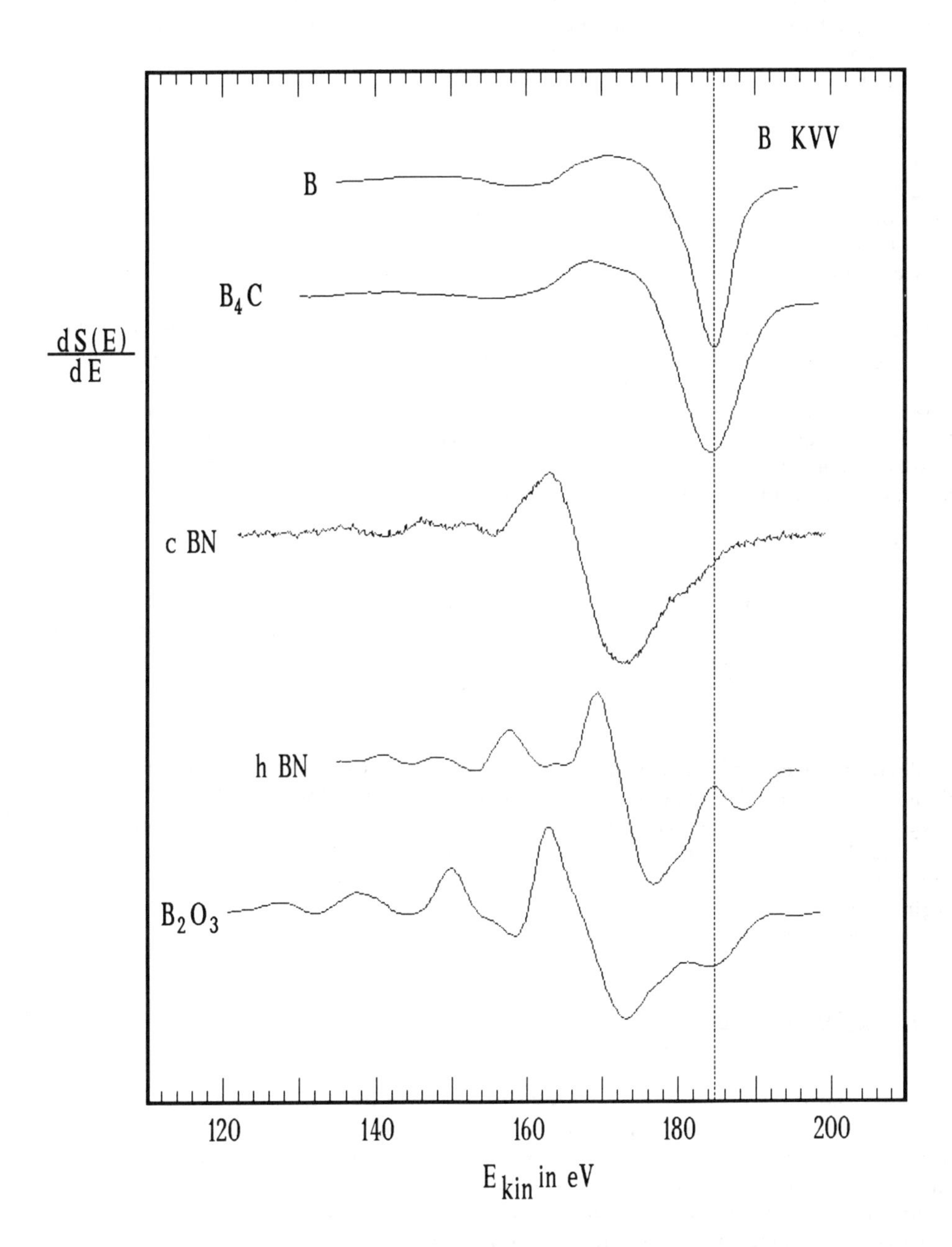

Figure 1 : First derivative KVV transition of boron, boron carbide, cubic boron nitride, hexagonal boron nitride and boron oxide.

finite instrumental resolution and inelastical energy losses was removed [6], using a Fourier transform technique. This treatment was performed to ensure that the observed structures are all due to Auger transitions rather than to inelastic processes following an Auger decay. All the energies presented are taken with respect to Fermi level and they are accurate to 1 eV or better.

The polycristalline samples of B and B_4C were cleaned by alternating argon ion etching (30 μA, 3 keV) and subsequent annealing at 1800 K. Only small amounts of Ca and N for B as well as N, O and Ar for B_4C could be detected as remaining impurities on the samples. The boron oxide specimen was produced by a reactive sputtering of clean boron. The results were quite similiar to those of J. W. Rogers Jr. et al. [7]. A certain amount of elemental boron remained even after prolonged sputtering.

The h-BN a white powder sample was prepared by adhesion on a carbon substrate. The powder particles got a sufficiently good electrical contact with the conducting carbon surface. Therefore no charging effects could be observed. This method is comparable to the preparation of pressing powders into soft metal foils like In [8]. But it has the advantage that carbon shows only one Auger decay channel, which is not superposing the boron transition being examined, and furthermore, its well known peak position can be used for energy calibration, too.

The c-BN consisted of crystals from a quality used in grinding and honing tools. They had an average grain size of 2 - 4 μm. To obtain Auger spectra from these insulating samples quite a different technique than those being reviewed by Wei [8] had to be developed. The small crystals were mounted on a carbon substrate , which was built by a colloid solution of graphite in water. Short argon ion etching (30 μA, 6 keV) was practised for cleaning the surfaces of the samples. The electron excited Auger spectra were measured under a permanent ethan exposure of about $1 \cdot 10^{-8}$ Torr.

The electron beam induces damage of this carbon-hydrogen compound producing a carbon overlayer in the area where it is striking the specimen [9]. Since the beam diameter is about 10^3 times the grain size of the c-BN crystals the produced layer is in an electrical contact with the carbon substrate screening the electrostatic fields. So the carbon coat in the direct vicinity of the excited, Auger relaxing atoms avoids charging effects. By this treatment the Auger lineshape of the B KVV transitions became stable and reproducible within 5 minutes.

3. EXPERIMENTAL RESULTS

In figure 1 the differentiated boron KVV spectra in pure boron, in boron carbide, in hexagonal and cubic boron nitride and in boron oxide are presented. Obvious is the influence of the chemical environment on the Auger transitions of the excited boron atoms. The pure element and its carbide show only one peak, while the multiplet for the nitrides and the oxide indicates extra decay channels. The transitions of the compounds are shifted to lower kinetic energies with respect to the pure element.

Especially for c-BN and h-BN there are differences in energy and lineshape of the multiplets.

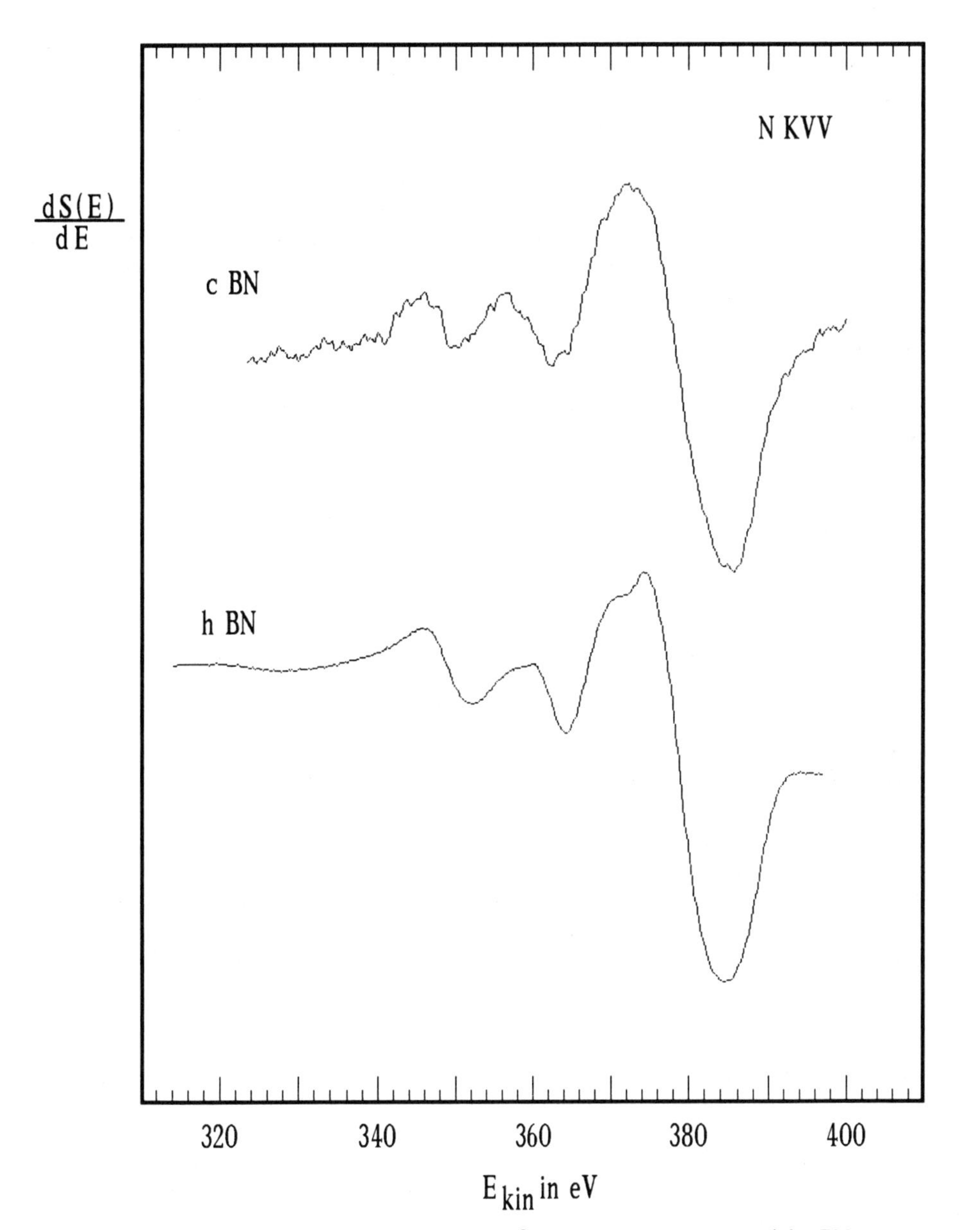

Figure 2 : First derivative KVV transition of nitrogen in c-BN and h-BN.

Observations during the preparation of boron oxide indicate that the high energy peak at 184 eV, which is marked by a dashed line, is caused by remaining amounts of the pure element.

The spectra of boron and boron oxide are in a good agreement with data published by Joyner and Hercules [10] and by Rogers and Knotek [7]. We could not find any accordance to the spectra of Dagoury and Vigner [4]. We are unable to explain the discrepancies in energy and lineshape especially for h-BN. Qualitative agreement with our data was found in their spectra of boron bombarded with oxygen.

In figure 2 the nitrogen KVV spectra from c-BN and h-BN are shown. They are very similiar in shape and energy, while the transitions at their respective electropositive partner show larger differences.

The important influence of the chemical environment of the boron atoms is not only reflected by the Auger spectra, but by the K level binding energies, too. From our XPS measurements we found this energy E_b to be 187 eV for boron and 186.1 eV for the carbide.

sample	B	B_4C	h-BN	c-BN	B_2O_3
E (B 1s)	187	186.1	191	191	193
E (VB_{max})	3	1	9.5	8.5	11
E (VB_{sat})	-	-	20.8	20.5	26.4
E (exciton)	-	-	-0.7	-0.7	-0.1
ΔE_{Gap}	1.5	0.5	5.6	6.0	7.0

TABLE 1: Binding energies relative to Fermi level in eV of the boron K level E(B 1s) from XPS measurements and the dominant maximum E(VB_{max}), the low energy satellite E(VB_{sat}) and the excitonic state E(exciton) of the valence band structure from SXS data for boron, B_4C, h-BN, c-BN and B_2O_3. The binding energy of c-BN has been assumed to be equal to that of h-BN, after [11]. The bandgap ΔE_{Gap} of these solids is added.

In h-BN and the oxide E_b is shifted to 191 eV and 193 eV. The binding energy of c-BN has not been measured. Fomichev and Rumsh [11] assume, that the degree of ionicity of the bonding alters little between the two BN modifications and therefore they consider that the K - level of boron in c-BN remains the same as in h-BN. All the binding energies of the boron K level from the respective compounds relative to Fermi level are listed in table 1.

The band gap is changing from 1.5 eV for boron [12] and 0.5 eV for its carbide [13] to the insulating h-BN with 5.6 eV [10] and c-BN with 6.0 ± 0.5 eV [14]. The boron oxide has a gap of 7 eV [12].

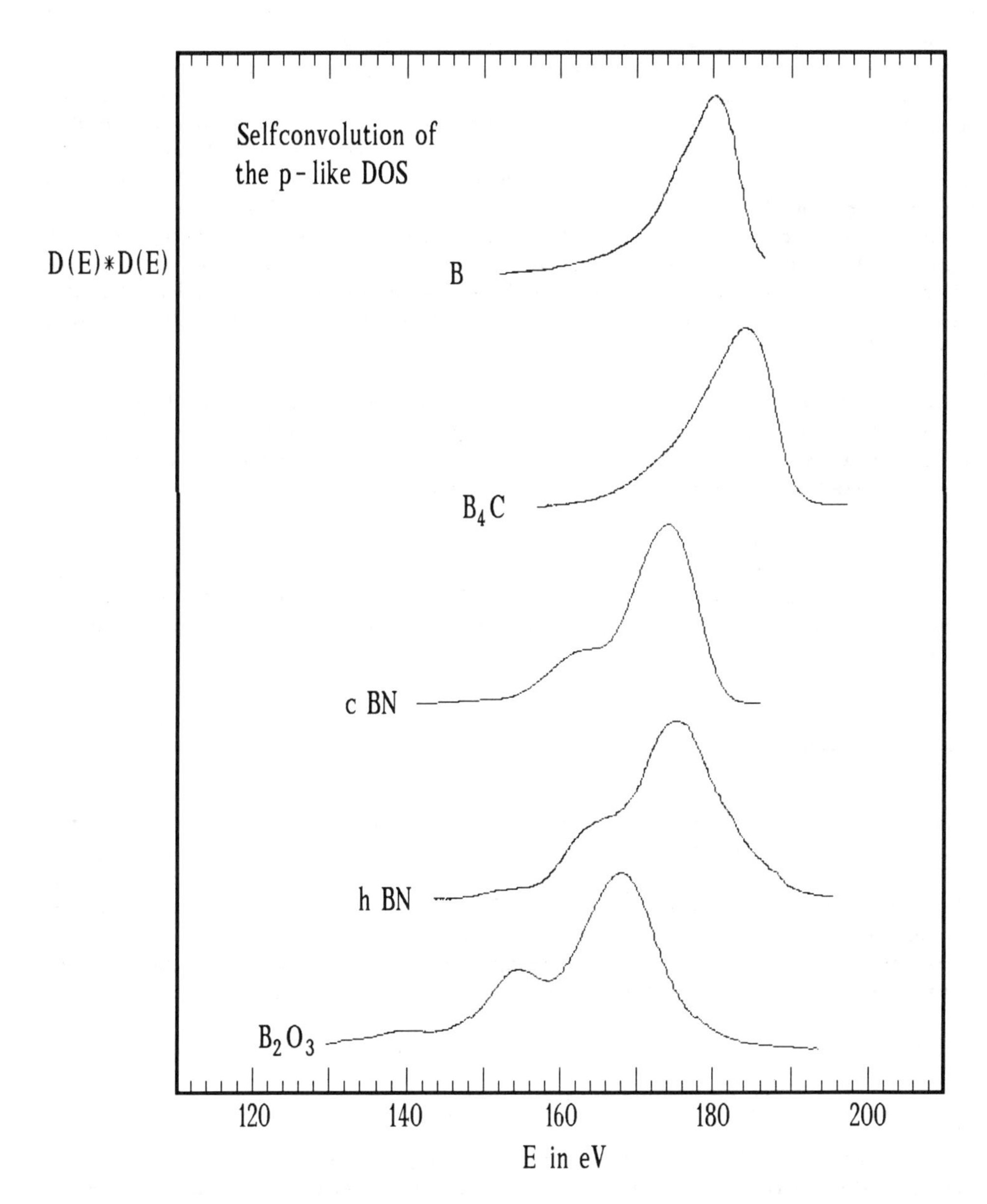

Figure 5 : Selfconvolution of the p-like DOS from SXS data [15] for boron, boron carbide, cubic boron nitride, hexagonal boron nitride and boron oxide.

The soft x-ray spectra of the boron K emission from reference [15] are presented in figure 3. Similar results are published for boron [16], boron nitride [11, 17, 18] and boron oxide [17,18]. These data from the radiative relaxation being in competition to the non-radiative Auger process are reflecting the influence of the chemical bond, too. So the maximum of the p-like DOS is shifted to higher binding energies. At the lower energy side of the boron oxide, the cubic and the hexagonal boron nitride spectra a second peak exists. This splitting in the local density of states at the side of the electropositive element depends on the binding energy of the 2s level of the electronegative partner of boron in the respective compound. From MO theory a mixing between the 2s orbital of the ligands (C, N and O) and the boron electron orbitals is predicted with the result of a characteristical seperation energy between the upper valence band and the ligand's 2s level. The average seperation energies are 8 eV for the carbides, 13 eV for the nitrides and 16 eV for the oxides [19, 20]. Because of the half-width of 9 eV of the valence band in boron carbide this splitting cannot be observed in the emission spectrum.

The DOS of the boron nitrides and oxide have a special feature at the high energy side being located within the band gap above Fermi level. These additional features marked with a line in figure 3 have been interpreted as the radiative decay of a core exciton localized at the side of the boron atom [11].

By the binding energies of the boron K level from XPS measurements the SXS data can be transformed into binding energies relative to Fermi level, too. These energies for the valence band maximum, the separated peak caused by the ligand's 2s level and the excitonic transition are listed in table 1 together with the band gap in the respective solids.

4. DISCUSSION

To demonstrate that the observed multiplets of the boron KVV transitions are not due to energy losses, the Auger spectra shown in figure 1 have been treated by the described integration, backgroundsubtraction and deconvolution techniques. The results are given in figure 4. No features in the spectra are caused by inelastic losses, Auger electrons may have suffered on their way from the emitting atom in the solid compounds into the vacuum. In the following discussion the Auger transitions in the different compounds are labeled as given in figure 4.

The energy E_{Auger} of the Auger transitions can be approximated as

$$E_{Auger} = EB(1s) - E(k) - E(l) - U_{eff} \qquad (1)$$

where EB(1s) is the boron's core electron binding energy. E(k) and E(l) are the energies of the electron levels involved in the Auger relaxation. U_{eff} is the effective interaction energy of the hole-hole Coulomb repulsion in the final state. This energy is a priori unknown.

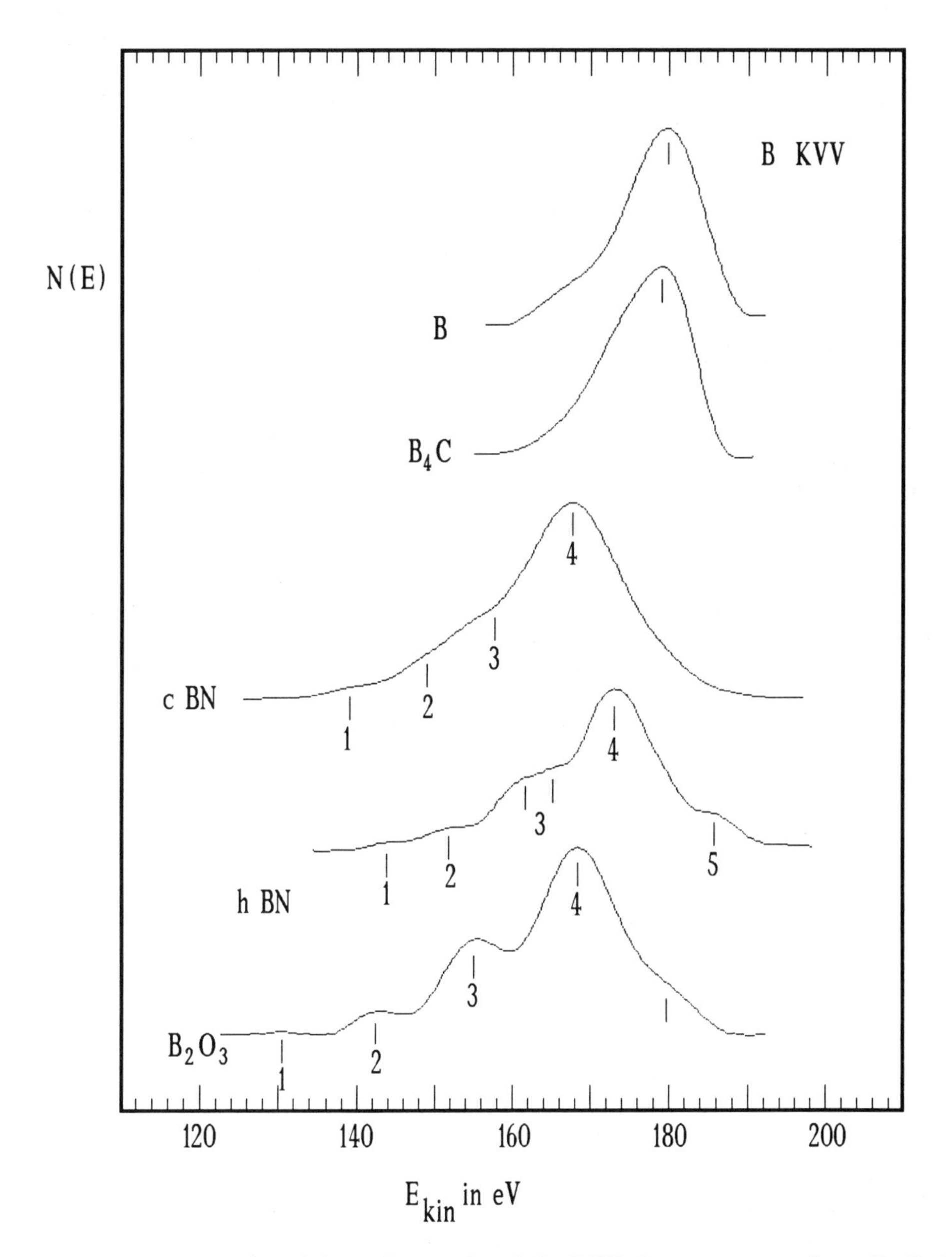

Figure 4 : Integrated and loss deconvoluted B-KVV Auger spectra from B, B_4C, c-BN, h-BN and B_2O_3.

The self-convolution of the density of states suggested by Lander [21] is a model to describe CVV Auger transitions from solids having a broad valence band. The range of validity of this model was limited by Cini [22] and Sawatzky [23] by comparison of the effective hole-hole-interaction energy U_{eff} with the width V of the valence band. If U_{eff} is small compared to the bandwidth the lineshape will be band-like, while it is atomic-like if $V \ll U_{eff}$. Atomic and band-like parts contribute to the lineshape if U_{eff} is of the order of V.

According to these models the main features of the Auger lines labled 2,3,4 in figure 4 will be discussed in terms of a transition model based on the local DOS, while the extra lines 1 and 5 need a different interpretation. Moreover the linewidth is not exactly reproduced leading to an extended model for boron and its carbide.

4.1 MAIN FEATURES

Assuming that the bandwidth V is large compared to the hole-hole- interaction U_{eff} we carried out the self-convolution of the p- like local density of states (DOS) from the SXS data given in figure 3. The results of this treatment are presented for boron and its respective compounds in figure 5.

The comparison with the corrected N(E) spectra from figure 4 shows, that the self-convolution is reflecting certain parts of the experimental data. So B and B_4C only have one transition, while the boron nitrides and the oxide show three peaks in figure 5. The energies of these main structures are listed in tables 2 and 3. The comparison of the transition energies from the N(E) mode with the energies calculated in this selffold model provides an estimation for the magnitude of the effective hole-hole interaction. The different values of U_{eff} are between 1.4 and 6.5 eV, reflecting that the valence band with a bandwidth of 15 eV and more causes a high degree of delocalisation and therefore only a minor interaction energy.

sample	B	B_4C	B_2O_3				
dS(E)/dE	184.8	184.3	184.0	173.2	156.0	144.0	132.0
N(E)	179.5	178.7	179.0	168.3	155.0	142.3	129.8
model	181	182		170	150	140	
peak	KVV	KVV		KL_1L_1	KL_1V	KVV	

TABLE 2: Energies of the boron Auger transitions in eV for boron, its carbide and oxide. dS(E)/dE: energy of the minima in the first derivative dS(E)/dE (figure 1). N(E): energy of the maxima of the N(E) mode (figure 4). model: energy calculated with equation(1) assuming $U_{eff} = 0$. peak: Description of the decay channel.

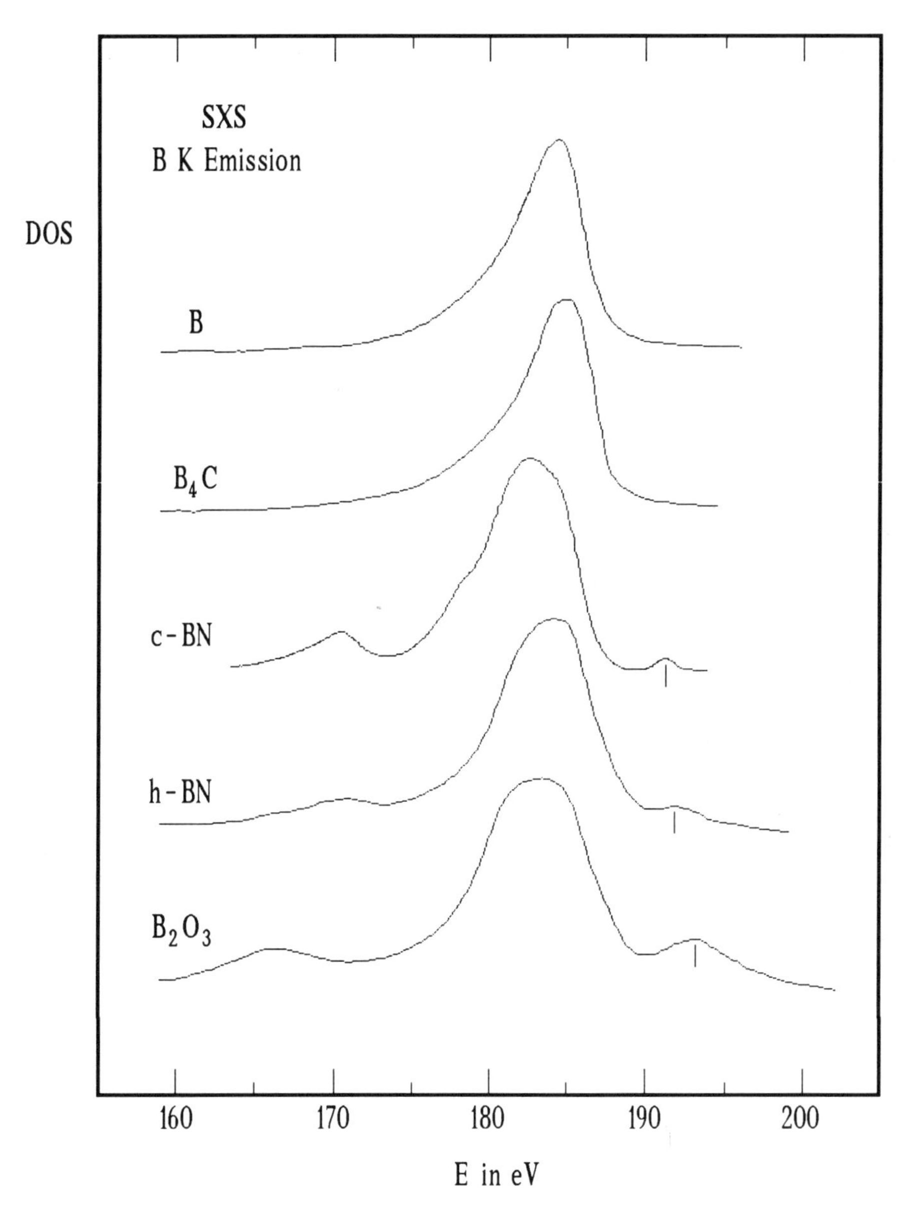

Figure 3: Soft x-ray spectra of boron, boron carbide, cubic and hexagonal boron nitride and boron oxide reproduced from reference [15].

h-BN				c-BN		
dS(E)/dE	N(E)	model	transition	dS(E)/dE	N(E)	model
188.5	185.0					
176.8	173.0	174.4	KVV	172.8	167.5	174.0
165.3	166.1					
162.7	160.6	163.1	$KL_1(N)V$	155.4	155.6	162.0
153.0	152.6	151.8	$KL_1(N)L_1(N)$	149.5	148.2	150.0
145.0	144.6			141.3	139.6	

TABLE 3: Energies of the boron Auger transitions in eV from h-BN and c-BN. dS(E)/dE: Energy of the minima of the first derivative (figure1). N(E): Maxima maxima of the N(E) mode (figure 4.). model: Energy calculated with equation 1 assuming U_{eff} = 0. transition: Description of the Auger decay channel.

So the transitions channels labled 2, 3 and 4 are attributed to the KL_1L_1, KL_1V and KVV Auger transitons. The energy separation of these lines varies with different electronegative partners in the respective compounds. This effect is controlled by the L_1 levels of the binding partner carbon, nitrogen and oxygen [2]. They open extra decay channels with their local DOS at the side of the excited boron atom. This observation supports the model that the splitting of the Auger multiplet is mainly caused by the binding partner in a compound. This is a further evidence for the advantage of AES as a probe for the local chemical environment of an atom.

But there are discrepancies in width as well as in the number of peaks between the lineshape of the integrated Auger spectra from measurements in figure 4 and the respective selfconvolution of the local p-like DOS in figure 5 .

4.2 LINEWIDTH

So in the case of boron and boron carbide the selfconvolution causes a smaller line-width (full mean at half maximum FWHM: 9.4 for B, 10.9 for B_4C) than it is observed in the measurements (FWHM: 16.4 for B, 13.3 for B_4C). For the pure lithium the Auger lineshape is well reproduced by the selfconvolution of the p-like DOS from SXS-data [25]. The Auger transitions could not be described by this model in a satisfactory way. The reason is the lack of the s-parts to this transition not being reflected by SXS. Its consideration in the selfconvolution causes contributions from s*p and s*s parts besides p*p. Assuming the well known configuration of the valence electrons in ground state of $s_{0.68}\,p_{1.32}$ still a deviation exists to the experimental lineshape. The introduction of a s-like screening of the beryllium core hole, which causes an electron configuration of the initial state of $s_{1.68}\,p_{1.32}$ at the side of the relaxing atom, produces quite a good agreement between the selfconvolution and the experimental data [26].

Since the influence of the s-like DOS to KVV-transition increases with increasing atomic number [25,27,28] such s-like screening effects in the initial state can lead to a better description of the B and B_4C -lineshape. Unfortunately the complex structure of boron and boron carbide hinder satisfactory band structure calculations in agreement with experimental data [29,30,31]. Complemantary in the case of boron carbide the L_1 level separated by an energy of 8 eV immerses into the wide valence band leading to an additional broadening of the Auger peak.

4.3 HIGH ENERGY PEAK

In the selfconvolution spectra of c-BN, h-BN and B_2O_3 given in figure 5, only three peaks appear being discussed in chapter 4.1. The Auger spectra of these compounds given in figure 4 show further decay channels. First the transition on the high energy part of the spectra is discussed.

B_2O_3 has a high energy peak at 179 eV. Following the oxidation process described in chapter 2 the Auger transition of the pure boron never is completely vanishing. Therefore we attribute the high energy peak in boron oxide to large amounts of non-oxidized boron [5]. This was reported by Rogers et al. [7], too. As the high energy peak in the h-BN spectrum (labled 5 in figure 4) is at an energy of 185 eV it cannot be caused by small amounts of pure boron.

First indications to the origin of this h-BN transition can be found in the electron loss spectra (ELS) from figure 6. Compared to the primary beam at $\Delta E = 0$, the low energy losses are amplified by a factor of 2, the region of the core and ionisation losses of the boron K level by a factor of 64.

The low energy losses are caused by plasmons for boron ($\hbar\omega$ = 22.6 eV) and boron carbide ($\hbar\omega$ = 24 eV). For boron nitride and oxide these losses are interpreted as interband transitions from the valence into the conduction band [7, 32]. In the region of the core losses double structures are clearly resolved for B_2O_3 , h-BN [5] and the new data from c-BN. The energies of these losses for c-BN, h-BN and B_2O_3 are listed in table 4 compared to the respective energy values won from quantum yield (QYS) and soft x-ray spectra (SXS) [11,18]. Using binding energy levels of the B 1s-level from XPS spectra and the SXS data the lower energy core losses are localised within the band gap, while the higher losses reflect transitions from the boron K level into the conduction band. Only one loss from excitation into the

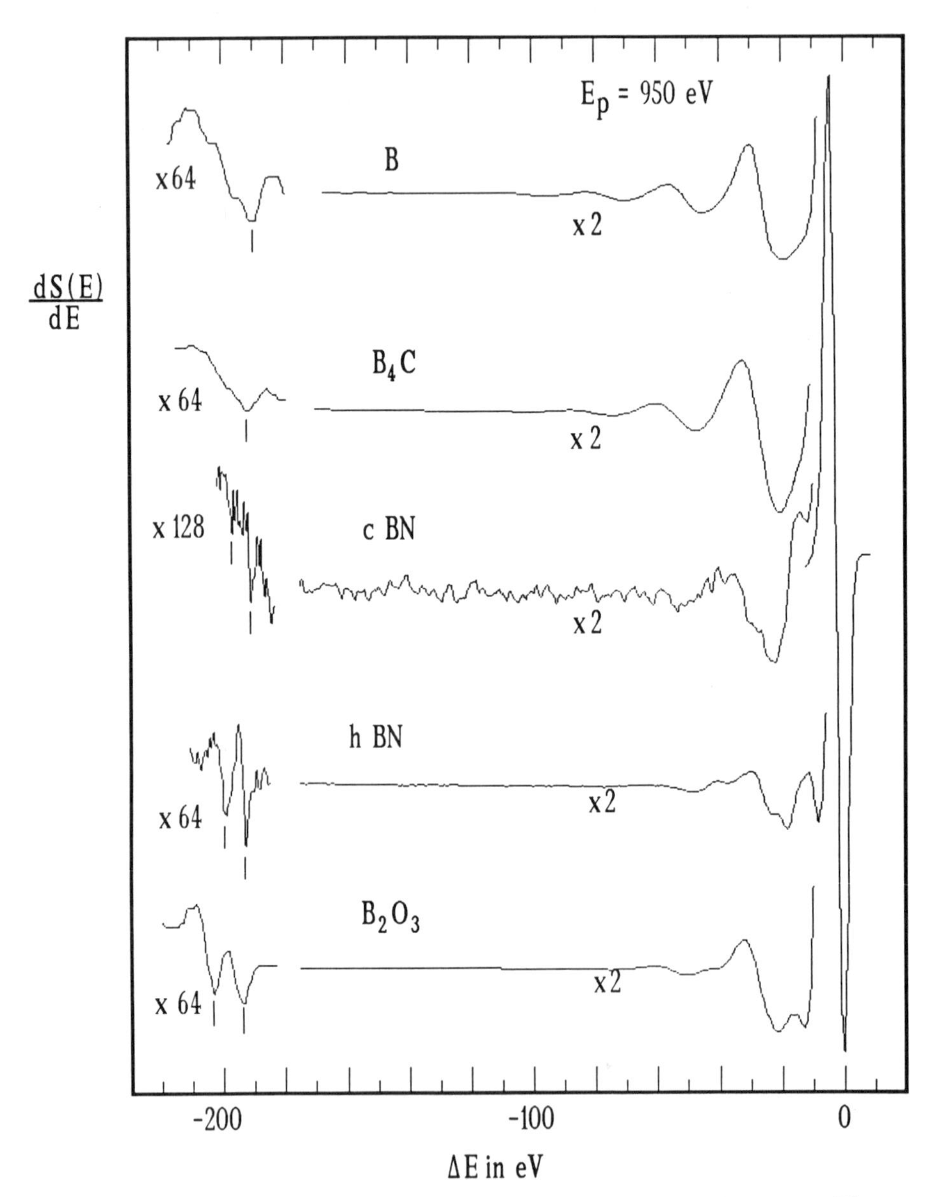

Figure 6 : Electron loss spectra of B, B_4C, c-BN, h-BN and B_2O_3 . The core losses are magnified by a factor of 64 (c-BN by a factor of 128). The energy losses are magnified by a factor of 2.

	exitonic level			conduction band	
sample	ELS	QYS	SXS	ELS	QYS
c - BN	190.6	191.8	191.3	197.4	197.5
h - BN	192.6	191.8	191.3	198.8	198.0
B_2O_3	194.0	194.0	193.5	203.0	202.7

TABLE 4: Comparison of the core losses from electron loss spectroscopy (ELS) with quantum yield (QYS) and soft x-ray spectra (SXS) for c - BN, h - BN and B_2O_3 [11,18].

conduction band is found in pure boron (190.3 eV) and its carbide (192.5 eV).

In the boron nitrides and the oxide the loss lifting a core electron into the band gap can be interpreted as an excitation of a core exciton. We do not expect them for boron and boron carbide because of their narrow band gaps of 1.5 and 0.5 eV [12,13]. We suppose that excitonic features in the loss spectra of the electropositive partner of a compound appear if the ionic part of the band and the gap increases [5]. This is the case in the series of boron, its carbide, nitrides and oxide. It is supported by similiar results for lithium and beryllium nitride [33] and for beryllium oxide [33, 34]. These compounds show such a double structure, too. Based on these observations we assume that in strongly ionic bonds the electron hole interaction is large since the core hole is not so effectively screened by the valence electrons being located mainly at the electronegative partner in these binary compounds. For alkali halides a similiar interpretation has been reported by Pantalides [35].

As mentioned in chapter 3 the SXS spectra of c - BN, h - BN and B_2O_3 are pointing to these excitonic state, too. According to table 1 their energies are 0.7, 0.7 and 0.1 eV above Fermi level. This reflects that such an excitation can decay in a radiating transition. There is a certain probability for a nonradiative relaxation, too. This was shown by Barth et al. [36] for h - BN. The authors were exciting only this core exciton with synchrotron radiation. In their experiment they determined the ratio of the direct nonradiative recombination to the KVV transitions to 1 : 2. By a Gaussian curve fit of the electron excited, integrated h-BN Auger spectrum from figure 4 we obtained a ratio of 1 : 11, reflecting that primary electrons at an energy of Ep = 2.5 keV are not exclusively populating this excitonic level.

By the direct recombination of the electron from the core exciton level with its K hole the energy will be transferred to a valence electron being emitted with a higher kinetic energy than the normal KVV Auger electrons. We estimated this energy to be 183.7 eV, which is in good agreement with the measured 185 eV of the high energy peak in table 3.

In the case of c-BN and B_2O_3 the direct recombination energies in this model are estimated to 184.3 and 181.8 eV. In the c-BN spectrum we did not find any hints to this transition. In the case of boron oxide this direct recombination peak is overlapping with the line of the remaining pure boron. Therefore it is not clear whether such a decay exists in this compound, too.

4.4 LOW ENERGY TRANSITIONS

The boron nitrides and the oxide from figure 4 show an additional peak labled 1 at the low energy side of their Auger spectra. Such low energy transitions are observed in the Auger multiplets of the electropositive element in the oxides and nitrides of lithium and beryllium, too [2].

Introducing a model of a partial localisation of the two holes in the final state the existance of the low energy transition can be explained. The two holes can be located either at the same or at different atomic sides. In this model the effective interaction energy U_{eff} would be large if the final state holes are localized at the same and small if they are at different atomic sides. According to equation 1 different U_{eff} is leading to different Auger energies.

We are assuming that the low energy peak is a final state of high localisation of the KL_1L_1 transition in the selfconvolution model.

Such localization effects have been discussed by Ramaker [37,38] and could be proved in the Auger spectra of the tetrahedral halides and hydrides of carbon and silicon [39, 42 , 43].

The splitting of KL_1V transition can also be attributed to partial localization effects. A similar interpretation has been proposed for Li_2O [25] following Auger data of LiF [40].

5. COMPARISON OF c-BN WITH h-BN

The integrated and loss corrected nitrogen Auger spectra of cubic and hexagonal boron nitride are presented in figure 7. Except for the energy difference of 3 eV in the KL_1V decay channel there is no great distinguishing feature in lineshape between the c-BN and h-BN data. Our N(KVV) transition in h-BN is very similar to that of Rye et al. [41] except for the energies. The source of this discrepancy can be the shift of their $(BN)_x$ spectrum they performed to come to a visually align with their N(KVV) lineshape of borazine.

Obviously the more important differences between c-BN and h-BN do not appear in the N(KVV) lines but in the boron transitions from figure 4. This is not surprising

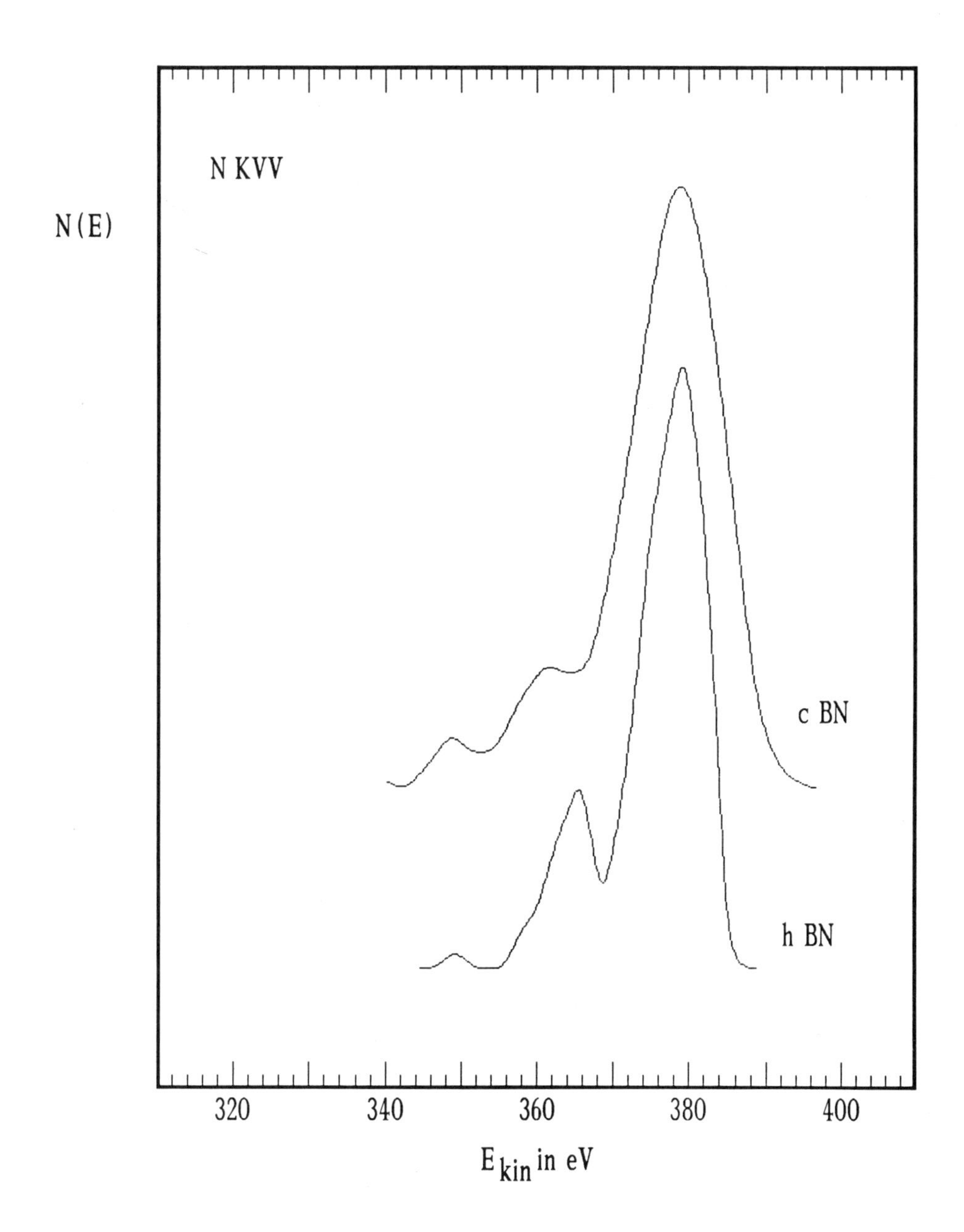

Figure 7 : The integrated and loss deconvoluted Auger spectra of nitrogen in cubic and hexagonal boron nitride.

because the more dramatic effects are always found at the side of the electropositive partner in the compound, which can be explained by a charge transfer model [1,44,45].

Depending on the electronegativity of the bonding partners in a compound there is more or less transfer of electron charge from the electropositive to the electronegative element. Therefore Auger recombination at the side of the electronegative atom is not so much effected by the chemical environment, because the holes of the final state will mainly be localized at the side of the initial state hole. Whilst the Auger decay at the electropositive element is ruled by its chemical surrounding built up by the anionic partners in the bond. In the final state the holes can be localized at the side either of the relaxing electropositive or of the electronegative atoms in the direct neighbourhood. So the B-KVV Auger spectra are more affected by the local symmetry - the hybridization - of the boron atom.

For the further discussion we should like to draw attention especially to these boron KVV transitions of c-BN and h-BN in figure 4. These spectra are different in energy and lineshape. Similar differences exist between the isoelectronic and iso- structural carbon modifications diamond and graphite [3, 46,47]. In both sp^3 hybrides the main transition is shifted to lower kinetic energies, 5eV for diamond and 4eV for c-BN compared to the respective sp^2 compounds graphite and h-BN. For c-BN and h-BN the structures from the selfconvolution model of the p-like DOS in figure 5 are remarkable similar. From chapter 4 we know that localization effects are determining the deviations from the self- fold model of the DOS. Therefore differences in the Auger spectra for cubic and hexagonal boron nitride do not arise from differences in the one electron DOS but from different localization effects caused by different hybridization.

Cini [22] and Sawatzky [23] limited the range of validity of the selfconvolution model. Dunlap et. al. [48] generalized these ideas to covalent systems with different elements in compound introducing criteria for localization onto atomic (AO), band (BO) and group orbitals (GO). Ramaker is assuming that bond orbital localization is dominant for BN [46].

The main difference between the Auger spectra of hexagonal and cubic boron nitride in figure 4 is the peak labled 5. It only exists in the h-BN transition and is explained in chapter 4.3 as the nonradiative decay of an exciton.

Like in graphite the boron and nitrogen atoms are sp^2 hybridized. With their bonds they are building up layers of symmetrical hexagons with alternating boron or nitrogen atoms in the corners. Whereby each boron is surrounded by three nitrogen atoms and vice versa. Additional there are π- orbitals at each element. From SXS and quantum yield data [11] it is well known that the π- orbital at the side of the nitrogen atoms is occupied, while it is empty at the boron side. From the anisotropic emission of the N K-emission band [49], the angular dependence of soft-x-ray absorption near the boron K threshold [50] and the angular distributions in electron energy loss spectra [51] it can be concluded, that these orbitals are perpendicular orientated to the hexagonal layers built up by the boron and nitrogen atoms via the

sp^2 - bonds. The narrow half width of 0.65 eV from photoemission yield at the boron K-edge [36] points to a sharp localization of this orbital at the boron atom.

The core losses in our ELS measurements (figure 6) show for h-BN as well as for c-BN a double structure with the lower energy loss being an excitation into the band gap. Why do we not observe such a Auger decay of a core exciton in the c-BN spectrum?

The SXS and QYS data from Fomichev and Rumsh [11] show that the selective maximum being significant for h-BN clearly decreases while the shape itself of the emission spectra changes very little. Also in the core losses of graphite a double structure is observed [52,53] while the respective spectra from diamond only have one main loss [53,54]. In diamond spectra hints to a state in the gap exist, but it was identified as a unoccupied surface state being present if hydrogen is absent from surface [54].

So we assume that the contribution of such an excitonic state to the Auger process only is found in the sp^2 and not in the sp^3 hybrides. As mentioned diamond has no excitonic features [53,54]. In the isostructural and isoelectronic c-BN it is observed because of the different ionicity of the two elements in the compound causing a charge transfer from boron to nitrogen. This transfer is leading to unoccupied states at the boron's and additional occupied states at the nitrogen's side.

Both graphite and h-BN show an excitonic feature in spectroscopy. In boron nitride it is localized exclusively at the boron atom. The nitrogen core losses [52] and the SXS data [11] do not point to such a state. Depending on the ionicity it is localizied contrary to the graphite only at one atomic side, at the boron atom. The spreading of this π -like state over surrounding carbon atoms in the graphite hinders its contribution to Auger decay. But in hexagonal boron nitride the core hole as well as the exciton are localized at the same atom opening the possibillity for the observed Auger recombination.

6. CONCLUSION

The presented Auger spectra of c-BN and h-BN show several decay channels. In comparison with the spectra of boron, boron carbide and boron oxide and with respect to published SXS data the main transitions can be understood in terms of a selfconvolution model of the p-like DOS.

Deviations from this model in energy, line shape and number of lines lead to the introduction of localization effects helping to explain the existance of an extra peak at the low energy side of the B-KVV transitions of the nitrides and the oxide. Supported by our ELS measurements the high energy peak in h-BN is interpreted as the decay of a core exciton.

The different arrangement of the atoms in c-BN having diamond and in h-BN having graphite structure causes differences in the Auger spectra, which are not expected by regarding their quite similiar local density of states from SXS data. The differences are attributed to sp^3 and sp^2 hybridization of c-BN and h-BN leading to distinct localizations. Expression of this structure effect on the B-KVV Auger decay is first a shift of the c-BN spectrum to lower kinetic energies compared to h-BN , equivalent observations exist for carbon transition in diamond and graphite. Second only for hexagonal boron nitride a high energy transition is observed. The special structure together with the different ionicity of the two elements leads in h-BN to the sharp localization of the exciton at the atomic side of the core hole. So the core exciton can relax via an Auger decay.

Our new data of c-BN together with the h-BN Auger lines will help in comparison with graphite and diamond spectra to come to a better understanding of localization and the influence of structures on the Auger decay from solids.

REFERENCES

1) Weissmann, R. and Müller, K.: Surface Sci. Rept., 1981, 1, 251
2) Hanke, G. and Müller, K.: Surface Sci., 1985, 152/153, 902
3) Mizokawa, Y., Miyusato, T., Nakamura, S., Geib, K.M. and Wilmsen, C.: Surface Sci., 1987, 182, 431
4) Dagoury, G. and Vigner, D.: Le Vide, 1977, 187, 51
5) Hanke, G., and Müller, K.: J. Vac. Sci. Technol., 1984, A2, 964
6) Mularie, W.M. and Peria, W.T.: Surface Sci., 1971, 26, 125
7) Rogers Jr., J.W. and Knotek, M.L.: Appl. Surf. Sci., 1982, 13,352
8) Wei, W.: J. Vac. Sci. Technol., 1988, A6, 2576
9) Kramer, M. and Müller, K.: submitted to J. Vac. Sci. Technol. A
10) Joyner, D.J. and Hercules, D.M.: J. Chem. Phys., 1980, 72, 1095
11) Fomichev, V.A. and Rumsh, M.A.: J. Phys. Chem. Solids, 1968, 29, 1015
12) Strehlow, W.H. and Cook, E.L.: J. Phys. Chem. Ref. Data, 1973, 2, 163
13) Werheit, H. and de Groot, K.: Phys. Status Solidi b, 1980, 97, 229
14) Zunger, A. and Freeman, A.J.: Phys. Rev. B, 1978, 17, 2030
15) Faessler, A.: Landolt-Börnstein, 6th ed., edited by Hellwege K.H., Springer, Berlin, 1955, I/4, 769 - 808
16) Hoffmann, L., Wiech, G. and Zöpf,E.: Z. Physik, 1969, 229, 131
17) Fomichev, V.A.: Sov. Phys. Solid State, 1971, 13, 754
18) Nakhmanson, M.S. and Baranoskii: Sov. Phys. Solid State, 1971, 12, 1966
19) Åberg, T.: X-Ray Spectra and Electronic Structure of Matter, Vol. 1, Eds. Faessler, A. and Wiech, G., Frank, München, 1973, pp. 1-37
20) Fischer, D.W.: Advances in X-Ray Analysis, Vol. 13, Plenum, New York, 1970, pp. 159-181
21) Lander, J.J.: Phys. Rev. 1953, 91, 1382
22) Cini, M.: Solid State Commun., 1976, 20, 605

23) Sawatzky, G.A.: Phys. Rev. Lett., 1977, 39, 504
24) Ramaker, D.E.: Appl. Surf. Sci., 1985, 21, 243
25) Madden, H.H. and Houston, J..: Solid State Commun.,1977, 21, 1081
26) Jennison, D.R., Madden, H.H. and Zehner D.M.: Phys. Rev. B, 1980, 21, 430
27) Madden, H.H. and Zehner, D.M.: Bull. Am. Phys. Soc., 1978, 23, 3357
28) Jennison, D.R.: Phys. Rev. B, 1978, 18, 6865
29) Perrot, F.: Phys. Rev. B, 1981, 23, 2004
30) Wertheit, H. and deGroot, K.: Phys. Status Solidi b, 1980, 97, 229
31) Yamazaki, M.: J. Chem. Phys., 1957, 27, 746
32) Büchner, U.: Phys. Status Solidi b, 1977, 81, 227
33) Hanke, G.: Thesis, Universität Erlangen-Nürnberg, 1984,
edited copy center 2000, Erlangen
34) Lukirskii, A.P. and Brytov, J.A.: Sov. Phys. Solid State, 1964, 6, 33
35) Pantelides, S:T.: Phys. Rev. B, 1975, 11, 2391
36) Barth, J., Kunz, C. and Zimkina, T.M.: Solid State Commun., 1980, 36, 453
37) Ramaker, D.E.: Phys. Rev. B, 1980, 21, 4608
38) Ramaker, D.E.: Chemistry and Physics of Solid Surfaces IV,
Eds. Vanselow, R. and Howe, R., Springer, Berlin, 1982, pp.19-50
39) Rye, R.R. and Houston, J.E.: J. Chem. Phys., 1983, 78,4321
40) Gallon, T.E. and Matthew, J.A.D.: Phys. Status Solidi, 1970, 41, 343
41) Rye, R.R., Kelber, J.A., Kellogg, G.E., Nebesny, K.W. and Lichtenberger, D.L.:
J. Chem. Phys., 1987, 86, 4375
42) Rye, R.R., Jennison, D.R. and Houston J.E.: J. Chem. Phys 1980, 73, 4867
43) Jennison, D.R., Kelber, J.A. and Rye, R.R.: J. Vac. Sci. Technol. 1981, 18, 466
44) Matthew, J.A.D. and Kominos, Y.: Surf. Sci., 1975, 53, 716
45) Fuggle, J.C.: Electron Spectroscopy: Theory, Techniques and Applications, Vol. 4,
Eds. Brundle, C.R. and Baker, A.D., Academic Press, London, 1981, pp.86-152
46) Ramaker, D.E.: Appl. Surf. Sci., 1985, 21, 243
47) Lurie, P.G. and Wilson, J.M.: Surf. Sci., 1977, 65, 476
48) Dunlap, B.I., Hutson, F.L. and Ramaker, D.E.: J. Vac. Sci., 1981, 18, 556
49) Tegeler, E., Kosuch, N., Wiech, G. and Faessler, A.: Phys. Status Solidi b, 1977, 84, 561
50) Davies, B.M., Bassani, F., Brown, F.C. and Olson, C.G.: Phys. Rev. B, 1981, 24, 3537
51) Leapman, R.D. and Silcox, J.: Phys. Rev. Lett., 1979, 42, 1361
52) Nassiopoulos, A.G. and Cazaux, J.: Surf. Sci., 1986, 165, 203
53) Koma, A. and Miki, K.: Appl. Phys. A, 1984, 34, 35
54) Pepper, S.V.:Surf. Sci., 1982, 123, 47

Materials Science Forum Vols. 54 & 55 (1990) pp. 229-260

THE PROPERTIES OF LOW PRESSURE CHEMICAL VAPOR DEPOSITED BORON NITRIDE THIN FILMS

S.S. Dana
IBM Research Division
Thomas J. Watson Research Center, Yorktown Heights, NY 10598, USA

ABSTRACT

Thin films containing boron, nitrogen and hydrogen have been prepared by Low Pressure Chemical Vapor Deposition. The principles of deposition and the various experimental deposition parameters investigated are described. The mechanical properties of the deposited films have been characterized by measuring the internal stress and the Young's modulus, the electrical properties by measuring the dc electrical conductivity, and the optical properties by using UV-visible spectroscopy.

The chemical composition of the films has been analyzed by ion beam techniques, hydrogen content by resonant nuclear reaction, and the boron-to-nitrogen atomic ratio by Rutherford backscattering, as well as electron microprobe. The microstructure of the films, particularly the bonding and distribution of hydrogen, is estimated from the NMR, infrared absorption spectroscopy and evolved gas analysis techniques. Data establishing correlations between the physical properties of the films and their chemical composition are presented.

Finally, the application of boron nitride films as substrates for x-ray lithography masks is discussed. Since radiation damage is the main limitation for their usage, the stability of the various film properties upon x-ray radiation is examined.

INTRODUCTION.-

Boron nitride films have attracted a great deal of interest for their unique physical properties, which makes them advantageous over other materials. The purpose of this chapter is to describe the properties of B-N-H films deposited by low pressure chemical vapor deposition (LPCVD). The results presented in this chapter are the work of many authors from different institutions, and are properly referenced.

Chemical vapor deposition (CVD) is the material synthesis method in which a mixture of gaseous reactants are transported in the vapor phase to a heated substrate surface on which a chemical reaction or pyrolysis (thermal decomposition) occurs to form a solid film deposit. Thin films of boron nitride have been grown by various types of CVD, including atmospheric-pressure [1-15], plasma-enhanced (PE-CVD) [13, 15-27], and metal-organic (MO-CVD) [28].

Reduced pressure CVD has been explored using a variety of precursors: pyrolysis of borazine $B_3N_3H_6$ at 300°-650°C [29, 30] and of β-trichloroborazole $B_3N_3H_3Cl_3$ at 700°-1100°C [3, 13, 31-35], the reactions between decaborane $B_{10}H_{14}$ and ammonia NH_3 at 350°-1150°C [36], boron trichloride BCl_3 and ammonia at 600°-900°C [15, 29, 37-43], BF_3 and ammonia [40, 96], or diborane B_2H_6 and ammonia at 250°-600°C [39, 44-46]. Ammonia is the usual source of nitrogen, since N_2 itself reacts much more slowly [37]. Films deposited below about 800°C are quasi-amorphous [42]. Stoichiometric films with a B/N ratio of 1 have been achieved at higher temperatures [1, 7, 35, 41, 94-95]. Films produced from BCl_3 as a starting material often react with atmospheric moisture to form crystalline boron oxide [36, 39, 41], while those produced from the pyrolysis of borazine at low temperatures dissolve in water [29].

Unless otherwise specified, we will only consider in this chapter the characterization of those LPCVD films formed by reacting diborane and ammonia. Because these films are nonstoichiometric, it has been proposed to call them "Boro-hydro-nitride" films [44].

I- FILM PREPARATION.-

1. *LPCVD process:*

In the low pressure (LP-CVD) technique the reaction is thermally driven, and the use of sub-atmospheric pressure permits a relatively low deposition temperature. The net effect of the higher gas diffusivity (inversely proportional to pressure) and of the increase of the boundary layer across which the reactants must diffuse (the layer is inversely proportional to the Reynolds number), is an increase of the gas-phase mass transfer of reactants by more than an order of magnitude as the pressure is lowered from its atmospheric value to 1 Torr [47]. Therefore the rate determining step at low pressures is the chemical reaction rate at the surface, so that a uniform temperature leads to excellent uniformity of film thickness and composition. Also the increased mean-free-path of the reactant gas molecules at lower pressure allows the substrates to be stacked vertically, and to be closely spaced for increased throughput. The use of a hot wall reactor and the vertical mounting minimize the formation and co-deposition of homogeneous gas phase nucleated particulates respectively, resulting in a lower defect density.

2. *Deposition equipment:*

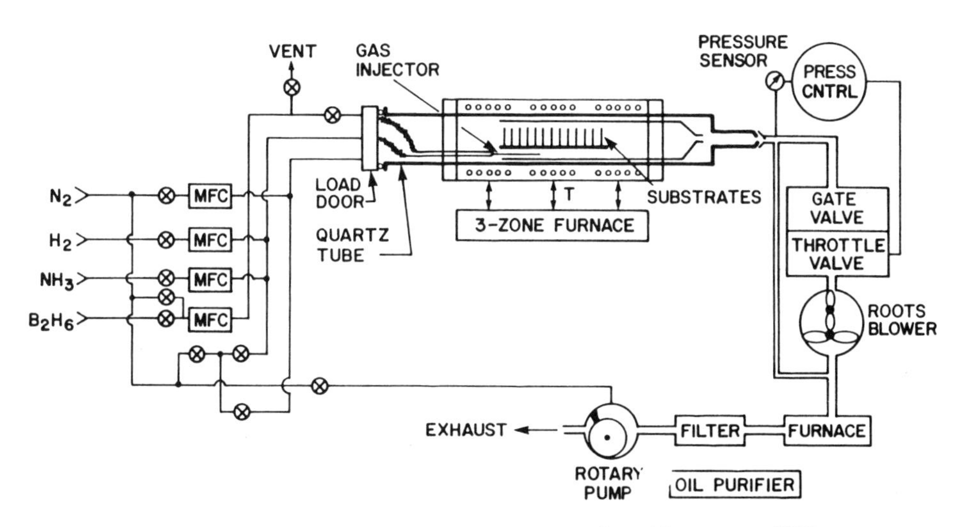

Fig. 1 - Schematic diagram of the low pressure deposition reactor [88].

A schematic of a typical apparatus [17, 39, 44, 48] is shown in Figure 1. The reaction takes place in a horizontal heavy wall round quartz tube (160 cm long with an inside diameter of 12-15 cm) resistance-heated with a three-zone furnace. The source gases, high purity ammonia and diborane diluted to 3-15% by volume in inert (nitrogen or helium) or hydrogen carrier gas, are introduced separately through inlet quartz tubes, and are mixed in the hot zone of the reaction chamber to prevent the formation of solid borohydride compound in the gas lines [1, 39, 44]. The inlet gas flow rates are measured by means of automatic mass flow controllers, and the total pressure is continuously monitored with a capacitance manometer located at the exhaust end of the reaction tube. The pressure is controlled automatically and independently of the total inlet gas flow rate, using a feedback loop from the pressure sensor output to a butterfly valve [44]. A Roots blower is used to maintain high flows at low pressure. A high conductance trap (stainless steel wool) is inserted in series with the mechanical pump, and a nitrogen gas stream was maintained to avoid hydrocarbon backstreaming from the pump oil. Also a recirculating filter for the pump oil prevented pump failure from particulates [44].

Prior to deposition, the silicon substrate wafers were cleaned by a sequence of acidic and basic hydrogen peroxide solutions [49], and located vertically on a quartz carrier inside the tube furnace, perpendicular to the gas flow. During temperature stabilization and outgassing, the reactor was evacuated down to a base pressure of about 1 mTorr and then backfilled with nitrogen. The diborane was introduced last and turned off first. During the deposition, the film thickness was monitored using the interference in the growing film of a He-Ne laser beam [44]. After deposition, all gases were turned off, the reactor was again evacuated and backfilled with N_2 to atmospheric pressure.

3. *Deposition parameters and mechanisms:*

The basic variables which govern the deposition process are the temperature of deposition T_{dep}, the total pressure P_{dep} , and the ratio of the flow rates of the reactant gases of NH_3 to B_2H_6, $R_{dep}=NH_3/B_2H_6$. The ranges of temperature T_{dep} and pressure P_{dep} were respectively 300°-600°C and 0.3-1 Torr. The value of R_{dep} was determined by calculating the net volumes of reactant gases flowing into the reactor, and was varied between 0.1 and 7; the flow rate of diborane was varied from 4 to 40 sccm while the flow

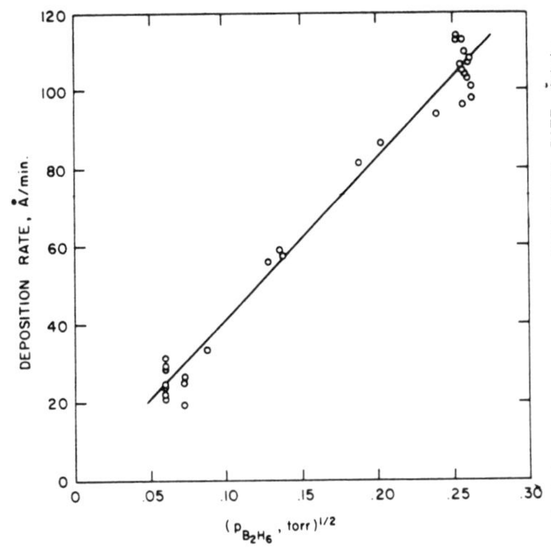

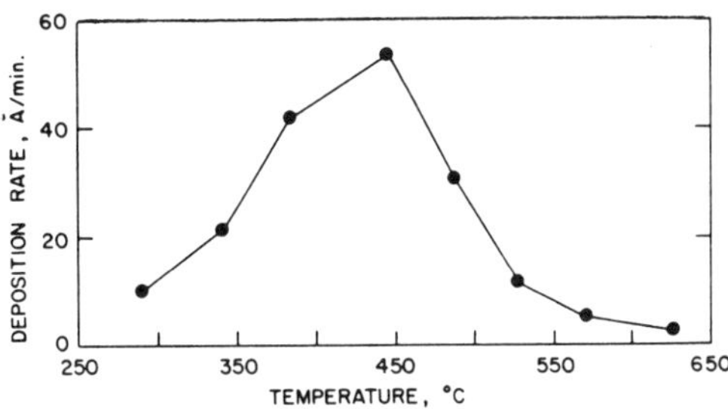

Fig. 3 - Deposition rate vs. the deposition temperature. Pressures are: diborane, 0.005; ammonia, 0.006; total P_{dep}=0.5 Torr [39].

Fig. 2 - Deposition rate vs. the square root of the diborane partial pressure for films deposited at T_{dep}=340°C, total pressure P_{dep}=0.5 Torr [39].

of ammonia was maintained at 10 sccm [44].

The deposition rate of the films as a function of the diborane partial pressure and of the temperature, is shown on Figures 2 and 3, respectively. The rate depends on the square root of the diborane partial pressure (for ammonia partial pressures between 0 and 0.1 Torr), and is independent of the total pressure (at constant reactant partial pressures) [39]. At low temperatures, the rate increases with increasing temperature, and an Arrhenius behavior is observed. The rate reaches a maximum at about 430°C, and then decreases rapidly with additional increases in temperature because of the depletion of reactants [39]. An Arrhenius plot is given in Figure 4 showing data obtained at four different diborane partial pressures. The activation energies range from 20 to 26 kcal/mole, and a concentration dependence was suggested [39]. An activation energy of 15.7 kcal/mole was calculated from the slope of the logarithmic plot of deposition rate versus 1/T for films deposited at P_{dep}=0.7 Torr, R_{dep}=7 and p_{NH3} =0.3 Torr.

Although many reactions are taking place simultaneously, the films appeared to be a compound with a reasonably well-defined stoichiometry [39], rather than an intimate mixture of boron nitride BN and elemental boron B, as was proposed for films deposited at atmospheric pressure [1, 7, 14, 36]: BN formed from the over-all reaction of B_2H_6 and NH_3 ($B_2H_6 + 2NH_3 = 2$ BN + 6 H_2), and B from the nearly independent decomposition of B_2H_6 ($B_2H_6 + 2NH_3 = 2$ B + 3 H_2),

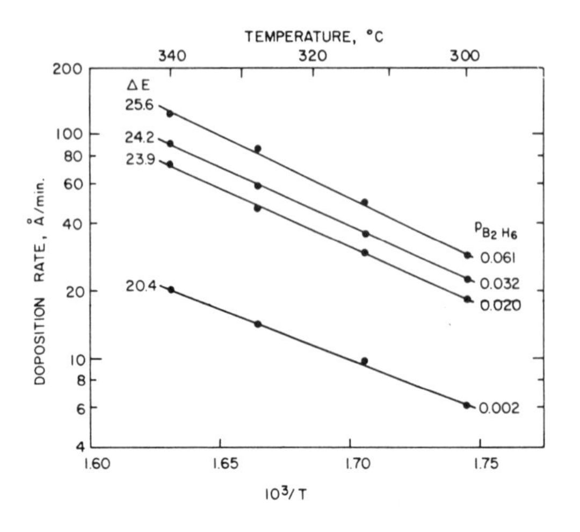

Fig. 4 - Arrhenius plot (P_{dep}=0.5 Torr, R_{dep}=2) [39].

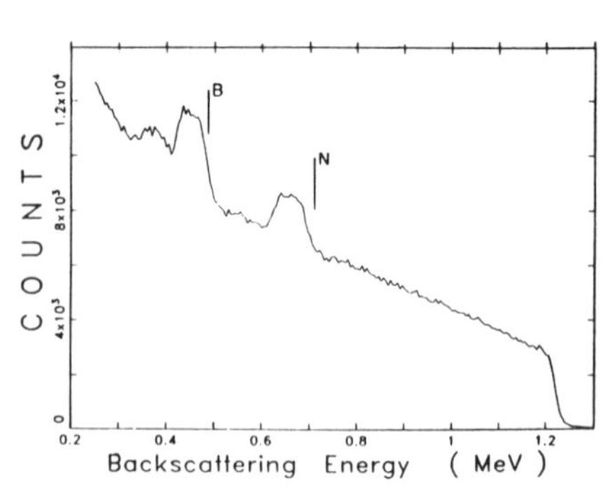

Fig. 5- Rutherford backscattering spectrum [88].

II- CHEMICAL PROPERTIES.-

1. *Chemical composition:*

The atomic density (per cm^2) of boron and nitrogen in the films was analyzed both by Rutherford backscattering and beam channeling to reduce the background from the silicon substrate (a typical spectrum is shown on Figure 5), as well as by electron probe x-ray microanalysis techniques [44]. The measured atomic ratio of boron to nitrogen is plotted in Figure 6 as a function of the gas flow ratio R_{dep} (Note that the straight lines should only be used as visual aid, and not to show a linear dependence). Although the electron microprobe data are only qualitative, they show the same behavior as the Rutherford data, i.e. a decrease of the atomic ratio in the film as the reactant gas ratio increases at a given temperature [44].

The B:N atomic ratio is essentially constant through the thickness of the film, as indicated by the depth profiles obtained by Auger analysis (shown on Figure 7), and by the data of Figure 6 (the film thicknesses of the samples for the Rutherford and microprobe analysis were respectively 0.1 and 1 μm) [88].

Hydrogen is invisible to most analytic methods (methods based on Auger emission are unsuitable because of hydrogen's low Z, RBS because of its light mass, and other

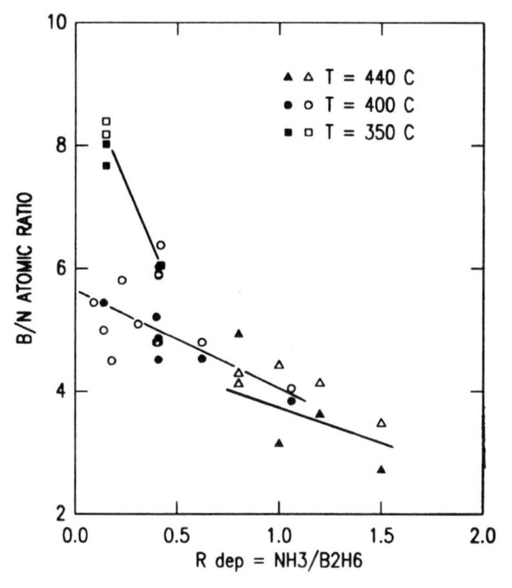

Fig. 6 - Boron to nitrogen atomic Ratio vs. gas Ratio R_{dep} for three deposition Temperatures at P_{dep}=0.36 Torr, obtained by electron microprobe (open symbols) and channeled Rutherford backscattering (solid symbols) [44].

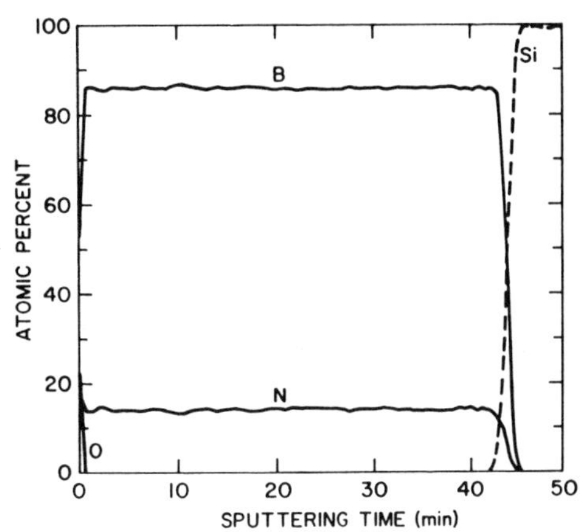

Fig. 7 - Depth profile, Auger analysis [88].

other methods because of hydrogen's nearly universal presence as a contaminant). The hydrogen concentration in the films was analyzed quantitatively [44] using the resonant nuclear reaction :

$^{15}N + H \rightarrow \ ^{12}C + {}^{4}He + \gamma\text{-ray (4.43 MeV)}$.

In this technique, a beam of accelerated $^{15}N^{2+}$ ions impinges upon the film and penetrates the silicon substrate. There is appreciable probability for reaction with hydrogen only at the precise energy 6.385 MeV [50]. Also, as the ^{15}N ions lose energy while they penetrate the films, the beam reaches the resonance energy at some depth. The hydrogen concentration profile vs. depth thus obtained was found [44] to remain essentially constant throughout the 1 micron depth (corresponding to the approximate range of the maximum energy (8 MeV) of the nitrogen ions used in the measurements), indicating that the films were homogeneous in hydrogen.

The atomic percent of hydrogen in the films was calculated from the measured nuclear resonance data, and plotted on Figure 8 as a function of the gas ratio R_{dep} for three

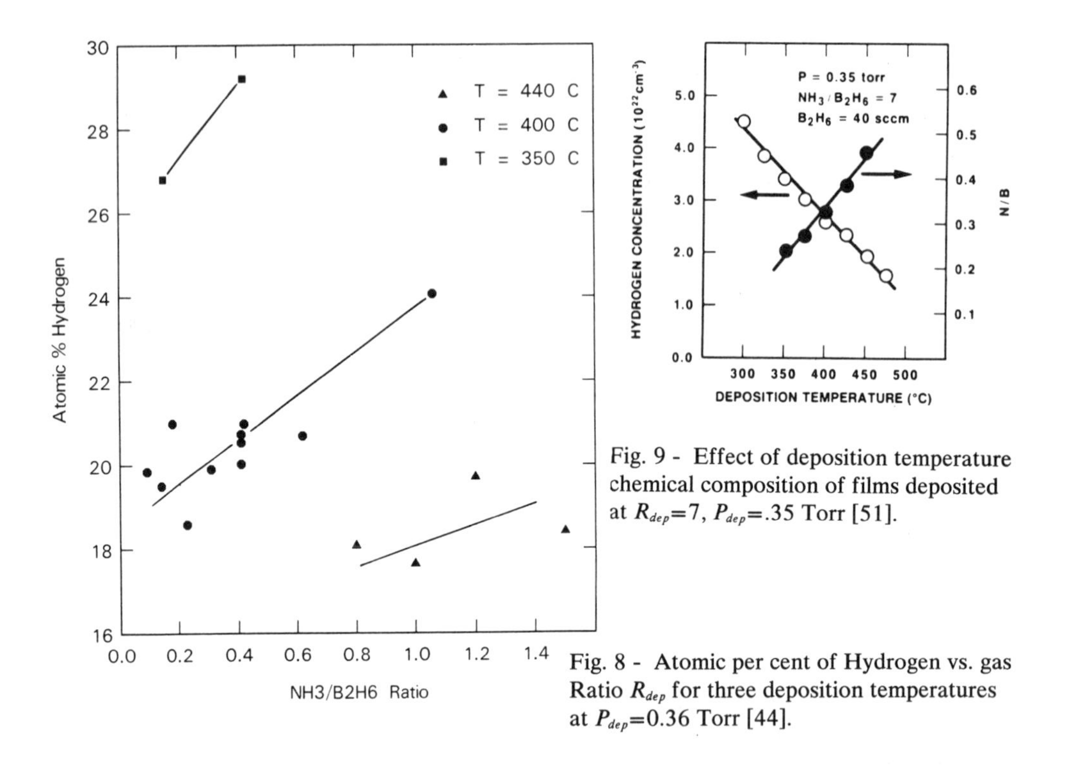

Fig. 9 - Effect of deposition temperature chemical composition of films deposited at R_{dep}=7, P_{dep}=.35 Torr [51].

Fig. 8 - Atomic per cent of Hydrogen vs. gas Ratio R_{dep} for three deposition temperatures at P_{dep}=0.36 Torr [44].

deposition temperatures. It was observed that at a given deposition temperature, the hydrogen content in the films increases with the gas ratio R_{dep}. Also for a given value of the gas ratio, the hydrogen content decreases with increasing deposition temperature [44].

That qualitative behavior was confirmed by other workers with films deposited at R_{dep}= 7 and P_{dep}=0.35 Torr [51]. The effect of T_{dep} on the boron to nitrogen ratio and on the hydrogen concentration in the films is shown on Figure 9. However at P_{dep} = 0.7 Torr, R_{dep} seems to have no effect on the hydrogen concentration in the films [51], as opposed to the dependence shown in Figure 8 at lower pressures P_{dep}. The effect of the B_2H_6 flow rate had little effect on the hydrogen composition or the stress, as shown in Figure 10.

2. ***Etching:***

The chemical stability of the films to various chemicals and gaseous ambients is excellent. The oxygen-free boron nitride films do not change after long times at room ambient. They are extremely resistant towards both acidic and basic aqueous solutions,

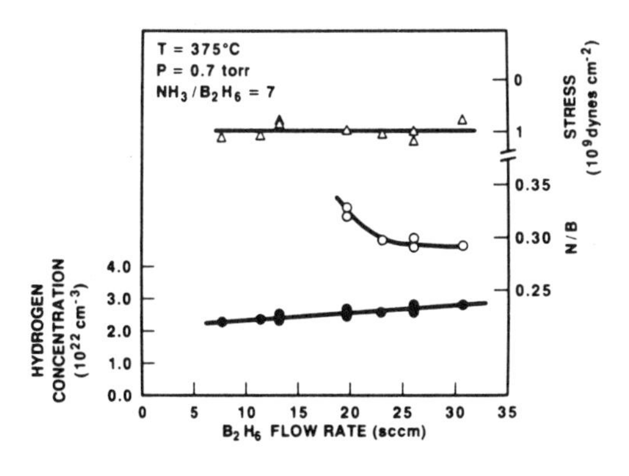

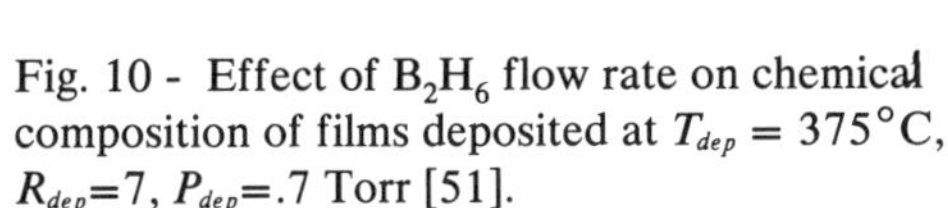

Fig. 10 - Effect of B_2H_6 flow rate on chemical composition of films deposited at T_{dep} = 375°C, R_{dep}=7, P_{dep}=.7 Torr [51].

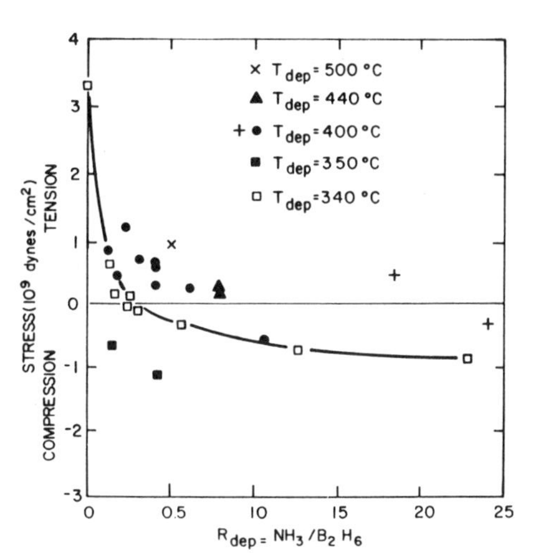

Fig. 11 - Film stress vs. gas Ratio R_{dep}: Open symbols for films deposited at P_{dep}=0.5 Torr, T_{dep} =340°C [39]; Closed symbols at P_{dep}=.36 Torr, T_{dep}=350, 400 and 440°C [44]; + and X [80].

being hydrophobic and insoluble in all etching solutions investigated [39, 44]. In addition, the films can be easily dry etched in a $CF_4 + O_2$ plasma with an etch rate of 0.3 μm/min [39, 44].

III- MECHANICAL PROPERTIES.-

1. *Stress:*

The average stress in the films was determined using Stoney's equation by measuring the radius of curvature of the silicon substrate before and after deposition [93], with an optically levered laser beam [39] or by x-ray diffraction [44]. The stress versus the gas ratio R_{dep} is plotted in Figure 11 for various deposition temperatures. The open symbols are for films deposited at 340°C and 0.5 Torr [39], the full symbols are for films deposited at 350°C, 400°C, 440 °C and 0.36 Torr [44]. Both data are in qualitative agreement, the stress changes from tensile to compressive as the gas ratio R_{dep} is increased.

The effect of the total pressure P_{dep} on the stress of the films is shown on Figure 12, and is independent of the B_2H_6 flow rate over the range of 7-40 sccm investigated.

2. ***Young's modulus:***

The value of $(E/1-\nu)$, where E is the Young's modulus of the films and ν is the Poisson's ratio, was found to be $2.0x10^{12}$ dyne/cm^2 for a film of composition $B_9N_2H_3$ measured using the vibrating-reed method [45, 53], $1.7x10^{12}$ dyne/cm^2 for B_3NH films [46] using the pressure-bulge technique [54, 55], and $2.4x10^{12}$ dyne/cm^2 using the acoustic resonant frequency method [55].

The fracture strength of a 1 inch square membrane 1.7 μm thick was $3.6x10^9$ dyne/cm^2, as obtained by the bulge technique for films deposited at R_{dep}=0.4, T_{dep}=400°C [56], and > $5x10^9$ dyne/cm^2 for a 4 μm thick membrane [55].

3. ***Adhesion:***

The adherence of the films, to silicon and SiO_2 substrates, has been entirely satisfactory [1, 39, 44]: loss of adhesion was not observed even after various chemical treatments, after etching patterns in a plasma, or after repeated temperature cycling between 25°C and 450°C. Peeling and cracking did occur [39] only when the films were heated for several hours in steam at 2 atm pressure and 125°C (due to a reaction at the film- silicon interface), or annealed in vacuum at temperatures above 500°C.

Films thicker than 0.1 μm are pinhole-free. The film surfaces are quite smooth. The only features observable were spherical hillocks ranging in size up to 1 μm in diameter; their density increased as the deposition temperature was decreased [1, 39].

IV- STRUCTURAL PROPERTIES.-

1. ***Microstructure:***

The microstructure of the as-deposited films was amorphous, as determined by transmission electron diffraction patterns [39, 44, 51] and x-ray diffraction analysis [45]. The grain size was estimated to be less than 250 nm [39].

The measured average x-ray transmission of the films for x-rays around 1 nm was about 90% for films of composition $B_{10}N_2H_3$ and 1.5 μm thickness [44], i.e. 20% higher than the calculated x-ray transmission for stoichiometric BN. This value is in agreement with the measured film densities: 1.7 g/cm^3 [44], 1.74 g/cm^3 [55], 1.63 g/cm^3 [46], 1.6 g/cm^3 [53] and 1.9 g/cm^3 [30] (the BN bulk density is 1.40-2.14 g/cm^3, increasing with T_{dep} [31, 33, 41, 43]).

X-ray photoelectron (XPS) and Auger (AES) electron spectroscopies of the films have been reported [57]. Point defects in amorphous boron nitride were observed by electron paramagnetic resonance (EPR) in as-deposited films [58], although no ESR signal was detected on B_3NH films [59].

2. ***Bonds:***

Evolved gas analysis techniques have been used to measure the rate and relative amount of evolution of hydrogen as a sample was heated and weaker bonds were broken. Based on a clear separation of the temperatures at which hydrogen was evolved, hydrogen was shown to exist in several bonding configurations: a weakly bonded state which is removed at relatively low temperatures, and much more strongly bonded states which require temperatures in excess of 800°C before the bonds are broken [46]. The higher temperature sites have been correlated with IR spectra and correspond to hydrogen bonded mainly to B and to N [59].

Infrared absorption spectroscopy showed large differences in the chemical bonding of hydrogen to both boron and nitrogen as BH, NH_2 and possibly NH [15, 29, 36, 39, 44]. The broadness and frequency shifting of the BN band also depends on experimental conditions. Peaks were seen at 3400 cm^{-1}; 3220 cm^{-1} and 1540 cm^{-1}; 2500 cm^{-1}; strong and asymmetric at 1400 cm^{-1}; and weaker and sharper at 800 cm^{-1}. These absorption bands were respectively attributed to the characteristic vibrations of N-H; NH_2; B-H bond stretching; B-N bond stretching and to B-N-B bond bending [1, 8, 22, 36, 39, 51, 60, 61]. The spectrum was insensitive to variations in the ratio R_{dep} (for $0.5 < R_{dep} < 5$), and sensitive to the deposition temperature (Figure 13) [39]. Similar spectra were observed for films deposited at temperatures greater than 550°C by pyrolysis of borazine [29].

There was no evidence from the ^1H NMR spectra of molecular H_2 trapped in the

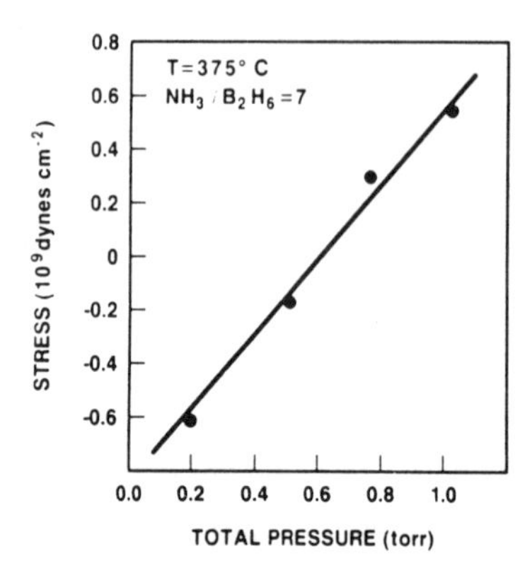

Fig. 12 - Effect of total pressure on stress for films deposited with T_{dep} = 375°C, R_{dep}=7 [51].

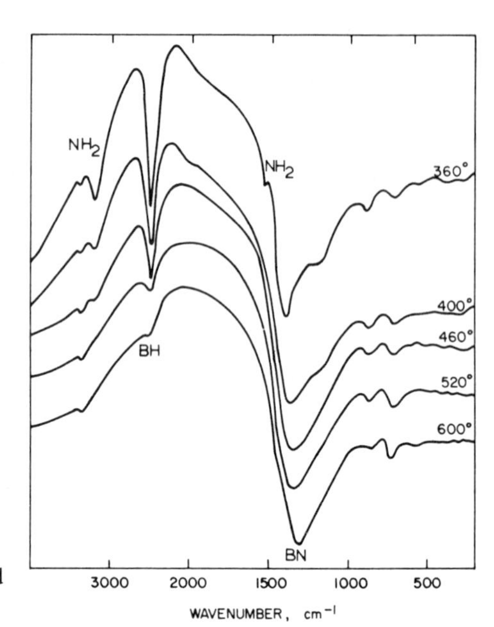

Fig. 13 - Percent transmitance for films deposited at different temperatures, with R_{dep}=1.3 [39].

films; hydrogen clusters and dihydride species (-BH_2 or -NH_2) were ruled out as the predominant bonding arrangement of hydrogen [59]. Hydrogen was found to be essentially homogeneously distributed throughout the films.

3. ***Film anneal:***

Annealing of the films in inert gas at temperatures as high as 440°C did not appreciably change their hydrogen content [44]. However hydrogen could be removed from the films when heated in vacuum at temperatures above 450°C [39, 46]. The kinetics of hydrogen desorption and the resulting effects on stress are illustrated in Figure 14; at stress values ≥ 3x10⁹ dynes/cm², film delamination prevailed. The N-H peaks progressively weaken and thoroughly disappear at 700°C, at which temperature the B-H peak initiates an apparent decrease in intensity and vanishes by 1100°C. Onset of phase separation was observed within the film at 900°C, with precipitation of an elemental boron-rich phase; phase separation was accompanied by severe darkening of the film and a reduction in stress to final values ≤ 2.0x10⁹ dynes/cm², thus insuring mechanical integrity of the films [51].

Although increases in tensile stress were shown to result from annealing in vacuum, reversible changes could be accomplished by annealing in a hydrogen ambient, as re-

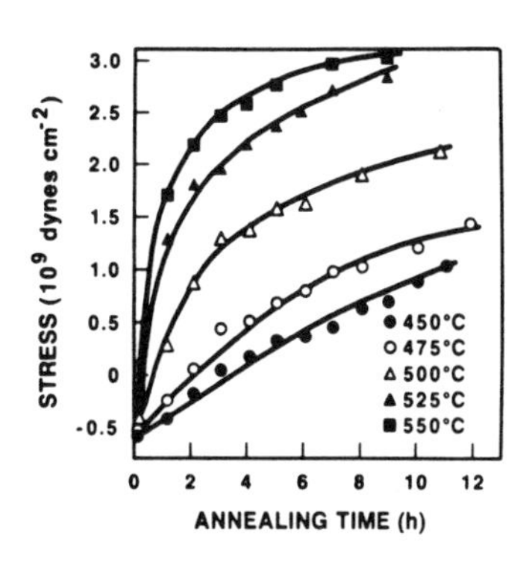

Fig. 14 - Effect of thermal treatment in vacuum (10^{-3} Torr) on stress of B_3NH films [51].

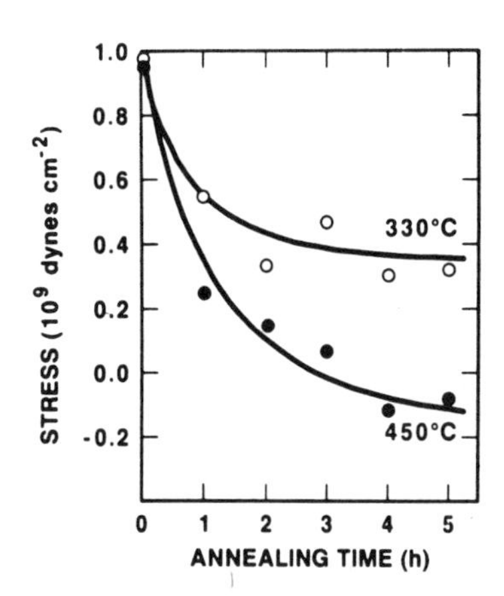

Fig. 15 - Effect of thermal treatment in hydrogen on stress of B_3NH films [51].

vealed in Figure 15 [51].

V- OPTICAL PROPERTIES.-

The films are colored due to their non-stoichiometry. On increase in the boron content in the films, they change from a light yellow to a dark brown (the stoichiometric boron nitride BN is colorless and transparent, because of its wide band gap).

The dispersion in the data presented in the literature confirms the critical role that the particular film deposition technique exerts on the optical properties [60]. Amorphous boron nitride with a B/N ratio of 1 has a refractive index of 1.8-1.9 at a wavelength of 633 nm [35] and 1.7-1.8 at 541 nm [7, 8, 15], while polycrystalline films have a slightly higher index of 1.9-2.0 [60], and cubic boron nitride has an index of 2.12 [63]. The absorption edge E_g reported was approximately 5 eV for amorphous BN [22, 64], 5.90 eV [28, 36] for stoichiometric boron nitride BN, and 5.83 eV [60] and 3.84 eV [1].

The refractive index n and the extinction coefficient k, which is related to the absorption coefficient α through the equation $\alpha = 4\pi k/\lambda$, were measured by ellipsometry [62], or calculated from the positions of the extrema of the transmission spectrophotometer data [44]

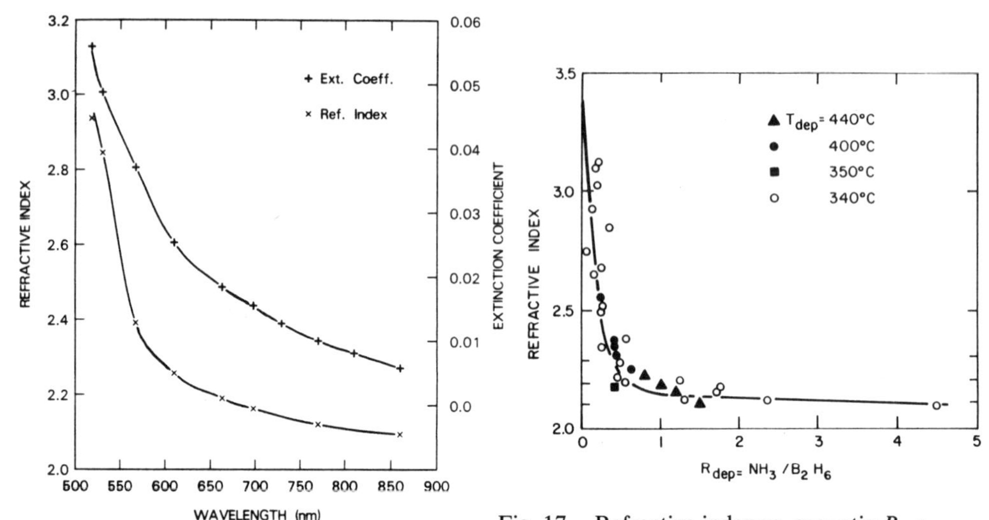

Fig. 16 - Refractive index and extinction coefficient vs. Wavelength (nm), measured on film deposited at T_{dep}=400°C, R_{dep}=0.41 and P_{dep}= 0.36 Torr [44].

Fig. 17 - Refractive index vs. gas ratio R_{dep}: Closed symbols measured at 600 nm for three deposition temperatures, at P_{dep}=0.36 Torr [44]; Open symbols measured at 546 nm on films deposited at T_{dep}= 340°C and P_{dep}=0.5 Torr [39].

(see Figure 28). Values of n and k are shown in Figure 16 as a function of wavelength λ, for a given set of deposition parameters (T_{dep} = 400°C, R_{dep} = 0.41, P_{dep} = 0.36 Torr). The maximum absorption occurs in the ultraviolet with a long tail extending throughout the visible region.

The refractive index and extinction coefficient for B_3NH films at 632.8 nm are respectively 2.0 and 0.005, and correspond to a 65 % transmission value for a 4 μm thick film [46]. They decrease rapidly as R_{dep} increases from 0 to 0.5, and are nearly constant when $R_{dep} > 1.5$. They are plotted on Figs. 17 and 18 as a function of the gas ratio R_{dep}, for various deposition temperatures.

The temperature dependence is more complex. The absorption coefficient increases with increasing deposition temperature, as shown on Figure 19, while the index goes through a broad maximum at about 400°C [39]. Figure 19 compares to the data for films formed by pyrolysis of borazine [29]. The index is the same at deposition temperatures greater than 550°C. However, at lower temperatures, the refractive index is much higher. This is a strong indication that the films are not the same.

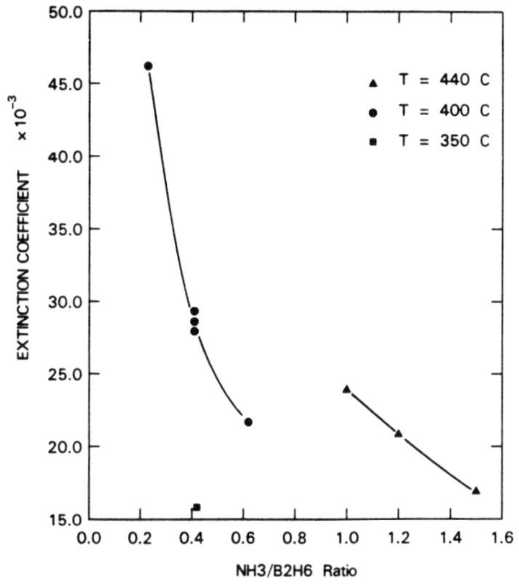

Fig. 18a - Extinction Coefficient measured at 600 nm vs. gas ratio R_{dep} for three deposition temperatures, at P_{dep}=0.36 Torr [44].

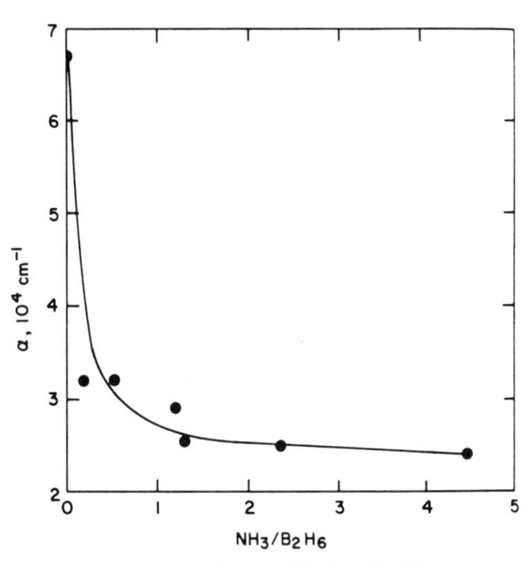

Fig. 18b - Absorption coefficient for films deposited at T_{dep}=340 °C and P_{dep}=0.5 Torr [39].

VI- ELECTRICAL PROPERTIES.-

Electrical transport properties of the films have been studied by measuring their current-voltage characteristics at room temperature [1, 45, 57].

In [45], MIS capacitor structures were made by evaporating aluminum electrode dots onto the BN films 1 to 3 μm thick deposited on silicon substrates. The current intensity was recorded at room temperature while the voltage was linearly ramped from 0 to 100 Volts at 1 Volt/sec. To compare films with different thicknesses, the current density at an electric field of E_0=2.3 x10^5 Volt/cm was used to calculate the ratio of the field over the current density, which was called the 'electrical resistivity' [45].

The 'electrical resistivity' was found to be strongly dependent on the film deposition parameters [45]. The influence of the gas ratio R_{dep} at a temperature of T_{dep}=400 °C, shown in Figure 20, and of the deposition temperature T_{dep} at a gas ratio R_{dep}=1, were determined [45]. At this temperature, the film's 'electrical resistivity' increases as the gas ratio R_{dep} is increased. At R_{dep}=1, the film's resistivity decreases as the deposition temperature T_{dep} is increased [45].

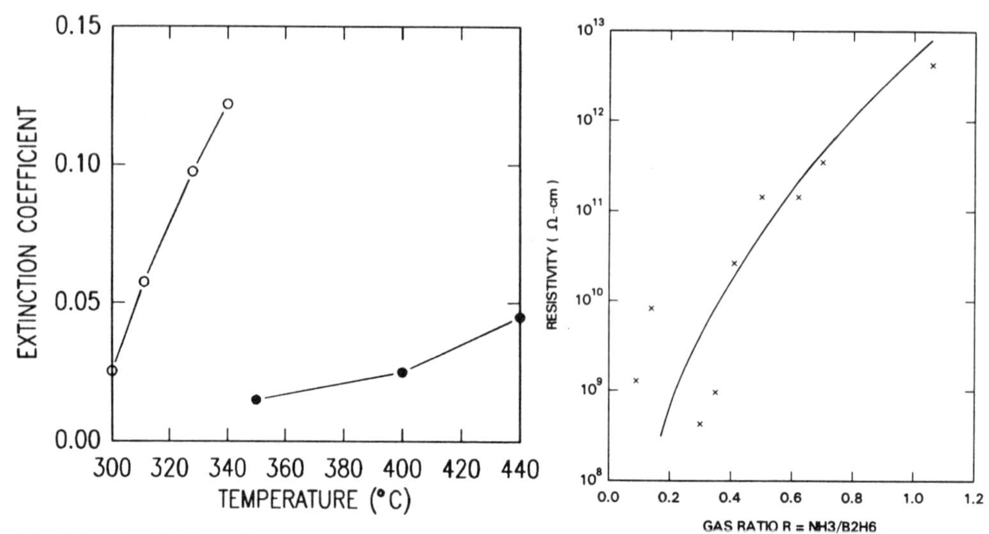

Fig. 19 - Qualitative effect of the deposition Temperature on the extinction coefficient at 600 nm: Open symbols for films deposited with R_{dep}=1.3 at pressure P_{dep}=0.5 Torr [39]; closed symbols for films deposited with R_{dep}=0.5 at pressure P_{dep}=0.36 Torr [44].

Fig. 20 - Effect of the deposition gas ratio R_{dep} (at T_{dep}=400°C, P_{dep}= 0.36 Torr) on the film's electrical resistivity [45].

It has been observed that the films are conducting even at low voltages [45], indicating that the dominant conduction mechanism is most likely to be electronic charge hopping through traps [1, 45]. The current density plotted on a logarithmic scale is linearly dependent on the square root of the applied electric field E. The electronic conductivity decreases when a B_3NH film is annealed at 450°C in a H_2 ambient, reflecting a reduction in the number of available traps caused by the hydrogen tieup of dangling bonds [51]. The conduction through the BN films in high electric fields can be well explained by the Poole-Frenkel type mechanism, which is controlled by the field-enhanced thermal excitation of trapped electrons into the conduction band [1, 51, 57, 65]:

$$\sigma = \sigma_0 \exp(\sqrt{(e^3E/\pi k_{op}\varepsilon_0)}\ /kT),$$

where e is the magnitude of the electronic charge, k_{op} is the dynamic dielectric constant of the material, ε_0 is the permittivity of free space, k the Boltzmann constant, and T the absolute temperature.

The breakdown field strength was found to be more than 3×10^6 V/cm [57], 5×10^6 V/cm [1, 66], 10^7 V/cm [5], and about 2×10^7 V/cm [3, 34]. At low field (below 10^5 Volt/cm), the conductivity was ohmic, with an electrical resistivity strongly increasing as T_{dep} increases, with reported values of 10^{10} to 10^{15} Ω.cm [67], of > 10^{16} Ω.cm [57], and about 10^{17} Ω.cm

[3, 34].

The calculated value for the dynamic permittivity $\varepsilon_i = k_{op}\varepsilon_0$ was consistent with that determined from capacitance measurements [51]. Typical capacitance versus gate voltage characteristics were performed at 1 MHz [45, 57]. From this data, the film thickness and the capacitor area, the static dielectric constant was estimated to be 5.8 for $B_{10}N_2H_3$ films [45], in agreement with the values 5.75 [57] or 5.3-7.7 for polycrystalline BN [16], but larger than the obtained values of 4.4 for microcrystalline films [34], 3.3-4.0 for high T_{dep} amorphous films [1, 5] or 3.7 for PECVD deposited films [13].

VII- CORRELATIONS-

We have so far recognized that the tensile stress, the optical transmission and the electrical resistivity of the films depend very critically on the deposition conditions.

It was deduced from Figs. 11 and 18a that as the gas ratio R_{dep} is increased at a given temperature T_{dep}, the stress in the films becomes more compressive and the films become optically more transparent. The same effect can be observed in Figure 19, as the temperature is decreased for a constant value of the gas ratio.

The fact that films under both tension and compression were found with approximately the same B:N atomic ratio [44], indicates that the stress does not depend substantially on the boron content of the films, as previously stated for atmospheric pressure CVD boron nitride films [7]. Furthermore the effect of deposition parameters on the composition indicates that the stress remains constant while the B:N ratio varies between 2 and 3 [44, 51]. Also, as T_{dep} is decreased from 450°C to 300°C, both the index decreases (Figure 21) and the B/N atomic ratio increases (Figure 6), in contradiction with other reports that boron rich films have a higher refractive index [14].

However, the observed behavior can be explained in terms of the effect of the hydrogen content both on the stress and the optical properties [44, 46], and is shown explicitly in Figures 22 and 23. Figure 23 shows that at a given temperature, the optical transparency increases with hydrogen concentration, while Figure 22 shows that the stress of the films depends on the hydrogen content in the opposite way [45].

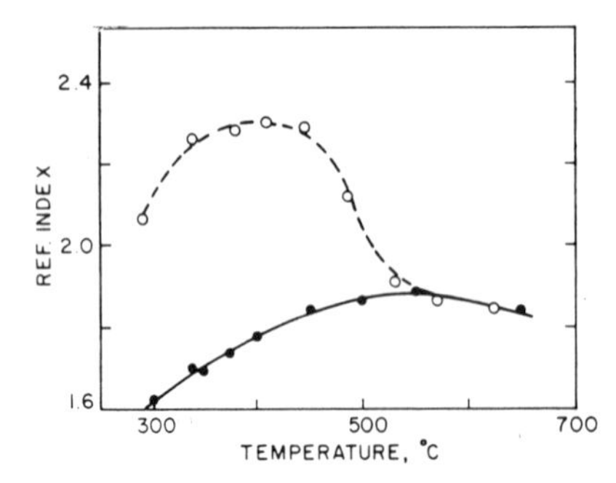

Fig. 21 - Refractive index measured at 546 nm:
Solid symbols: Borazine, P_{dep}=0.3 Torr [29].
Open symbols: Diborane, P_{dep}=0.5 Torr, R_{dep}=1.3 [39].

We also deduce from Figure 22, that when the hydrogen content is less than about 21 atomic % the stress is tensile, and that for hydrogen concentrations above 24 at.%, the stress is compressive. A plot of the measured stress vs. hydrogen concentration, shown in Figure 24 [46, 51] for various annealed films confirms the correlation reported previously [44] between these two parameters. This relation between stress and total hydrogen concentration was used to tailor the stress of the films, either by adjusting the deposition conditions [39] or by an annealing process [46, 51, 68].

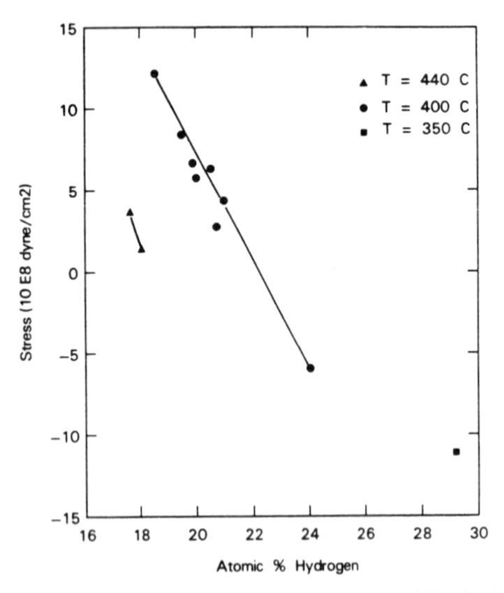

Fig. 22 - Stress vs Atomic per cent of Hydrogen for three deposition temperatures, at pressure P_{dep}= 0.36 Torr (Each data point corresponds to films deposited with a different set of experimental parameters) [44].

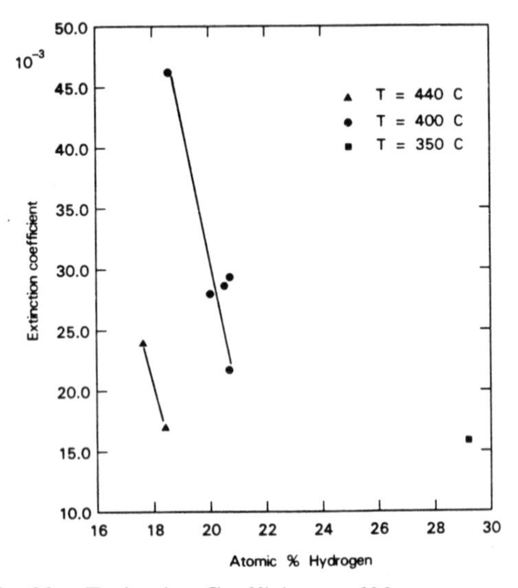

Fig. 23 - Extinction Coefficient at 600 nm vs. Atomic per cent of Hydrogen for three deposition temperatures, at pressure P_{dep}=0.36 Torr (Each data point corresponds to films deposited with a different set of experimental parameters) [44].

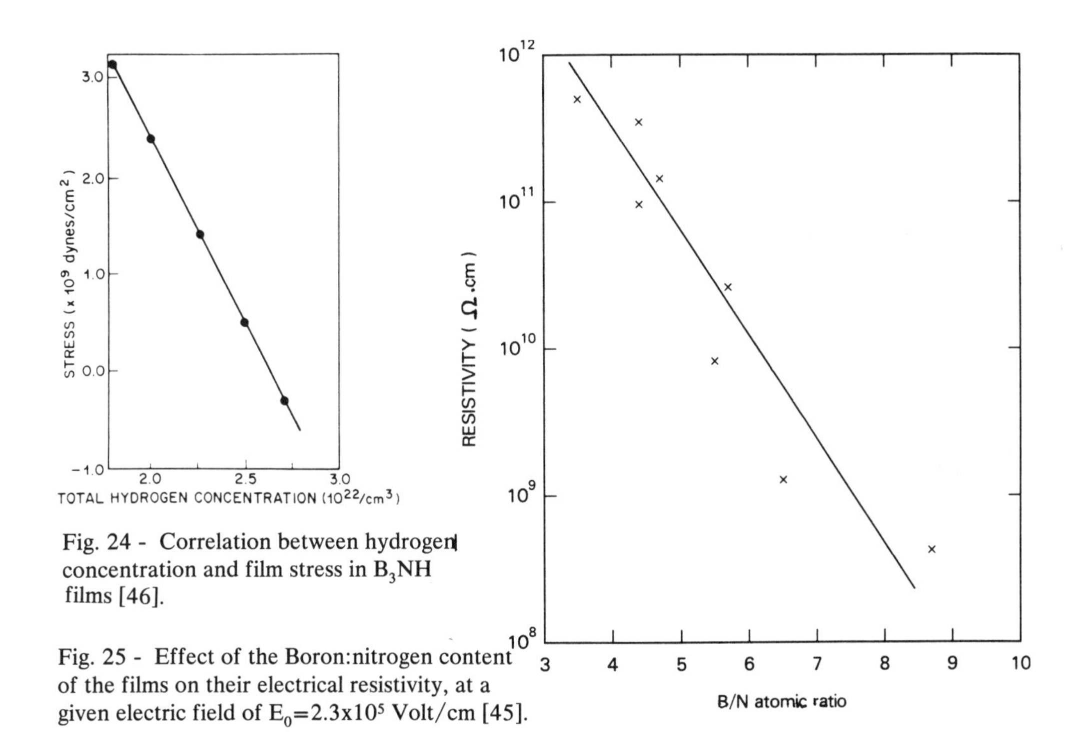

Fig. 24 - Correlation between hydrogen concentration and film stress in B_3NH films [46].

Fig. 25 - Effect of the Boron:nitrogen content of the films on their electrical resistivity, at a given electric field of E_0=2.3x10^5 Volt/cm [45].

The effect of the chemical composition on the electrical resistivity of films deposited at 400°C is shown in Figures 25 and 26. The less resistive films are those with the highest boron content, and also the ones with the lowest hydrogen content [45].

The film stress and electrical resistivity were measured in each sample of different stoichiometry, and the results are plotted on Figure 27. The resistivity is also plotted on Figure 28 versus the optical absorption coefficient of films deposited under the same conditions. These two plots show an inverse correlation between electrical resistivity and both mechanical stress and optical transparency: the least resistive films have both the highest tensile stress and the lowest optical transparency [45].

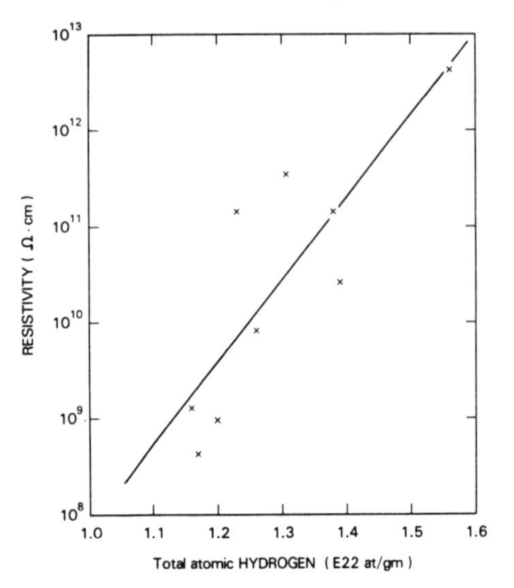

Fig. 26 - Effect of the hydrogen content of films deposited at T_{dep}=400°C, P_{dep}=0.36 Torr on their electrical resistivity at a given electric field of E_0=2.3x10^5 Volt/cm [45].

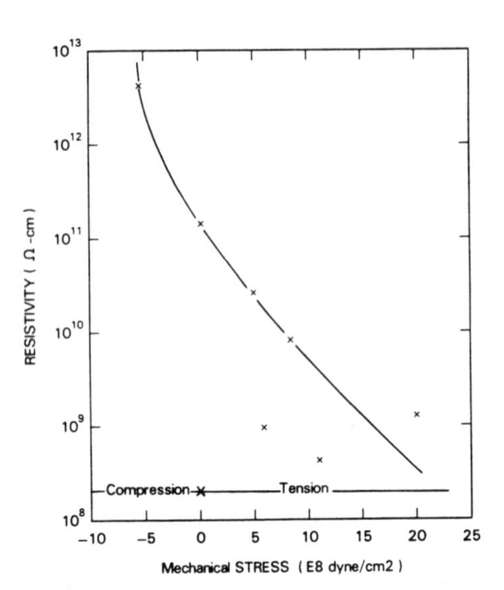

Fig. 27 - Relation between electrical and mechanical properties [45].

VIII- EFFECTS OF RADIATION ON FILM PROPERTIES.-

1. *Effect on Optical Properties:*

Upon x-ray radiation, the optical transmission of a membrane decreased, as can be observed in Figure 29, which shows the spectra of the same membrane before and after synchrotron irradiation with total consecutive incident energy densities of 700 and 1100 J/cm². After irradiation of half of a membrane area, the optical transmission in both the adjacent irradiated and non-irradiated areas was measured with a spectrometer attached to an optical microscope [45]. The relative change in the optical absorption coefficient was calculated from the ratio of the corresponding measured intensities, and plotted on Figure 30 as a function of irradiation dose for three different film compositions.

The optical degradation of B_3NH films 4 μm thick, approached a saturation transmission value of around 40 % at 632.8 nm after a dose of about 200 kJ/cm³ (it was > 65% before irradiation [51]). Although there was some tailing off in the rate of

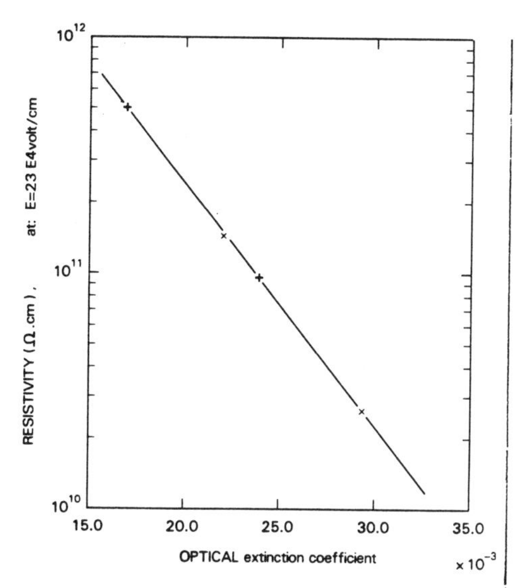

Fig. 28 - Relation between electrical and optical properties [45].

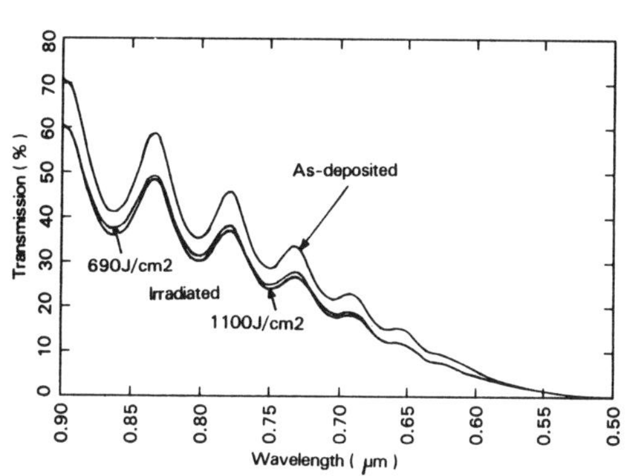

Fig. 29 - Optical transmission spectra of membrane, as-deposited and after total incident energy densities of 690 and 1100 J/cm² [45].

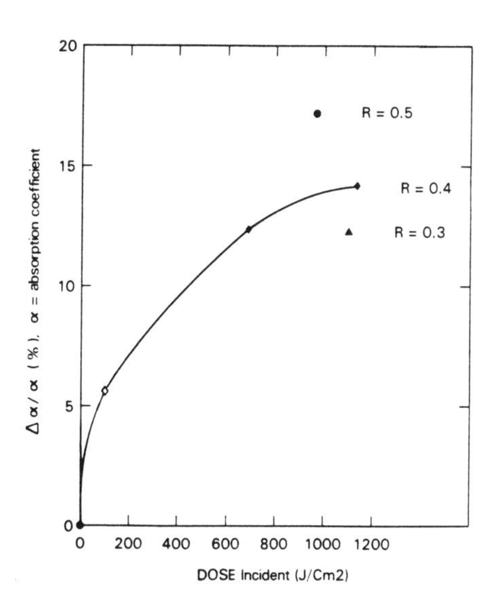

Fig. 30 - Effect of radiation on the optical absorption [45].

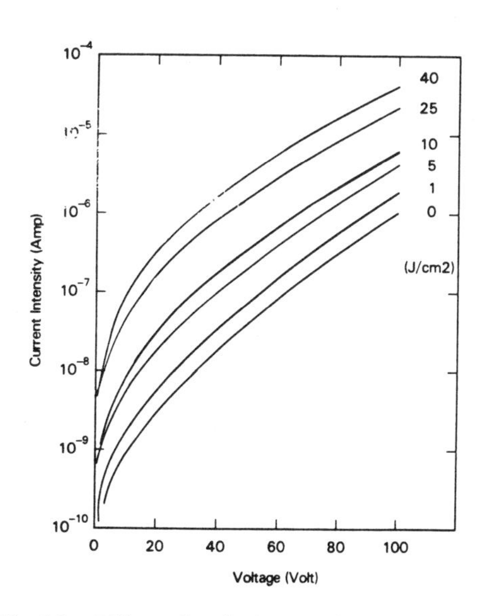

Fig. 31 - Effect of radiation on the electrical resistivity of film deposited at T_{dep}=400°C and P_{dep}=0.36 Torr [45].

increase of the absorption coefficient, saturation was less apparent in a film of composition B_4NH, where the transmission fell from 50% to 20% after absorbing 250 kJ/cm^3 [30].

Films deposited at lower values of R_{dep} [45] or at higher temperatures T_{dep} [30], were less affected by the absorbed radiation. Also there was no evidence of optical degradation on near-stoichiometric BN films after an exposure dose of 2x10^3 kJ/cm^3.

The reduced transmission as a function of absorbed dose in the film was found to be primarily dependent upon the cumulative absorbed dose, and independent of the x-ray flux or source [46]. One mechanism cited by these authors is the breaking of chemical bonds as x-rays are absorbed. The dangling bonds formed within the film will create energy levels in the bandgap, traps or color centers, which affect the optical transparency of the films.

2. ***Effect on Mechanical Properties:***

The maximum strain change induced into the irradiated area was 269 ppm for a film with B_3NH composition at an absorbed dose of 130 kJ/cm^3 [46], 60 ppm for a B_4NH composition at an absorbed dose of 3000 kJ/cm^3 [69], and less than 80 ppm for a $B_{10}N_2H_3$ composition at 1100 kJ/cm^2 [45]. The induced distortion is much smaller in films deposited either at higher temperatures T_{dep} [30] or at higher values of R_{dep} [45]. Also a B_3NH film which had been annealed at 1050°C for 3 hrs in a N_2 ambient to remove its hydrogen, was mechanically stable under synchrotron radiation after an exposure of 4.3x10^3 kJ/cm^3.

Moreover the total strain change increases with the cumulative absorbed dose, but also tends to saturate at a value dependent upon the incident x-ray flux [30, 46].

The strain change corresponds to a decrease in the film tension, since the elastic constant is unchanged after the sample is irradiated [45, 46]. The magnitude of the tensile stress change per unit of absorbed dose (kJ/cm^3) is 1.2x10^4 dyne/cm^2 for a B_4NH membrane deposited at T_{dep}=500°C [30, 69], 3.85x10^6 dyne/cm^2 for a B_4NH membrane [30], 3x10^6 dyne/cm^2 for a B_3NH membrane deposited at T_{dep}=400°C [46], and 6.8x10^6 dyne/cm^2 for a film deposited at T_{dep}=400°C and R_{dep}=1.25 [53].

3. ***Effect on chemical composition and structure:***

The boron-to-nitrogen ratio and the hydrogen content were measured before and after irradiation, for three values of the deposition gas ratio R_{dep}. Within experimental error, no change in composition could be detected after x-ray irradiation with incident energy densities up to 1100 J/cm^2 [45].

Infrared transmission measurements of the membranes did not show any significant structural change after irradiation. Even in the worst case of radiation damage where the membrane wrinkled, the infrared spectra of exposed and unexposed regions show that no chemical bonds disappeared and that no new bonds were formed [45]. Measurements of the boron and nitrogen near-edge x-ray absorption spectra showed no resolvable difference, suggesting that the bonding between the majority of boron and nitrogen atoms is largely unaffected by the radiation [30].

From NMR it was concluded that the damage sites are not correlated to the majority hydrogen sites, i.e. the hydrogen does not aggregate to damage sites. The ^1H NMR linewidth was essentially unchanged from as-deposited to irradiated, which suggests that hydrogen spatial distribution was unaffected [59].

ESR spectroscopy of irradiated films indicated that the resonance consists of a single symmetric line centered at g=2.005 with a peak-to-peak derivative linewidth of about 37 G [58], and 33 G [59]. An increase of as much as a factor of three was measured on films irradiated to 1 kJ/cm^2 [58]. The ESR signal intensity yields a spin density of $7.3 \times 10^{19}/g$ in films irradiated with a dose of 30 kJ/cm^3, i.e. approximately 1 unpaired spin per 175 nominal B_3NH units [59].

A damage mechanism was proposed for the films irradiated at sufficiently high doses and dose rates [30, 46]. The observed mechanical instability was explained by the multiple bonding configurations for hydrogen [59]. Upon irradiation the bonding configurations of the hydrogen are modified such that more of the hydrogen becomes weakly bonded, causing the films to become less tensile and relax.

4. ***Effect on Electrical Properties:***

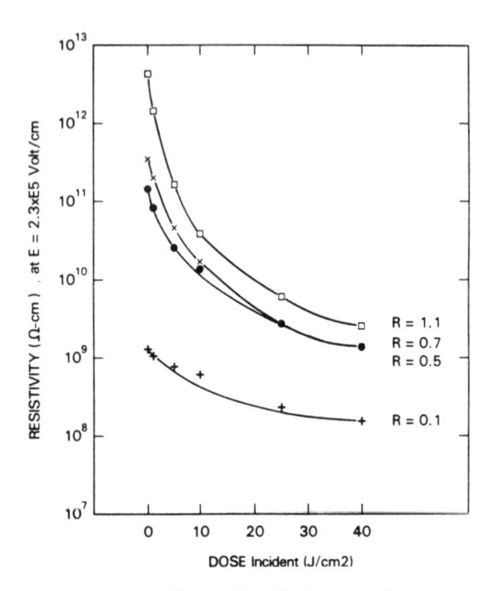

Fig. 32 - Effect of radiation on the electrical resistivity of various films deposited at 400°C (R=R_{dep}) [45].

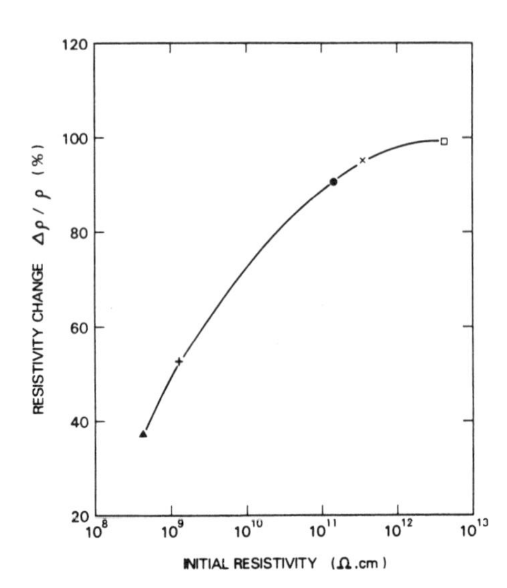

Fig. 33 - Influence of the initial film resistivity on the electrical damage after an irradiation dose of 10 J/cm² [45].

In contrast to similar measurements in SiO_2 films, no gross shift in the C-V curve was observed before and after irradiation with incident energy densities of 10 and 100 J/cm² [45].

However, the electrical resistivity was observed to decrease after irradiation, as shown in Figure 31 for successive exposures to total incident energy densities of 1, 5, 10, 25 and 40 J/cm² [45]. It is interesting to note that electrical damage can be detected at very small doses compared to mechanical or optical damage; this behavior may be used to advantage in the monitoring of the effects of radiation on the deposited films.

The fact that the film becomes more conductive upon irradiation seems to be due to the generation of electrically-active defects. Indeed, a strong electron spin resonance signal reveals the radiation damage as an increase of the radiation-induced defect concentration. At a constant bias, the enhancement in conductivity observed with increases in absorbed dose is direct evidence for the progressively higher density of traps generated [51].

Figure 32 shows the influence of the gas ratio R_{dep} on the extent of the damage; the relative change in resistivity after an incident energy density of 10 mJ/cm^2, is shown in Figure 33. These results show that the rate of change of the electrical resistivity with dose, increases with the value of the film's initial (as deposited) resistivity [45].

IX- BORON NITRIDE X-RAY LITHOGRAPHY MASK SUBSTRATES.-

1. ***Boron nitride x-ray mask membranes:***

X-ray lithography is a proximity lithographic technique to replicate patterns with feature dimensions of 0.5 μm or less, and is being investigated for the production of very large scale integrated circuit (ULSI) devices. An x-ray mask consists of a thin-film mêmbrane which is held in tension by a supporting frame, and that supports the x-ray absorber which makes up the circuit pattern. One-to-one masks are required. A cross section of a typical mask structure is shown in Figure 34.

Boron nitride films have been widely selected for use as a membrane material in x-ray lithography mask substrates [19, 44, 51, 55, 70-83]. LPCVD boron nitride films were most desirable for their key properties: very low x-ray absorption coefficient (low atomic number material), controllable tension to achieve membrane flatness and me-

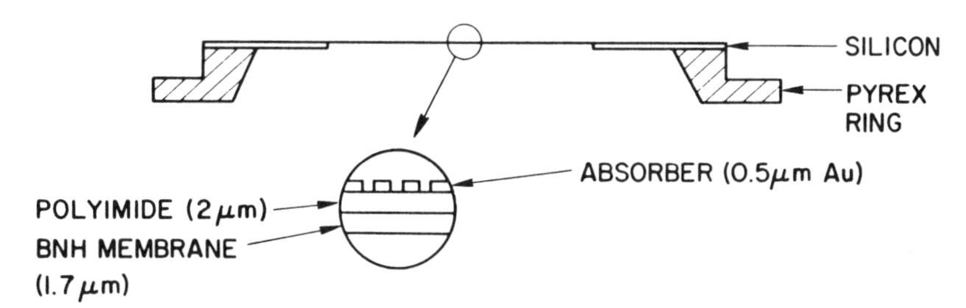

Fig. 34 - X-ray lithography mask structure [88].

chanical strength, high Young's modulus that provides the dimensional stability required of less than 6 ppm in order to minimize the mechanical distortion induced by the stress in the absorber, adequate optical transparency required for alignment of mask to wafer during registration, thermal expansion coefficient close to that of the silicon frame to minimize the variation in tension, low defect density, ease of fabrication and chemical inertness which are advantageous throughout all phases of mask processing.

2. ***Fabrication:***

For adequate x-ray transmission and image contrast, the membranes were required to be about 4 μm thick when utilizing a conventional x-ray source [85], and about 2 μm thick with the soft x-ray spectrum of an electron storage ring source [44, 84]. The mechanical tension and optical transparency of the films were optimized in one of two ways: either by tailoring the deposition parameters [44] or by an anneal process after deposition to adjust the hydrogen content [68]. It was demonstrated that membranes approximately 2 μm thick and of large area (8 cm in diameter) could be easily fabricated [44, 46, 51, 55, 77, 79].

One fabrication sequence used to fabricate the x-ray masks is outlined in Figure 35, and is as follows: the boron nitride film was deposited by LPCVD on both sides of a sacrificial silicon wafer 10 cm in diameter, of (100) orientation, and polished on both sides. After defining a central exposure area and several alignment windows on the back of the wafer by reactive ion etching, the silicon substrate was etched chemically in a solution of potassium hydroxide in these areas to a thickness of about 10 μm, using the B-N-H film as an etch mask. Final etching was completed in a mixture of hydrofluoric-nitric-acetic 1:1:1 acid solution. The two-step etching was required to control the size of the square alignment windows. The silicon substrate was then bonded onto a Pyrex ring to ensure structural strength during handling. A completed x-ray mask blank is shown on Figure 36. Finally, the substrate was overcoated with a two micron thick polyimide layer for additional durability, and ready to be patterned by the use of electron beam lithography.

3. ***Stability:***

Such membranes have performed satisfactorily with conventional electron impact sources, and were routinely fabricated for the earliest pilot-line applications of full-field x-ray lithography with palladium sources [16, 70, 79, 80, 85-87], such as 1-Mbit bubble memory [77]. Long term stability of the mask with these low power sources was measured to be better than +- 0.03 μm (1 σ) [55, 80].

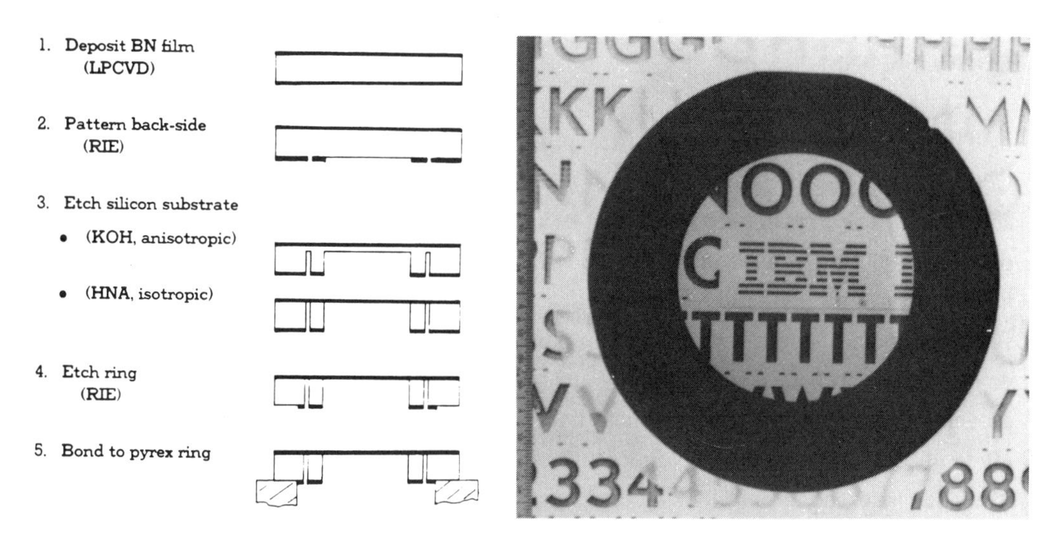

Fig. 35 - Boron nitride x-ray mask fabrication processing sequence [44].

However damage appeared in the membrane after long-term exposure to a synchrotron source, as evidenced by both a mechanical relaxation of the film, which caused a geometrical distortion of the written pattern, and by optical darkening of the films, which is of concern when in-field optical alignment systems are used [30, 45, 46]. Surface crystallites of boric oxide were also observed when the films were irradiated in oxygen-containing ambients where surface reactivity was enhanced due to the surface dangling bonds of boron [46]. The cumulative damage effects reduce the lifetime of the mask and depend upon the exposure process, i.e. the energy spectrum, the dose and incident flux intensity varying from 1 to 100 mW/cm^2 depending on the x-ray source (x-ray tube with rotating or fixed anode, plasma or synchrotron), and the exposure ambient (O_2 or He).

Exposures using a Pd source appeared to saturate at lower values of the strain changes than those obtained for exposures to the high radiation flux of a storage ring [46]. Furthermore, experimental evidence suggests that the radiation damage in membranes prepared from "boro-hydro-nitride" films is less pronounced for films with the lowest as-deposited initial electrical resistivity [45], than for films deposited at higher temperatures T_{dep} [30]. A composite membrane with a very low hydrogen content was also claimed to provide the required mechanical stability and optical transparency under synchrotron radiation [51].

CONCLUSIONS.-

The mechanical, optical and electrical properties of Boron Nitride films deposited by the LPCVD method, differ from each other considerably depending on the various deposition conditions. The physical properties can be tailored through two main control parameters, the deposition temperature and small values of the gas ratio in the reactants (they are nearly independent of the gas ratio for $0.5 < R_{dep} < 5$).

The mechanical stress and optical transparency of the films exhibit a correlation with the dc electrical resistivity. This is mainly attributed to the important role of hydrogen. Hydrogen is present in multiple configurations in the as-deposited films, and is essentially randomly distributed as monohydrides and with a homogeneous concentration. The tensile stress decreases monotonically with the hydrogen content, whereas the optical transparency is found to increase.

The effects of x-ray radiation in the range of 1 to 3 keV on the film properties were reviewed. In particular, no change in chemical composition after x-ray irradiation was detected for the films studied. The breaking of chemical bonds within the film as x-rays are absorbed, affects the optical transparency of the films. Evidence suggests that the hydrogen spatial distribution, and the bonding between the majority of boron and nitrogen atoms, are largely unaffected by the radiation. The observed mechanical instability of the boron nitride membranes can be explained by the modification of the bonding configurations of the hydrogen; upon irradiation, more of the hydrogen becomes weakly bonded, causing the films to become less tensile and relax.

It is the hope of the author that the results presented here will be useful to select the range of deposition parameters which will provide the required film properties to suit each specific application.

Acknowledgements:

The author would like to thank Juan Maldonado for many discussions during the course of the work, Raul Acosta for his thorough comments on this chapter, and Jim Stathis and Alan Wilson for reviewing the manuscript.

REFERENCES:

1. M. J. Rand and J. Roberts: J. Electrochem. Soc. 115 (1968) 423.
2. C.F. Powell et al.: Vapor deposition, Wiley, New York (1962) 382, 663.
3. R. R. Haberecht, R. J. Patterson and R. D. Humphries: Proc. Conf. on Electrical Insulation, National Research Council, Washington, DC no 1238 (1964), p 50.
4. H.O. Pierson: J. Compos. Mater. 9 (1975) 228.
5. M.Hirayama and K. Shohno: J. Electrochem. Soc. 122 (1975) 1671.
6. T. Kimura, K Yamamoto and S. Yugo: Jpn. J. Appl. Phys. 17 (1978) 1871.
7. S.P. Murarka, C.C. Chang, D.N.K. Wang, and T.E. Smith: J. Electrochem. Soc. 126, No. 11 (1979) 1951.
8. T. Takahashi et al.: J. Crystal Growth 53 (1981) 418, and 47 (1979) 245.
9. K. Shohno, T. Kim and C. Kim: J. Electrochem. Soc. 127 (1980) 1546.
10. M. Sano and M. Aoki: Thin Solid films 83 (1981) 247.
11. G. Constant and R. Feurer: J. Less-Common Met. 82 (1981) 113.
12. S. Motojima et al.: Thin Solid Films 88 (1982) 269.
13. W. Schmolla and H. L. Hartnagel: Solid-State Electronics 26 (1983) 931.
14. C. Kim, B. Sohn, and K. Shono: J. Electrochem. Soc. 131 (1984) 1384.
15. S. P. S. Arya, A. D'Amico: Thin Solid Films, 157 (1988) 267.
16. S. B. Hyder and T. O. Yep: J. Electrochem. Soc. 123 (1976) 1721.
17. R.S. Rosler: Solid State Technology 20 (1977) 63.
18. O. Gafri, A. Grill, D. Itzhak, A. Inspektor and R. Avni: Thin Solid Films 72 (1980) 523.
19. D.L. Brors: Proc. SPIE 333 (1982) 111.
20. W. Schmolla and D.H. Hartnagel: J. Phys. D, 15 (1982) L95.
21. W. Schmolla and D.H. Hartnagel: J. Electrochem. Soc. 129 (1982) 2637.
22. H. Miyamoto, H. Hirose and Y. Osaka: Jpn. J. Appl. Phys. 22 (1983) L216.
23. T.H. Yuzuriha et al.: J. Vac. Sci. Technol. A3 (1985) 2135.
24. T.H. Yuzuriha and D.W. Hess: Thin Solid Films 140 (1986) 199.
25. O. Matsumoto et al.: Electrochem. Soc. Proc., Tenth International Conf. (1987), p 552; and Denki Kagaru 56 (1988) 478.

26. A. Chayahara et al.: Applied Surface Science 33/34 (1988) 561.

27. M. Yamada et al.: Proc. Electrochem. Soc. 88-8 (1988) 178.

28. K. Nakamura: J. Electrochem. Soc. 133 (1986) 1120.

29. A. C. Adams: J. Electrochem. Soc. 128 (1981) 1378.

30. P.L. King, L. Pan, P. Pianetta, A. Shimkunas, P. Mauger and D. Seligson: J. Vac. Sci. Technol. B6 No 1 (1988) 162.

31. A.E. Stock: Ber. 56B (1923) 1463.

32. R. Kiessling: Acta. Chem. Scand. 2 (1948) 707.

33. R. Francis and E.R. Flint: US Army report WAL 766.41/1.

34. R.J. Patterson, R. D. Humphries and R. R. Haberecht: Meetings of the Electrochemical Society (1963): Pittsburgh, PA, Abstract 103; and New York, Abstract 197.

35. R.N. Singh: Electrochem. Soc. Proc., Tenth International Conf. (1987) 543.

36. K. Nakamura: J. Electrochem. Soc. 132 (1985) 1757.

37. M. Basche, U.S.P. 315006 (1964).

38. G. Clerc and P. Gerlach: Proc. 5th Int. Conf. on CVD, Slough, edited by J.M. Blocher et al. (1975) 777.

39. A.C. Adams and C.D. Capio: J. Electrochem. Soc. 127 (1980) 399.

40. H. Hannache, R. Naslam and C. Bernard: J. Less-Common Met. 95 (1983) 221.

41. T. Matsuda et al.: Proc. 5th European Conf. on CVD, Uppsala (1985) 420.

42. T. Matsuda, N. Uno, H. Nakae, and T. Hirai : J. Mater. Sci. 21 (1986) 649.

43. T. Matsuda, H. Nakae and T. Hirai: J. Materials Sci. 23 (1988) 509.

44. S. S. Dana and J. R. Maldonado: J. Vac. Sci. Technol. B4 No1 (1986) 235.

45. S. S. Dana, J. Batey, J. R. Maldonado, O. Vladimirsky, R. Fair, R. Viswanathan: Proceedings of the International Conference on Microlithography, "Microelectronic Engineering" 6 (1987) 233.

46. W.A. Johnson, R.A. Levy, D.J. Resnick, T.E. Saunders, H. Betz and H. Oertel: J. Vac. Sci. Technol. B5 No 1 (1987) 257.

47. W. Kern and V. Ban: Thin film process, Academic Press, III-2 (1978 257.

48. A.C. Adams and C.D. Capio: J. Electrochem. Soc. 126 (1979) 1042.

49. W. Kern and D.A. Puotinen: RCA Rev. 31 (1970) 187.

50. W.A. Landford: Solar Cells 2 (1980) 351.

51. R.A. Levy, D.J. Resnick, R.C. Frye, A.W. Yanof, G.M. Wells and F. Cerina: J. Vac. Sci. Technol. B6 No 1 (1988) 154.

52. G. Stoney: Proc. R. Soc. London Ser. A 87 (1909) 172.

53. M. Nakaishi, M. Yamada and K. Nakagawa: Electrochem. Soc. Proc. (1988).

54. E.I. Bromley, J.N. Randall, D.C. Flanders and R.W. Mountain: J. Vac. Sci. Technol. B 1 (1983) 1364.

55. M. Karnezos: J. Vac. Sci. Technol. B 4 (1986) 226.

56. C. E. Uzoh, J.R. Maldonado, S. S. Dana, R. Acosta, I. Babich and O. Vladimirsky: J. of Vacuum Science and Technology B6(6) (1988) 2178.

57. E. Yamaguchi and M. Minakata: J. Appl. Phys. 55, No8 (1984) 3098.

58. J. Stathis and S. S. Dana: to be published.

59. T.M. Duncan et al.: J. Appl. Phys. 64 (1988) 2990.

60. W. Baronian: Mat. Res. Bull. 7, No 2 (1972) 119.

61. E.G. Brame et al.: J. Inorg. Nucl. Chem. 5 (1957) 48.

62. R.J. Archer: J. Opt. Soc Am. 52 (1962) 970.

63. P.J. Gielisse et al.: Phys. Rev. 155 (1967) 1039.

64. M.D. Wiggins and C.R. Aita: J. Vac. Sci. Technol. A2 (1984) 322.

65. J. Frenkel: Tech. Phys. USSR 5 (1938) 685.

66. K. Shohno et al.: Electrochem. Soc. Proc. 79 (1979) 234.

67. C. Kim and K. Shono: Japan. J. Appl. Phys. 20,No 10 (1981) 1901.

68. H.J. Levinstein et al.: US Patent Nber 4,522,842 (1985).

69. M. Karnezos: Solid State Technology Sept (1987) 151.

70. D. Maydan, G.A. Coquin, H.J. Levinstein, A.K. Sinha and D.N.K..Wang : J.Vac.Sci. Technol. 16 (1979) 1959; and Proc. 15th Symp. EIPBT (1979).

71. A.C. Adams et al.: US Patent Nber 4,171,489 (1979).

72. M.P. Lepselter et al.: US Patent Nber 4,253,029 (1979).

73. D. Maydan: J. Vac. Sci. Technol. 17 (1980) 1164.

74. K.E. Bean: Thin Solid films 83 (1981) 173.

75. B.B. Triplett and S. Jones: Proc. SPIE Microlithography VII, 333 (1982) 118.

76. P.D. Blais: Semicon West Tech. Prog. Proc., San Mateo, CA (May 1982) 49.

77. B.B. Triplett, R. Hollman: Proc. of the IEEE 71 No 5 (1983) 585.

78. P. B.: Semiconductor Int. 17 (May 1983).

79. A.P. Neukermans: Proc. SPIE 393 (1983) 93; and Solid State Technology Sept (1984) 185.

80. A.R. Shimkunas: Solid State Technology 27, No 9 (1984) 192; and Electronics Sept. (1985) 48.

81. R.P. Jaeger, M. Karnezos and H. Nakano: SPIE 537 (1985) 75.

82. A.W. Yanof, D.J. Resnick, C.A. Jankoski and W.A. Johnson: Proceedings of SPIE 632 (1986) 118.

83. L. Hanlon, M. Greenstein, W. Grossmann and A. Neukermans: J. Electrochem. Soc. B4 (1986) 305.

84. R.P. Halbich, J.P. Silverman, W.D. Grobman, J.R. Maldonado and J.M. Warlaumont: J.Vac. Sci. Technol. B1 (4), p 1262 (1983).

85. M.P. Lepselter: Proc. IEEE 71, No. 5 (1983).

86. P. Burgraff: Semiconductor International April (1985) 92.

87. P. Tisher: From Electronics to Microelectronics, W.A Kaiser and W.E. Proebster eds., Amsterdam North-Holland (1980) 47.

88. S.S. Dana: 29th International Symposium on Electron, Ion and Photon Beams, Portland, Oregon, May 1985.

89. Spec. Publ. Chem. Soc. 30 (1977) 167.

90. C. Capio, H. Levinstein, A. Adams, D. Wang, A. Sinha: PCT Intern. Appl. 80-00634 (1980).

91. A.D. Wilson: Proceedings SPIE 537 (1985) 85.

92. T.M. Duncan et al.: J. Appl. Phys. 64(6) (1988) 2990.

93. A. Segmuller and M. Murakani, in Analytical techniques for thin films, edited by K.N. Tu and R. Rosenberg, Academic Press.

94. T. Matsuda: J. Materials Science 24 (1989) 2353.

95. A.W. Moore et al.: J. Appl. Phys. 65, No. 12 (1989) 5109.

96. O. Dugne et al.: J. de physique, Colloque C5, (1989) 333.

Materials Science Forum Vols. 54 & 55 (1990) pp. 261-276

BORON-NITROGEN-HYDROGEN THIN FILMS

M. Karnezos
Hewlett Packard Laboratories
1501 Page Mill Road, Palo Alto, CA 94304, USA

ABSTRACT

Refractory thin films of boron-nitrogen-hydrogen (B-N-H) are characterized by very high yield strength and unusual resistance to wet chemicals. These properties combined with standard semiconductor thin film growth techniques, allow the fabrication of defect free membranes with high yield and wide range of thickness. The controlled tensile film stress, the low mass density and low atomic number make these membranes ideal candidates for vacuum window applications with excellent transmission to soft radiation and electron beams. The superb adhesion to substrates like silicon, coupled with orientation dependent etching of the substrate, allows the fabrication of complex structures used in sensors or plain passivation of surfaces. The thin film growth technique is described and the dependence of the film properties on the growth parameters is discussed. Measurement and control of film parameters is very essential to the success of most of the applications. Some of the measurement techniques, though not unique to this material are described as they were used. The characteristics of some published applications are given and suggestions for others are made.

INTRODUCTION

The latest interest for B-N-H thin films was started by research efforts at various laboratories for the development of masks for x-ray lithography [1]. The requirements for the mask membrane are numerous. Low x-ray absorption coefficient of 0.1/um (Figure 1) is desired for high mask contrast in the wavelength range of 4-12 Angstrom. A membrane thickness of 2-5 um to allow at least 50% transparency to He-Ne laser light is necessary for optical alignment. Dimensional stability [2] of about 5 ppm over few square centimeters, under a stress of 2x10^8 dyne/cm^2 from the metal (Au or W) x-ray absorber, is required for mask-to-mask overlay accuracy. Defect free surface of <1 defect/cm^2 is a must to the feasibility of masks in the sub-micron regime and to minimize the cost of repair. Ability to withstand exposure to large doses (1x10^6 Joule/cm2) of x-ray radiation without damage is essential for high volume chip manufacturing. These are rather stringent requirements which led the materials development and, B-N-H has met to a certain degree.

The very desirable properties of B-N-H films are, the high material strength, low mass density, its compatibility with semiconductor processing and the fact that can be grown with very low defect density. The weak point of LPCVD grown B-N-H is the fact that it is not stoichiometric and contains hydrogen. The hydrogen content highly depends on the growth parameters and determines the film stress. High doses of radiation distort the membrane. This is believed to happen by the release of hydrogen which in turn changes the local stress of the membrane. B-N-H is grown using Low Pressure Chemical Vapor Deposition (LPCVD). Equipment have been developed which can grow very clean and uniform thickness films over 150 mm silicon wafers with controlled stress and composition. Although the films are labeled as boron nitride the composition Bx-N-H varies with x=2-6, but the hydrogen content is not determined very accurately.

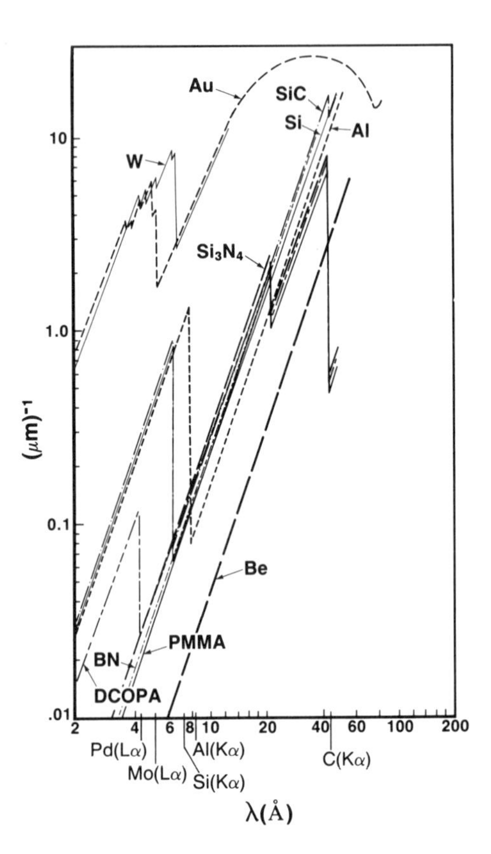

Figure 1. X-ray absorption coefficients of some of the most absorbing and most transparent materials used for masks at wavelengths appropriate to x-ray lithography.

The growth process development has required measurement of film parameters like thickness, membrane flatness, film stress and particulate contamination. Material properties like Young's modulus, thermal conductivity, dielectric constant and index of refraction have been measured for certain compositions. Although by all indications the films look amorphous, there is no detailed crystallographic study.

The need for the development of thin film B-N-H came from x-ray masks but there are applications in the field of vacuum windows transparent to radiation and electrons, that have attracted attention. These and others will be discussed in the applications section.

THIN FILM GROWTH

EXPERIMENTAL - The films are grown [3] in a LPCVD system employing a three zone furnace schematically shown in figure 2. The reaction takes place in the central 20 cm part of the round quartz tube where the temperature profile is flat to +/- 1 deg.C. The ends of the tube are sealed with o-rings kept at about 10 deg.C above ambient without water cooling. The Si wafers are placed back-to-back in pairs vertically on a quartz boat, at a spacing of 15 mm. The boat is entered into the tube from the front with a cantilever mechanism without touching the tube walls. The reaction gases, diborane B2H6, ammonia NH3 and hydrogen H2 are entered individually at the front through the flange to prevent premature reaction and formation of solid components in the gas lines. The gas flow is metered through individual flow controllers and diverted to the pump by-passing the tube, until the desired total flow of 500 standard cubic centimeters per minute (sccm) is reached. The flow then is directed into the tube at temperature of 470 deg.C. The

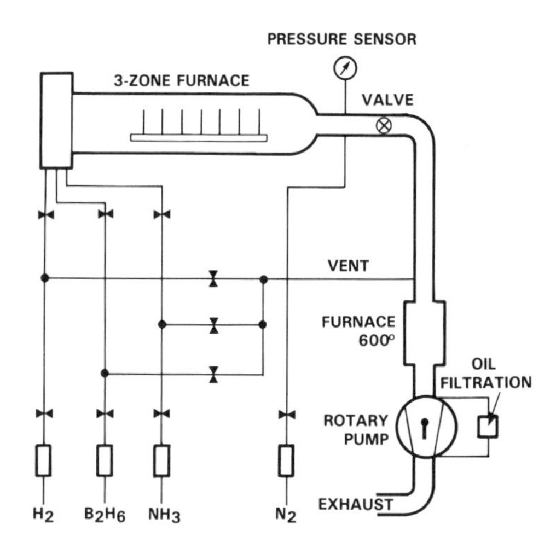

Figure 2. Schematic diagram of the LPCVD reactor.

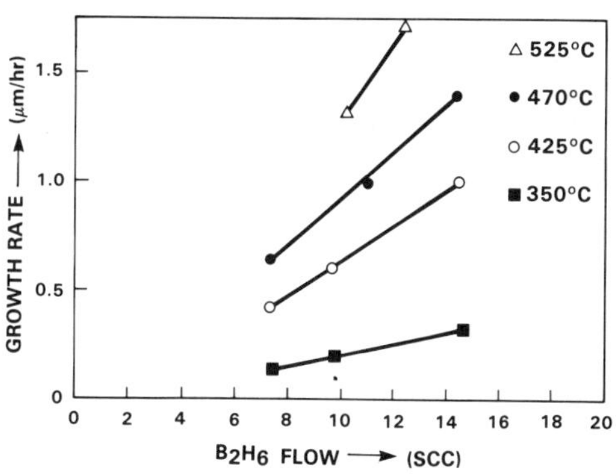

Figure 3. Growth rate vs. the B2H6 flow at various temperatures. The pressure is kept constant at P=0.9 Torr and the total flow 500 sccm by varying the flow of H2 to compensate for the changes of the B2H6 flow.

pressure in the tube is maintained at 900 mTorr with a rotary pump at the back-end of the tube. To prevent rapid corrosion of the pump from the ammonia and the reaction by-products, an after-burner furnace operating at ~600 deg.C is installed before the pump. Also, the use of silicone based pump oil with an oil filtration unit provides trouble free operation and a long pump life-time.

DEPOSITION - The films are grown on 100 mm Si wafers by reacting diborane and ammonia in the hydrogen carrying gas. The diborane is always received in hydrogen carrying gas 10% B2H6, 90% H2 and the ammonia is liquid. Not all the reactant gases combine in the tube to form the films. Unreacted diborane and other species are observed at the end of the tube. The growth rate of solid films increases predictably with increasing diborane flow (Figure 3). It also increases linearly as the mixture gets reacher in diborane at constant growth temperature (Figure 4) and can reach 1.5 um/hr before the mixture gets depleted of diborane. However, the films appear to have a textured surface covered with ~1um high hillocks when the ammonia flow reaches 25 sccm. Below that the films are perfectly smooth with no observable surface texture. Figure 5 shows increasing growth rate as the temperature increases to cover the growth range of this setup with the remaining growth parameters remaining constant. The most uniform depositions across the wafer and

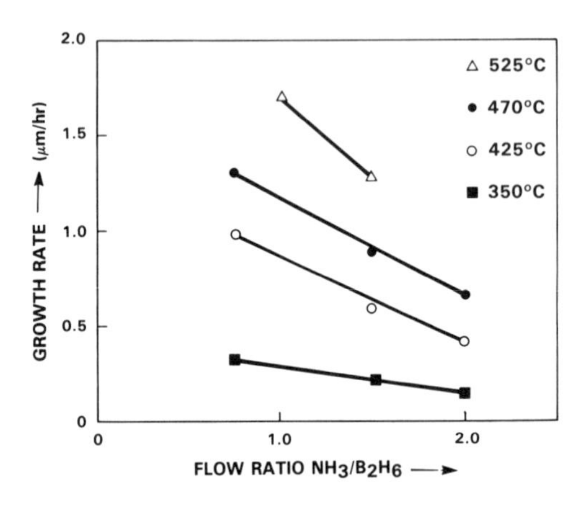

Figure 4. Growth rate vs. flow ratio of NH3/B2H6 at various temperatures. The pressure is kept constant at P=0.9 Torr, the total flow of NH3+B2H6 at 25 sccm and the total gas flow 500 sccm.

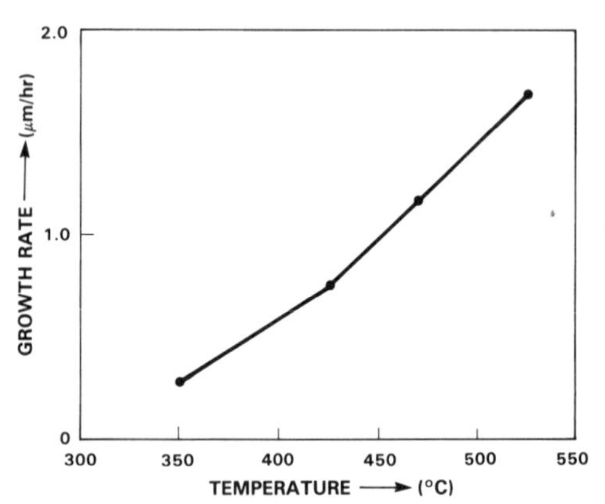

Figure 5. Growth rate vs. temperature at P=0.9 Torr, reactive gas flow NH3+B2H6=25 sccm flow ratio NH3/B2H6=1 and total gas flow 500 sccm.

across the boat were achieved in the 425-470 deg.C temperature range; varying the pressure and the reactant gas flow would probably improve the uniformity at other temperatures. At 525 deg.C the deposition zone is very narrow since the deposition rate is high resulting in rapid depletion of the reactant gases. Typical depositions are done at 470 deg.C, 0.9 Torr pressure, NH_3/B_2H_6 flow ratio of 0.75, reactant gas flow $NH_3+B_2H_6$ of 25 sccm and total gas flow of 500 sccm. The film thickness uniformity also depends on the uniformity of gas flow in the tube.

Although the gas pressure is low, <1 Torr, the flow has gradients which can be favorably affected by the boat design. Film thickness uniformity of 5% across 100 mm Si wafers can be easily achieved by presenting an axially symmetric boat and aiding the flow uniformity with a symmetric gas entry at the cold section of the tube in the front.

Films with low defect density are very desirable. Protruding defects on the film are duplicated in subsequent films like metals or dielectrics deposited by evaporation or, sputtering or spin-on like photoresist, etc. These in turn will translate into defects on any patterns formed lithographically. X-ray masks require membranes with defect density about 0.1 defect/cm2 of 0.1 um diameter of less. This requires Si wafers which have lower defect densities and a deposition process at least as good. The best results of <1 defect/ cm2 [4] were achieved by loading the Si wafers immediately after opening the box as received from the vendor under a better than class 10 clean room environment. Tweezers were avoided and the wafers were handled only from the edges. Contact of either surface of the wafer with any other object will result in defects.

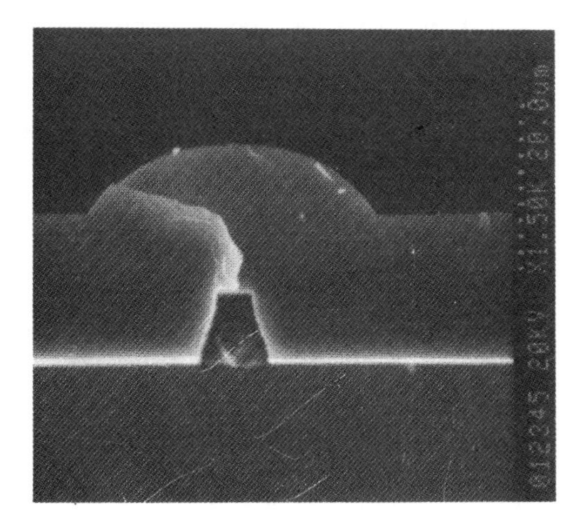

Figure 6. Copper trace 10 um thick etched on Si wafer. B_4NH film 18 um thick deposited on top. Excellent adhesion and step coverage with no voids.

The contact area of the wafer with the boat was also minimized. During the loading of the boat it is very essential that the wafer does not scratch the boat in any way. This will result in a spray of particles of Si or boat material (quartz or B-N-H on boat from previous depositions). It is also very essential that the boat is loaded and unloaded into the furnace tube with a cantilever mechanism to avoid rubbing the boat on to the tube walls to generate particles. The reaction gases are filtered and the mechanical pump is never turned off without isolating the tube first, to prevent back-streaming of oil vapors into the reaction chamber. The furnace is vented with N_2 at 350 deg.C and it is also back purged during unloading to prevent O2 and water vapor from possibly oxidizing the films. The boats and the reaction tube can be used for multiple depositions, up to a total deposited films thickness of about 30 um. Thicker films start flaking-off under their residual stress, causing a serious source of defects. Also thermal cycling of the boat and tube between the operating temperature and ambient causes flaking of the deposited films. This is minimized by replacing the boat and tube often and maintaining the tube at 350 deg.C when not operating. No wet chemical has been identified capable of stripping the B-N-H film from the boat or tube.

B-N-H films were deposited on a variety of substrates, including Si, SiO2, Si3N4, quartz, Pyrex, W, Ti, Cu, Cr and stainless steel. The adhesion to these substrates is very good, but flaking may occur after repeated thermal cycling in the cases where significant mismatch of the B-N-H to substrate thermal expansion coefficient exists. The primary requirement for the deposition is that the substrate does not react with the gases and can stand the growth temperature without degradation or material changes.

The low pressure of the process allows diffusion of the gases and deposition even in very restricted spaces, but at lower deposition rates. The Si wafers although loaded back-to-back and held together during the deposition, there is a thin film of about 0.1-0.2 um deposited on the backs. The step coverage of the deposition process is excellent. Deposition on substrates with patterned films show complete and uniform coverage resulting in sealing of the features with B-N-H (Figure 6), even though the step is 10 um high.

FILM PROPERTIES

STRUCTURE AND COMPOSITION - The films are amorphous and appear very smooth under scanning electron microscope. If any texture exists it would be less than 50 Angstrom. The B:N ratio was measured with Auger spectroscopy and Rutherford backscattering (RBS) to be 4:1 for the films grown under the usual growth conditions of 470 deg.C, flow ratio NH3/B2H6=0.75, pressure= 0.9 Torr and total gas flow of 500 sccm. The hydrogen concentration [5] was established with nuclear reaction analysis. The resulting approximate composition is B4NH.

The density of B4NH was measured on uniform films deposited on Si wafers by measuring the thickness and the added mass. Measurements on many wafers grown under the same conditions yielded density d=1.74 gr/cm3.

STRESS - The film stress was maintained above the compressive state for all the applications. Figure 7 shows increasing stress with decreasing NH3/B2H6 flow ratio. This results in films with higher hydrogen content as shown by R.A. Levy et.al [5] Increasing temperature also results in increased stress, though not as dramatic compared to the NH3/B2H6 ratio

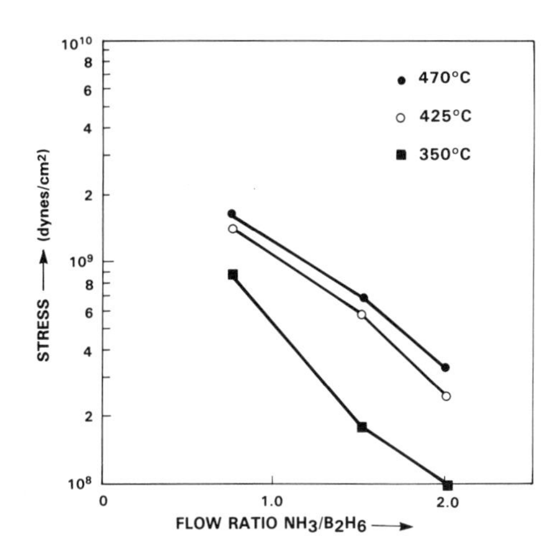

Figure 7. Stress of B4NH film vs. flow ratio NH3/B2H6 at constant pressure P=0.9 Torr, total reactive gas flow NH3+B2H6=25 sccm total gas flow 500 sccm. All films had the same thickness t=0.5 um.

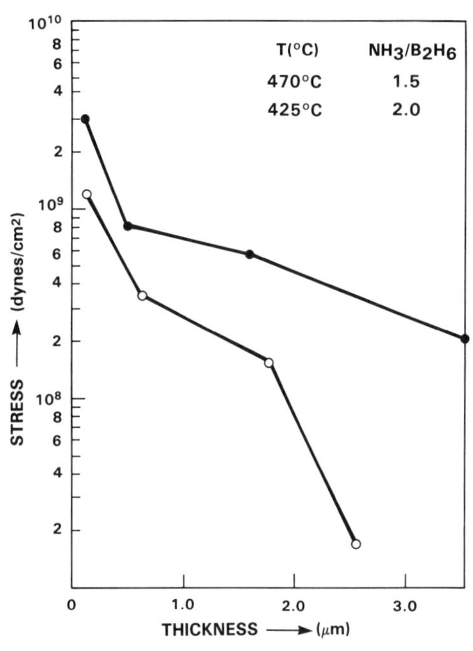

Figure 8. Tensile film stress vs. the total film thickness at two growth temperatures. The pressure P=0.9 Torr and reactive gas flow NH3+B2H6=22 sccm were kept constant.

dependence. Most of the film stress appears to arise from a lattice mismatch to the Si substrate. Figure 8 shows the stress dependence on the total film thickness with all the growth conditions remaining constant. The decrease in stress as the thickness increases is dramatic indicating that a thin layer of about 0.05 um carries most of the stress, while the remaining film thickness is well matched to this thin layer. This was verified by subsequent depositions on wafers already covered with B4NH and the total stress did not increase appreciably. The range of measured stress varies, 1x10^7-3x10^9 dynes/cm2. Although the film adhesion appears to be excellent, at stress levels above 3x109 dynes/cm2 and for films few micrometers thick, the Si wafers have been observed to crack under the stress and occasional film delamination occurred at the edges.

The stress is monitored using three different techniques [4]. At the wafer level a single beam interferometer, as shown in Figure 9 is used to measure the bow h of the B4NH wafers. The stress σ_o is given by [6].

$$\sigma_o = (h/3a^2)(E/1-\nu)(ts^2/tf)$$

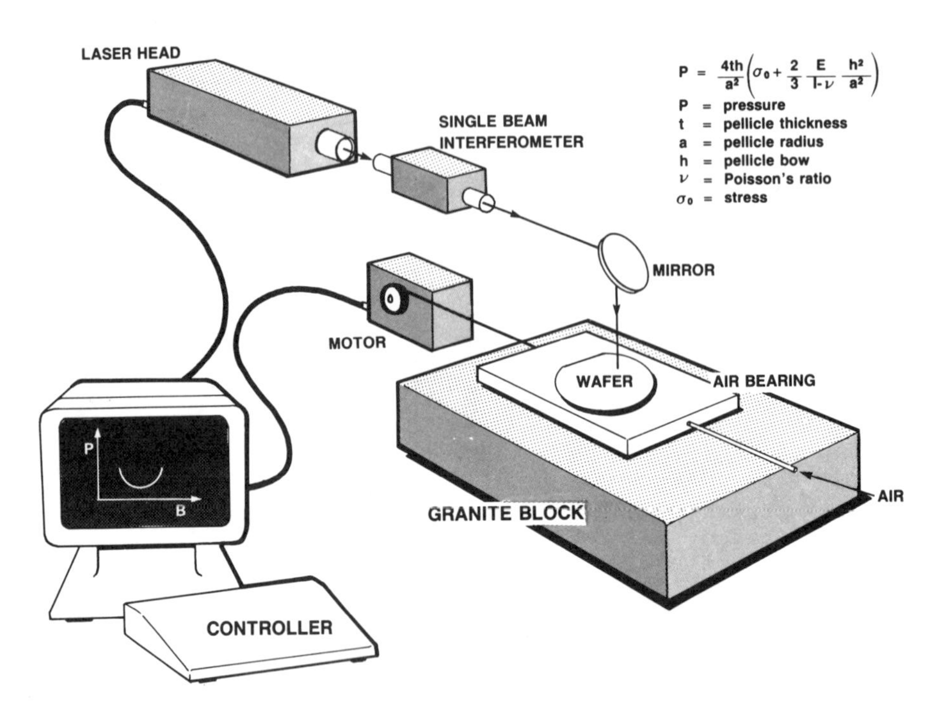

Figure 9. Single beam line interferometer used to measure wafer and membrane flatness. Resolution of the system is +- 0.1 um.

where tf and ts are the thickness of film and substrate, h and a are the bow at radius a, and $E/1-\nu$ = 1.805x10^12 dynes/cm2 for [100] silicon [7]. This method is fast and accurate (+\- 3%) and is routinely used for process control. Typical data are tf =4 um, h=95 um, and σ_0 = 1.5x 10^9 dyn/cm2 on 100 mm Si wafers. A similar setup using a capacitive gauge instead of the laser was used to measure non-reflecting films. It needs calibration as the output depends on the probe-wafer gap.

After the Si is etched and the membrane is formed the above setup can be used to measure the stress by applying a positive pressure to the membrane. This "bulge" method [8] provides a direct measurement on the membrane with increased sensitivity and accuracy. The maximum deflection h is measured as a function of the differential pressure P, applied across the membrane. The stress σ_0 and $E/1-\nu$ for B4NH are obtained by a least squares fit of

$$P = (4tf/a^2)(\sigma_0+2/3(E/1-\nu)h^2/a^2)h \qquad (1)$$

to the data (Figure 10). These two parameters can be varied separately by changing the growth conditions mentioned earlier. A faster and less cumbersome method measures the resonant

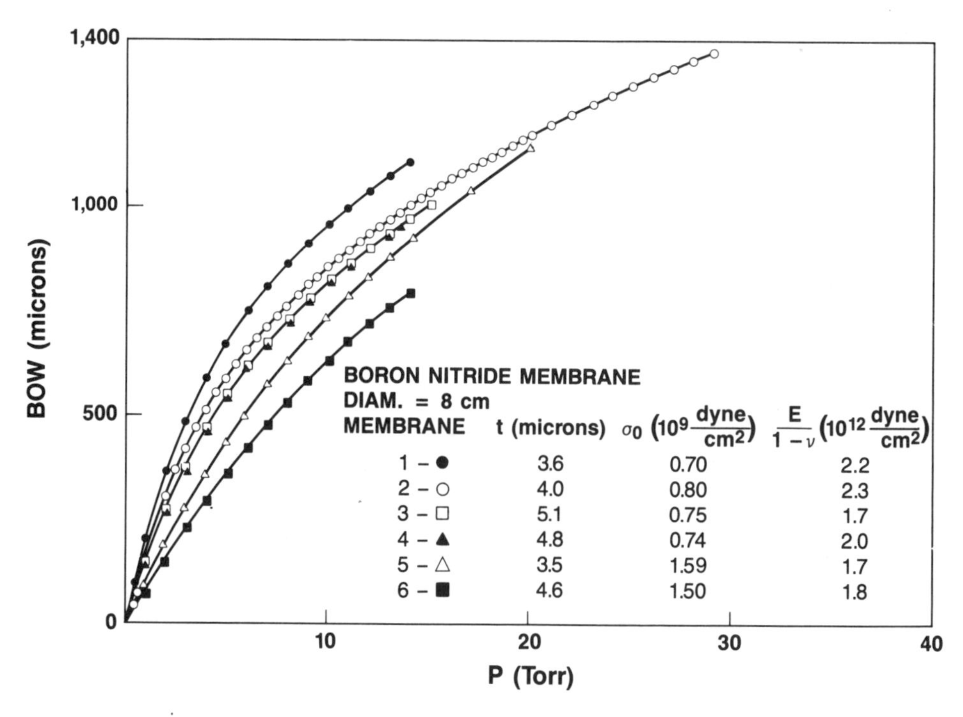

Figure 10. Membrane bow vs. pressure data from the "bulge" method. Solid line represents the least squares fit of Eq(1) to the data. Stress σ_o and $E/1-\nu$ are the fitted parameters.

frequency [9-11] of the membrane using a setup illustrated in Figure 11. The measurement is done at atmospheric pressure and the results are corrected for air loading on the membrane to obtain the vacuum resonant frequency, νvac:

$$\nu vac = \nu air[1+1.34(a\ da/tf\ df]^{1/2}$$

where da is the density of air, a, tf, and df are the radious, thickness, and density of the membrane, respectively. The stress σ_o can be obtained from

$$\sigma_o = df(2.61a\ \nu vac)^2$$

A membrane with a=3.94 cm, tf=5.5 um, df= 1.74 g/cm3 and σ_o = 1.0x 10^9 dynes/cm2 will resonate at approximately 850 Hz. Stress measurements from the above three methods agree to +/- 10%.

DIELECTRIC CONSTANT - The dielectric constant of B4NH was measured on free standing membranes. On a 80 mm diameter membrane two thin film aluminum plates 0.4 um thick, 50 mm diameter were evaporated on both sides of the membrane forming a capacitor with the B4NH membrane as the dielectric. The dielectric constant was calculated from the measured capacitance at 10 KHz, but little change was observed for lower frequencies down to 120 Hz. The measured dielectric constant of typical run is $\varepsilon/\varepsilon_o$ =7 although material grown at different temperature and NH3/B2H6 flow ratio can have a value down to $\varepsilon/\varepsilon_o$ =5.5 as Table I shows.

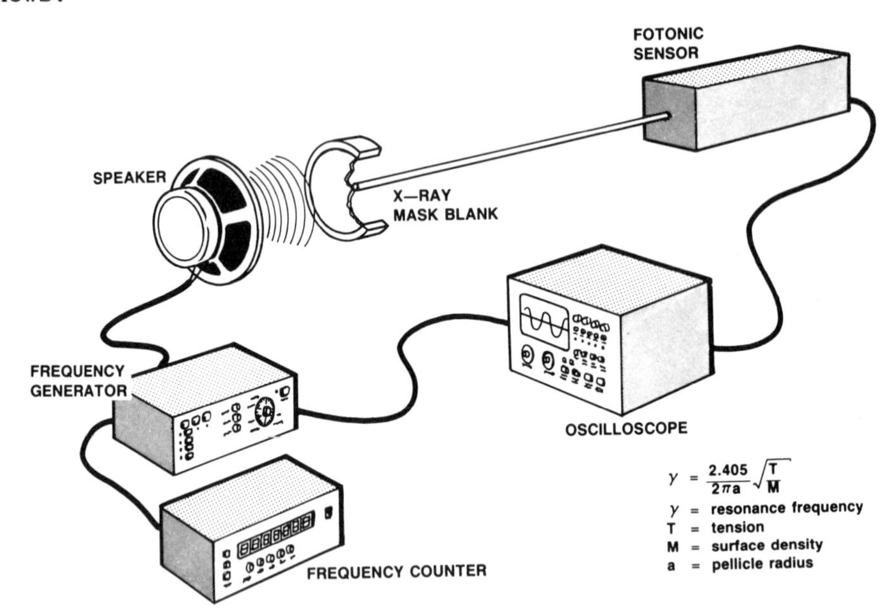

Figure 11. X-ray mask blank acoustic resonance measurement setup. Diverging light beam from the optical fiber is reflected by the membrane and collected back, thus sensing the gap change. Membrane acoustic resonance frequence and displacement are measured from the oscilloscope trace.

Table I

Growth Conditions			Film Properties			
Temp. (C)	Flow NH3/B2H6	Flow (sccm) NH3+B2H6	Thickness (um)	σ_o (dyne/cm2)	C (nF)	$\frac{\varepsilon}{\varepsilon_o}$
425	2.04	22.0	2.6	1.8x10^7	41.0	5.86
425	2.02	38.5	3.5	3.5x10^8	28.8	5.54
470	0.76	25.6	4.7	1.2x10^9	27.1	7.00
470	1.50	25.0	3.4	7.3x10^7	31.5	5.92
525	1.50	25.0	3.8	3.2x10^6	29.1	6.11
525	1.00	25.0	4.3	8.4x10^8	30.2	7.07

In the process of measurement a soft breakdown of the dielectric

$\varepsilon/\varepsilon_0$ =7 was observed at approximately 1.2x10^7 V/m but a hard breakdown was not evident up through 1.2x10^8 V/m (Figure 12).

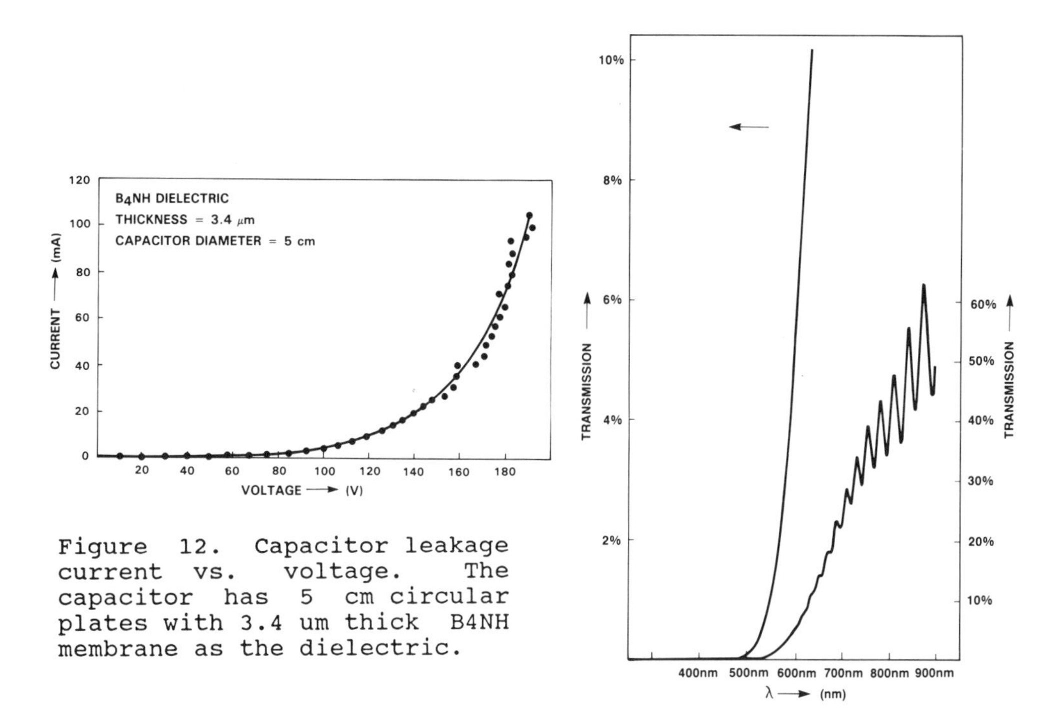

Figure 12. Capacitor leakage current vs. voltage. The capacitor has 5 cm circular plates with 3.4 um thick B4NH membrane as the dielectric.

Figure 13. Optical transmission of 5.07 um thick B4NH membrane in the visible range.

<u>OPTICAL PROPERTIES</u> - The optical transmission of B4NH films was measured on free standing membranes using a spectrophotometer. The optical transmission decreases as the content of boron [12] increases. Figure 13 shows that below 500 um there is total absorption, but at about 900 um there is about 60% transmission. The index of refraction was measured by ellipsometry and the thickness using reflectance spectroscopy. The measured index of refraction was n=2.3. Continuous films were grown down to 0.02 um thick on Si wafers. 0.1um thick pin-hole free membranes were formed over 7 mm. Thick films up to 20 um were also grown on Si without delamination or wafers braking.

<u>THERMAL PROPERTIES</u> - The thermal conductivity and heat capacity were measured on 1 um thick B4NH films deposited on Si under standard growth conditions. Heaters and temperature sensors were lithographically printed on the film for the measurements. The measured thermal conductivity was k=0.012 W/cm.K and heat capacity c=2.4 Joules/cm3.K. When compared to thermally grown SiO2, which has the same values as fused quartz, the

conductivity is 13% less and the volume heat capacity is 41% greater. This makes B4NH quite a good thermal insulator.

The thermal expansion coefficient [13] was measured for films deposited on Si wafers by measuring the wafer bow at different temperatures in the range of 20 deg.C - 270 deg.C. The wafer bow is caused by the differential thermal expansion between Si and B4NH. The expansion coefficient a=2.2x10^-6/K was calculated from the bow and the known expansion coefficient of Si.

APPLICATIONS

B4NH thin films have been used in a variety of applications. The high film strength and low density lend the film for radiation window applications while its low dielectric constant combined with semiconductor processing make it desirable as a thin dielectric. The inertness and resistance to wet chemicals helps in the use of B4NH as a passivation coating. Some of these applications will be described below.

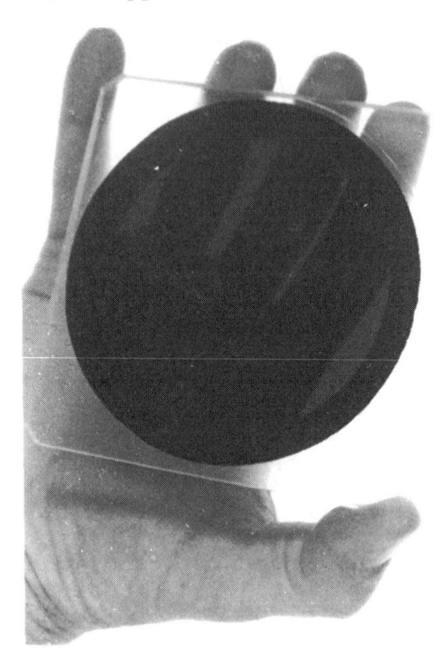

Figure 14. X-ray mask blank. Pyrex frame 10x10 cm, B4NH membrane 4 um thick, 8 cm diameter. The membrane is anodically bonded to the Pyrex frame.

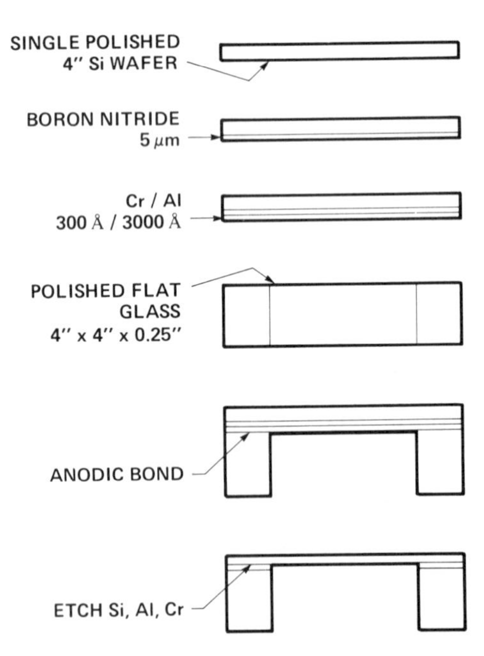

Figure 15. X-ray mask blank fabrication process.

<u>X-RAY MASK BLANK</u> - An x-ray mask blank consists of a 4-5 um thick B4NH membrane bonded to a Pyrex frame as shown in Figure 14. The fabrication process is shown in Figure 15. This structure has certain characteristics very desirable for a mask blank. The front surface of the film that carries the growth defects is not used for the mask pattern. Instead the back side of the film which was in contact with the Si wafer is used for the subsequent mask absorber film deposition. By its nature this surface is free of protruding defects and very desirable for fine line lithography. The anodic bonding [14] layer of Cr/Al is very thin and uniform in thickness. Therefore the bonding frame planarity is maintained through the bonding process. This results in a membrane under tension, typically 1.3×10^9 dynes/cm2 and <0.2 um flatness over the entire aperture of the mask. These membranes are very durable, surviving immersion in boiling sulfuric peroxide for cleaning as well as scrubbing with a wafer scrubber. The x-ray absorbing metal, Au or W, are deposited directly on the membrane using a thin layer of Ti for improved adhesion. These metal films are patterned and etched using standard semiconductor processes. In the process of reactive ion etching of the W film with SF6, it was demonstrated that the B4NH membrane can also be etched to form submicron patterns [15] with high fidelity (Figure 16). In general B4NH can be etched in

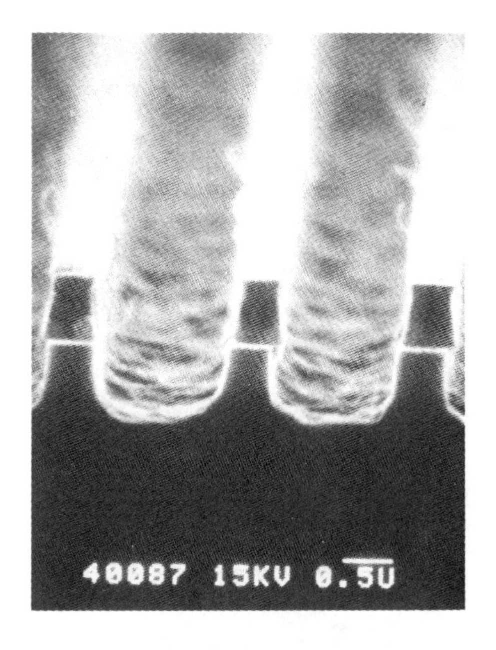

Figure 16. Submicron lines of 0.5 um thick W patterned on a B4NH membrane. The etching continued into the membrane replicating the pattern.

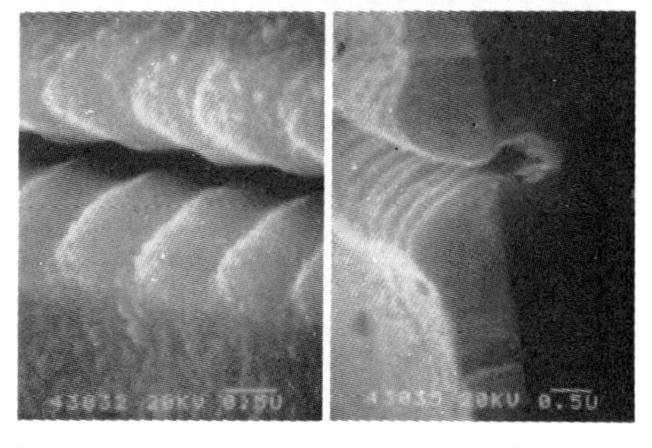

Figure 17. Argon-ion laser induced etching of W in an atmosphere of oxygen. 0.6 um thick W on 5 um B4NH. The beam is stepped at 1 um intervals. Note the etching into the B4NH.

fluorine plasma from gases like CF4, SF6 or others. Local etching [16] can also be achieved using an Argon-ion laser in an atmosphere of O2 (Figure 17) or using an excimer laser (Figure 18). A variety of x-ray windows and electron windows have been fabricated using B4NH membranes and processes similar to that of an x-ray mask blank. Electron windows [17] of 1 cm diameter and 3-4 um thick have transmission up to 65% at 10 KeV and hold pressure of many hundreds of atmospheres. Windows of 2 mm diameter and 0.2 um thick have 30% transmission at CKa radiation (44.7 Angstrom), 50% transmission at BKa (67.6 Angstrom) and hold pressure of more than an atmosphere and maintain very high vacuum. Sample holders for TEM measurements have also been fabricated using 400 Angstrom thick free standing membranes.

Figure 18. 18 um thick B4NH film deposited on Si wafer. The via is etched using a focused eximer laser beam. The walls are near 90 degrees and very clean and the bottom of the via free of residue.

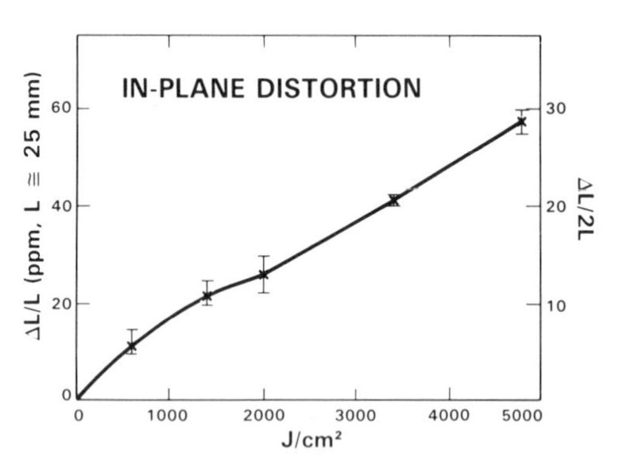

Figure 19. Strain vs. incident dose for a B4NH membrane exposed to synchrotron radiation. Exposed area 25x8 mm, 4 um thick, L=25 mm.

Distortions of the B4NH membrane from high doses of x-ray radiation [18,19] have been observed. These are attributed to the release of hydrogen resulting in a decrease of the membrane stress locally. The resulting in plane distortions are cumulative with the absorbed dose as shown in Figure 19. They are measured using a diffraction technique [20] with high resolution, <0.1 um. The nonuniform absorption of radiation, with depth, can cause stress gradients and locally buckle the membrane out of plane. Out of plane distortions >0.1 um were measured using a He-Ne laser interferometer.

BURIED CAPACITOR DIELECTRIC - An application of the B4NH as a dielectric film is in capacitors as they are used in high density interconnect substrates. These substrates consist of alternating layers of metals like Cu or Al and dielectrics like polyimide, deposited on Si wafers. The metal layers are patterned photolitographically to form the interconnection of various integrated circuits mounted on the substrate. Capacitors are used as charge reservoirs for the switching output drivers. It is very desirable to locate them very close to the switching drivers to minimize the total inductance of the connection. The capacitance required is of the order of 20 nF/cm2. These are fabricated directly on doped Si wafers and subsequently are covered (i.e. buried) by the metal/polyimide layers. The requirement on the capacitor dielectric is high dielectric constant and very thin films free of pin holes. In this application 0.2 um thick B4NH films were used over 20 cm2 area to produce capacitors operating at voltages less 10V. The capacitors met the specifications and survived numerous anneals up to 450 deg.C without leakage or shorts.

PRESSURE TRANSDUCERS - Thin membrane based pressure transducers are used for sensing low pressures in the range 0-100 Torr. The change in the membrane deflection exposed to pressure is sensed as capacitance variation of a capacitor formed by the membrane and a fixed plate. Capacitive sensors of 1 mm diameter with 0.5 um thick membrane and 10 um plate separation were fabricated. The zero pressure capacitance was <1 pF and the sensitivity about 2000 ppm/mTorr. This method is susceptible to RFI/EMI disturbances if the sensing circuitry is away from the sensor. This problem can be circumvented if an optical means of sensing the membrane deflection is employed. The membrane is mounted at the end of a optical fiber and laser light is sent through the fiber to reflect on the membrane exposed to pressure. The reflected intensity is a function of the membrane deflection. Such sensors were built using 0.5 mm diameter membrane at 4 um thick. Sensitivity of ~ 350 ppm/Torr was measured in the range of 40-70 Torr where the response is very linear. The sensitivity can be greatly improved by using thinner membranes.

SUMMARY

The following is a summary of the measured B4NH membrane parameters.

Density = 1.74 g/cm^3
$E/1-\nu$ = 2x10^12 dynes/cm^2
Thermal Coefficient of Expansion = 2.2x10^-6/K
Heat Conductivity = 0.012 W/cm-K
Specific Heat = 2.4 J/cm^3-K or 1.38 J/g-K
Dielectric Constant = 5.5 - 7.0
Film Thickness = 0.02 - 20 um (continous)

Film Stress = 1x10^6 - 3x10^9 dynes/cm^2 , controlled
Thickness uniformity = <10% accross 100 mm
Defect density = <1 defect/cm^2
Optical Transmission = 10% at He-Ne line with 4 um film
X-ray Transmission = 90% Pd La (4.4 Angstrom), 4 um film
30% C Ka (44.7 Angstrom), 0.17 um film
50% B Ka (67.6 Angstrom), 0.17 um film
Composition = B4NH , approximate
Structure = Amorphous
LPCVD grown on : Si, SiTi, SiO2, Si3N4, Quartz, S.S.Steel, Pyrex
Cu, W, Ti

REFERENCES

1. D.L. Spears, H.I. Smith, Electron. Lett. 8 102 (1972).
2. M. Karnezos, Solid State Techol., p.151, Sept. 1987.
3. A.C. Adams, C.D. Capio, J. Electrochem. Soc., 127, 399 (1980).
4. M. Karnezos, J. Vac. Sci. Technol. B4, p.226 (1986).
5. R.A. Levy, D.J. Resnich, R.C. Frye, A.W. Yanof, J. Vac. Sci. Technol., B6 154 (1988).
6. R. Glang, R.A. Holmwood, and R.L. Rosenfeld, Rev. Sci. Instrum. 36, 7 (1965).
7. W.A. Brantley, J. appl. Phys. 44, 534 (1973).
8. J.W. Beams, in Structure and Properties of Thin Films, edited by C. Neugebauer, J.B. Newkirk, D.A. Vermilyea (Wiley, New York, 1959).
9. Lord Rayleigh, Theory of Sound (Dover, New York), Vol.I.
10. Morse and Feshback, Methods of Theoretical Physics (McGraw-Hill, New York, 1953), pp. 1452-1455, 1565.
11. T.M. Yarwood, Acoustics (McMillan, London,1953).
12. S.S. Dana, J.R. Maldonado, J. Vas. Sci. Technol., B4, 253 (1986).
13. T.F. Retajczyk, Jr., A.K. Sinha, Apply. Phys. Lett., 36 162 (1980).
14. U.S. Patent 4,632,871 (1986).
15. M. Karnezos, R. Ruby, B. Heflinger, H. Nakano, R. Jones, J. Vac. Sci. Technol., B5 283 (1987).
16. W. M. Grossman, M. Karnezos, J. Vac. Sci. Technol., B5, 843 (1987).
17. L. Hanlon, M. Greenstein, W. Grossman, A. Neukermans, J. Vac., Sci. Technol., B4 305 (1986).
18. W.A. Johnson, R.A. Levy, D. J. Resnick, T.E. Saunders, A.W. Yanof, H. Betz, H. Huber, H. Oertel, J. Vac., Sci. Technol., B5 257 (1987).
19. G.M. Wells, G. Chen, D. So, E. L. Brodsky, K. Kriesel, F. Cerrina, M. Karnezos, J. Vac. Sci. Technol., B6 2190 (1988)
20. R. Ruby, D. Dalwin, M. Karnezos, J. Vac. Sci. Technol., B5, 272 (1987).

Materials Science Forum Vols. 54 & 55 (1990) pp. 277-294
Copyright Trans Tech Publications, Switzerland

CHARACTERIZATION OF BN THIN FILMS DEPOSITED BY PLASMA CVD

Y. Osaka (a), A. Chayahara (b), H. Yokoyama (a), M. Okamoto (a), T. Hamada (a), T. Imura (a) and M. Fujisawa (c)
(a) Dept. of Electrical Engineering, Hiroshima University
Saijo, Higashi-Hiroshima 724, Japan
(b) Government Industrial Research Institute
Osaka, Midorigaoka, Ikeda 563, Japan
(c) Synchrotron Radiation Lab., Institute for Solid State Physics
The University of Tokyo, Midori-cho, Tanasi, Tokyo 188, Japan

ABSTRACT

Cubic BN thin films are formed in RF discharge in B_2H_6 and N_2 at low pressures under a magnetic field to confine the plasma, for negatively self-biased substrate electrodes. Ion bombardment on the growing surface is suggested to play an important role in the formation of cubic BN. The deposited films are characterized by infrared absorption spectroscopy and transmission electron microscopy which shows that they are composed of microcrystals of cubic BN with 100-200 Å grain size. Optical reflectance measurements of the films were carried out at room temperature in the photon energy range 5-25 eV. Structure in the imaginary dielectric function spectra was in good agreement with that predicted for crystalline BN by previous energy band calculations.

§1. INTRODUCTION

Boron nitride is chemically and thermally stable and highly insulating. Cubic boron nitride (c-BN) having the zinc-blende type structure is an especially hard material comparable to diamond. However, BN is usually crystallized in the graphite-like hexagonal form. Synthesis of c-BN in equilibrium requires very high pressures and temperatures [1]. Recently, new processes for the deposition of c-BN thin films from vapor phase at lower pressures have been investigated [2-7]. In these techniques, it seems that energetic ions impinging on the growing surface play an essential role in the synthesis of c-BN.

We have deposited BN films at relatively low temperature with only the RF discharge and investigated the effect of substrate self bias to accelerate ions toward the growing surface on the structure of BN thin films [8]. In this experimental investigation we have used two types of discharges with RF substrate bias, that is, RF discharge in a magnetic field and microwave discharge at the electron cyclotron resonance (ECR) condition. In the case of only RF discharge in a magnetic field, it is possible to use the more simple deposition apparatus. However, in these apparatuses, the effect of self bias to accelerate ions is not separated from that on plasma conditions such as plasma density and plasma potential. In this sense, the method using ECR plasma and RF substrate bias at the same time is appropriate to investigate the effect of self bias to accelerate ions, because the plasma conditions can be varied by ECR plasma. The RF discharge in the magnetic field and ECR plasma can be kept at a low pressure [9], so ions can be accelerated efficiently. A negative self bias is supplied by RF power added to the substrate electrode in order to avoid dielectric breakdown of the depositing films in dc bias.

The deposition techniques are described in §2. Deposition conditions for creating c-BN films are discussed in §3, where a growth mechanism of c-BN is discussed. The deposited films are characterized by infrared absorption spectroscopy and transmission electron spectroscopy (TEM). The experimental results are described in §4. Optical reflectance spectra of the films were measured in the energy range 5-25 eV with synchrotron radiation. The spectra were compared with those for sintered c-BN and pyrolytic h-BN. The experimental results are described in §5, where the qualitative interpretation of reflectance spectra is made.

From the reflectance measurements of c-BN, the well-known Kramers-Kronig analysis was performed to determine the imaginary (ε_2) part of the dielectric function. We compare these results with a calculated ε_2 spectrum of c-BN in §6.

§2. DEPOSITION TECHNIQUES

Figure 1 and 2 shows the apparatus to deposit BN thin films from the vapor phase of B_2H_6 and N_2. It is evacuated by a diffusion pump through a throttling gate valve. A base pressure is 2×10^{-7} Torr. Figure 1 is the apparatus using both ECR plasma and RF substrate bias. Microwave power is introduced through a rectangular waveguide and a fused quartz window into the plasma chamber of stainless steel with inside diameter 11 cm and height 16 cm cooled with water. Magnetic field intensity is 1000 G at the center of coils. In Figure 2 the arrangement using only RF power is shown. The magnetic field direction is changed to obtain uniform thin films.

The substrate electrode is coupled to RF generator through an impedance-matching network with a blocking capacitor. The RF power is applied between the grounded chamber and the substrate electrode. A negative dc self-bias is generated basically due to the higher mobility of electrons as compared to ions in a megahertz glow discharge. The self-bias is measured using oscilloscope connected to the substrate electrode through a choke coil to block RF power. The temperature of the substrate holder is measured by a thermocouple behind the electrode. The deposition conditions are summarized in Table I. We used single crystal silicon and fused quartz as substrates.

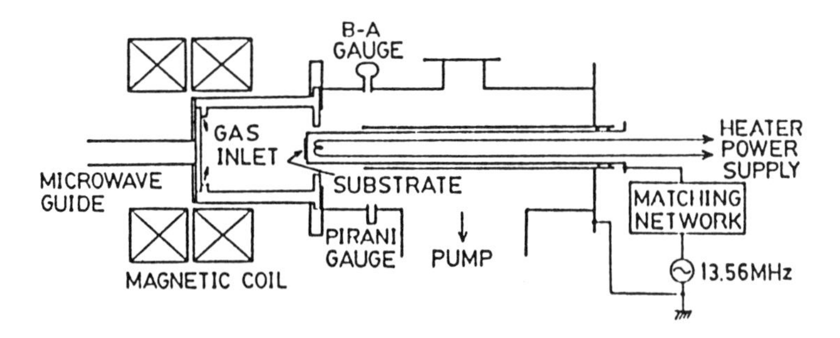

Figure 1 The apparatus to deposit BN thin films in which ECR plasma and RF substrate bias are used.

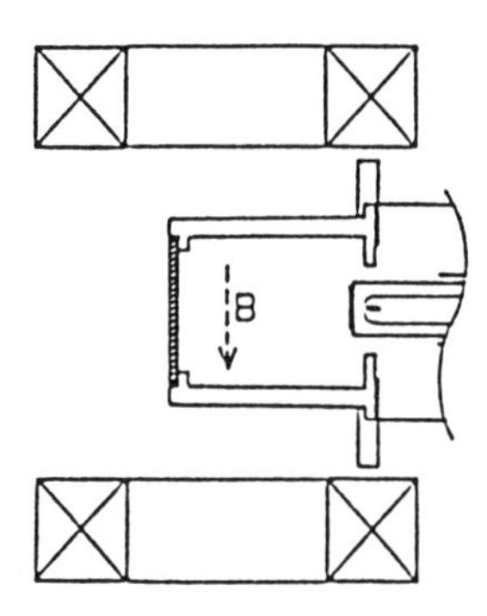

Figure 2 The magnet coil arrangement for RF plasma CVD. Microwave guide is disconnected.

Table I Deposition conditions

Flow rate		
B_2H_6 (Ar base 10 %)	12 - 10	SCCM
N_2	1.2 - 10	SCCM
Pressure	$3x10^{-4}$ - $2x10^{-3}$	Torr
Substrate holder temp.	R.T. - 600	°C
Microwave power (2.45 GHz)	0, 160	W
Magnetic field (max.)	1000	G
RF bias power (13.56 MHz)	0 - 200	W
Negative self bias	0 - 400	V
Deposition rate	0 - 1	Å/s

Deposition rate and the composition of the films at a typical deposition condition shown in Table II is described.

Table II Deposition conditions

Flow rate		
B_2H_6(Ar base 10 %)	10	SCCM
N_2	0.2-5	SCCM
Pressure	$4x10^{-4}$	Torr
Substrate holder temp.	70	°C
Microwave power (2.45 GHz)	320	W
Magnetic field (max.)	1000	G
Negative self bias	0	V

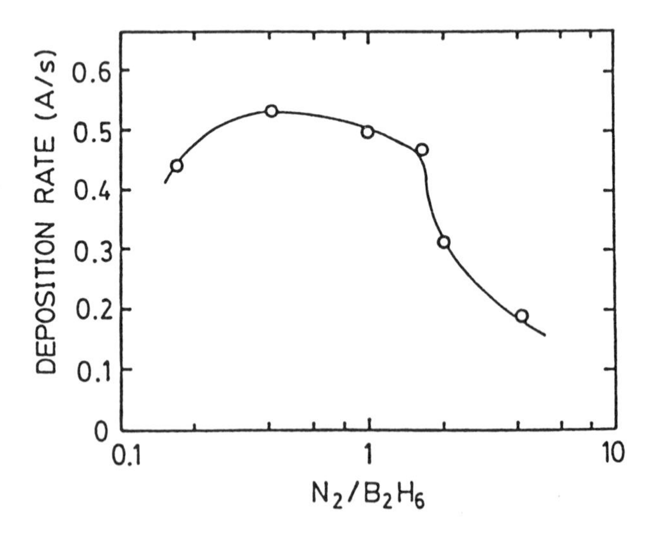

Figure 3 Deposition rates of h-BN films as a function of the N_2/B_2H_6 ratio.

Deposition rates are shown in Figure 3 as a function of N_2/B_2H_6. Deposition rate (maximum 0.53 Å/s) is between that of thermal CVD and spattering method. Prepared films are uniform on both Si substrate and SiO_2. The decrease of deposition rate with increasing N_2/B_2H_6 may be caused by etching by nitrogen.

The composition of the obtained films was determined by use of the XPS analysis technique. Figure 3 shows the B/N ratio in the films as a function of the N_2/B_2H_6 ratio. As the N_2/B_2H_6 ratio increases, the nitrogen content in the film increases. Stoichiometric boron nitride films have been obtained at N_2/B_2H_6=1. In the following, this condition is fixed to prepare the BN films except for §4.

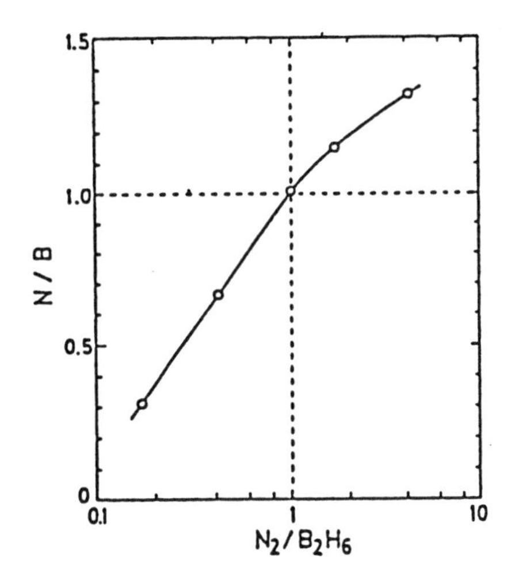

Figure 4 The B/N ratio in the films as a function of the N_2/B_2H_6 ratio.

§3. DEPOSITION CONDITIONS FOR CREATING c-BN FILMS (GROWTH MECHANISM OF c-BN)

Figure 5 shows the infrared absorption spectra of BN films deposited without a microwave power on silicon substrates at the pressure: 3×10^{-4} Torr, flow rate: N_2 4 SCCM, B_2H_6(Ar based, 10 %) 10 SCCM. No absorption due to B-H or N-H stretching vibration was observed. The absorption bands at 1370 cm^{-1} and about 800 cm^{-1} are due to in-plane and out-of-plane lattice vibrations of graphite-like hexagonal BN, respectively [10]. The absorption band near 1070 cm^{-1} is assigned to the TO mode of cubic BN [11]. Figure 5 shows that the cubic phase is formed only at the pertinent self-bias between -200 and -300, whereas in the regions below 200 V and above 400 V, only the hexagonal phase grows. At the N_2 flow rate of less than 4 SCCM, the hexagonal phase increases.

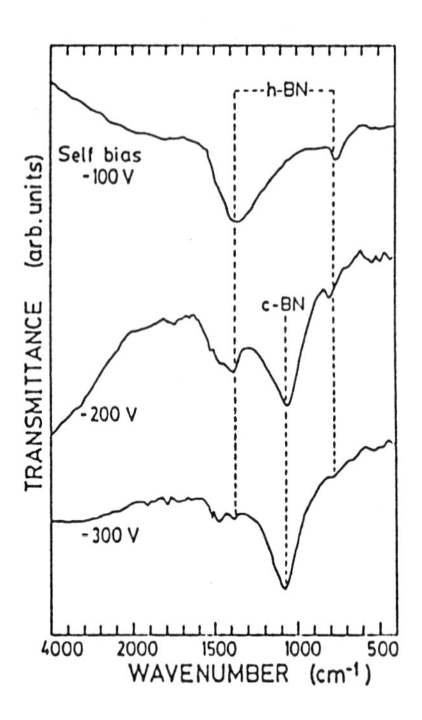

Figure 5 Infrared absorption spectra of deposited BN films with several negative self-bias voltage.

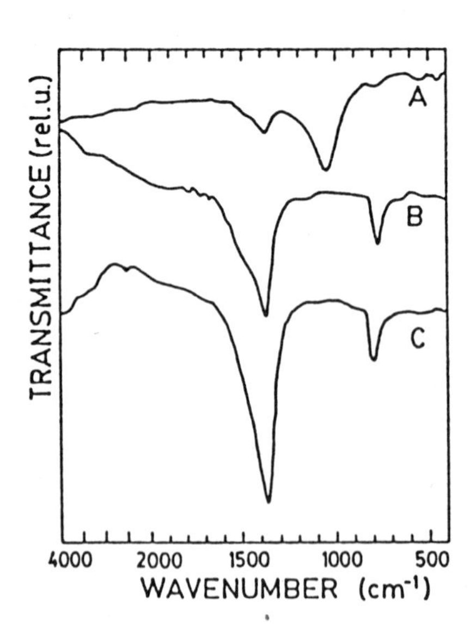

Figure 6 Infrared absorption spectra of BN thin films deposited using ECR plasma (160 W) and following self bias; (A) 50 V, (B) 30 V, (C) 0 V.

Figure 6 shows the infrared absorption spectra of the BN thin films for three values of substrate bias using ECR plasma (Microwave power: 160 W). At lower substrate bias the structure of deposited BN films becomes hexagonal. As the self-bias increase, the film thickness decreases due to sputter etching so that the absorption strength becomes weaker. The absorption strength of the in-plane lattice vibration decreases especially in comparison with that of the out-of-plane lattice vibration. These facts show that the ions accelerated by self-bias to the substrate sputter selectively the hexagonal grains which have c-axis alignment and prevent the growth of such a hexagonal phase. The cubic phase appears at the self-bias of greater than 40 V. Thus, the structure of deposited BN is sensitively influenced by a negative self-bias, and the cubic phase can be synthesized at the pertinent self-bias, energetic ions sputter selectively the weakly bonded atoms located in the hexagonal phase on the growing surface and increase the volume fraction of the cubic phase.

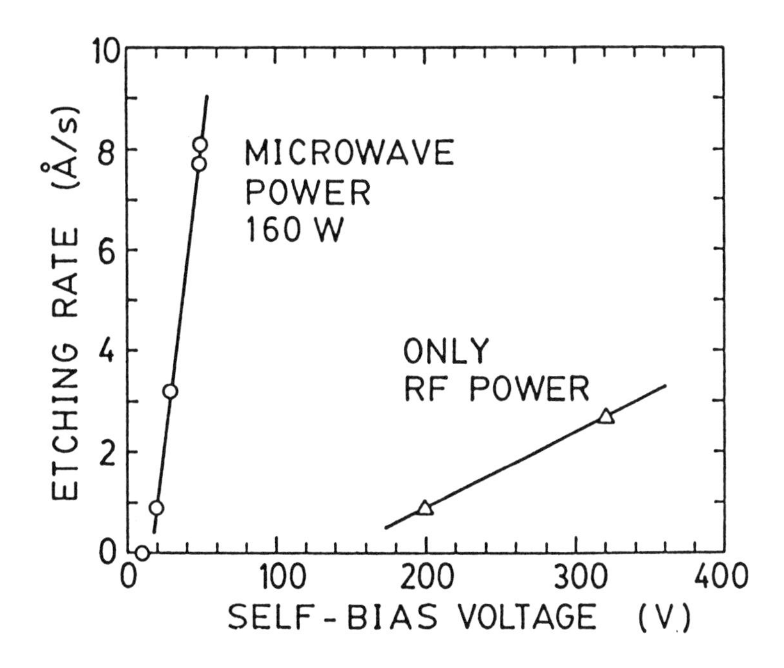

Figure 7 Etching rate of silicon (100) in Ar plasma (Pressure $4x10^{-4}$ Torr, Magnetic field 1000 G and substrate temperature 100 °C).

The threshold value of the self bias for the formation of cubic phase in the case using only RF discharge is greater than that in ECR plasma. This threshold value seems to depend on the plasma conditions, but cubic phase can not have been synthesized without the substrate bias in this experiment up to now. The actual ion energy impinging the growth surface for cubic phase formation in the plasma CVD is determined by plasma potential and self-bias. The degree of ionization in the ECR plasma is higher than that in RF plasma. This fact is ascertained by Figure 7. For same self-bias voltage, etching rate of silicon (100) in Ar ECR plasma is larger than that in Ar RF plasma, which shows that ion density in ECR plasma is larger. Consequently, assuming that the formation of cubic phase necessitate a critical value of the energy flux of collecting ions on the growth surface, the ion energy necessary to the synthesis of the cubic phase seems to become lower.

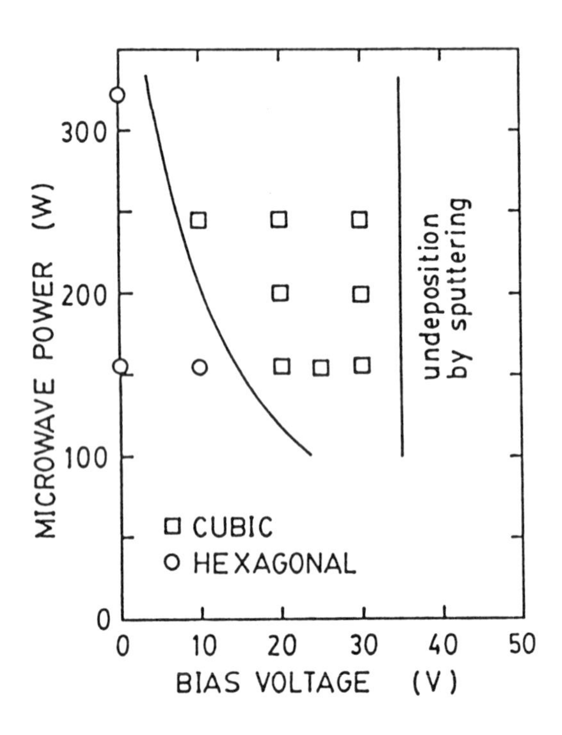

Figure 8 Schematic illustration of deposition conditions for ECR plasma (Pressure $4x10^{-4}$ Torr and Magnetic field 1000 G)

For ECR plasma (Pressure $4x10^{-4}$ Torr), the deposition conditions for creating c-BN phase in films are schematically shown in Figure 8.

We measured current density-voltage characteristics of a probe in ECR Ar plasma (Pressure $4x10^{-4}$ Torr). As a probe, the substrate electrode was used. In Figure 9, the current density-voltage characteristics for negative bias are shown for various microwave power. From this Figure, we can see that the ion current density in plasma is essential for the creation of c-BN phase.

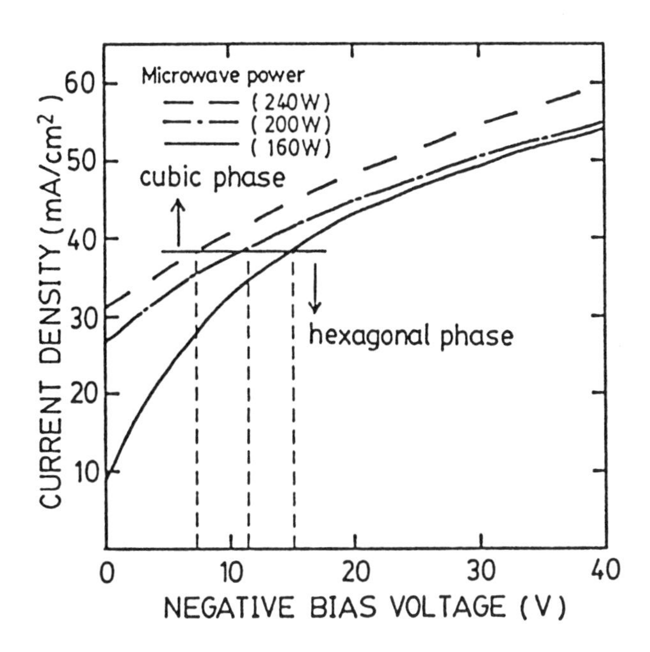

Figure 9 Current density-voltage characteristics for various microwave power in ECR Ar plasma ($4x10^{-4}$ Torr).

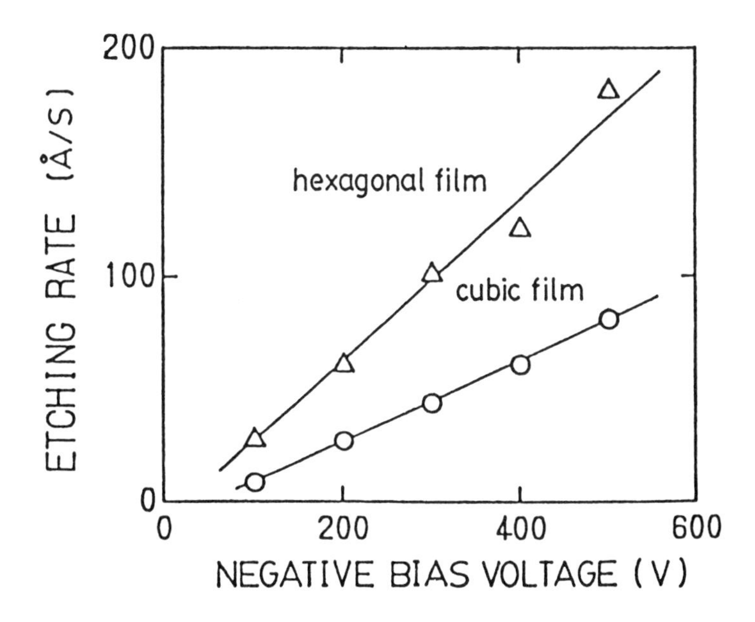

Figure 10 Etching rates of hexagonal and cubic films.

The etching rate of films prepared in RF Ar plasma (Pressure $4x10^{-4}$ Torr) were measured. Figure 10 shows the etching rate of hexagonal and cubic films. The volume fraction of this cubic film is nearly 100 %. This shows that the etching rate of hexagonal phase is about twice of one of cubic phase. From these results, the following growth mechanism on cubic phases in films is suggested : I) the formation of cubic phase has the threshold value (~ $38mA/cm^2$ in Figure 9) of ion current density. II) above this threshold value, the hexagonal phases on the growing surface are selectively etched and the volume fraction of cubic phase is increased with the increase of ion current density.

The initial growth process of films was examined by infrared absorption spectra. Figure 11 shows the dependence of the absorbance of cubic (~ 1070 cm^{-1}) and hexagonal IR band (~ 1370 cm^{-1}) on film thickness. This film consist of hexagonal and cubic phases. It must be noted that the volume fraction of cubic phase in films is nearly independent of the film thickness, which suggests that the interface layer on silicon has the cubic phase.

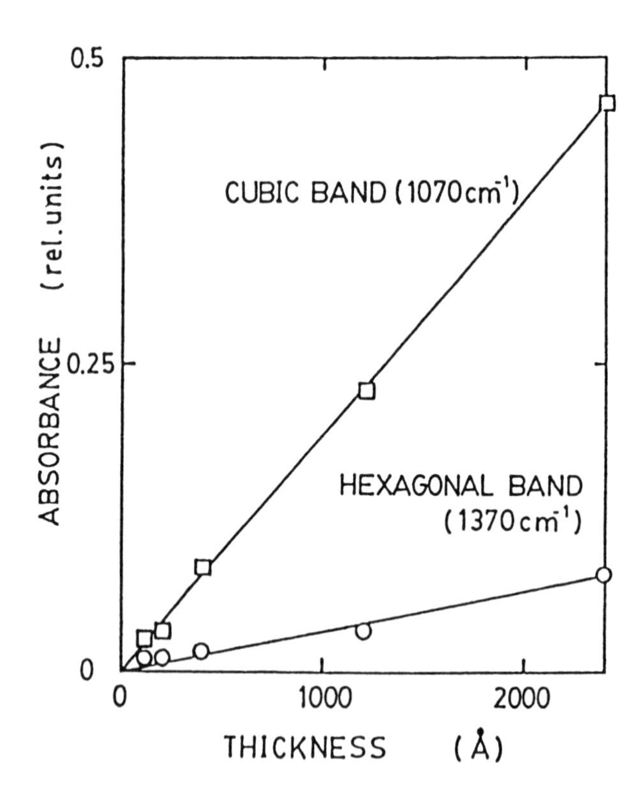

Figure 11 Dependence of the absorbance of cubic and hexagonal IR bands on film thickness.

§4. TEM AND IR

The bright and dark field images of transmission electron microscopy show that the deposited films consist of microcrystals of BN with about 100 - 200 Å grain size. Measured d spacings in the deposited cubic and hexagonal BN thin films are tabulated and compared with the d values of the respective bulks in Table III. These results clearly indicated that cubic BN and hexagonal BN can be synthesized using the plasma CVD technique.

Infrared spectra of films show a strong band near 1370 cm^{-1} and a weak band near 800 cm^{-1} of hexagonal phase, and a strong band near 1070 cm^{-1} of cubic phase. Near N-H and B-H stretching region at 3430 cm^{-1} and 2500 cm^{-1} there are no absorption, and thus our films contain no bonding hydrogen. For the deposition condition shown in Table II, the dependence of a crystalline quality of hexagonal film on N_2/B_2H_6 ratio was studied. Figure 12 show infrared spectra under oblique (45 degree) incidence against the film surface. New weak band near 1610 cm^{-1} is longitudinal optic (LO) mode [9] and the intensity of this mode becomes strong with increasing N_2/B_2H_6 ratio. The ratio of absorbance between LO and TO modes is maximum at N_2/B_2H_6=1. Since the intensity of the LO mode corresponds to a crystallinity of films, this shows a best crystalline quality of hexagonal film at N_2/B_2H_6=1, which is consistent the observation that the electron diffraction pattern of TEM has the sharpest ring at N_2/B_2H_6=1.

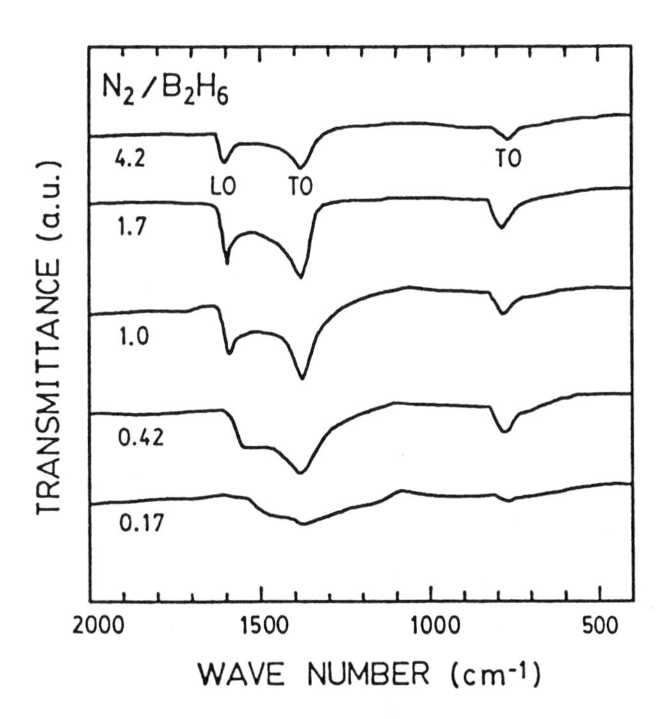

Figure 12 Infrared spectra under oblique (45 degree) incidence for various N_2/B_2H_6 ratio.

Table III The d values observed on the deposition films compared with that for a cubic boron nitride (ASTM 25-1033) and hexagonal boron nitride (ASTM 9-12)

cubic BN			hexagonal BN		
hkl	ASTM d (Å)	observed d spacing (Å)	hkl	ASTM d (Å)	observed d spacing (Å)
(111)	2.088	2.051	(002)	3.33	3.326
(220)	1.2785	1.290	(100)	2.17	2.148
(311)	1.0901	1.096	(101)	2.06	
(331)	0.8297	0.808	(102)	1.817	
			(004)	1.667	1.663
			(103)	1.552	
			(104)	1.322	
			(110)	1.253	1.234
			(112)	1.173	1.178

§5. REFLECTANCE SPECTRA

Reflectance spectra of our microcrystalline films are compared with that of bulk samples. Reflectance spectra in the photon energy range 5-25 eV were measured by using the SOR-RING Beam Line 1 at The Institute for Solid State Physics of the University of Tokyo with a 1-m Seya-Namioka-type monochromator. The incident angle was near-normal, 15 ° from the normal axis of the sample surface. The reflected photons were detected by a sodium-salicylate-coated photomultiplier in the wavelength region 440-2000 Å and by a normal photomultiplier in 1900-3000 Å.

The samples for reflectance measurements were a sintered c-BN plate, microcrystalline c-BN thin films, a p-BN plate and h-BN thin films. The sintered plate of c-BN produced at a high pressure and temperature was offered by Sumitomo Electric Industries, Ltd. This sample (5x5x0.5 mm^3) was characterized as cubic by using X-ray diffraction and Raman scattering. A plate of p-BN prepared by high-temperature CVD was obtained from Shin-Etsu Chemical Co., Ltd. X-ray diffraction and Raman-scattering measurements showed that p-BN has a long-range c-axis stacking order and a good c-axis alignment among the crystallites, though it is polycrystalline in the hexagonal basal plane. The reflectance spectra of c-BN films are shown in Figure 13.

The reflectance spectra are shown in Figure 13. The spectrum of the deposited c-BN thin film by the plasma CVD (b) is similar to that of the c-BN sinter synthesized under high pressure and temperature (a). Both spectra show peaks at 11.7 and 14.0 eV, and a broad peak around 18 eV. The 11.4 and 14.0 eV peaks seem to correspond qualitatively to the E_1 and E_2 peaks of the zinc-blende-type semiconductor. By use of the Harrison's theory [12], the E_2 peak was calculated to be around 14.2 eV, which nearly agrees with the observed value.

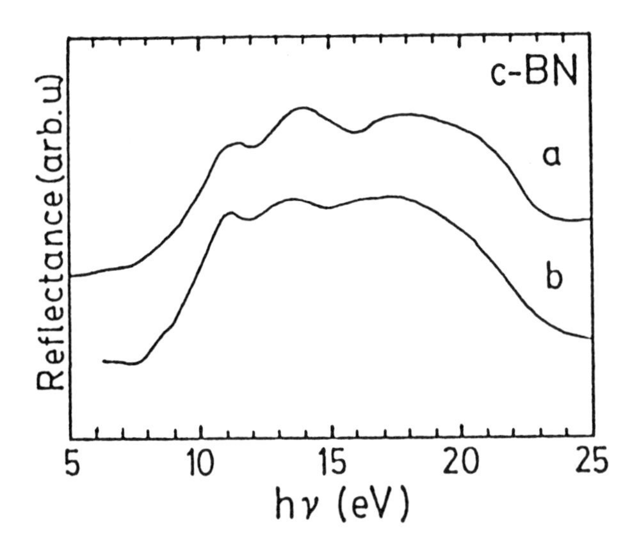

Figure 13 Reflectance spectra of c-BN. (a) c-BN sintered, (b) microcrystalline c-BN.

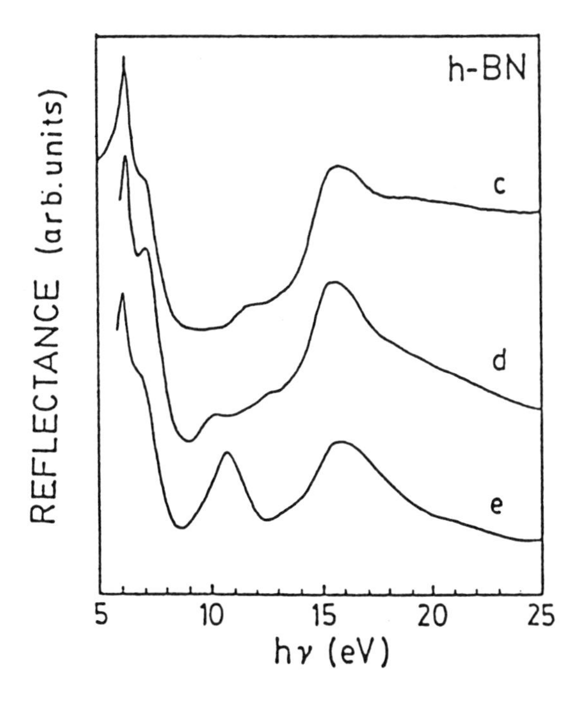

Figure 14 Reflectance spectra of h-BN. (c) p-BN, (d) h-BN thin film without self bias and (e) h-BN with 100 V self bias.

Figure 14 shows the reflectance spectra of h-BN. The corresponding IR spectra are shown in Figure 15.

The comparison of these figures yields that the peak at 6.2 eV is due to the π (bonding) $\rightarrow$ π^* (antibonding) transition and the peaks around near 10 eV and 16 eV are due to the σ (bonding) $\rightarrow$ σ^* (antibonding) and to the σ (bonding) $\rightarrow$ π^* (antibonding), respectively. The detailed discussion for this assignment is given in reference [13].

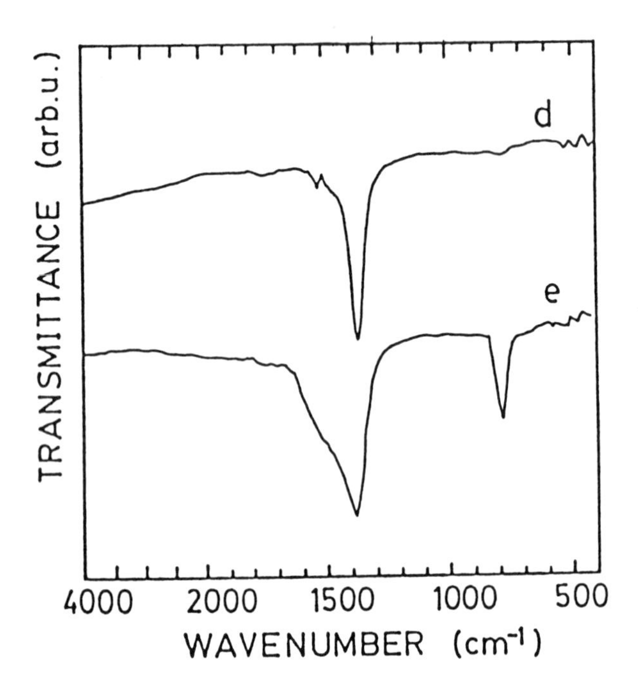

Figure 15 Infrared absorption spectra of microcrystalline h-BN. (d) h-BN thin film without self bias and (e) h-BN with 100 V self bias.

§6. IMAGINARY DIELECTRIC CONSTANT OF c-BN

The electronic structure of c-BN has been studied by optical reflectivity and absorption, and X-ray emission spectroscopy. There is little experimental data available. Philipp and Taft mentioned 'crude' reflectance measurement in their paper on the optical properties of diamond, reporting a peak near 14.5 eV [14]. Chrenko measured the ultraviolet absorption spectra of c-BN crystals and deduced a minimum value of the indirect optical gap as 6.4 ± 0.5eV [15].

Fomichev and Rumsch studied the spectra concerning the core levels by means of X-ray emission spectroscopy and deduced a value of about 6 eV for this gap [16].

Many theoretical studies of the band structure of c-BN have been reported [17-24]. Their results must be compared with the experimental data.

Figure 16 shows the results for ε_2 (imaginary dielectric function) obtained by well-known Kramers-Kronig transformation from reflectance spectrum. The $\varepsilon_2(E)$ (E:photon energy) spectrum for the sintered c-BN sample shows peaks at 9.05 eV and 11.7 eV and shoulders near 13.2 eV and 16.7 eV. In similar to this, the $_2$ for the microcrystalline c-BN has a broad structure with a peak at 11.7 eV. In Figure 16, theoretical curve for ε_2 calculated by Tsay et al.[20] is also shown.

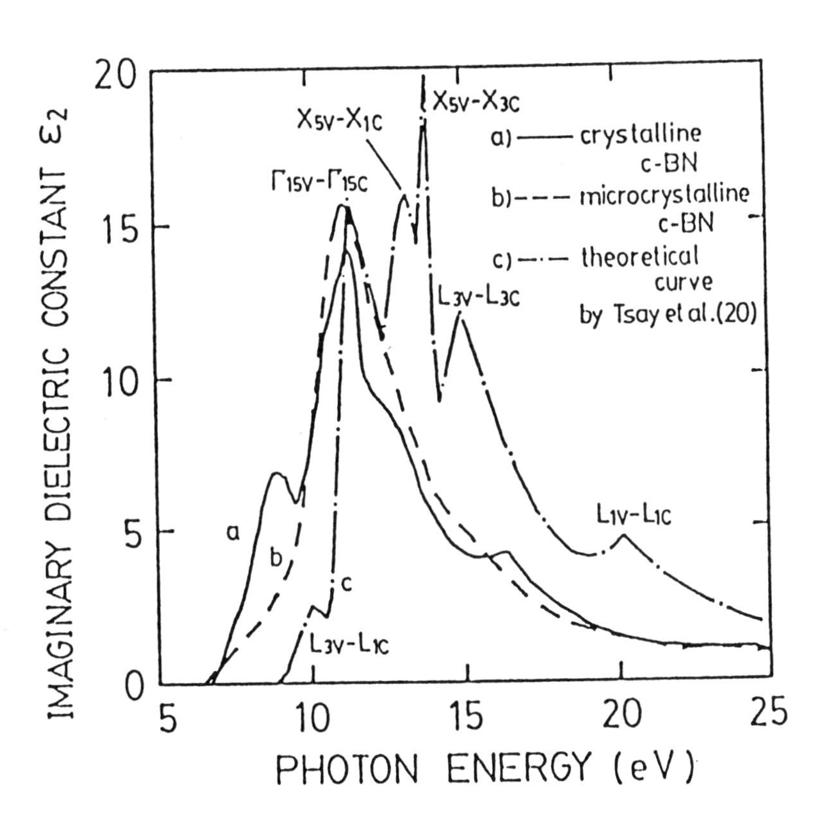

Figure 16 The $\varepsilon_2(E)$ spectra of c-BN. The full and broken curves represent the ε_2 for crystalline and microcrystalline c-BN, respectivity. The dotted curve is the theoretical one calculated by Tsay et al.[20]

Their calculation on the K X-ray emission spectra agrees with the experimental result [15], quantitatively. The ε_2 of crystalline c-BN is in semi-qualitative agreement with the theoretical result if our result is displaced to the higher energy side by about (2 ~ 3) eV. This seems to be related to that the experimental energy gap of c-BN is about (6.0 ~ 6.4) eV [14-15] but the theoretical value is 8.0 eV [20]. Other band calculations have also been reported [16-19,21-23]. Interband transition energies at symmetry points in several theoretical results [16-20] are compared on Table II in the reference [20]. In these calculations, the transition which has the energy difference near 13.7 eV is due to the direct transition $X_{5V} \rightarrow X_{3C}$. In other calculations [21-23], energies associated to the transition X_{5V} X_{1C} are about 13 eV. These results suggest that the peak of ε_2 at 11.7eV is due to the direct transition at the X point. Assuming that the peak of ε_2 at 11.7 eV is due to the transitions $X_{5V} \rightarrow X_{3C}$ and $X_{5V} \rightarrow X_{1C}$, from Figure 16, we can speculate that energies associated to the transitions $\Gamma_{15V} \rightarrow \Gamma_{15C}$, $L_{3V} \rightarrow L_{3C}$, and $L_{1V} \rightarrow L_{1C}$ are 9.05 eV, 13.2 eV, and 16.7 eV, respectively.

For the ε_2 of microcrystalline c-BN, the peak at 9.05 eV and the shoulders at 13.2 eV and 16.7 eV are eliminated. This seems to be related to the fact that the microcrystalline c-BN does not grow epitaxially on silicon substrate and has the rough surface.

§7. SUMMARY

Our results are summarized as follows:

(1) Using plasma CVD technique with substrate negative self bias, cubic BN phase could be synthesized from B_2H_6 (Ar dilution) and H_2 gas mixtures.
(2) The substrate negative self-bias is essential to the formation of c-BN.
(3) The formation of cubic phase has the threshald value of ion current density on a growing surface.
(4) The images of TEM show that the deposited films consist of microcrystals of BN with about 100-200 Å grain size.
(5) The microcrystallines c-BN and h-BN have similar electric structure to the bulk BN material.

References

[1] Wentrof, Jr.R.H.: J.Chem. Phys. 1961,34,809
[2] Halverson,W. and Quinto,D.T.: J.Vacuum Technol. 1985,A3,2141
[3] Inagawa,K., Watanabe,K., Ohsone,H., Saitoh,K. and Itoh,A. in: Proc. 10th Symp. on ISIAT'86, Tokyo, 1986, p.381.
[4] Satou,M. and Fujimoto,F.: Japan.J.Appl.Phys. 1983,22,L171
[5] Szmidt,J., Jakubowski,A., Michalski,A. and Rusok,A.: Thin Solid Films 1983,110,7
[6] Weissmantel,C., Bewilogua,K., Breuer,K., Dietrich,D., Ebersbach,U., Erler,H.J., Rau,B. and Reisse,G.: Thin Solid Films 1982,96,31
[7] Zhou,P.F., Mori,T. and Namba,Y.: Shinku 1985,28,581
[8] Chayahara,A., Yokoyama,H., Imura,T. and Osaka,Y.: Japan.J.Appl.Phys. 1987,26,L1435
[9] Matsuo,S. and Kiuchi,M.: Japan.J.Appl.Phys. 1983,22,L210
[10] Geick,R., Perry,C.H. and Rupprecht,G.: Phys.Rev. 1966,146,543
[11] Gielisse,P.J., Mitra,S.S., Plendl,J.N., Griffis,R.D., Mansur,L.C., Marshall,R. and Pascoe,E.A.: Phys.Rev. 1967,155,1039
[12] Harrison,W.A.: Electronic Strustures and the Properties of Solids (Freeman,W.H. & Co., San Francisco, 1980) p.108.
[13] Chayahara,A., Yokoyama,H., Imura,T., Osaka,Y. and Fujisawa,M.: Japan.J.Appl.Phys. 1988,27,440
[14] Philipp,H.R. and Taft,E.A.: Phys.Rev. 1962,127,)159
[15] Chrenko,R.M.: Solid State Commun. 1974,14,511
[16] Fomichev,V.A. and Rumsch,M.A.: J.Phys.Chem.Solids 1968,29,1015
[17] Kleinman,L. and Philips,J.C.: Phys.Rev. 1960,117,460
[18] Bassani,F. and Yoshimine,M.: Phys.Rev. 1963,130,20
[19] Wief,D.R. and Keown,R.: J.Chem.Phys. 1967,47,3113
[20] Hemstreet, Jr.L.A. and Fong,C.Y.: Phys.Rev. 1972,B6,1464
[21] Tsay,Y.F., Vaidyanathan,A. and Mitra,S.S.: Phys.Rev. 1979,B19,5422
[22] Zunger,A. and Freeman,A.J.: Phys.Rev. 1978,B17,2030
[23] Prasad,C. and Dubey,J.D.: Phys.Stat.Sol(b) 1984,125,629
[24] Park,K.T., Terakura,T. and Hamada,N.: J.Phys.C:Solid State Phys. 1987,20,1241

Materials Science Forum Vols. 54 & 55 (1990) pp. 295-312

A PROMISING BORON-CARBON-NITROGEN THIN FILM

K. Montasser, S. Morita (a) and S. Hattori (a)
Dept. of Electrical Engineering, Faculty of Engineering, Al-Azhar University
Nasr City, Cairo, Egypt
(a) Dept. of Electronic Mechanical Engineering, Nagoya University
Furo-cho, Chikusa-ku, Nagoya 464, Japan

ABSTRACT

The deposition of a hydrogenated B-C-N (Boron-Carbon-Nitrogen) thin film by a 13.56 MHz rf plasma CVD (Chemical Vapor Deposition) using a gas mixture of diborane (5% vol.% in nitrogen), ethane (or methane), and argon (or nitrogen) as carrier gases is presented. The deposition rate was 80-120 °A/min. The refractive index was 1.3-1.6. The Knoop microhardness of the films was 1825-3324 Kg/mm^2. In the near-UV region, the films had a direct band gap value of 13.7 eV. The stress of the films showed to be dependent on the type of the carrier gas. Suitable tensile stress values, for x-ray lithography; of $0.8x10^9$ and $6.5x10^9$ dyn/cm^2 could be obtained with argon and nitrogen discharges, respectively. It could be noticed that both film composition and microhardness depend on the deposition conditions. Besides, there is an interrelation between film composition and microhardness.

It should be mentioned that the deposition was carried out at room temperature and without heating the substrate.

I. INTRODUCTION

Boron compounds are wellknown to be useful in various technical fields, such as metallurgy, ceramics and electronics. Thin films of boron compounds have also attracted the attention of many workers for their physical and chemical properties. Boron nitrogen compounds(BN) are among them. They can be found to have special interest for their superficial analogies with their organic counterparts [1].

In view of BN good thermal conductivity, it is valuable as a heat sink and mounting material in value and transition circuits which can replace the tonic and less readily worked beryllia. Another importance, but still with some conservations; is that it can be used as a distortion-free radiation mask for its structurally stability and free of distortion over the range of common processing conditions [2].

However, all the above mentioned applications have encouraged us to synthesize a new compound which may have some unique properties. The starting point of this research work stemmed from the idea of combining the properties of both boron nitrogen and carbon films. This led us to form a hard and transparent B-C-N. On the mean time, the research work concentrated at first on plastic coating and low-z coating material in a pellet target for laser fusion targets [3].

Recently, x-ray lithography is widely believed to occupy an important future lithographic niche in the vicinity of 0.5 micron and below. It will be technically easier to achieve resolution and depth of focus with x-ray than with optical steppers in this lithographic system. It will be much less expensive to print large volume runners like Dynamic RAMs, microprocessors, or CODECs with x-ray than with electron beam direct write [4].

The main key point in order to manage such a precise lithographic technique is the x-ray mask substrate. The mask should be capable to meet the main requirements of the practical x-ray mask substrate, such as ease of formation technique, suitable mechanical properties, transparency and resistance to x-ray radiation, stability, optical transparency, etching resistivity and low defect density.

Such a challenging point also encouraged us, in an advanced stage of research to concentrate on trying to get a good B-C-N thin film capable to meet the main requirements for the x-ray lithography.

The formation of a transparent BN thin film deposited outside the discharge region is presented in section II. The films were formed by both rf plasma CVD at 13.56 MHz and microwave plasma at 2.45 GHz. In section III, the new B-C-N thin film deposited by low temperature plasma CVD at a frequency of 13.56 MHz is introduced. The films could be obtained with a gas mixture of diborane (5 vol.% in nitrogen), ethane (or methane), and argon (or nitrogen) as carrier gases. The properties of the B-C-N thin film are presented in section IV, such as the refractive index, microhardness, plasma etch rate, film band gap, and stress. The film composition and microhardness were found to be dependent on the deposition conditions, as presented in section V. The effect of film composition on microhardness is also presented.

II. BORON NITROGEN COMPOUNDS

Boron nitrogen compounds have attracted the attention of a lot of researchers to work with for main two factors [1]. First, the BN moiety is isoelectronic with C-C. Second, the sum of the covalent single-bond radii of boron and nitrogen is very similar in magnitude to that of two carbon atoms. Consequently, there are superficial analogies between certain boron nitrogen compounds and their organic counter parts.

Potential applications of boron nitrogen films have been proposed such as:

1- high temperature dielectrics [5].
2- heat-dissipation coating [6].
3- passivation layers [7].
4- diffusion sources of boron [8].
5- sodium barriers [7].

According to the formation conditions, boron nitrogen can be formed in different structures by plasma CVD. For example, hexagonal boron nitrogen (h-BN) can be formed with very close analogy to graphite, and is close to its coplanar C-C bond length (1.45 °A) by treatment at temperatures below 1800 °C. Cubic BN can be formed with very high degree of hardness similar to diamond at temperature over 2000 °C.

A. A HISTORICAL REVIEW

There has been a considerable work on BN thin film formation specially during the last 15 years. Rand and Roberts [7] deposited a vitreous film of BN up to 600 nm thick on a variety of substrates at 600-1000 °C by a reaction between diborane and ammonia in hydrogen or inert carrier gas. Hirayama and Shohno [8] reported that the films deposited below 1000 °C are amorphous. However, when the deposition temperature is over 1000 °C, the film is polycrystalline with a hexagonal structure. Hyder and Yep [9] grew thin films of BN by reactive plasma deposition using the ammonia-diborane reaction at a high substrate temperature (over 1000 °C) on graphite, Si and BN substrates. The deposition gave smooth and transparent films and they found that they had better crystalline quality than those deposited by NH_3:B_2H_6 plasma CVD alone at a high temperature.

Murark et al. [10] deposited BN films by reacting ammonia and diborane in an inert carrier gas in a temperture range of 400-900 °C. They investigated the properties of BN films as function of deposition conditions. They also reported that the film properties can be controlled by the flow rate of reactant gases. Adams and Capio at Bell Labs. [11] deposited films of BN compound with a composition of approximately B_6NH_x at reduced pressure by reacting diborane and ammonia at 250-600 °C. The films were in a ring structure, and hydrogen bonded as BH, NH_2 and possibly NH. Gafri et al. [12] obtained amorphous film containing a small amount of h-BN using BCl_3 at a relatively low substrate temperature of 550-620 °C. But the authors did not mention anything about the transparency of the films. Sano and Aoki [13] presented transparent and smooth thin films of BN on fused silica and sapphire at a substrate temperature of 1000-1100 °C by reacting BCl_3 with NH_3. The films were chemically inert and adherent to the substrate. Morita et al. [14] also obtained a transparent film, which was supposed to have a polymeric structure; by plasma CVD using ammonia-diborane.

According to the above mentioned review, it can be noticed that the crystallization of the films is strictly related to the substrate temperature and it decreases proportionally with it.

However, no publications have appeared concerning the deposition of a hard and transparent BN thin film at room temperature and without heating the substrate. We happened to be interested in amorphous BN, especially as deposited at low temperature for using as x-ray mask substrate. In the following subsection, the formation of a transparent BN thin film is introduced. The deposition was carried out outside the discharge region at room ambients.

B. THE FORMATION OF TRANSPARENT BN THIN FILM AT ROOM TEMPERATURE

Two types of reactors are used, namely rf and microwave ones. Figure 1 illustrates the inductively coupled rf plasma reactor at 13.56 MHz.

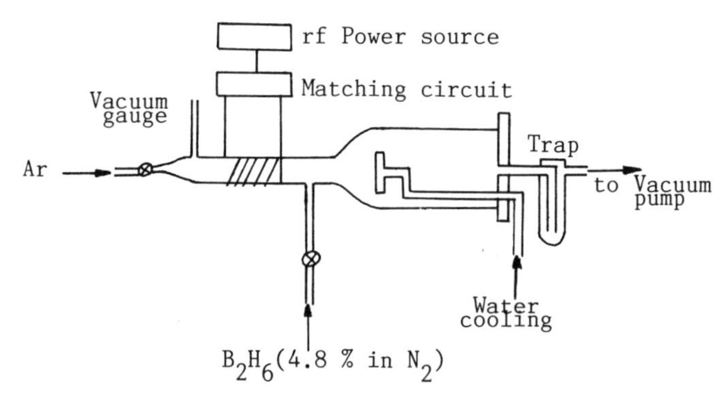

Figure 1. Schematic diagram of the inductively coupled rf plasma reactor.

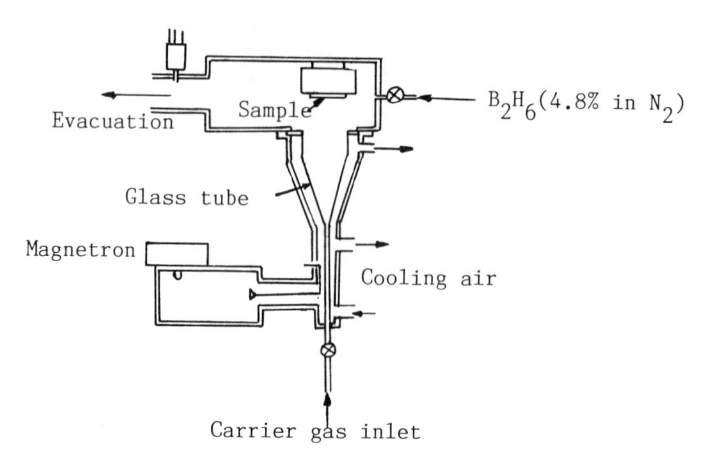

Figure 2. Schematic diagram of the 2.45GHz microwave plasma reactor.

A schematic diagram of the microwave reactor is given in Figure 2. A 2.45 GHz microwave generated by a magnetron is transmitted through a rectangular waveguide and converted to a travelling wave through a concentric waveguide. The full details of the two reactors can be found elsewhere [15-17].

Only diborane gas (B_2H_6), diluted in nitrogen at 4.8 vol.%; was used as the reactant gas. The deposition was performed outside the discharge region at room ambient. Table I illustrates the deposition conditions and some results.

Table I. Deposition Conditions and Results

Film No.	1	2	3	4	5
Deposition method	rf plasma	rf plasma	rf plasma	rf plasma	Microwave
Carrier gas flow rate(ml min^{-1})	Ar 11	Ar 11	Ar 11	N_2 20	N_2 550
Diborane(4.8 vol% in N_2) flow rate (ml min^{-1})	10	16	19	10	275
Total pressure(Pa)	32.5	39	54.5	56	149
Discharge power(W)	30	30	30	10	800
Deposition rate (nm min^{-1})	16.7	18.3	---	---	---
Film thickness(nm)	500	550	---	---	---
Film description	Hard, stable, and transparent	Hard, stable, and transparent	Hard, stable and transparent	Hard, stable	Soft and colored

Three kinds of substrates were used: NaCl, glass(Corning 7059) and mirrior-polished n-type Si(100) with resistivities of 20-45 Ω.cm. A spectrophotometer at a wavelength of 1000-200nm was used to measure the transparency of the films deposited on the glass substrate. The molecular structure of the deposited films on NaCl crystal and Si substrates was investigated by infrared(IR) absorption spectroscopy in the range 4000-600 cm^{-1}.

Both the film thickness and refractive index for some films were determined by an ellipsometer at a wavelength of 632.8 nm by an He-Ne laser source.

C. RESULTS

The infrared spectra of the films deposited by both rf and microwave plasma are shown in Figures 3 and 4, respectively. In the case of rf plasma deposition, it is clear that there appears a main absorption at 1400-1300 cm^{-1}, a relatively deep absorption at about 3200 cm^{-1}, and another deep absorption at 2500 cm^{-1} for the nitrogen gas discharge. While those for microwave plasma deposition are at 3200, 1450, 1190 and 780 cm^{-1}.

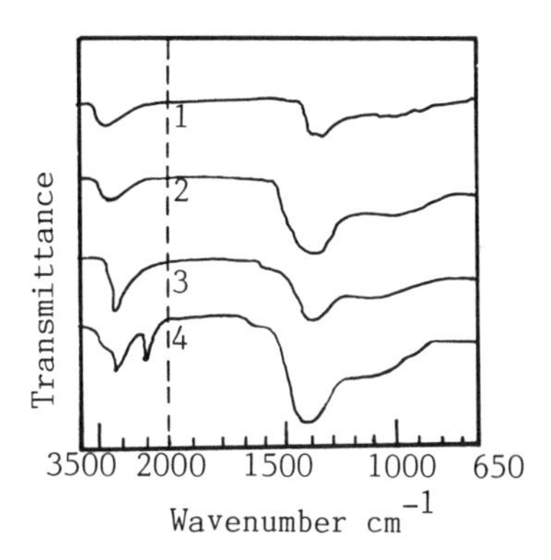

Figure 3. Infrared spectra for the films deposited by rf plasma (1-3) with argon as carrier gas and (4) with N_2 as carrier gas.

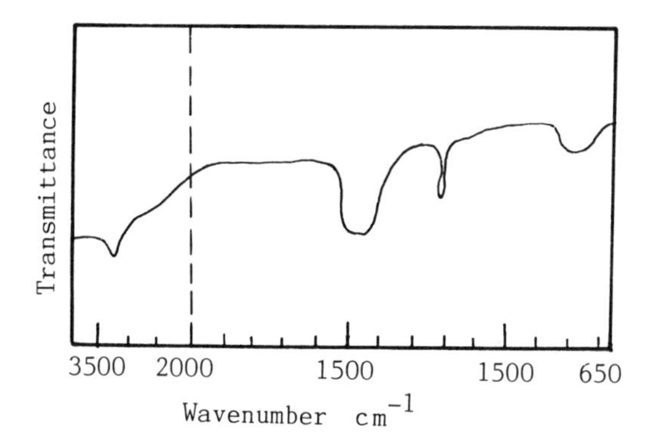

Figure 4. Infrared spectrum of film No. 5 deposited by microwave plasma.

The relatively deep absorption at 3200 cm^{-1} can be assigned to either NH_2 groups (3240-3210 cm^{-1}) [11], and/or N-H bond (3500-3000 cm^{-1}) [18]. The absorption at 2500 cm^{-1} can be assigned to B-H stretch (2520-2510 cm^{-1}) [7,11]. This absorption disappeared in the case of rf deposition with argon, as a carrier gas; at a relatively low pressure of about 32.5 Pa. This means that the films deposited with N_2 or argon as carrier gasa at higher pressures are formed in a polymeric structure. When trying to determine the absorption peak at 1450 cm^{-1}, it can not exactly be referred to any chemical bond of BN compound. But the cyclic BN compound showed a peak at 1415-1400 cm^{-1} [11]. Whereas the absorption peak at 1450 cm^{-1} will be expected to be a highly crosslinked BN bond structure as the absorption peak of cyclic BN bond was shifted to a shorter wavelength. The absorption peak at 1400-1310 cm^{-1} will be referred to BN, and/or B-O bond structures [19]. The absorption at 1190 cm^{-1} may be assigned to BH_2 absorption (1205-1140 cm^{-1}) [20]. While the absorption at about 780 cm^{-1} can be adopted to B-N-B bending vibration [13] or B-B bonds [7,13].

The refractive index value was 1.67-2.7 in a total pressure range of 65-240 Pa at a discharge power level of 850 W with the microwave discharge.

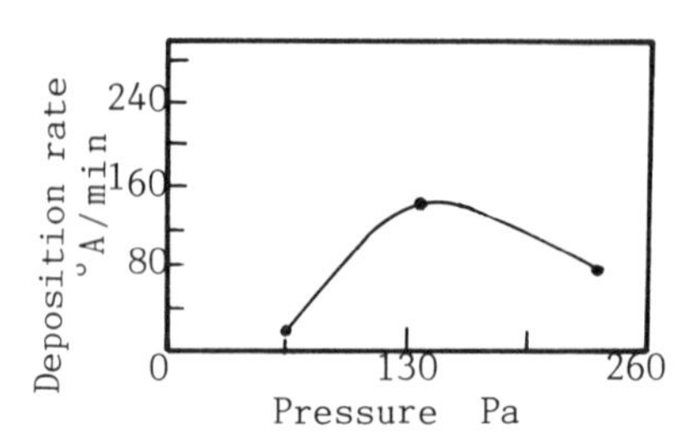

Figure 5. Deposition rate against the total pressure for an N_2/B_2H_6 ratio=18 and an argon flow rate of 50 ml min^{-1} by microwave plasma.

Figure 5 represents the deposition rate against the total pressure at an argon flow rate of 50 ml min^{-1} and a discharge power level of 1000 W with the microwave discharge. The deposition rate increases gradually with increasing the total pressure up to a maximum value at about 130 Pa. The same behaviour was reported by Gafri et al. [12] for BN thin film deposition by rf plasma at a substrate temperature of 550-620 °C. Also, it was noticed for the deposition of SiC thin film by Katz et al. [20].

It was found that both the refractive index and deposition rate depend on the argon flow rate. They all decrease with increasing argon flow rate up to a limited value of about 50 ml min^{-1}, after which they all increase.

D. COMPARISON BETWEEN RF AND MICROWAVE PLASMA DEPOSITION RESULTS

According to the infrared spectra of the two deposition methods, the absorption peak at 1400-1330 cm^{-1}, for rf plasma deposition; was shifted to 1450 cm^{-1} in the case of microwave plasma. This frequency shift usually occurs at a high deposition temperature, which results from the microwave field application.

The relatively deep absorption peak at 3200 cm^{-1} was detected in both methods. This absorption was referred to either NH_2 group (3240-3210 cm^{-1}) [11], or N-H bonds (3500-3000 cm^{-1}) [16]. It can be noticed that this absorption peak is relatively small in the case of microwave plasma deposition. This is due to too high gas temperature resulting from the microwave field which increases the dissociation of hydrogen.

The absorption peak at 2500 cm^{-1}, which was assigned to B-H stretch (2520-2510 cm^{-1}) [1,11], disappeared in the deposition with microwave plasma. This is also referred to the above mentioned reasons.

The absorption at about 780 cm^{-1}, that was adopted to B-N-B bending vibration or/and B-B bonds, appeared only in the microwave plasma deposition. This may also be due to the high gas energy that would lead to a high degree of dissociation of diborane, and in accordance to a high boron content. This absorption peak appeared in some films deposited by rf plasma at a relatively higher argon flow rates of about 20 ml min^{-1} or more. That is because the increase of argon flow rate leads to high electron temperature in the discharge region and consequently to higher boron content. It was noticed that such a high boron content films were of higher refractive index values. In this portion, it can be mentioned here that the value of the refractive index depends on many other factors, such as the total pressure inside the reactor, argon flow rate, N_2/B_2H_6 ratio and the position of the wafer inside the reactor.

It must be noted that the deposition in both methods was carried out outside the discharge region. This would keep the deposited films away from the electron bombardment that usually happens in the discharge region. Also, the films deposited in the discharge region lost their transparency and became whitely fogged. This is possibly due to the existence of some N-H groups, or a high boron content in the films.

III. TRANSPARENT BORON-CARBON-NITROGEN THIN FILMS

Compounds containing only boron and nitrogen have some special properties, such as being electrically insulating, chemically inert, thermally stable, and resistible to corrosion. Besides, they have desirable mechanical properties and reasonable wide bandgap. In the field of x-ray lithography, they have been used as good x-ray mask substrates. However, the control of the internal stress and stability of the mask films are still the main discouraging points for a wide range of applications.

In the first stages of this research work, the interests concentrated on adding some of the properties of carbon to those of boron-nitrogen compounds. During that stage, the main target was to form a suitable B-C-N (Boron-Carbon-Nitrogen) thin film for plastic coating and as a low-z layer in a pellet target for laser-fusion experiments. However, in an advanced stage of study the need for an x-ray mask turned over our interests to get a good thin film that can meet the main requirements for the x-ray lithography.

A. FORMATION METHOD

Two kinds of rf reactors at 13.56 MHz were used. The first was the same previous inductively coupled reactor but with two reactant gases inlets. The second was a capacitively coupled one, Figure 6.

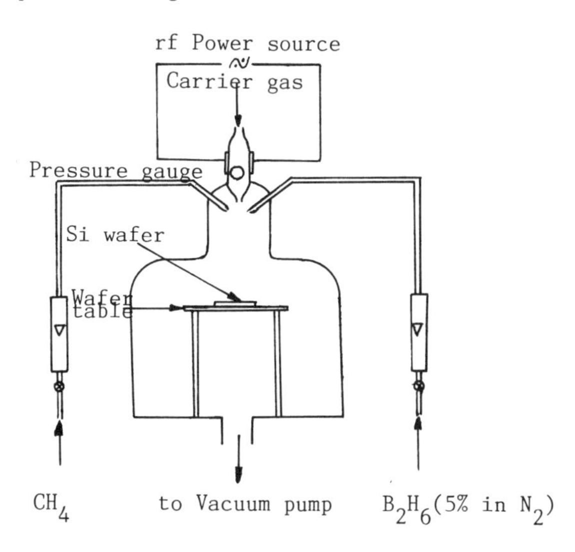

Figure 6. Schematic diagram of the capacitively coupled rf apparatus.

Both reactors gave nearly the same results. Two kinds of carrier gases, namely argon and nitrogenwere used alternatively. Diborane (B_2H_6) diluted in nitrogen to about 5 vol.%, and methane (or ethane) were used as reactant gases. The full details of the two reactors can be found in references [21-23].

Table II. Deposition Conditions and Results

Film No.	Ar flow rate (ml min^{-1})	B_2H_6:C_2H_6 wt. ratio	Total pressure (Pa)	Film thickness (nm)	Comments
1	6	1.35	48.1	730	Transparent Stable
2	12	0.87	48.1	600	Transparent Stable
3	30	0.14	48.1	---	Transparent Unstable
4	45	1.30	117	---	Whitely Fogged
5	6	1.30	117	---	Whitely Fogged
6	6	2.00	48.1	710	Transparent Stable
7	6	2.42	48.1	---	Transparent Stable

Some examples of the deposition conditions and general features of the films, obtained by the inductively coupled reactor; are given in Table II.

B. DEPOSITION

The films were formed at room temperature and without heating the substrate. Most of the films deposited in this work were found to be stable and adherent to the substrate in room ambients for long time [24]. Besides, they could be obtained repeatedly with the same results. The films containing boron at atomic percentages of more than 10% were transformed by atmospheric changes after six weeks and gradually lost transparency [25]. For such films the atomic percentage of oxygen increased proportionally with the increase of boron content. The most stable films were those containing carbon at atomic percentage of more than 45% and nitrogen at around 25%. Deposition at pressures less than 52 Pa (with the capacitively coupled rf reactor) gave transparent films with a moderate deposition rate of 80-120 °A/min. The films deposited at pressures higher than 52 Pa were increasingly fogged, appearing white. The transparencies of the films on glass substrates were investigated by spectrophotometry at wavelengths of 10000-2000 °A. It was noticed that the films obtained at lower pressures (less than 52 Pa) showed a very degree of transparency without any noticeable absorption. The deposition rates and thicknesses were as high as 150 °A/min and 7000°A, respectively. In more advanced stages of deposition, films of about 2 μm could be obtained.

C. INFRARED SPECTRA

It was very difficult to determine the molecular structure of such a complex film only from its IR spectra. The spectra were dependent on the ratios of reactant gases, and both the kind and flow rate of the carrier gas [24]. Figure 7 illustrates IR spectrum of the films deposited at a pressure of 20 Pa (with the capacitively coupled rf reactor), discharge power level of 30 W, and $B_2H_6:CH_4$ ratio of 0.15.

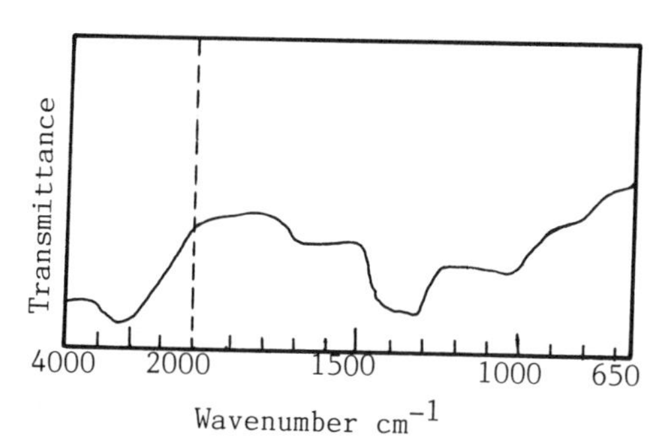

Figure 7. Infrared spectrum of the films at 20 Pa pressure, discharge power level of 30 W and $B_2H_6:CH_4=0.15$.

This film was supposed to contain mainly the groups B-N, C-N, B-O, and B-CH_3 in the range 1400-1250 cm^{-1}. O-H stretching modes and N-H absorption are still remaining in the films but not so strong, at 3400 and 3200 cm^{-1}, respectively. The shoulder at 1100-1000 $cm^{-1}$was assigned to B-C stretching. B-H-B bridge groups were represented by the shoulder at 1700 cm^{-1}.

D. COMPOSITION

According to the quantitative analysis of the films through the ESCA (Electron Spectroscopy for Chemical Analysis) data and IR spectra, they contain boron, nitrogen, carbon, hydrogen and oxygen. The ESCA analysis was carried out by an ESCA MK II (VG Scientific) apparatus, with ion etching by argon at 8 kV and 200 $\mu A/cm^2$ for some films. Quantitative analysis of the hydrogen was not available.

Oxygen was not mixed with the reactant gases, but it appeared in the films at different atomic percentages. It was noticed that this value is a function of the boron content in the film, and it may be arisen from one of the following reasons [26]. First, it may be due to some leaked air into the reactor because of the high activity of boron through the evacuation rotary pump. Second, as boron has a high reactivity in the discharge region, it may react with the glass wall of the reactor tube. The second reason is supposed by the appearance of very small amount of silicon of about 1 at.% or less in some samples. In spite of this oxygen content, the films were stable and adherent to the substrate in room ambients.

E. A DISCUSSION OF THE DEPOSITION MECHANISM

By investigating the molecular structure of the deposited films, regarding the IR spectra; the deposition mechanism can be suggested to be occuring according to the following chemical reactions [27]:

$$B_2H_6 \longrightarrow 2BH_3 \longrightarrow 2BH_2 + H_2 \longrightarrow 2BH + 2H_2$$

$$C_2H_6 \longrightarrow C_2H_4 + H_2 \longrightarrow C_2H_2 + 2H_2$$

The above mentioned reactions are expected as in the glow discharge, it is wellknown that the electron temperature is high compared to the gas temperature. Such a phenomenon would lead to an activation and decomposition of both diborane and ethane. It is also supposed that decomposition reactions are assisted by the activated argon.

Nitrogen was supposed to be activated in the glow discharge as nitrogen atom had been detected in the deposited film. The possibility of nitrization is expected to be small due to the large ionization energy of nitrogen.

Generally speaking, the formation of B-C-N compounds would take place on the surface of the substrate according to the deposition conditions. Under the irradiation of ionized particles and activated and decomposed molecules, various chemical bonds can be formed, such as B-B, B-H-B, B-C, B-N and C-N. It was noticed that at low pressures of about 52 Pa, the density of B-C is high.

IV. THE PROPERTIES OF B-C-N THIN FILMS

A. REFRACTIVE INDEX

The refractive index of the film was found to be in the range 1.3-1.6, as measured by an ellipsometer with He-Ne laser source at wavelengths of 632.8 nm. The higher value of refractive index was obtained at higher discharge power level and pressure. In the case when the same diluted diborane in nitrogen was only used as the reactant gas, see section II; the refractive index was 1.67-2.75. It is expected that this value changes in accordance with the content of both boron and carbon, and the film structure. Boron-rich films have relatively higher refractive index values.

B. MICROHARDNESS

An E. Lietz Wetzlar Micro Vickers hardness tester was used to measure the hardness of the films on silicon substrates. The measurements were performed according to the following formula:

$$HV = \frac{1854\ P}{d^2}$$

where:

HV = hardness in Kg/mm^2.
P = load in grams.
d = average length of the diagonal formed on the thin film by the load, in μm.

The microhardnesses of the films were in the range 3324-1825 Kg/mm^2. Whitely foggy films with values as high as 4500 Kg/mm^2 could also be obtained. It can be expected to get higher refractive index values with different deposition conditions. A full detailed characterization of the microhardness is given in the next section.

C. PLASMA ETCHING RATE

Low pressure plasma etching of 0.53 Pa at a discharge power level of 50 W was carried out by using different etching gases, namely CF_4, O_2 and H_2 [19]. Figure 8 illustrates the depth of etching against the duration. It is clear that at the initial stages of etching, the etching rates are 125, 55 and 28 nm min^{-1} with the etching gases CF_4, O_2 and H_2, respectively. It is clear that CF_4 gas gave the highest etching rate, while the lowest value was for H_2.

In the case of plasma etching with O_2, the etch depth increases with the etch duration up to a value of about 500 nm and then it saturates regardless of the etch duration. This indicates that there has been a balance between the etch rate and the oxidation of the film.

D. FILM BAND GAP

The film band gap is required when studying the electrical properties for some applications, such as laser optical diode. The film band gap value is dependent on the method to be used for the measurements [13].

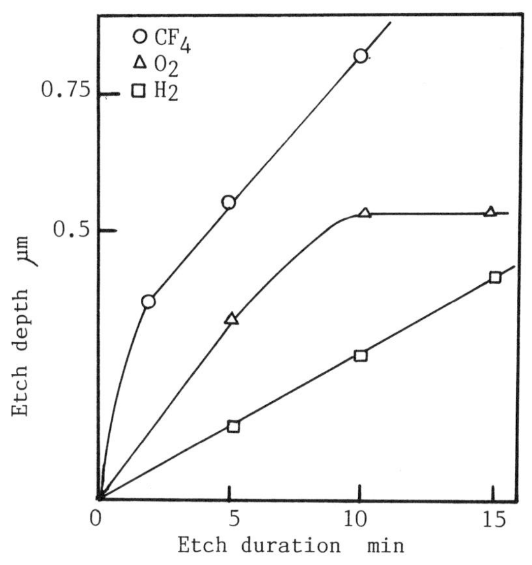

Figure 8. Etch depth versus etch duration with CF_4, O_2 and H_2 gases.

Table III gives some values of the band gap for different compounds [28 ,29].

Table III. Energy Band Gap for Different Compounds (eV)

	Diamond (cub.)	Boron nitride (cub.)	Boron nitride (hex.)	Silicon (cub.)	Silicon carbide (cub.)	Silicon carbide (hex.)	AlN	GaN	Ge (cub.)
Direct	7.4	10-13	5.8	1.15	6.0		6.28	3.39	0.805
Indirect	5.47	7-10			2.5	2.99			0.664

An UVSOR (Ultraviolet Synchrotron Orbital Radiation) apparatus was used to measure the film band gap [30]. An ultraviolet radiation of 68 °A/mm was applied. The B-C-N films were first deposited on LiF substrates to thicknesses of 200 and 1000 °A.

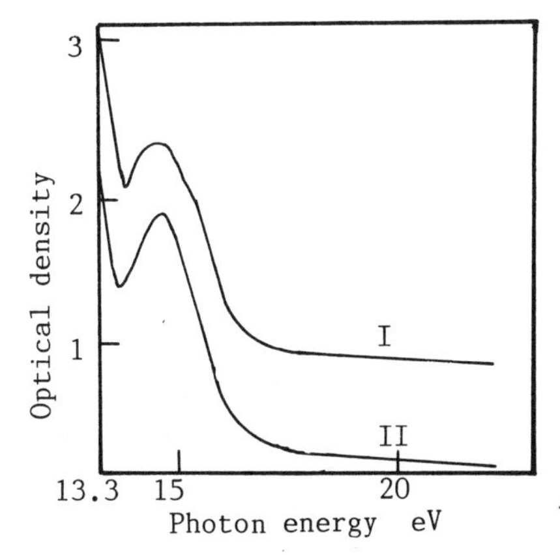

Figure 9. The optical density of the films against the photon energy for: I) 200 °A and II) 1000 °A, thickness.

Figure 9 represents the optical density of the films as a function of the photon energy, in eV . In the near-UV region the main absorption peak was observed at 14.7 eV and a sharp drop occured near 13.7 eV. The sharp drop is attributed to the direct band gap of the film. This value indicates that the B-C-N film is an insulator.

E. MECHANICAL PROPERTIES

As a matter of fact, a suitable ruggedness of optical mask is needed as they must be handled by shop personnel in a production environment. The microhardness of the obtained B-C-N films, deposited at room temperature and without heating the substrate; is in the range 3324-1825 Kg/mm^2. These values seem to be reasonable when they are compared with those of the other mask substrates.

Another factor which has a special importance is the film stress. If the film has a compressive stress, it will shrivel when the base silicon substrate is selectively etched away from the back side in the membrane fabrication process. On the other hand, when the stress in the film is highly tensile, the film will be susceptible to cracking. Therefore, a film for an x-ray mask substrate having an appropriate tensile stress should be prepared for practical x-ray lithography. The tensile stress of a membrane can be determined conveniently with a "bulge" chamber by measuring the displacement of the membrane as a function of the differential press [31]. Membrane stress may also be measured using a resonance frequency technique [32]. The "wafer bow" method is one of the most famous ways to measure the stress of a film deposited on a wafer, Figure 10.

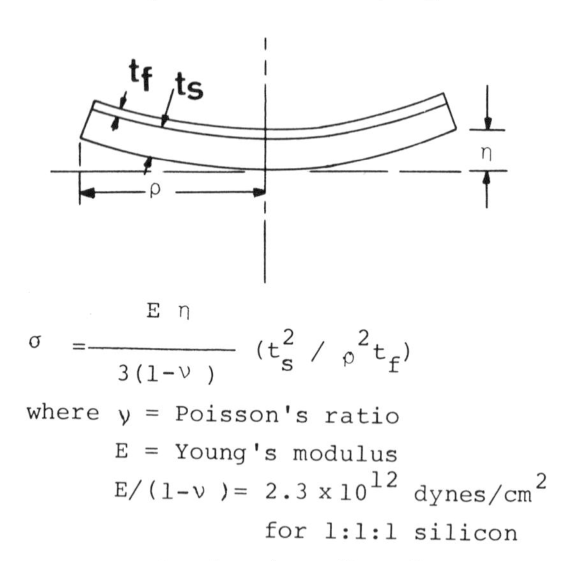

Figure 10. The stress measurement using the radius of curvature, "wafer bow" method; Ref. [33,34].

In this method, the bowing of the wafer is measured before and after coating [33,34]. This method was used to measure the stress of the B-C-N films. The substrate curvature radius was measured using the Newton rings interference technique [33]. The stress of the films showed to be dependent on the type of the carrier gas, namely argon and nitrogen. However, both kinds of discharges led to the required tensile stress needed for the x-ray mask substrate. The tensile stress was 0.8×10^9 and 6.5×10^9 dyn/cm^2 for the argon and nitrogen discharges, respectively. When referring to the literature, it can be noticed that the first stress value is reasonable to fulfill one of the main requirements for a self-standing x-ray mask membrane [34].

V. FILM CHARACTERISTICS

A. RELATION BETWEEN FILM COMPOSITION AND DEPOSITION CONDITIONS

The film composition depends mainly on the total pressure inside the reactor, discharge power level and reactant gases ratio.

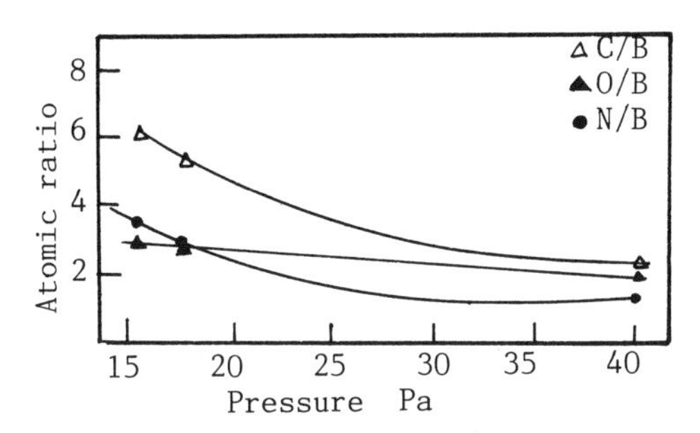

Figure 11. The atomic ratio dependence on the total pressure inside the reactor, at discharge power level of 30 W and $B_2H_6:CH_4$ ratio of 0.5.

Figure 11 displays the atomic ratios C/B, O/B and N/B,at a discharge power level of 30 W and $B_2H_6:CH_4$ ratio of 0.5;as a function of the total pressure inside the reactor. It is clear that these ratios decrease with increasing pressure. At lower pressures, C/B ratio is higher than the others. Then they all become nearly the same at higher pressures. This indicates that the change in C/B at the low pressure is the most prominent factor.

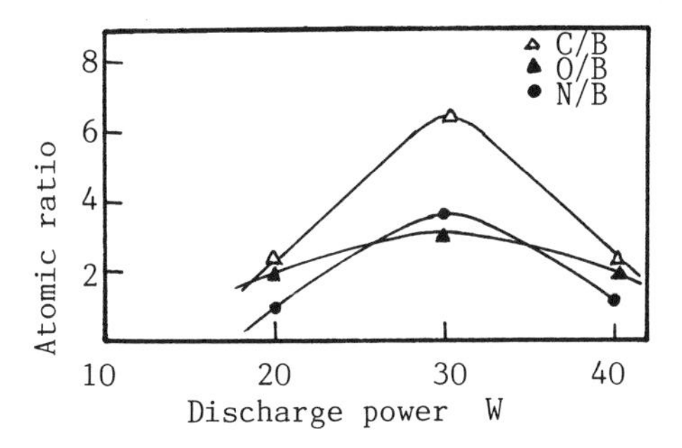

Figure 12. The atomic ratio versus discharge power level, at 20 Pa pressure and $B_2H_6:CH_4$ ratio of 0.5.

The dependence of the same atomic ratios on the discharge power level, at a pressure of 20 Pa and $B_2H_6:CH_4$ ratio of 0.5; is given in Figure 12. It can be noticed that there is a peak value for all the ratios after which they all decrease with the increase of the discharge power level.

Figure 13 represents the same atomic ratios against the reactant gases ratio ($B_2H_6:CH_4$), at a discharge power level of 30 W and a pressure of 20 Pa. It is clear that the C/B ratio is inversly proportional to $B_2H_6:CH_4$, while the two other ratios increase proportionally. Also, the oxygen content increases with increasing the diborane flow rate. It may be concluded that there is an interrela-

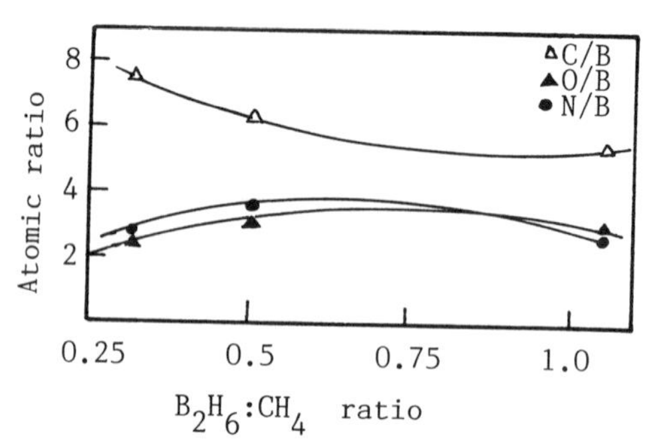

Figure 13. The atomic ratio against B_2H_6:CH_4 ratio, at a discharge power level of 30 W and 20 Pa pressure.

tion between all the elements in both gas and solid phases rather constant in comparison with the atomic ratio in row material gas that in the produced film is greatly reduced.

B. THE DEPENDENCE OF MICROHARDNESS ON THE DEPOSITION CONDITIONS

The Wetzler Micro Vickers hardness tester gives Knoop microhardness values in Kg/mm^2. The relation between the microhardness and the total pressure inside the reactor is presented in Figure 14.

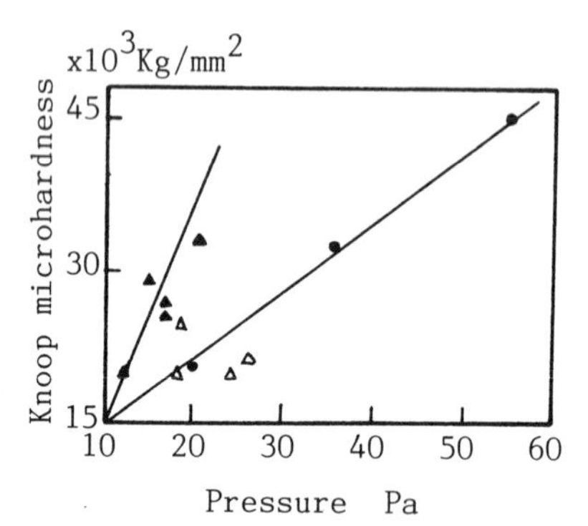

Figure 14. Knoop microhardness versus the total pressure, for 5000 °A-thick films at B_2H_6:CH_4 ratios of about ●: 0.15 and ▲: 0.6 with argon discharge, and △: 0.6 with N_2 discharge, and at discharge power level of 30 W.

In case of argon glow discharge, the microhardness can be divided into two groups. The first is for a monomer gases (B_2H_6:CH_4) ratio of about 0.6, and the second for the ratio of about 0.15. It is clear that the microhardness increases linearly with the total pressure. It must be mentioned that very high microhardness values could be obtained, but at the expense of the transparency of the film. Some values for the nitrogen glow discharge deposition are also illustrated on the same graph. It can be noticed that these values are smaller than those for argon glow discharge. This may be due to the difference in the ionization energy of the two carrier gases.

Figure 15 shows the dependence of the microhardness on the monomer gases ratios, with nitrogen glow discharge at a discharge power level of 30 W and a total pressure of 20 Pa. It is clear that the microhardness reaches a maximum

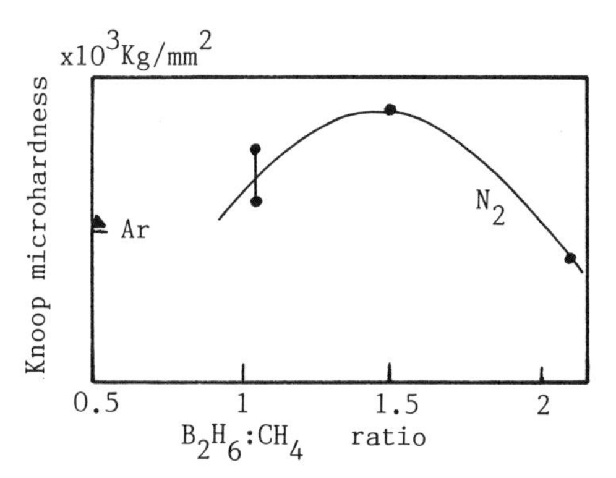

Figure 15. Dependence of microhardness on B_2H_6:CH_4 ratio at discharge power level of 30 W and 20 Pa pressure.

value at a monomer gases ratio of about 1.5, and then it decreases.

C. EFFECT OF FILM COMPOSITION ON MICROHARDNESS

In this subsection, the effect of the different elements contained in the film on its microhardness is to be investigated.

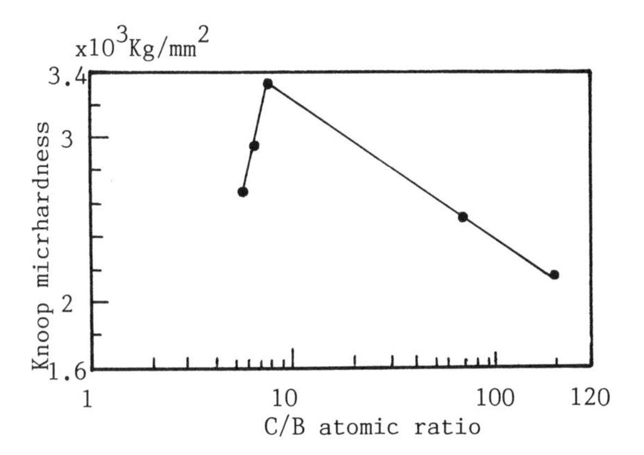

Figure 16. Knoop microhardness versus C/B atomic ratio for 5000 °A-thick films

Figures 16-18 illustrate the microhardness as a function of C/B, N/B and O/B, for films of 5000 °A thicknesses, respectively. It is clear, from Figures 16 and 17; that the microhardness has a maximum value of about 3324 Kg/mm^2, after which it decreases with the decrease of boron content in the film. As for the dependence on the O/B atomic ratio, it is clear that the microhardness is inversely proportional to the oxygen content in the film. However, this dependence is not clear at O/B ratio less than unity so far, as this work could clarify. This may be attributed to some weak oxygen bonds in the films specially for the low pressure high carbon films, i.e.; more polymer-like ones

Transparent films with high Knoop microhardness values of 3324 Kg/mm^2 were obtained at an atomic ratio of C:N:O:B equals 7.6:2.85:2.7:1. Such films can not be expected to have a polymeric nature as they have relatively high microhardness values when compared with those of polymers.

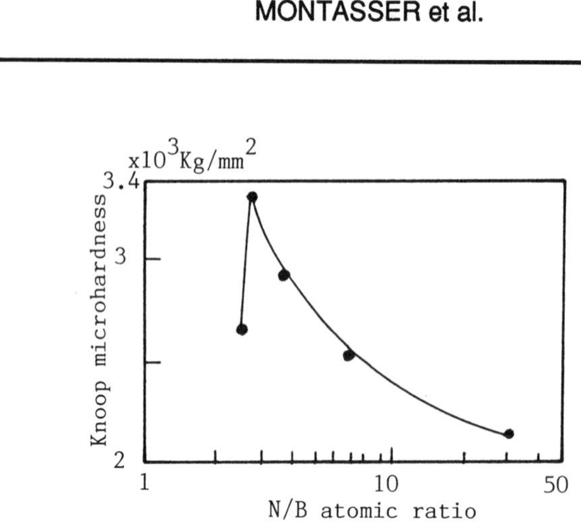

Figure 17. Knoop microhardness against N/B atomic ratios for 5000 °A-thick films.

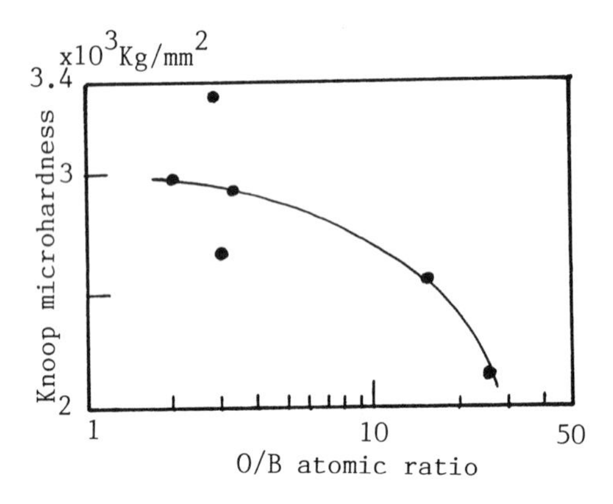

Figure 18. Knoop microhardness against O/B atomic ratios for 5000 °A- thick films.

VI. CONCLUSIONS

A new hydrogenated B-C-N compound could be obtained as a try to combine the properties of both carbon and BN compounds. The deposition was carried out at room temperature and without heating the substrate. Nitrogen was supposed to be activated in the glow discharge as nitrogen atom could be detected in the films.

A moderate deposition rate of 80-120 °A/min could be obtained. The refractive index of the film was 1.3-1.6 when measured with an ellipsometer operated by He-Ne laser source at wavelength of 632.8 nm. These values are relatively low when compared with those of BN thin films (1.67-2.75). The refractive index value is mainly a function of both carbon and boron content in the film.

The degree of transparency was very high without any noticeable absorption at wavelengths of 10000-2000 °A. Most of the films showed a very high degree of reproducibility. In addition, they were stable and adherent to the substrates at room ambients. It was noticed that the stability of the films depended on the boron and carbon atomic percentages.

The Knoop microhardness of the films were 1825-3324 Kg/mm^2. Whitely foggy films with values as high as 4500 Kg/mm^2 could also be obtained. In the near-UV region, the films had a direct band gap value of 13.7 eV, which indicates that they are insulators.

The stress of the films showed to dependent on the type of the carrier gas. Suitable tensile stress values of $0.8x10^9$ and $6.5x10^9$ dyn/cm^2, reasonable for the x-ray mask substrate; could be obtained with argon and nitrogen discharges, respectively.

The film composition and microhardness depended on the deposition conditions, such as the total pressure inside the reactor, monomer gases ratio, and discharge power level. Besides, there is a concrete interrelation between the film composition and microhardness. A transparent film with a Knoop microhardness of 3324 Kg/mm^2 was obtained at C/B, N/B, and O/B atomic ratios of 7.6, 2.85, and 2.7, respectively.

In fact, it should be concluded that, we were able to form B-C-N thin films which could realize some of the main requirements for a suitable x-ray mask substrate. The main results are ease of fabrication method, as plasma CVD deposition was used; resistance to x-ray radiation by the ultraviolet synchrotron orbital radiation (UVSOR) in a wide wavelength range of 1000-5 °A, stability, optical transparency, needed for the mask alignment; and the required slightly tensile stress value [35].

REFERENCES

1) Bailar, J.C., Emeleus, H.J., Nyholm, S.R., and Dickenson, A.F.: "Compressive Inorganic Chemistry", Pergamon, Oxford, 1973.

2) Adams, A.C., Capio, C.D., Levinstein, H.J., Sinha, A.K., and Wang, D.N.: Radiation Mask Structure, U.S. Patent 4, 1979, 171, 489.

3) Takigawa, A., Yoshida, M., Morita, S., Hattori, S., and Montasser, K.: Japanese Patent, Jan. 1985, Open File No. Showa 60-16814.

4) Herzog, H.J., Hersener, J., and Strohm, K.M.: "Stress Analysis of Absorber Metals for Silicon x-ray Masks", Microcircuit Eng. 84, Heuberger, A., and Beneking, H. eds., Academic Press, 1985, 317.

5) Carpenter, L.G., and Kirley, P.J.: J. Physics D 15, 1982, 1143.

6) Simpson, A., and Stuckes, A.D.: J. Physics C 4, 1971, 1710.

7) Rand, M., and Roberts, J.: J. Electrochem. Soc., 1968, 115, 423.

8) Hirayama, M., and Shohno, K.: ibid, 1975, 122, 1671.

9) Hyder, S.B., and Yep, T.: ibid, 1976, 123, 1721.

10) Murarka, S.P., Chung, C., Wang, D., and Smith, T.: ibid, 1979, 126, 1951.

11) Adams, A.C., and Capio, C.D.: ibid, 1980, 127, 399.

12) Gafri, O., Grill, A., and Itzhak, D.: Thin Solid Films, 1980, 72, 523.

13) Sano, M., and Aoki, M.: ibid, 1981, 83, 247.

14) Morita, S., Ishibashi, S., Hattori, S., and Ieda, M.: The 28th Meeting of J. of "Phys. Soc. of Japan" (Oyo Buturi Gakkai), Nagoya, 1981, 129P-E-5.

15) Montasser, K.: Ph.D. Graduation Thesis, "Boron Containing Compound Films by Plasma CVD", Dept. of Elect. Eng.& Electron., Nagoya University, Japan, 1987.

16) Hattori, S., Chinen, S., and Ishida, H.: J. of Phys. Elect. Sci. Instro., 1971, 4.

17) Montasser, K., Tamano, J., Morita, S., and Hattori, S.: Plasma Chem. and Plasma Proc., 1984, 4, 251.

18) Tibbitt, J.M., Bell, A.T., and Shen, M.:"Plasma Chem. of Polymers", Shen, M. eds., Dekker, New York, 1976.

19) Montasser, K., Morita, S., and Hattori, S.: The 8th Sympo. on Ion Sources and Ion-Assisted Tech. (ISIAT), Tokyo, 1984, Session VI-3.

20) Katz, M., Itzhak, D., Grill, A., and Avni, R.: Thin Solid Films, 1980, 72, 497.

21) Kato, I., Wakana, S., Hara, S., and Kezuka, H.: Jpns. J. of Applied Phys., 1982, 21, L470-L472.

22) Morita, S., Montasser, K., and Hattori, S.: The 8th Meeting of the Jpns. Group of Vacuum Sci., Nagoya, 1984, 1.

23) Montasser, K., Morita, S., and Hattori, S.: The 46th Meeting of J. of "Phys. Soc. of Japan" (Oyo Buturi Gakkai), Kyoto, 1985, 3PR4.

24) Montasser, K., Morita, S., and Hattori, S.: Thin Solid Films, 1984, 117, 311.

25) Montasser, K., Morita, S., and Hattori, S.: J. Appl. Phys., 1985, 58, 3185.

26) Montasser, K., Morita, S., and Hattori, S.: The 7th Intern. Sympos. on Plasma Chemistry (ISPC7), Eindhoven, The Netherlands, 1985, P-5-5.

27) Vossen, J., and Kern, W.:"Thin Film Processing", Academic Press, 1978.

28) Sobolev, V.V., Kroitorn, S.G., Sokolov, E.B., and Chengov, V.P.: Sov. Phys. Solid State, 1978, 20, 12.

29) Larach, S., and Shrader, R.E.: Phys. Rev., 1956, 104, 68.

30) Watanabe, M., Uchida, A., Matsudo, O., Sakai, K., Takai, K., Katayama, T., Yoshida, K., and Kihara, M.: IEEE Trs. Nucl. Sci., 1981, NS-28, 3175.

31) Bromely, E.I., Randall, J.N., Flanders, D.C., and Mountain, R.W.: J. Vac. Sci. Technol., 1983, B1, 1364.

32) Acosta, R.E., Maldonado, J.R., Towart, L.K., and Warlaumont, J.R.: Pro. SPIE, 1983, 448, 114.

33) Ghate, P.B., and Hall, L.H.: J. Electrochem. Soc.: Solid State Sci. Tech., 1972, 119, 491.

34) Aiyer, C.R., Gangal, S.A., Montasser, K., Morita, S., and Hattori, S.: Thin Solid Films, 1988, 163, 229, The 7th Intern. Conf. on Thin Films, New Delhi, Dec. 7-11, 1987, 10DT02, 212.

35) Montasser, K., Yamada, H., Morita, S., and Hattori, S.: Jpns. J. Appl. Phys., 1987, 26(4), L237-L239.

Materials Science Forum Vols. 54 & 55 (1990) pp. 313-328

CUBIC BORON NITRIDE PN JUNCTION MADE AT HIGH PRESSURE

O. Mishima
National Institute for Research in Inorganic Materials
1-1 Namiki, Tsukuba, Ibaraki 305, Japan

ABSTRACT

A temperature difference method was successfully applied to grow large crystals of cubic boron nitride (cBN) at high pressure. The reported existence of a semiconducting cBN was reconfirmed and a functionable *pn* junction diode of cBN was fabricated by growing an *n*-type crystal epitaxially on a *p*-type seed crystal. Rectification characteristics of the diode were observed from room temperature to ~650℃. Injection luminescence of UV light was observed from the diode. Thus, the cBN can be a good potential candidate as an electronic and optoelectronic material. However, improvement in the synthetic method and ohmic contact as well as the study of the properties of cBN are necessary for broad applications.

1. INTRODUCTION

Although there is an obvious difference between the B-N and C-C groupings because of the difference in the electronegativity between boron and nitrogen, the extreme importance of carbon in many fields aroused interest in the analogy between C-C and B-N bondings in hope of significant properties from B-N compounds. In this article, one may obtain a glimpse of the similarities and differences between cubic carbon (diamond) and cubic boron nitride (cBN).

The diamond was synthesized in 1955 by researchers of General Electric Company by converting graphite in a high-pressure and high-temperature apparatus [1]. Just after the success in the diamond synthesis, in 1957, the cubic boron nitride was synthesized from hexagonal boron nitride (hBN) under the similar extreme conditions by Robert H. Wentorf, Jr.[2,3]. He was one of the GE-diamond team and was inspired by the structural resemblance between carbon and boron nitride and by the analogy to other tetrahedrally bonded III-V compounds like AlN. He used the high-pressure high-temperature apparatus which was developed for the diamond synthesis.

It may be natural that one takes an interest in the semiconducting properties of this simplest III-V compound on the analogy of GaAs, as well as the mechanical behavior regarding the analogy of diamonds. Wentorf doped selected impurities and found that both *p*- and *n*-type semiconductors of cBN were obtainable [4] while only the *p* type was made for diamonds [5]. He attempted further to make a *pn* junction of cBN with what is called the film method under high pressure [4]. The *pn* composite bodies he obtained with this method were, however, too small to be examined. No meanful measurement was possible [4] and the formation of the *pn* junction was unconfirmed. The problem of fabricating the *pn* junction at high pressure remained to be solved.

The size of the cBN crystals produced with the film method, in which the hBN dissolved in a molten solvent and the cBN precipitated due to the solubility difference between the two materials, was usually less than ~0.5 mm. The smallness of the cBN crystal caused not only experimental difficulties regarding the study of the properties of cBN but also limitations as regards the application of this material. A large cBN crystal was desired.

The temperature difference method at high pressure was successfully applied to obtain large gem-quality single crystals of diamonds [6,7]. The supersaturation necessary for the growth was obtained by the solubility difference due to the temperature difference between the nutrient and the seed in a solvent bath. Nowadays, high-quality single crystals of diamonds ~15 mm in size are available using this method. However, there appeared to be no convincing paper on the crystal growth of cBN with the temperature difference method, which might be due to the relatively poor understanding of the phase diagram of a BN-solvent system compared with a carbon-solvent system. Although the solvent of BN was studied to some extent [8], the growth condition of the large cBN crystal was not found. The growth of the large crystal has been a problem for the last two decades or more.

We have started to make large cBN single crystals by the temperature difference method a few years ago at Tsukuba. By confirming that the present understanding of crystal morphology in relation to supersaturation was applicable even under a very high pressure and by pursuing a suitable growth condition, we have succeeded in growing large cBN crystals [9]. We have further attempted to make large semiconducting cBN crystals and to fabricate the *pn* junction with the temperature difference method, feeling the interest probably similar to what Wentorf might have. Consequently, the functionable electronic device, a *pn* junction diode of the cBN, was produced at high pressure [10]. Moreover, the cBN diode showed outstanding properties as a high temperature diode [10] and as a light-emitting diode [11]. The present article reviews our recent work on the growth of the cBN crystals and on the fabrication and the properties of the cBN *pn* junction diode.

2. SYNTHESIS OF LARGE SINGLE CRYSTALS OF CUBIC BN [9]

Since the cBN is stable only at high pressure and metastable at 1 bar, it is generally produced at high pressure. Fortunately, the pressure and the temperature of the synthesis of cBN are nearly the same as those of diamonds, and one can appropriate the relatively established technique of the diamond synthesis in high-pressure technology. Our high-pressure equipment and assemblies in a pressure chamber for the growth of cBN crystals were initially made for the diamond synthesis [12] and modified slightly to withstand high temperatures because the growth temperature of cBN was slightly higher than that for diamonds.

The principle of the temperature difference method is dissolution at high temperature and precipitation at low temperature in the growth cell located in the high-pressure oven. The growth cell, made of molybdenum with a cylindrical shape, was compressed in the high-pressure chamber and heated with the surrounding graphite-tube heater located also in the pressure medium in the chamber.(Fig. 1) BN powder (cBN or hBN) was placed at the high-temperature end in the cell as a nutrient. The solvent used was usually lithium calcium boron nitride and occasionally lithium boron nitride. These solvents are solid compounds with respective chemical formulas at room temperature ($LiCaBN_2$ and Li_3BN_2). Although the formulas seem complicated, it may be realized that these compounds include adequate B and N atoms in Li and/or Ca to provide sufficient supersaturation of BN efficiently. Mg and Sr are also known to be the solvent of BN. These compounds react easily with the moisture in the air. Although the compounds were treated with normal care, oxygen might exist in the present solvent.

From preliminary experiments using the $LiCaBN_2$ solvent and the cell, having two (top and bottom) solvent baths arranged vertically and symmetrically in the heater tube, the top bath was found to be more suitable to suppress nucleations and to grow large cBN crystals. This may be due to the circumstance that the density of cBN is higher than that of the solvent; the cBN nutrient located at the top side in the bottom bath may fall and provide many seeds in the solvent. The suitable cell assembly may, thus, depend on the density of the individual solvent. In the present work with

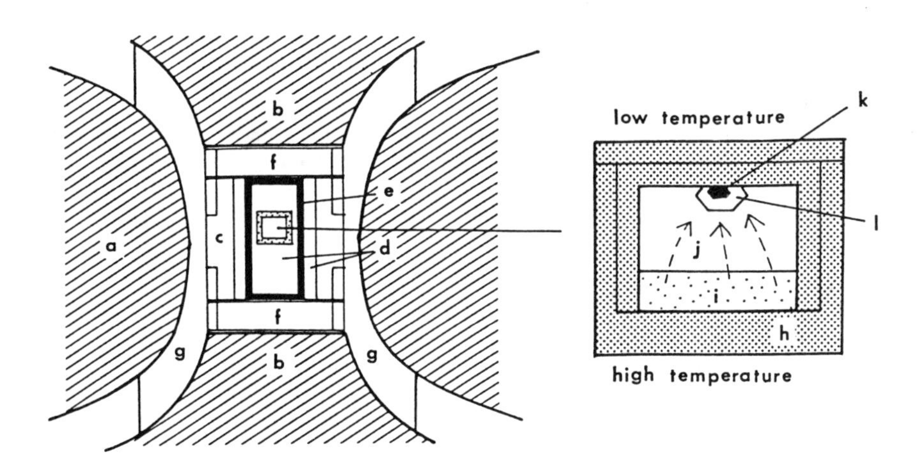

Fig. 1. Assembly for the growth of the cBN crystal. The growth cell was compressed in a modified belt-type high-pressure apparatus [12]. The inner diameter of the cell depended on the size of the apparatus and was ~4 mm (using a 25-mm i.d. cylinder and a 800-ton press) or ~11 mm (using a 75-mm i.d. cylinder and a 14000-ton press). a) WC or steel cylinder b) top and bottom WC anvil c) NaCl pressure medium d) hBN pressure medium e) graphite tube heater f) ZrO_2 thermal insulator g) pressure gasket h) Mo growth cell i) BN nutrient j) solvent k) cBN seed crystal l) growing cBN crystal

Li_3BN_2 or $LiCaBN_2$, the growth cell was always located at the upper side in the tube heater to grow cBN at the top end of the cell.

To change the temperature difference in the cell, we employed two simple methods. One is to move the growth cell in the heater tube to change the axial temperature difference in the cell. Another is to insert a thin partition sheet with a hole between the solvent and the BN nutrient in the cell and to change the diameter of the hole to change the radial temperature difference.

The growth cell was first compressed to ~50 kbar at room temperature with a hydraulic press. Electrical power was then supplied to the heater and kept constant at a certain value, corresponding to the estimated cell temperature of 1700 ~ 1800℃, for 10 ~ 70 hrs while the press load was fixed.

Precise control of pressure, temperature and temperature difference in the growth cell is still difficult for the present high-pressure technology, partly due to the lack of simple and effective monitors of pressure and temperature. Experimental errors caused the uncertainty of reproducibility of pressure and temperature, estimated ±~2 kbar and ±~30℃, and caused lack of reproducibility of the growth of cBN.

It was found that cBN crystals could grow by spontaneous nucleation in the temperature difference method. To grow high-quality large single crystals, one may be able to apply the existing growth theory, which describes the growth mechanism changes from the dislocation-controlled growth to the surface-nucleation-controlled growth and further to the unstable deposition on a rough surface as supersaturation increases [13]. The crystal shape grown in these mechanisms are expected to change from small bulky and polyhedral shape to hollow or skeleton shape and further to dendritic shape as the increase of supersaturation [13]. The question was whether the present understanding of crystal morphology in relation to supersaturation was applicable even under very high pressure and high temperature. Although the theory was expected to be applicable under moderate pressure by considering abundant morphological observations of minerals, it might not be correct in general under extremely high pressure. There seemed, to our knowledge, to be no experimental inspection. Then, the relation between crystal morphology and supersaturation was briefly studied in this cBN work. As a result of the observation of the cBN crystals grown under various growth conditions, the applicability of the theory at ~50 kbar and ~2000℃ was nearly confirmed experimentally, which provided technical guidance to growing good crystals under very high pressure.

Figure 2 shows the shape of the grown cBN crystals. In the figure, h and r, the experimental parameters, are the distance between the centers of the growth cell and the heater and the diameter of the hole of the partition sheet in the cell, respectively. These values were thought to correspond approximately to the temperature difference between the nutrient and the growing crystal and to the supersaturation of boron and nitrogen near the growing crystal. When h and/or r are large, the supersaturation is large. As shown in the figure, no obvious growth occurred if the supersaturation was small. If the temperature is fixed, tiny bulky crystals and then large hollow or skeleton crystals with inclusions of solvent appeared as the supersaturation increased. The observed trend of shape and size of the grown cBN crystals agreed with what was expected from the growth theory if the experimantal error was taken into account.

The increase in temperature caused apparently the same effect on the growth shape as the increase of supersaturation did, which was explicable due to the increase in the growth rate at high temperature.

The morphology of the synthetic diamond crystal was usually considered to relate only to pressure and temperature. Once we realized the growth theory was applicable at ~50 kbar and ~2000℃, the morphology of a

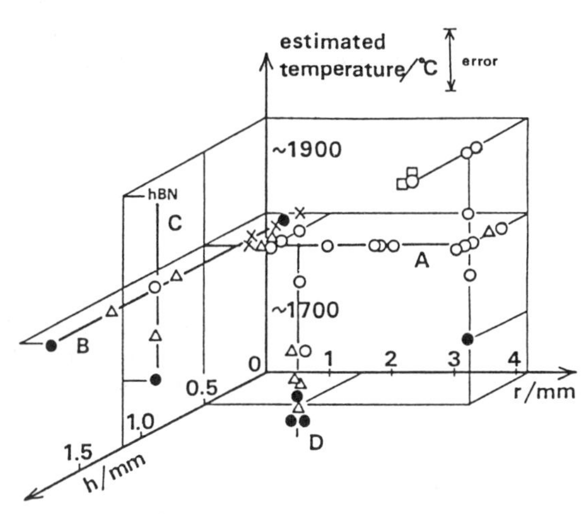

Fig. 2. Appearance of cBN crystals, grown under different conditions of h, r, and temperature. ×) no growth of cBN △) bulky and polyhedral cBN ○) cracked and hollow crystal with inclusions of solvent □) large, cracked, and polycrystalline cBN ●) no growth of cBN (Temperature is too low.)

synthetic diamond would also relate to supersaturation. The *bunching* effect might be also expected when impurities were added in the solvent of a diamond or cBN.

To grow a high-quality large single crystal quickly, the experimental conditions should be set between those for a small bulky crystal and a large hollow crystal. When the temperature difference is made too small, there will be no growth even though the temperature is increased. It seemed experimentally easier to adjust the temperature than to control the temperature difference.

By finding a suitable growth condition according to the growth theory, we could grow cBN crystals of 3 ~ 5 mm in size (Fig. 3). When a cBN seed crystal was placed at the low temperature end in the cell, epitaxial growth on the seed was possible, as confirmed by the X-ray diffraction analysis, which was common in the diamond growth. Much larger cBN crystals should be made in a larger growth cell in a larger high-pressure apparatus and by growing crystals for a longer time.

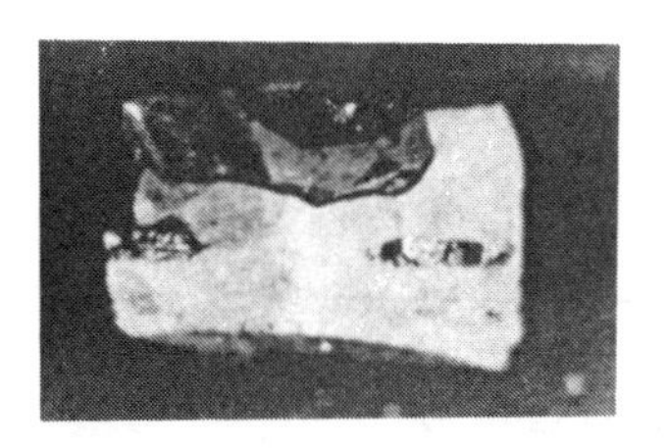

(a) 1 mm

Fig. 3. The cBN crystal grown at the low temperature end in the growth cell with ~4 mm i.d.

The imperfections of the grown cBN crystal are not studied in detail yet. The observed imperfections were impurities of Mg, Si and Mo, and twins, as well as inclusions of the solvent and cracks. The impurities of O, Li, Ca, and C and vacancies were considered as other imperfections in the present crystals. The color of cBN was usually honey, yellow, amber or brown and occasionally almost colorless. The color might be darker when the crystal was grown at a lower temperature.

Different from diamonds, the cBN is a polar material along the ⟨111⟩ axis. The effect of orientation of the cBN seed on the growth was found. On the analogy of the similar effect in other polar crystals [14], it was suggested that the shiny and predominant surface which was normal to the ⟨111⟩ axis of the cBN crystal corresponded to the N-terminating surface [9]. However, the recent Ratherford backscattering measurement [15], in which the cleavaged (110) surface of the cBN crystal was bombarded with He ions to determine the polarity of the cBN, indicated the opposite result. A more detailed study on the relation between the crystal habit and polarity is needed.

When the cBN crystals were recovered in the $LiCaBN_2$ solvent at a slow cooling rate, erosive patterns were observed on the surface of the crystal [9]. This fact suggested the complexity of the phase diagram of the BN-$LiCaBN_2$ system and indicated the existence of a peritectic relation, as discussed in the case of diamonds [16].

The large cBN crystals were obtained by finding the suitable growth condition. These crystals were used for measurement of Raman [17] (Fig.4), reflectance [18], luminescence and so on. Productivity of large and high-quality crystals is, however, still limited due to the experimental difficulties and improvements in the growth techniques at high pressure is greatly desired.

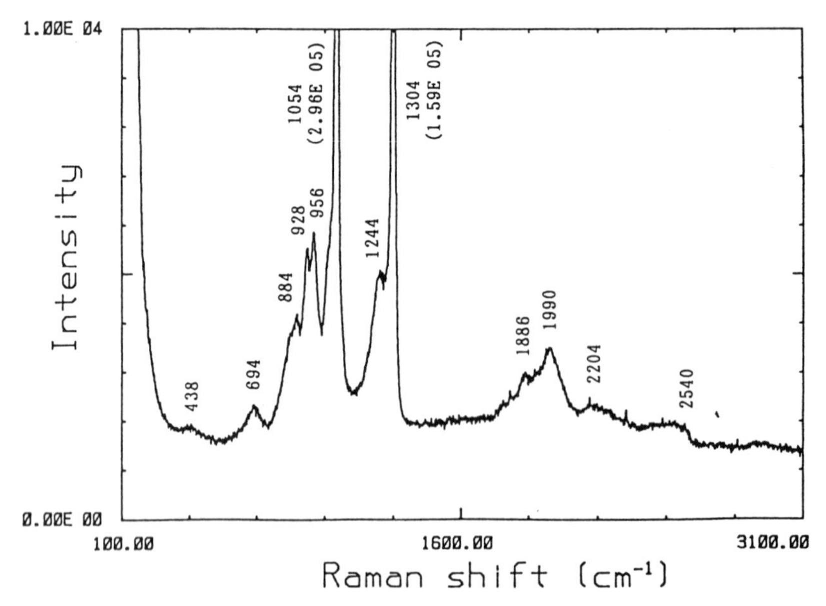

Fig. 4. Raman spectra of the cBN single crystal (~1.5 mm) obtained with SPEX 1403 double-monochromator at 1 bar and 295 K. (Backscattering measurement with unpolarized light) [17]

3. CUBIC BN SEMICONDUCTING CRYSTALS [10]

The cBN is insulator at room temperature. Wentorf reported the preparation of the *p*- and the *n*-type semiconducting cBN using the film method [4]. Beryllium was doped to form the *p* type; and substances such as S, Si and KCN were doped to form the *n* type. Semiconducting properties were shown by measuring thermoelectric power, resistivity and temperature dependence of the resistivety. The *p*-type crystals were blue in color and appeared the same as the *p*-type semiconducting diamond while the *n*-type cBN crystals were yellow, brown, or reddish-brown. We have applied the developed temperature difference method to prepare large semiconducting cBN crystals by adding Be or Si to the solvent [10]. The results obtained so far agreed with Wentorf's observation.

When a small amount of Be metal (<~1 wt%) was added to the solvent, blue, polyhedral and electrically conductive crystals were grown. As Wentorf observed, the larger the amount of Be in the cell, the darker the blue color of cBN. A very small amount of Be was enough to make the blue color. The addition of too much Be made it difficult to grow the crystal. Lack of uniformity of the blue color in the crystal was usually observed.

By adding Si powder (<~5 wt%) to the solvent, yellowish-orange, transparent, polyhedral and electrically conductive crystals were made. Existence of Si in the cBN crystal was confirmed using the EPMA method.

No method for making ohmic contact with the semiconducting cBN has been developed yet. Although we usually adopted the conventional Ag-painted electrode, which was made on the cBN surfaces cleaned with acid, the electrode caused large and unstable contact resistance ($10^4 \sim 10^6$ Ω at room temperature). When a large current was passed through, the increase in temperature at the electrode worsened the property of the electrode and finally induced spark discharges; causing the electrode to need repairs. When the cBN crystal was mechanically squeezed with needles or plates of metal, the current could be more stably applied.

Voltage (V) - Current (I) characteristics of the doped and undoped crystals (1 ~ 2 mm in size) were measured by using Ag-painted electrodes. When the two-point proof method was applied, results differed greatly depending on conditions of the contact, places and polarity of electrode and the crystals which were even made in the same growth cell. The difference may be caused by the non-ohmic contact, the doping amount in an individual crystal and the lack of uniformity of doping amount in the crystal. The nonlinear V-I relation was observed by means of the two-point proof method and showed occasionally the character of a Shottoky diode.

The Be-doped and Si-doped crystals were, no doubt, conductive, as Wentorf observed, while the undoped crystals had high resistivity (>~10^6 Ω).

The four-point proof method was applied to the rather conductive Be-doped and Si-doped crystals and the undoped crystals. The Ag-painted electrodes were carefully made and the distances between the Ag termimals were 0.2 ~ 0.5 mm. The electric power supplied to the sample was limited according to the conditions of the electrodes, to avoid the increase in temperature. A direct current was flowed for a few seconds.

Linear V-I characteristics were observed by using the four-point proof method. The resistance of the Be-doped and Si-doped crystals at room temperature was $10^2 \sim 10^3$ Ω and $10^3 \sim 10^4$ Ω, respectively, while the undoped crystals had a large resistance. Resistivity of the Be- and the Si-doped crystals was tentativily estimated to be $1 \sim 10^2$ Ω·cm and $10^2 \sim 10^3$ Ω·cm, respectively, if the effect of the lack of uniformity of the resistance in the crystals was neglected. Mobility and the amount of doping were not estimated.

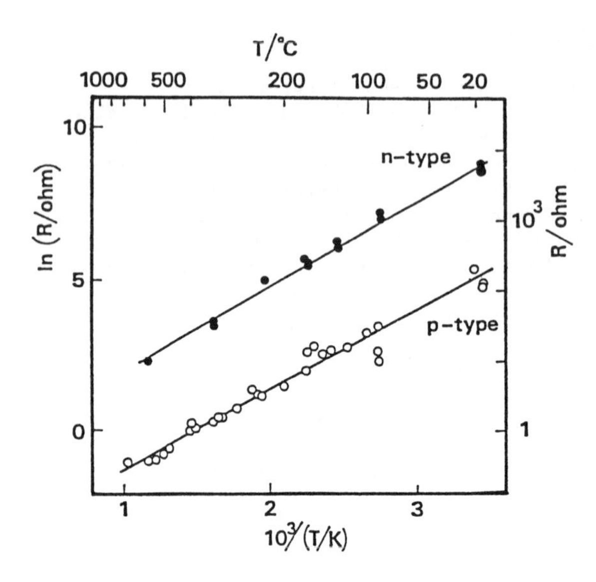

Fig. 5. Temperature dependence of resistance of cBN crystals: O) the Be-doped *p*-type crystal ●) the Si-doped *n*-type crystal

Results of the electric measurement from room temperature up to ~600℃ are shown in Fig. 5. The resistance of the semiconducting cBN crystals decreased almost lineally with a decrease in $10^3/T$ and became 1 ~ 10 Ω at ~600℃. Activation energy of the measured Be-doped crystal, estimated from slope in the figure, was 0.23 ± ~0.02 eV and that of the Si-doped crystal was 0.24 ± ~0.03 eV. In Ref. 10, a deep activation center with an activation energy of 1.11 eV was discribed. Origin of the center has been found later to be not of the cBN but of the ZiO_2-base sample holder which had low resistivity at high temperature [19].

According to the work of Wentorf [4], surfur may become an effective donor of the cBN. At present, we have not succeeded in doping S in the cBN, probably due to undesired reaction between S and Mo of the growth cell.

Though we have not yet measured the hole effect of the semiconducting cBN, our results of the Be- and the Si-doped cBN suggest validity of Wentorf's study of the existence of the *p*- and the *n*-type cBN. An interesting question may be why the cBN can be made into both the *p* and *n* type in contrast to diamonds which cannot easily be made into the *n*-type semiconductor. Nitrogen in diamonds cannot become an effective donor due to large activation energy. Although there was an attempt to dope phosphorus in diamond, no appreciable conductivity was obtained so far. It may be considered that Si-N bonding can be more easily formed than C-P bonding.

4. *PN* JUNCTION OF CUBIC BN [10]

The conductive *p*- and *n*-type cBN crystals, being of a substantial size, could be produced with the temperature difference method at a high pressure, even though producibility and quality of the semiconducting cBN were not optimum and the ohmic contact has not been developed yet. It may be a consecutive and reasonable desire to fabricate the *pn* junction with the temperature difference method and to examine diode action of the *pn* junction.

A few dark-blue Be-doped *p*-type crystals with dimensions of 0.5 ~ 2.0 mm were placed at the low-temperature end of the cell as seed crystals. By adding Si to the $LiCaBN_2$ solvent, the *n*-type crystals were grown on the *p*-type seeds.

Apparent pn composite cBN crystals (1 ~ 2 mm in size) could be obtained, which were looking like a tiny one-sided boiled egg in appearance consisting of a dark-blue *p*-type region and an yellowish-orange *n*-type region around it. The interface between the *p*- and *n*-type regions, which could be seen through the transparent *n*-type crystal, showed good growth on the seed crystal and was round because of slight melting of the seed crystal in the initial stage of the crystal growth. There were occasionally small pools of the trapped solvent and fine cracks in the *n*-type region near the *pn* interface.

The next subject was to examine the *pn* junction. Several *pn* composite crystals were shaped by grinding with a diamond wheel to expose both the *p*- and the *n*-type regions so that the *pn* junction could be tested. The undesirable solvent pools near the interface were eliminated as much as possible.

Figure 6 shows photos of the surface, which was cleaved almost vertically with respect to the *pn* interface. Such cleavage surfaces were sometimes made during a recovery of cBN from the high pressure-high temperature condition or during the grinding with the diamond wheel. The dark-blue right-hand side in Fig. 6a corresponds to the *p*-type region and the yellowish-orange left-hand side to the *n*-type region. The observation of the *pn* boundary with the SEM, which was resolvable to ~50 Å, revealed that a very smooth and very flat cleavage surface over the *pn* boundary was certainly obtainable, which proved that it was possible to make good epitaxial growth on the seed.

Formation of the *pn* junction was studied by means of the electron beam induced current (EBIC) measurement and electric measurement.

Figure 6b shows the EBIC picture taken of the surface of Fig. 6a at room temperature. A bright continuous line across the sample indicated the existence of a space-charge layer in the crystal and the location of this line corresponded to the boundary between the blue *p*-type and the yellowish-orange *n*-type regions. The EBIC picture showed also that the EBIC flowed from the *n* type to the *p* type. Thus, the observed space-charge later agreed with the *pn* space-charge layer, that we expected would exist if the junction was really formed, and the formation of the *pn* junction was, no doubt, confirmed. The intensity of the EBIC image varied along the pn boundary, indicating non-homogeneity in the crystal. The width of the present *pn* space-charge layer was less than ~200 nm.

The most distinctive function of the *pn* junction is the rectification characteristics. The V-I measurement was carried out repeatly on the several *pn* samples by using the four-point probe method. The location of the *pn* boundary on the surface of the crystal was determined with the EBIC method and the Ag electrodes were attached correctly to the sample. Several tens of volts at the current electrode were necessary to flow a few milliamperes through the sample due to the large contact resistance at the electrodes at room temperature. Direct current flowed for a few seconds. Although the V-I characteristics differed slightly depending on location and polarity of the electrodes, usual rectification characteristics were always observed when the current passed through the junction.

The formation of the *pn* junction of cBN was clearly confirmed and a cBN *pn* diode was made. Even though the producibility and quality of the diode were still low due to the difficulty of the high pressure synthesis, it was possible to fabricate a functionable electronic device at a very high pressure. The cBN diode operated as a rectifier and as a detector of the electron beam as shown in the EBIC measurement. A *pnp* or *npn* transistor of cBN may be obtainable in the same way although we have yet not to try it.

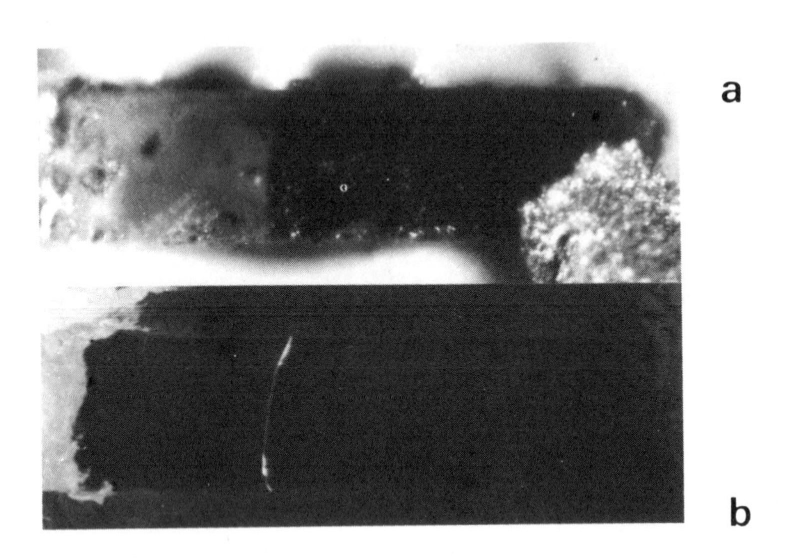

Fig. 6. The cBN *pn* junction diode. a) A photograph of the cleavage surface of the diode with electrodes for the EBIC measurement. The right- and left-hand side correspond to the *p*- and *n*-type region, respectively. b) The EBIC image of the same diode. The space-charge region is the bright and continuous line in the picture.

5. PROPERTIES OF THE CUBIC BN *PN* JUNCTION DIODE

The existence of the semiconducting cBN was confirmed and the functionable *pn* junction of cBN was made under an extreme condition. The study was motivated mainly from academic and experimental curiosity, since the cBN was the tetrahedrally bonded simplest III-V compound stable at a high pressure.

The next subject may be to investigate the properties of the diode and appraisal of potentiality of cBN as a practically useful electronic material. There may be peculiarities regarding the cBN device, since the cBN is the simplest III-V compound as recognized from the periodic table.

As to an electronic structure and an energy gap of cBN, there were several theoretical studies [20,21,22] while only a few experimental measurements were carried out due to the experimental difficulty caused by the hitherto small size of the cBN crystal. From theoretical calculation, the minimum energy gap was found to be an indirect type.

Exact value of the energy gap of cBN has not been determined yet. The theoretical value was scattered in a range of 3 ~ 10 eV. Experimental results of the measurement of UV-light absorption showed the gap was larger than 6.4 ± 0.5 eV [23] or >~6.6 eV [24]. The value of 6.0 ± 0.5 eV was obtained from an X-ray emission spectra [25]. The reflectance data, recently obtained by using UVSOR-light, indicated ~6 eV [18]. From these results, the value of the energy gap of cBN may be around 6.4 ~ 7 eV.

The expected energy gap of cBN is the largest among those of existing IV, III-V, and II-VI semiconductors, as shown in Table 1. Moreover, the cBN can be made into both the *p* and the *n* type. Because of these two properties, the cBN is expected to have extreme potentiality as a wide-gap semiconductor. We have examined preliminarily the properties of the present cBN pn diode although the optimum effect of the cBN device would be, of course, obtained by using much more refined cBN crystals.

Table 1. The energy gap of wide-gap semiconductors.

Semiconductor	Energy gap (eV)
BN (p,n)	~7 ?
AlN (n)	6.3
C (p)	5.5
ZnS (n)	3.8
GaN (n)	3.4
SiC (p,n)	2.9
ZnSe (n)	2.8
CdS (n)	2.6
AlP (p,n)	2.5
ZnTe (p)	2.4
GaP (p,n)	2.4
BP (p,n)	2.0

5-1. THE HIGH TEMPERATURE DIODE

Because of the large energy gap of cBN, an ideal cBN diode should operate up to 1300~1500℃, the temperature at which cBN becomes thermally unstable and changes into hBN at 1 bar. The present cBN diode, operating at room temperature, may be also functionable at a much higher temperature.

The rectification characteristics of the diode were examined from room temperature to about 800℃ by means of the four-point probe method with the Ag-painted electrodes. The hBN sample holder, which had high resistivity even at a high temperature, was used. The rectification characteristics were stably and reproducibly observed up to ~650℃ [19] (Fig. 7) and the measurement became unstable and a leakage current increased at ~700℃.

The observed operation temperature (~650℃) of the present cBN *pn* diode was comparable to that of the SiC transistor which was considered to be the highest temperature semiconductor device existing at present [26].

The cBN diode made at a high pressure was found to be a functional high-temperature diode. The large temperature dependence of conductivity of the cBN diode in the forward-bias condition may, however, cause some difficulty regarding practical use.

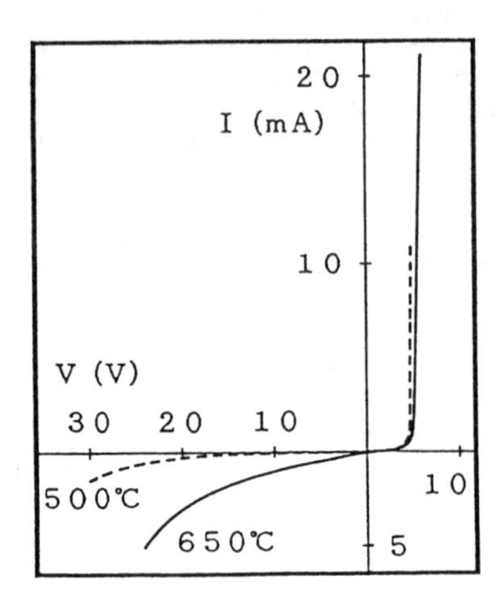

Fig. 7. The rectification characteristics of the cBN pn diode at high temperatures.

5-2. AN UV LIGHT-EMITTING DIODE [11]

Light may be emitted by recombination of the minority carriers injected through the *pn* junction of the present cBN *pn* diode. Even UV light may appear due to the large energy gap of cBN, which would be new and interesting. Thus, an optical measurement on the diode was carried out [11].

While a direct and stationary electric current was passed through the junction, luminescence from the diode was studied by means of a microscopic observation and spectrum measurement at room temperature. The electrodes were made by painting Ag on the cBN crystal or by squeezing the crystal mechanically with metal slabs or needles. Although the amount of the current through the diode was limited due to deterioration of the electrodes when the Ag painted electrodes were used, a stable large current, depending on the size of the diode, could be supplied when the mechanical-contact method was employed. Although the applied voltage on the electrodes was several tens of volts, the voltage over the *pn* junction was several volts as shown in the measurement of the rectification characteristics. A large current caused an increase in the temperature of the diode, which made it easier to supply the much larger current. A substantial reverse current could also be passed by applying reverse bias beyond breakdown voltage.

Several *pn* junction diodes have been examined so far and the results have agreed basically with each other. The formation of the *pn* junction of these diodes was confirmed from the existence of the *pn* space-charge layer, as measured by using the EBIC method, and/or the rectification characteristics.

In the forward-bias condition, a whitish-blue emission was clearly observed along the entire *pn* interface region as shown in the microscopic photograph of Fig. 8c. Appearances of the emission varied along the junction, corresponding seemingly to the intensity of the EBIC image (Fig. 8b). Brightness of the luminescence was stable and constant with the constant current and increased continuously from very faint bluish light to a

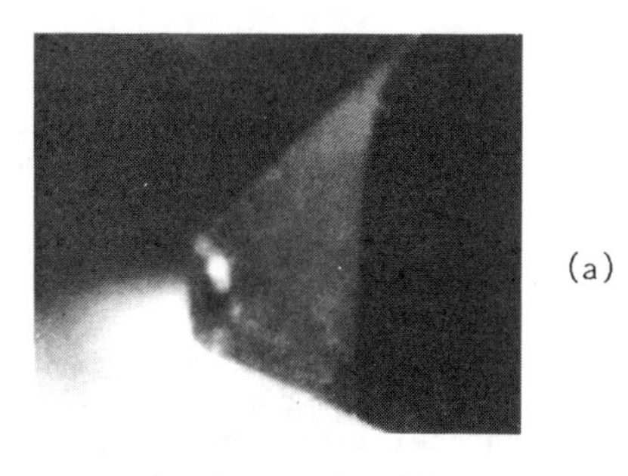

(a)

Fig. 8. The cBN *pn* junction. a) The microscopic picture. The right- and left-hand side correspond to the *p*- and *n*-type region, respectively. b) The EBIC picture confirming the formation of the *pn* junction. c) The luminescence from the *pn* junction in a forward-bias condition.

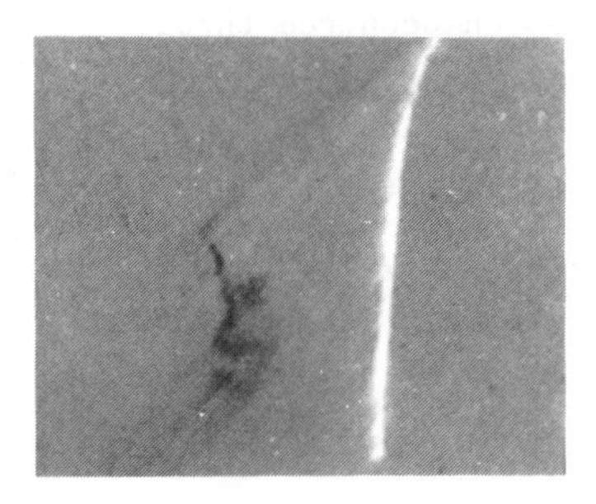

(b)

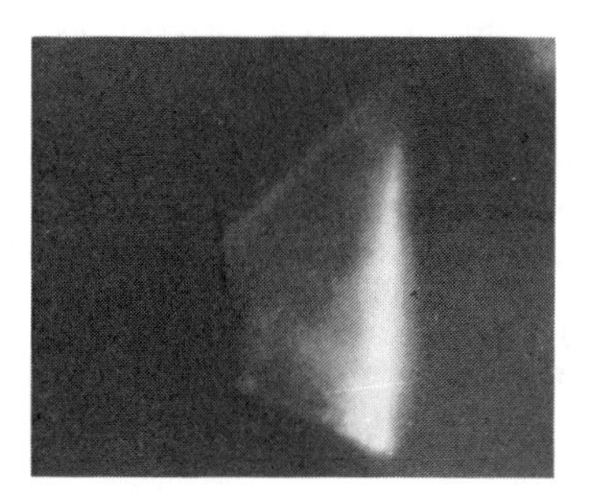

(c)

intense whitish light with an increase in the current. When the amount of the current was increased too much, the color of the whole *pn* diode changed reddish-orange, which may be due to the increased temperature of the diode. The whitish-blue emission was easily recognized under room light with the naked eye.

Unstable sparkling discharges were sometimes observed from the electrode. The induced color of cBN was more yellowish than that near the *pn* junction. The color and the location of the light of the discharges were apparently different from those of the emission at the *pn* junction.

When the reverse current flowed, the whitish-blue color almost disappeared and the faint orange color was observed near the junction.

Spectra of the emission from the diode were measured in a range from $\sim$185 to $\sim$650 nm. The measurement below $\sim$200 nm was done by allowing argon gas to flow in the optical path.

Spectral patterns in the forward-bias condition differed after the electrodes were repaired and the diode was reset in the instrument. Two typical spectra are shown in Fig. 9. The reason why different types of the spectra appear is explicable from lack of reproducibility of forming electrodes and the lack of uniformity in the crystal. When the spectra were corrected by using the preliminarily calibrated spectral sensitivity of the instrument, peaks were obtained at ~260, ~310 and ~490 nm (Fig. 9b). The observed shortest wavelength was ~215 nm (~5.8 eV).

In the reverse-bias condition, there were emissions mainly in the long wavelength region. The intensity of the emission was very weak compared with that in the forward bias condition.

The emission of the UV light from the well-confirmed *pn* junction diode was clearly observed when the current passed in the forward direction. Thus, the cBN *pn* junction diode was functional as a light-emitting diode (LED). The presence of the rather long-wavelength light, compared with that expected from the energy gap of cBN, implied that the emission occurred through levels of unknown crystalline defects and/or impurities. More detailed studies are in progress [27].

Short wavelength limit of the *pn*-type LED has been in the blue from a SiC *pn* diode. In the present study, the cBN *pn* LED emitted the UV light though the UV emission from a *ms*-type AlN diode had been reported [28]. The emission of the UV light supported again the existence of the both the *p*- and the *n*-type cBN. The cBN LED operates certainly at a high temperature as well.

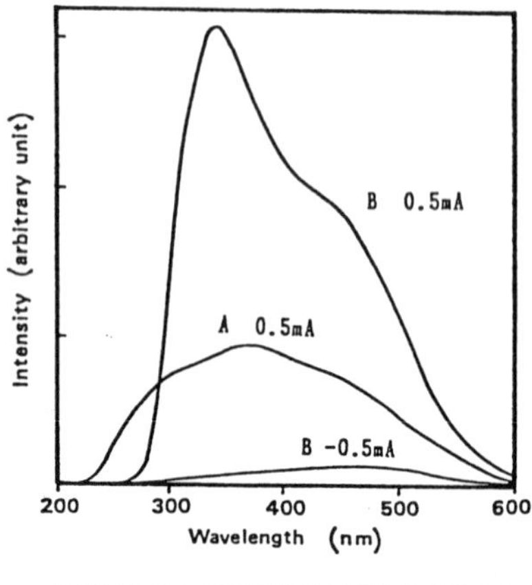

(a)

Fig. 9. Spectra of the cBN LED. A and B correspond to two different conditions of electrodes. a) The uncorrected spectra obtained with a SPEX 1402 double-monochromator and an EMI-6256S photomultiplier. b) The preliminarily corrected spectra.

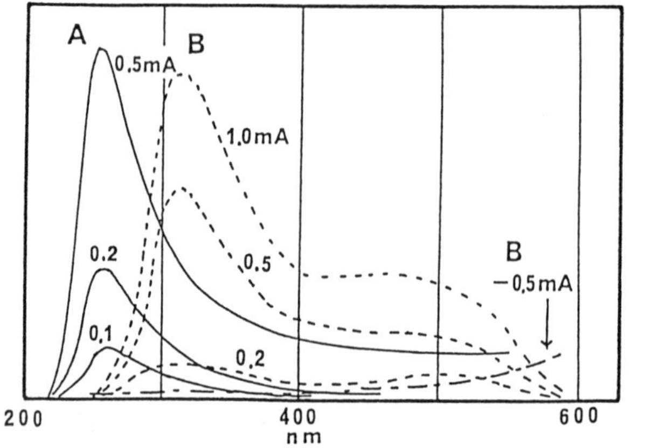

(b)

6. CONCLUSION REMARKS

The principal results of this cBN work may be summarized as follows:

(1) It was found that the large cBN crystals could be obtainable by the temperature difference method at a high pressure.

(2) The existence of the semiconducting cBN was reconfirmed.

(3) The functionable electronic device (diode) and the optoelectronic device (LED) were made of cBN for the first time.

(4) The high-temperature semiconductor device (diode) operating, probably, at the highest temperature was obtained.

(5) The *pn*-type LED emitting the shortest-wavelength light was obtained.

(6) It is probably the first time that the complicated electronic and optoelectronic device has been made under a very high pressure.

Although purity and perfection of the crystal were not optimum in the present work, the cBN *pn* diode demonstrated extraordinary character, which was, of course, due to the remarkable electronic properties of this simplest III - V compound.

The existence of the *p*- and the *n*-type cBN, reported by Wentorf, was strongly supported by the present study. The existence of the both type is noteworthy. As shown in Table 1, the cBN is the largest energy gap semiconductor with both the *p* and the *n* type among the known semiconductors. We may expect the existence of much larger energy-gap semiconductors in the relatively ionic materials like BeO. Since the wide-gap materials may form easily deep acceptor levels or deep donor levels, they may not become semiconductors with substantial conductivity at ordinary temperature. Thus, the cBN is probably the widest-gap semiconductor having the *p* and the *n* type, which can provide outstanding electronic properties and special applications, as seen partly in the present *pn* diode.

Bonding in the cBN crystals has covalent-aspects as diamonds which provide the tetrahedral-bonding structure, hard mechanical properties and semiconductive electronic properties. However, roughly speaking, because of the ionic properties of cBN, the cBN has the wider energy-gap and both the *p* and the *n* type compared with the diamond. The electronic properties of cBN are superior to those of the diamond and the cBN may be regarded as electronic "diamond" because of the outstanding properties as the wide-gap semiconductor.

The beautiful cBN crystal seems more difficult to obtain than the diamond crystal. However, the crystal growth and properties of cBN could be anyhow controlled under the extreme condition and the electronic device made of this cBN material could be actually obtained. One can see the pretty blue light emitted from the *pn* junction of the cBN diode.

Needless to say, the *pn* junction of cBN has very high potentiality of application [27]. Improvement in the synthetic method and the ohmic contact as well as the study of the properties of cBN should be achieved for broad applications.

7. ACKNOWLEDGMENTS

The work reported here was carried out with many collaborators of NIRIM. They are K. Era, O. Fukunaga, T. Ohsawa, J. Tanaka, M. Tsutsumi S. Yamaoka and Y. Wada. The auther is also grateful to Drs. M. Aono and T. Kobayashi of Riken, K. Aoki of Natl.Chem.Lab. and T. Hattori of Musashi Inst. Technol.

REFERENCES

1) Bundy, F.P., Hall, H.T., Strong, H.M. and Wentorf,Jr., R.H.: Nature, 1955, 176, 51.
2) Wentorf,Jr., R.H.: J.Chem.Phys., 1957, 26, 956.
3) Wentorf,Jr., R.H.: J.Chem.Phys., 1961, 34, 809.
4) Wentorf,Jr., R.H.: J.Chem.Phys., 1962, 36, 1990.
5) Wentorf,Jr., R.H. and Bovenkerk, H.P.: J.Chem.Phys., 1962, 36, 1987.
6) Wentorf,Jr., R.H.: J.Phys.Chem., 1971, 75, 1833.
7) Strong, H.M. and Chrenko, R.M.: J.Phys.Chem., 1971, 75, 1838.
8) DeVries, R.C. and Fleischer, J.F.: J.Cryst.Growth., 1972, 13/14, 88.
9) Mishima, O., Yamaoka, S. and Fukunaga, O.: J.Appl.Phys., 1987, 61, 2825.
10) Mishima, O., Tanaka, J., Yamaoka, S. and Fukunaga, O.: Science, 1987, 238, 181.
11) Mishima, O., Koh, E., Tanaka, J. and Yamaoka, S.: Appl.Phys.Lett., 1988, 53, 964.
12) Fukunaga, O., Yamaoka, S., Endo, T., Akaishi, M. and Kanda, H.: in *High-Pressure Science and Technology*, 1, edited by Timmerhaus, K.D. and Borber, M.S. (Plenum, New York, 1979) pp, 846-853.
13) Sunagawa, I.: Bull. Miner., 1981, 104, 81.
14) Barber, H.D. and Heasell, E.L.: J.Phys.Chem.Solids, 1965, 26, 1561.
15) Kobayashi, T.: private communication.
16) Kanda, H., Akaishi, M., Setaka, N., Yamaoka, S. and Fukunaga, O.: J.Mater. Sci., 1980, 15, 2743.
17) Aoki, K. and Mishima, O.: unpublished work.
18) Miyata, N., Moriki, K., Mishima, O., Fujisawa, M. and Hattori, T.: to be published.
19) Mishima, O., Tanaka, J., Yamaoka, S. and Fukunaga, O.: unpublished work.
20) Zunger, A. and Freeman, A.J.: Phys.Rev., 1978, B17, 2030.
21) Huang, M.Z. and Ching, W.Y.: J.Phys.Chem.Solids, 1985, 46, 977.
22) Wentzcovitch, R.M., Chang, K.J. and Cohen, M.L.: Phys.Rev., 1987, B34, 1071.
23) Chrenko, R.M.: Solid State Commun., 1974, 14, 511.
24) Degawa, J., Tsuji, K. and Yazu, S.: private communication.
25) Fomichev, A. and Rumsh, M.A.: J.Phys.Chem.Solid, 1968, 29, 1015.
26) Palmour, J.W., Kong, H.S. and Davis, R.F.: Appl.Phys.Lett., 1987, 51, 2028.
27) Era, K., Mishima, O., Wada, Y., Tanaka, J. and Yamaoka, S.: Proc. of Intnl. Workshop on EL, Tottori, 1988 (Springer, 1989), in press.
28) Rutz, R.F.: Appl.Phys.Lett., 1976, 28, 379.

Materials Science Forum Vols. 54 & 55 (1990) pp. 329-352

THIN FILM BORON NITRIDE FOR SEMICONDUCTOR APPLICATION

E. Yamaguchi
NTT Basic Research Laboratories
Musashino-shi, Tokyo 180, Japan

ABSTRACT

This chapter discusses the preparation of thin film boron nitride, its application to semiconductor devices, and its electronic properties. First, we briefly review the technologies of boron nitride film deposition, including chemical vapor deposition methods, plasma process methods, sputter deposition methods, and electron-beam evaporation methods. Then, we discuss its application to semiconductor devices, for example, as a protective coating, a boron diffusion source for Si devices and a gate insulator for III-V semiconductor metal-insulator-semiconductor (MIS) field-effect transistors. Finally, we discuss the theoretical considerations for the electronic structures of BN with crystalline and non-crystalline structures. Here, a theory for the electronic structures of non-crystalline BN/III-V semiconductor interfaces is presented to clarify the origin of the interface states in the MIS structures.

1 INTRODUCTION

Thin films of stoichiometric boron nitride, BN, have been successfully formed by various techniques[1]–[38]. Most of the deposited films have crystalline or non-crystalline structures with a hexagonal phase. Crystalline hexagonal BN, which is isoelectronic with graphite, is a good insulator with a band-gap of more than 4 eV. Non-crystalline BN is also highly insulating, chemically inert, and thermally stable. Therefore, thin film BN was expected to have potential applications to semiconductors as a protective coating[8], as a diffusion source[9], and as a gate insulator for metal-insulator-semiconductor field-effect transistors (MISFETs)[39,40].

The thin film boron nitride deposition process on semiconductors must meet several important requirements to maintain the quality of the resulting interface. The insulating thin films should be formed with relatively low substrate temperatures. Temperature limitations often exist with compound semiconductors where one of the components is very volatile. Moreover, the energy of film producing particles should be small enough not to cause the sputtering of semiconductor surface atoms. A low oxygen content should be essential for producing the best boron nitride film properties, because the resulting boron oxide is very sensitive to moisture with the reaction:

$$B_2O_3 + 3H_2O \rightarrow 2B(OH)_3.$$

This chapter discusses the preparation techniques of boron nitride thin films at low temperatures in section 2. Their application to semiconductor devices is discussed in section 3.

Cubic BN, which consists of tetrahedrally bonded zinc-blende structures, has the largest band-gap among IV and III-V semiconductors. It has almost the same high hardness and high thermal conductivity as diamond. Electronic devices, such as diodes and transistors, made of cubic BN are therefore likely to operate at high temperatures. Although thin film growth of single-crystalline cubic BN on semiconductors has not been successful yet, bulk cubic BN about 1 mm large with *p-n* junctions is now being made by the temperature-difference solvent method[41,42]. This is reviewed in a different chapter.

Section 4 discusses theoretical considerations for the electronic structures of BN with crystalline (hexagonal, zinc-blende, or wurtzite) and non-crystalline (tetrahedrally-bonded amorphous) structures. Here, a theory for the electronic structures of non-crystalline BN/III-V semiconductor interfaces will be also presented to clarify the origin of the interface states in the MIS systems.

2 PREPARATION OF BORON NITRIDE FILMS

Thin film boron nitride has been grown by the chemical vapor deposition (CVD) method, plasma process method, sputter deposition method, and electron-beam evaporation method. This section discusses these techniques of boron nitride film preparation for semiconductor applications. The measured band-gap, static dielectric constant, and refractive index of these pyrolytic films are summarized in Table 1.

Depositions at low substrate temperatures with low energy of particles producing the film are likely to result in *turbostratic* BN films. Namely, the B_3N_3 hexagons are deposited in layers parallel to the substrate but in a random order. Hexagonal BN, in both crystalline and *turbostratic* form, is slowly attacked by water vapor *etc.* This is probably related to the intercalation of hexagonal (or *turbostratic*) BN, like graphite intercalation.

Conversely, film preparation methods making use of high-energy plasma or a high-energy ionized beam can synthesize cubic BN micro-crystals, although these methods are not suitable for semiconductor application. This section also includes a brief discussion of these techniques.

2.1 Chemical Vapor Deposition Method

The CVD method, which makes use of thermally activated vapor-phase reaction at substrate surfaces, is one of the most suitable techniques for making a uniform insulating film with very few defects. Thin film boron nitride was firstly formed by the vapor-phase reaction of boron trichloride with a large excess of ammonia at about 1000°C or higher:

$$BCl_3 + NH_3 \rightarrow BN + 3HCl,$$

or the thermal decomposition of β-trichloroborazole at 700–1400°C [1,2,3]:

$$H_3B_3N_3Cl_3 \rightarrow 3BN + 3HCl.$$

Table 1: Comparison of the measured band-gap, static dielectric constant, and refractive index of thin films of pyrolytic boron nitride.

Reference	Band-gap (eV)	Dielectric const.	Refractive index
CVD			
[4]	5.8		1.9–2.0
[6]	5.8		
[8]	3.8	3.7	1.7–1.8
[9]		3.3–3.5	
[10]			1.7–2.8
[13]		5.5–6.0	1.7
[16]	5.9		
Plasma			
[18]		5.3–7.7	
[24]		2.9–4.3	
[25]	5.0	6.5	
Sputter			
[30]	4.9–5.2	3.8	
[32]	5.4–5.6		1.6–1.7
[33]	4.5–4.9		

The deposited films had a *turbostratic* structure. The static dielectric constant was 4.4, and the resistivity ranged from 10^{16} to 10^{17} Ω-cm at room temperature.

Using the same reaction of BCl_3 with NH_3, the CVD of boron nitride films at lower temperatures has been studied by Baronian[4] (570–800°C), Takahashi *et al.*[5] (800–1050°C), Sano and Aoki[6] (600–1100°C), and Motojima *et al.*[7] (250–700°C). Films deposited at temperatures below 450°C were unstable in moist atmosphere and they devitrified. This is because the BCl_3+NH_3 reaction produces intermediate complexes, Cl_2BNH_2 or $ClB(NH_2)_2$. In contrast, Films deposited at temperatures above 600°C were very stable. Films deposited at temperatures above 570°C had the optical band-gap of 5.8 eV.

Rand and Roberts[8] have formed boron nitride film from a reaction between diborane and ammonia in hydrogen by the CVD method. The reaction proceeds as follows:

$$B_2H_6 + NH_3 \rightarrow 2BN + 6H_2$$

with free energy equal to −165 kcal/mole at 900 K. Below about 300°C, the diammoniate of diborane is formed as follows:

$$B_2H_6 + 2NH_3 \rightarrow (NH_3)_2BH_2^+BH_4^-.$$

With an increase in heating temperature, this stable white powder rearranges into borazine $B_3N_3H_6$, then into some polymeric materials such as $(BNH_2)_n$ and $(BNH)_n$ and finally into BN. In their work, the reaction was carried out at 600°C to 1080°C, where the substrate was usually silicon. The deposited film was micro-crystalline (10–65 Å) with a hexagonal structure. It had a band-gap of 3.8 eV and a static dielectric constant of 3.7. These values are much lower than those of stoichiometric pyrolytic BN films (band-gap 5–6 eV and static dielectric constant 5.5–7.7), possibly indicating that they prepared boron-rich films.

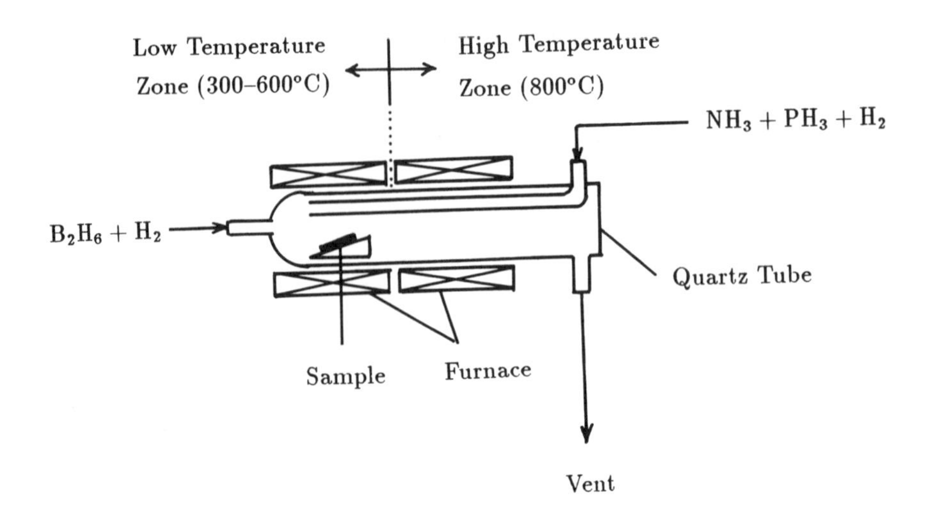

Figure 1: CVD equipment. From Yamaguchi and Minakata[12,13].

The boron nitride CVD using a B_2H_6–NH_3–H_2 system has been successively studied by Hirayama and Shohno[9], Murarka *et al.*[10], Adams and Capio[11], and Yamaguchi and Minakata[12,13]. In particular, Yamaguchi and Minakata have deposited phosphorus doped boron nitride films onto III-V semiconductor InP. The doping was carried out by injecting phosphine PH_3 into the reactant gas mixture. Figure 1 shows a typical deposition set-up given by Yamaguchi and Minakata. They used the two-temperature zone technique; namely, the deposition was carried out at low temperatures (300–600°C) to avoid surface degradation in the semiconductor and decomposition of B_2H_6 to boron. Two source gases, NH_3 and PH_3, are heated up to 800°C just before the reaction, in order to make N and P atoms reactive. This technique prevents both the formation of intermediate complexes and the volatilization of P atoms at the InP surface. Chemical depostion above about 500°C yielded a stoichiometric boron nitride, where the deposition rate was about 2×10^2 Å/min. in which the doped P concentration was less than 10^{-1} in units of N concentration. As the depostion temperature decreased, the [B]/[N] and [P]/[N] ratios increased. Sputter etching by Ar^+ ions showed that the etching rate for the boron nitride films was half or less that for other CVD films such as SiO_2, Si_3N_4, and Al_2O_3. The deposited films were insoluble in water, and in hydrochloric, hydrofluoric, sulfuric, and nitric acids of all concentrations. The stoichiometric boron nitride films had a refractive index of 1.72 and a static dielectric constant of 5.75 ± 0.28.

Boron nitride film preparation by CVD methods has also been investigated using other reaction systems[14,15,16]. Furukawa[15] has reported that significant reduction of deposition temperature to 330°C was achieved using a B_2H_6+NF_3 reaction. Nakamura[16] has developed a CVD in the molecular flow region ($< 10^{-3}$ Torr), and applied it to the preparation of boron nitride films using decaborane $B_{10}H_{14}$ and NH_3 gases at 300–1150°C. The obtained films were non-crystalline with the optical band-gap of 5.9 eV.

2.2 Plasma Process Method

The first attempt to grow thin film boron nitride using the plasma process method was

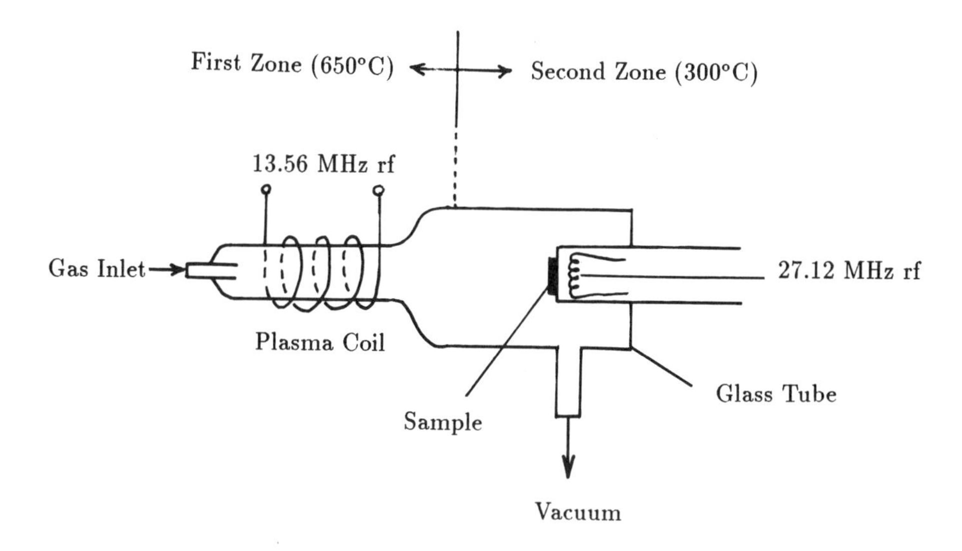

Figure 2: Double-plasma reactor for plasma MOCVD. After Schmolla and Hartnagel[23,24].

made by Alexander *et al*[17]. They made boron nitride film by the reactive plasma deposition at room temperature. Hyder and Yep[18] have also grown boron nitride film by the reactive plasma technique. They used the $B_2H_6+NH_3$ reaction at temperatures between 750°C and 1000°C under 13.56 MHz rf excitation (typically 4 W). The deposited boron nitride films on Si, graphite, and BN exhibited hexagonal structures. The boron nitride films deposited at lower temperatures (550–620°C) using the rf plasma reaction (0.5 MHz) of BCl_3 with NH_3 were mostly non-crystalline with small amounts of hexagonal BN[19].

Sokołowski *et al.* first succeeded in synthesizing micro-crystalline wurtzite BN film[20] and cubic BN film[21] by reactive pulse plasma deposition. They used a plasma accelerator; the plasma was generated and accelerated using a pulse discharge with a boron rod electrode in a N_2–H_2 gas mixture, where the peak inter-electrode voltage was 3.2–7 kV, the plasma temperature was 20,000 K and the substrate (Si, amorphous carbon *etc.*) temperature was 300 K. The obtained layers were extremely fine-grained with a crystal size of 20–40 Å. The resistance of the film was 10^{13} Ω. The interface properties of BN/Si systems prepared by the reactive pulse plasma technique were also investigated by Szmidt *et al*[22]. The density of interface states was about $10^{12}\text{eV}^{-1}\text{cm}^{-2}$.

As discussed in section 1, both temperature limitations and incident particle energy limitations exist with compound semiconductors. To avoid high substrate temperatures and high particle energies, Schmolla and Hartnagel[23,24] have developed a double-plasma system for the deposition of boron nitride on III-V semiconductors by metal-organic CVD (MOCVD). Starting materials were NH_3 and $BH_3N(C_2H_5)_3$ (borantriethylamin). The reactor (see figure 2) had two plasma zones. The liquid $BH_3N(C_2H_5)_3$ was evaporated and transported with Ar gas into the first plasma zone. Here, suitable products were obtained by the reaction with NH_3 under 13.56 MHz rf excitation (150 W), then transported into the second zone, where a 27.12 MHz rf electrode, positioned behind the substrate, generated the second plasma. Typical gas temperatures were 650°C in the first zone and 300°C in the second zones, where the reaction pressure was 2–3 Torr. The obtained film contained oxygen, but the [O]/[N] ratio decreased rapidly to 0.2–0.3 as the rf input power increased to 40 W. The static dielectric constant was 2.9–4.3. This may indicate that the films were not stoichiometric.

The glow discharge decomposition method is one plasma process technique where the substrate temperature can be greatly reduced. Miyamoto *et al.*[25] reported preparing boron nitride films at a temperature below 300°C by the glow discharge decomposition of a $B_2H_6+NH_3+H_2$ gas mixture. The glow discharge was excited by supplying 13.56 MHz rf power (3–40 W) through two external ring electrodes. X-ray diffraction and Raman spectroscopy measurements indicated that the boron nitride films deposited on silicon or quartz substrate were composed of non-crystalline phases with hexagonal-like bonds. The optical band-gap was about 5.0 eV and the static dielectric constant was 6.5. Very recently, Matsumoto *et al.*[26] have made micro-crystals of wurtzite or cubic BN using the same method for 20 hours.

The electron cyclotron-resonance (ECR) plasma technique has been applied to obtain micro-crystalline cubic BN[27,28,29]. The apparatus includes an external magnet which increases the path length of the electrons between cathode and anode, and hence enhances the plasma. Here, the source gases were vapor $B_3N_3H_6$[27], vapor $H_3BO_4+NH_3$[28], or $B_2H_6+NH_3$[29]. The obtained films contained micro-crystals of cubic BN.

2.3 Sputter Deposition and Electron-Beam Evaporation Methods

Reactive sputter deposition does not require the substrate to be at a high temperature. However, the incident particle energy is so high that the technique is not suitable for semiconductor applications. Noreika and Francombe[30] first reported boron nitride film prepared by reactive sputtering. The target was B slices. Films were deposited on Si or silica substrates in N_2 (5×10^{-3} Torr). Films deposited both on unheated substrates and on substrates heated to 900°C displayed similar micro-crystalline appearances (crystal size $\simeq$ 100 Å) with hexagonal phase. The band-gap was 4.9–5.2 eV for the stoichiometric boron nitride films.

The reactive sputtering method has been successively applied to the synthesis of boron nitride films by Tanaka *et al.*[31], Wiggins *et al.*[32], Yasuda *et al.*[33], and Seidel *et al.*[34]. Wiggins *et al.*[32] grew boron nitride films on water-cooled Si or Al_2O_3 (sapphire) substrates by sputter deposition from a BN target in Ar, Ar/N_2, and N_2 atmospheres. They found that films grown in gas containing 10% to 100% N_2 approached stoichiometry ([B]/[N]=1.1–1.4) with a band-gap of 5.4 to 5.6 eV. However, films grown in pure Ar were clearly nonstoichiometric ([B]/[N]=5.1) with a band-gap of 3.3 eV, as shown in figure 3; no long-range crystallographic order was detected in these films. Yasuda *et al.*[33] used the same method to prepare BN films, where the sputtering gas was a mixture of Ar (93 %) and H_2 (7 %) gases. They obtained slightly boron-rich non-crystalline films ([B]/[N]$\simeq$ 1.13) with a band-gap of 4.5–4.9 eV. Seidel *et al.* also reported a similar method of making boron nitride films, using rf reactive diode sputtering (650 W) with a high negative substrate bias of 900 V. From the transmission electron diffraction measurements, they estimated that nearly the whole film consisited of crystalline cubic BN micro-crystals with diameters in the range from 50 to 1000 Å.

The electron-beam evaporation technique is another powerful method which uses solid sources. The electron-beam evaporation of amorphous boron in a nitrogen plasma to form cubic BN micro-crystals has been reported by Weismantel *et al*[35,36]. The ionized species were accelerated with a bias of 500–3000 V in the plane of the substrates (NaCl, Si *etc.*) to obtain micro-crystalline cubic BN 10 nm in size. Satou and Fujimoto[37] used 30 keV N_2^+ ion beam bombardment on electron-beam-evaporated boron films to produce boron nitride

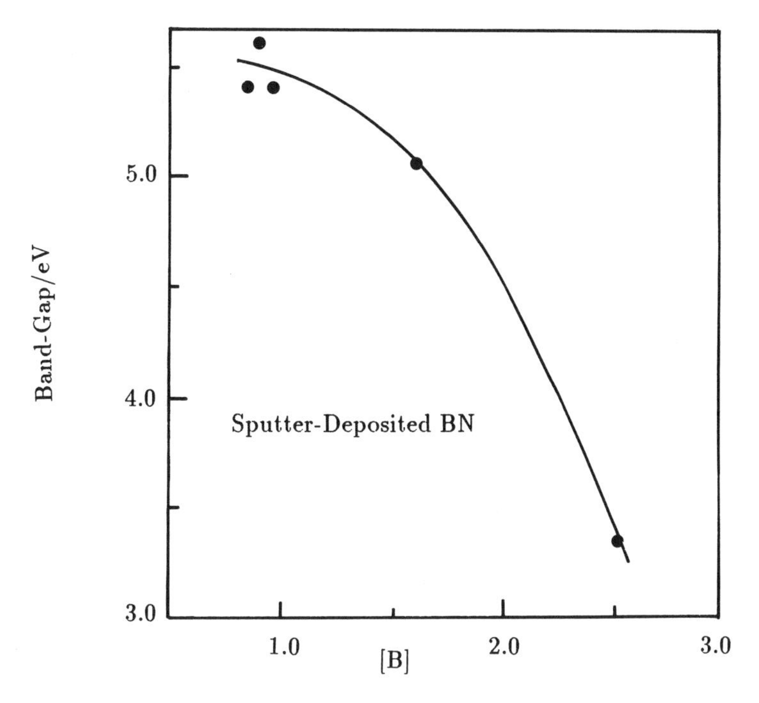

Figure 3: Band-gap of sputter deposited BN as a function of boron concentration relative to an external standard. After Wiggins *et al*[32].

films at room temperature. Zhou *et al.*[38] used almost the same technique as Weismantel, to make micro-crystals of cubic BN with a grain size of less than a few thousand angstroms.

3 SEMICONDUCTOR APPLICATIONS

As discussed in the previous section, BN films pyrolytically deposited at low temperatures have been shown to be mostly of hexagonal or *turbostratic* structures. They are layered structures with the B_3N_3 hexagons parallel to the substrate. Since it is very difficult for any ion to pass through the hexagons, the BN films were expected to work as a good protective coating. Furthermore, B is the most common doping element for p-type Si in current semiconductor technology. Therefore, BN films were also expected to be good diffusion sources for Si devices. Boron and nitrogen may not behave as dopants in the III-V semiconductors because they are of the same group in the periodic table as the semiconductor component. Hence, boron nitride film was expected to be particularly suitable for gate insulator application to III-V semiconductor MISFETs. This section discusses each possibile application for BN films.

3.1 Protective Coating

Rand and Roberts[8] have first investigated the properties of sodium drift and diffussion properties in pyrolytic boron nitride films. Their test was carried out with $^{22}NaCl$. At 400°C, with the surface bias plate at 4 to 10 V (field about 10^5 V/cm) for 1 hour, no sodium ion drift was detected in any boron nitride film in spite of the fact that the film was boron-rich. The profiles of thermal diffusion for 22 hours at 600°C showed that film deposited at 850°C exhibited a steep drop in Na ion concentration at a depth at 3000 Å. It also indicated a pileup of sodium at 2100 Å due to some sort of barrier. On the other hand, boron nitride deposited at 600°C was apparently not a barrier to sodium ion diffusion. In addition, all thin film boron nitride was highly resistant to all aqueous acids and bases.

They also found that all BN films were very hydrophobic. Water contact angles from 42° to 80° were measured, even after many days' exposure to atmospheric moisture. However, changes in chemical and physical properties with time were observed; these appear to depend on the deposition condition. In fact, freshly deposited boron nitride was not readily scratched by stainless steel tweezer tips. It remained so if stored in a desiccator. If stored in room ambient, it became soft enough to scratch in a few weeks. This has been found to be due to adsorption or absorption of molecular water. As denoted previously, this may be related to the intercalation between each layer of B_3N_3 hexagons. Further investigation is needed into BN intercalation.

3.2 Diffusion Source for Si Devices

Rand and Roberts also found that during deposition of the boron nitride film, boron diffused into the Si substrate; a boron nitride film deposited at 900°C or above acted as an infinite source of boron in Si. Hirayama and Shohno[9] have investigated the diffusion properties of boron in Si in more detail. The boron surface concentration *vs.* deposition temperature is shown in figure 4(a). The boron surface concentration, which was independent of deposition time, increased with an increase in the deposition temperature up to about 1000°C, and then suddenly decreased. In this temperature range, the film had a poly-crystalline hexagonal form. After the films were deposited on n-type Si at 700°C, the samples were heated in a hydrogen atmosphere. Values of boron surface concentrations are plotted in figure 4(b). The curve in this figure gives the solid solubility of boron in Si at each temperature[43], from which the activation energy for boron diffusion in Si was determined to be 88.9 kcal.

They applied these diffusion properties to the fabrication of a Si planar diode, as follows. First, a boron nitride film was deposited for 10 seconds at 700°C on n-type Si with windows opened in SiO_2. Next, the samples were heated in a nitrogen atmosphere at 1100°C for 120 minutes, to diffuse the boron in the Si and to decompose the BN in the windows. The diffusion depth of boron was about 3 μm, and the boron surface concentration was about 8×10^{18} cm^{-3}. Finally, nickel was deposited directly by electroless plating on the Si surfaces of the windows. Thus, they made planer diodes using only one photo-mask. The obtained device exhibited break down voltages of 115–120 V, as expected from the diffusion depth and resistivity of Si substrates.

3.3 Gate Insulator for Field-Effect Transistors

The first attempt to apply boron nitride film to the gate insulator for III-V semiconductor MISFETs was made by Yamaguchi *et al*[39]. They made enhancement-mode InP MISFETs with a gate insulator of BN film. The fabrication process of the device is shown in figure 5. The substrate was Zn-doped p-type InP (carrier concentration $\simeq 2 \times 10^{16}$ cm^{-3}) with a (100) surface orientation. First, the source and drain n^+ regions were formed by Si ion implantation with a dose of 2×10^{14} cm^{-2} at 200 keV, and subsequent annealing at 650°C for 15 minutes with an encapsulating film of CVD-SiO_2. After etching with HCl to form the channel regions, boron nitride film was deposited as a gate insulator at 400–500°C by the CVD method discussed in section 2.1. A typical thickness was 1000 Å. After the BN was

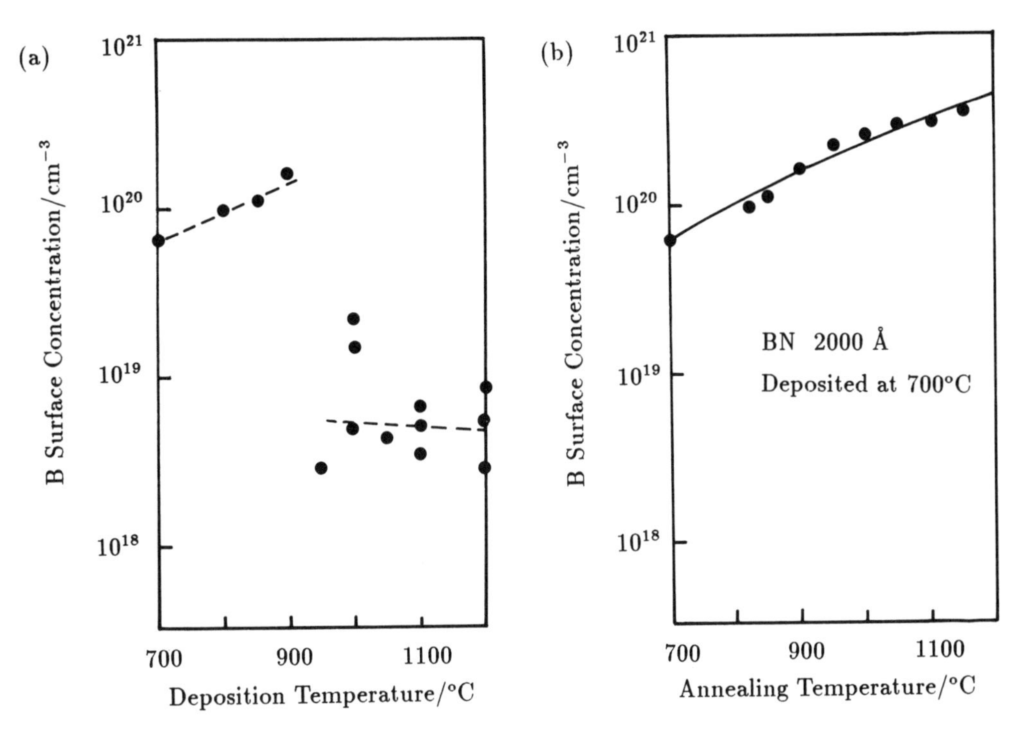

Figure 4: Boron surface concentration on Si as a function of (a) deposition temperature of CVD-BN and (b) annealing temperature for the BN film deposited at 700°C. After Hirayama and Shohno[9].

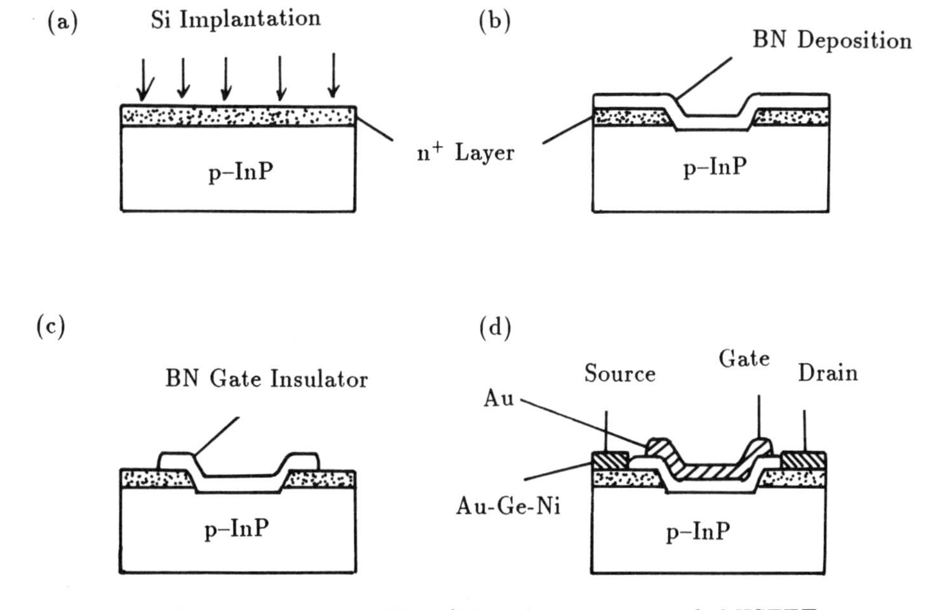

Figure 5: Fabrication process of InP/BN enhancement-mode MISFETs.

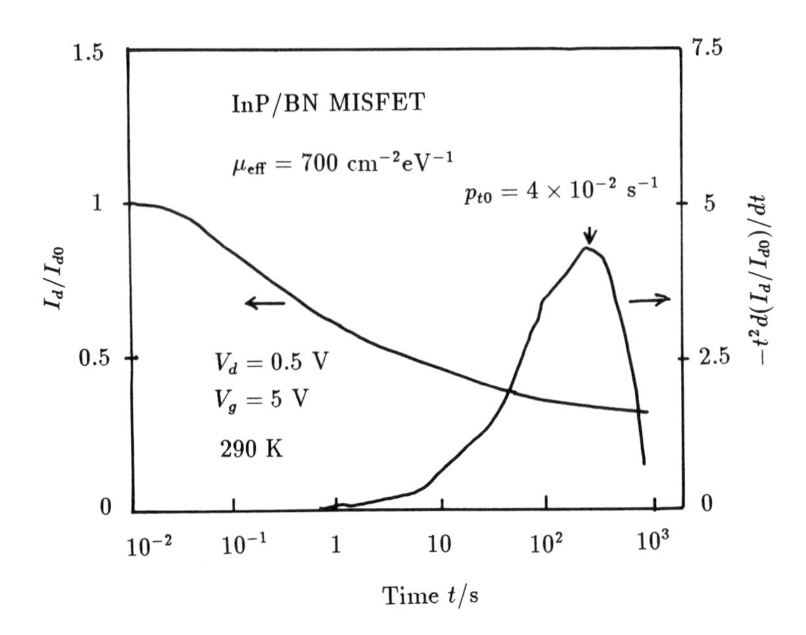

Figure 6: Typical characteristics of current drift and its spectrum in InP MISFETs with a gate insulator of BN deposited by CVD. From Yamaguchi *et al*[39].

etched by Ar^+ sputter in the region of source and drain electrodes, Au-Ge-Ni was deposited. Finally, Au was deposited as the gate electrode. The preliminary capacitance-voltage (C-V) analysis showed that the interface state density of n-InP/BN had a U-shaped distribution with the minimum of $\simeq 10^{10}$ cm^{-2}eV^{-1}[12,13]. The interface state density was too high near the valence-band edge to establish a two-dimensional hole gas. The insulator breakdown field was more than 3×10^6 V/cm. The fabricated inversion-mode InP/BN MISFET had a maximum effective mobility μ_{eff} of 700 cm^2/Vs with a typical threshold voltage of 0.7 V at room temperature, where field-effect mobility went through a maximum of 980 cm^2/Vs at 2.5 V and then decreased with increasing gate voltage due to defect scattering[44,45].

In common, the existence of interface traps in MIS systems gives a slowly decreasing drain-current with time to the MISFETs as well as a clockwise hysteresis in C-V curve to the MIS capacitors, due to charge injection into the insulator. On the other hand, the existence of mobile ions in the insulator causes a slowly increasing drain-current and an anti-clockwise hysteresis in C-V curves in MIS systems, because the mobile ions enhance the electric field at interface. In the present InP/BN MISFETs, the drain current I_d slowly decreases with time t under constant gate bias V_g and drain voltage V_d as shown in figure 6. Since this current drift depends on the temperature, they attributed this phenomenon to the tunneling of channel electrons to the localized trap states above the Fermi level in the insulating BN. Current transient spectroscopy analysis found a single trap level in BN above the Fermi level with a surface tunneling probability p_{t0} of 4×10^{-2} s^{-1} at 290 K (see figure 6). They have also fabricated n-channel GaAs/BN MISFETs; however, these MISFETs did not work at all, indicating the existence of a high interface state density near the conduction band edge of GaAs.

Schmolla[40] has also applied thin film boron nitride to the gate insulator for enhancement-mode InP MISFETs, where the film was deposited on semi-insulating InP using the double

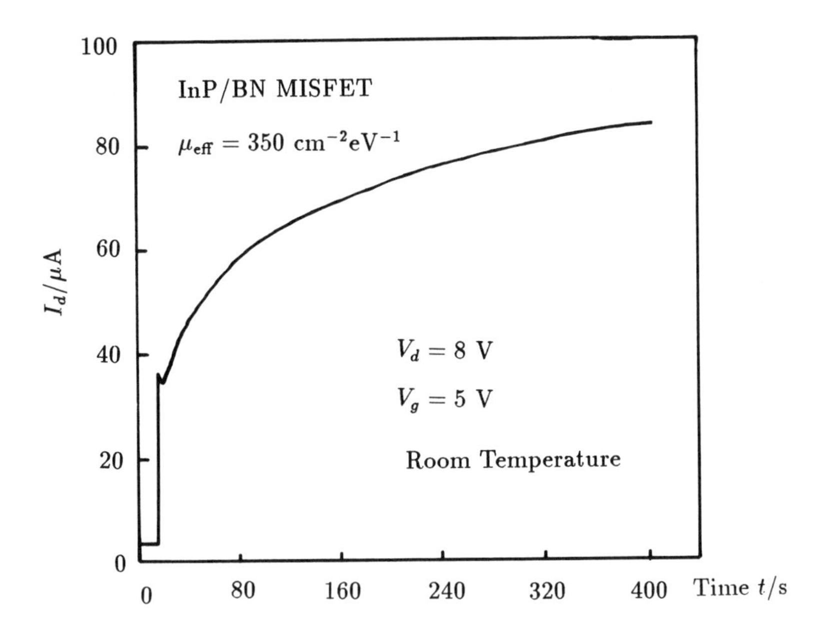

Figure 7: Typical characteristics of current drift in InP MISFETs with a gate insulator of BN deposited by plasma MOCVD. After Schmolla[40].

plasma method at 300°C as mentioned in section 2.2. They fabricated MISFETs with and without SiO_2 film. The fabricated MISFETs exhibited an effective mobility of about 350 cm^2/Vs with a typical threshold voltage of 0.8 V at room temperature. Without SiO_2 film, the drain current showed an increasing drift as shown in figure 7, indicating a large number of mobile ions in the deposited boron nitride film. Figure 7 also shows a decreasing drift in the early stage of transient current. It can be considered that this origin is the interface traps as found by Yamaguchi *et al.* On the other hand, for the MISFETs with an additional SiO_2 film, the drain current decreased with time and the time constant was smaller. They attributed this anomaly in the InP/BN/SiO_2 system to the transport of channel electrons through BN to the SiO_2 film. However, the resistivity of BN films is too high to yield the electron transport. Therefore, this is believed to be because the mobile ions were absent in the InP/BN/SiO_2 systems, due to the heat treatment during the SiO_2 depositions.

4 THEORETICAL CONSIDERATIONS

Hexagonal and cubic boron nitride are both ideally suited for testing a computational scheme designed for the study of periodic systems. Therefore, they have been the object of many electronic structure calculations performed with *ab initio* methods as well as semiempirical methods. This section discusses these electronic structure, and then discusses the electronic structure of non-crystalline BN by simulating tetrahedrally bonded amorphous BN with the Bethe lattice. Finally, it presents a theory of the electronic structure of non-crystalline BN/III-V semiconductor interfaces.

4.1 Electronic Structures of Crystalline BN

Hexagonal BN is isoelectronic with graphite. It has a layered and is characterized by strong intralayer bonds and weak interlayer interactions. However, the π-bonding and π-antibonding bands overlap at the Brillouin-zone boundary in graphite, making graphite a semimetal, while on the other hand, these bands are separated by an energy gap in hexagonal BN, making hexagonal BN an insulator. The energy band structure has been calculated by two-dimensional approximation[46,47,48], tight-binding calculations[49,50,51,52], the OPW method[53], the full-potential linearized APW method[54,55], and the numerical-basis-set LCAO plus the norm-conserving pseudopotential in the local-density-functional formalism[56]. The earlier calculations[46]–[53] concluded that hexagonal BN has a direct band-gap at the K- or H-point. However, recent first-principle calculations[54,55] have suggested that, on the contrary, the gap is indirect with the valence band maximum around the H-point and the conduction band minimum at the M-point. The calculated band energies and the experimental data are summarized in Table 2. An example of the calculated band structure is shown in figure 8.

Cubic BN is isoelectronic with diamond. Since 1960, many theoretical calculations have been reported for cubic BN[55,57]–[64]. These calculations have concluded that cubic BN has an indirect band-gap with the conduction band minimum at the X-point. However, they give widely different results for the values of band-gap and valence bandwidth. Kleinman and Phillips[57] predicted an indirect band-gap of 10.5 eV in their pseudopotential study. Bassani and Yoshimine[58], using the OPW method, found this gap to be about 3 eV. Zunger and Freeman[61] applied first-principle local-density formalism and found an indirect gap of 8.7 eV, while Dovesi *et al.*[63] predicted this gap to be 11.3 eV in their exact-exchange Hartree-Fock calculation. Prasad and Dubey[64] gave the value of 8.6 eV, using the composite wave variational version of the APW method. The most recent first-principle calculation for cubic BN has been made by Park *et al*[55], obtaining the indirect gap of 4.4 eV. They also predicted the indirect and direct gaps to be 4.9 eV and 8.2 eV for wurtzite BN.

This inconsistency in the energy band results of cubic BN has been attributed to the uncertainty in knowing the crystalline potentials due to the large electronic charge transfer from nitrogen sites to boron sites. The calculated results and the experimental results [65,66,67] are summarized in Table 3. An example of the calculated band structure is shown in figure 9.

The band structure of cubic BN can be well reproduced by the empirical sp^3s^* tight-binding method[68] as follows. For the wave vector $\vec{k}$ in the Brillouin zone, we construct the following Bloch sums, defined by the linear conbination of atomic orbitals $|\alpha i\rangle \equiv |\alpha b; \vec{R}\rangle$:

$$|\chi_b^\alpha(\vec{k})\rangle = N^{-1/2} \sum_{\vec{R}} e^{i\vec{k}\cdot(\vec{R}+\vec{r}_b)} |\alpha b; \vec{R}\rangle. \tag{1}$$

Here, N and $\vec{R}$ represent the number and position of the unit cells, $\alpha = s, p_x, p_y, p_z, s^*$, $i = \{b, \vec{R}\}$ and is the atomic site index, and r_b represents the atom position (b =anion or cation) in a unit cell.

Using the $|\chi_b^\alpha(\vec{k})\rangle$ as a basis set, the Hamiltonian $\widehat{H}$ for zinc-blende crystals is written as the following 10×10 matrix:

$$\widehat{H} = \begin{matrix} & a & c \\ a & H^a & H^{ac} \\ c & \text{h.c.} & H^c \end{matrix}, \tag{2}$$

Table 2: Comparison of energy gaps and valence bandwidths for hexagonal BN obtained by recent calculations and experiments (values in eV)

Reference	Band-gap
theoretical	
[49]	4.5 (H→H)
[51]	5.1 (K→K)
[52]	4.3 (K→K)
[53]	3.8
[54]	3.9 (H→M)
[55]	4.5 (H→M)
[56]	5.4 (H→H)
experimental	
[50]	4.3
[51]	5.8
[67]	4.5

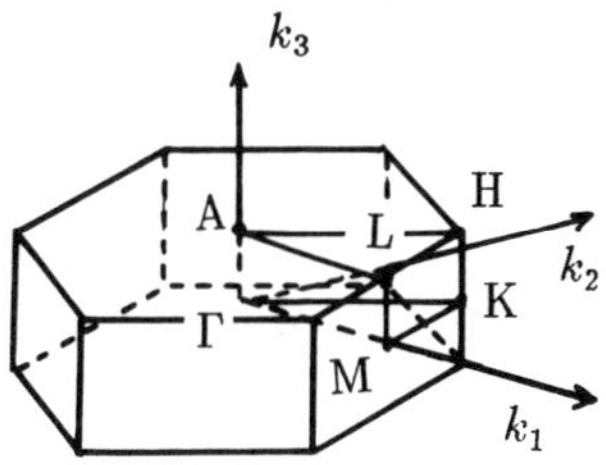

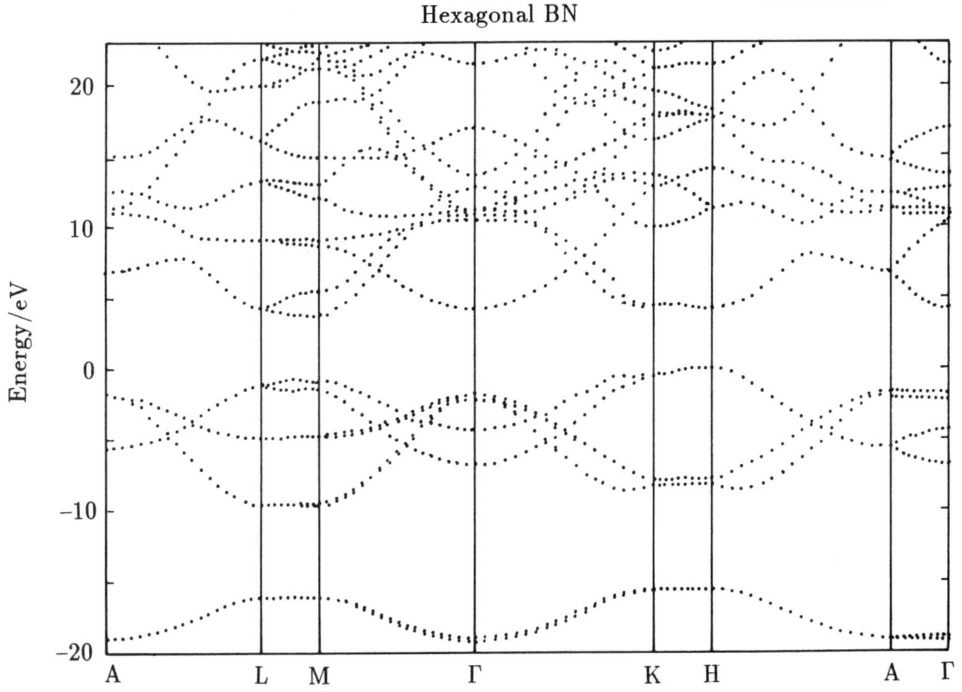

Figure 8: Band structure of hexagonal BN calculated by the full-potential linearized APW method. After Catellani *et al*[54].

Table 3: Comparison of important energy gaps and valence bandwidths for cubic BN obtained by different calculations and experiments (values in eV).

Reference	Indirect gap	Direct gap	Valence bandwidth
theoretical			
[57]	10.5	14.3	17.9
[58]	2.8	7.7	23.2
[59]	7.2	8.9	17.8
[60]	7.6	8.4	27.5
[61]	8.7	10.8	19.1
[62]	8.0	8.8	27.4
[63]	11.3		
[64]	8.6	10.9	16.7
[55]	4.4	8.8	20.1
experimental			
[65]	8.8		
[66]	6.4		22.0
[67]	6.0		15.4 to 22.0

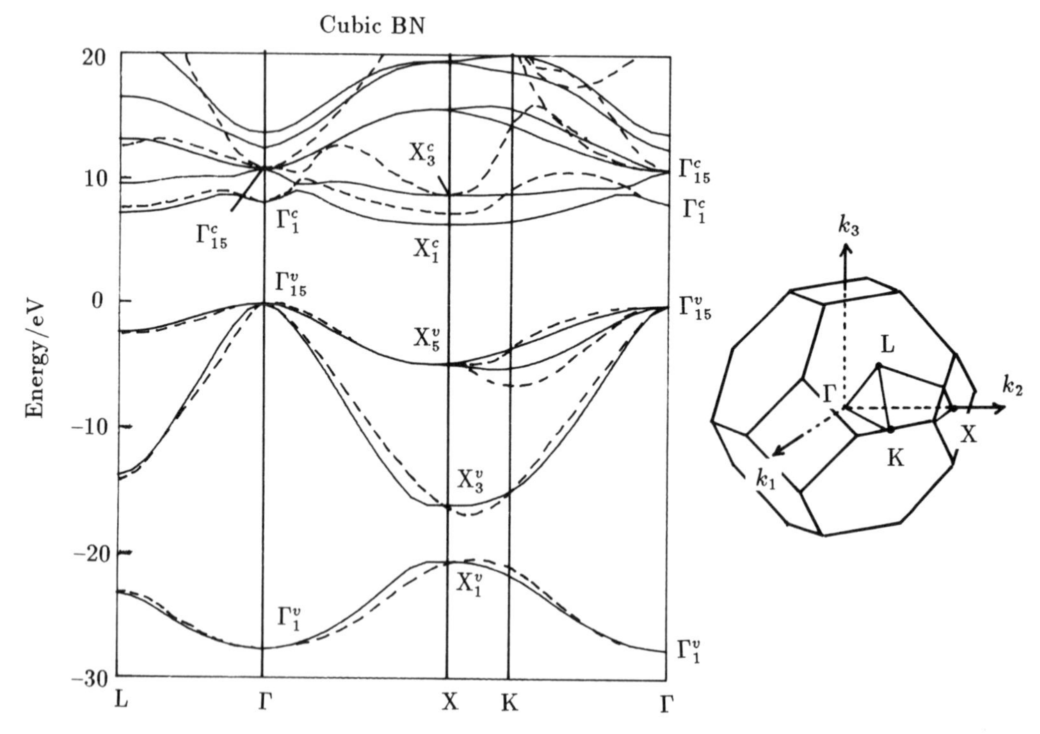

Figure 9: Band structure of cubic BN. Dashed lines represent the results of the non-local empirical pseudopotential calculation by Hamstreet and Fong[60]. Solid lines represent the results of the empirical sp^3s^* tight-binding calculation.

Table 4: The input band energies and tight-binding parameters for cubic BN. The energy is measured from the valence band maximum (Γ^v_{15}) in unit of eV.

Input band energies		Tight-binding parameters	
Γ^v_1	-27.69 [60]	$E(s_-)$	-14.71
Γ^c_1	8.06 [60]	$E(s_+)$	-4.91
Γ^c_{15}	10.75 [60]	$E(p_-)$	3.27
X^v_1	-20.64 [60]	$E(p_+)$	7.48
X^v_3	-16.13 [60]	$V(ss)$	-17.19
X^v_5	-4.84 [60]	$V(xx)$	4.95
X^c_1	6.40 [66]	$V(xy)$	10.00
X^c_3	8.66 [60]	$V(s_-p_+)$	12.55
$\Delta E(s)$	9.80 [68]	$V(s_+p_-)$	14.18
$\Delta E(p)$	4.21 [68]	$V(s^*_-p_+)$	7.16
$E(s^*_-)$	12.40 [68]	$V(s^*_+p_-)$	6.77
$E(s^*_+)$	13.68 [68]		

Here, H^a and H^c are intrasite interactions and are expressed as

$$H^a = \begin{array}{c} \\ s \\ p_x \\ p_y \\ p_z \\ s^* \end{array} \begin{array}{c} \begin{array}{ccccc} s & p_x & p_y & p_z & s^* \end{array} \\ \begin{pmatrix} E(s_-) & 0 & 0 & 0 & 0 \\ 0 & E(p_-) & 0 & 0 & 0 \\ 0 & 0 & E(p_-) & 0 & 0 \\ 0 & 0 & 0 & E(p_-) & 0 \\ 0 & 0 & 0 & 0 & E(s^*_-) \end{pmatrix} \end{array}, \tag{3}$$

$$H^c = \begin{array}{c} \\ s \\ p_x \\ p_y \\ p_z \\ s^* \end{array} \begin{array}{c} \begin{array}{ccccc} s & p_x & p_y & p_z & s^* \end{array} \\ \begin{pmatrix} E(s_+) & 0 & 0 & 0 & 0 \\ 0 & E(p_+) & 0 & 0 & 0 \\ 0 & 0 & E(p_+) & 0 & 0 \\ 0 & 0 & 0 & E(p_+) & 0 \\ 0 & 0 & 0 & 0 & E(s^*_+) \end{pmatrix} \end{array}. \tag{4}$$

In equation 2, H^{ac} is the nearest-neighbor interaction and is written as

$$H^{ac} = \begin{array}{c} \\ s \\ p_x \\ p_y \\ p_z \\ s^* \end{array} \begin{array}{c} \begin{array}{ccccc} s & p_x & p_y & p_z & s^* \end{array} \\ \begin{pmatrix} g_0V(ss) & g_1V(s_-p_+) & g_2V(s_-p_+) & g_3V(s_-p_+) & 0 \\ -g_1V(s_+p_-) & g_0V(xx) & g_3V(xy) & g_2V(xy) & -g_1V(s^*_+p_-) \\ -g_2V(s_+p_-) & g_3V(xy) & g_0V(xx) & g_1V(xy) & -g_2V(s^*_+p_-) \\ -g_3V(s_+p_-) & g_2V(xy) & g_1V(xy) & g_0V(xx) & -g_3V(s^*_+p_-) \\ 0 & g_1V(s^*_-p_+) & g_2V(s^*_-p_+) & g_3V(s^*_-p_+) & 0 \end{pmatrix} \end{array}, \tag{5}$$

where

$$g_0 = (e^{i\vec{k}\cdot\vec{\tau}_0} + e^{i\vec{k}\cdot\vec{\tau}_1} + e^{i\vec{k}\cdot\vec{\tau}_2} + e^{i\vec{k}\cdot\vec{\tau}_3})/4 \tag{6}$$

$$g_1 = (e^{i\vec{k}\cdot\vec{\tau}_0} + e^{i\vec{k}\cdot\vec{\tau}_1} - e^{i\vec{k}\cdot\vec{\tau}_2} - e^{i\vec{k}\cdot\vec{\tau}_3})/4 \tag{7}$$

$$g_2 = (e^{i\vec{k}\cdot\vec{\tau}_0} - e^{i\vec{k}\cdot\vec{\tau}_1} + e^{i\vec{k}\cdot\vec{\tau}_2} - e^{i\vec{k}\cdot\vec{\tau}_3})/4 \tag{8}$$

$$g_3 = (e^{i\vec{k}\cdot\vec{\tau}_0} - e^{i\vec{k}\cdot\vec{\tau}_1} - e^{i\vec{k}\cdot\vec{\tau}_2} + e^{i\vec{k}\cdot\vec{\tau}_3})/4 \tag{9}$$

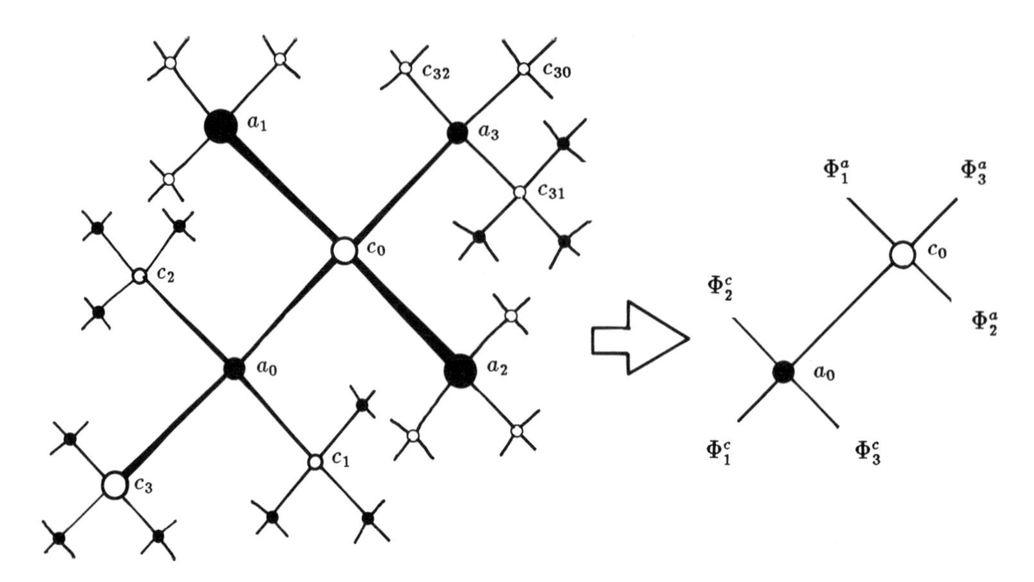

Figure 10: Portion of a Bethe lattice for a fourfold coordinated system. The Bethe lattice is represented by an anion (a_0), cation (a_1), and effective fields Φ_ν^a and Φ_ν^c.

with $\vec{\tau}_0 = (1,1,1)/4$; $\vec{\tau}_1 = (1,-1,-1)/4$; $\vec{\tau}_2 = (-1,1,-1)/4$; and $\vec{\tau}_3 = (-1,-1,1)/4$ in unit of lattice constant a_L ($a_L = 3.62$ Å for cubic BN[70]). In the above expressions, we ignore the second neighbor interactions.

In this method, band energies at the Γ-point and X-point are used as input data. The tight-binding parameters are determined so as to fit Γ- and X-point energies with the input data. Their analytic expressions are given in reference [69]. The input Γ- and X-point energies and the obtained tight-binding parameters for cubic BN are summarized in Table 4. Thus, the band structures can be obtained by diagonalizing the Hamiltonian $\widehat{H}$ for each $\vec{k}$. The calculated band structures for cubic BN are shown by the solid lines in figure 9, where the dashed lines represent the results of reference [60].

4.2 Electronic Structures of Non-Crystalline BN

As shown in the previous sections, two broad classes of disordered systems have been encountered in non-crystalline BN: *turbostratic* BN consisting of layered compound with disordered interlayer bonds, and amorphous BN consisting of a random network of tetrahedrally coordinated atoms. This subsection discusses the theoretical study of the electronic structures only for tetrahedrally-bonded amorphous BN, because theoretical models are only available for these systems.

There have been several theoretical models of tetrahedrally coordinated covalent solids in the amorphous phase[71]–[81]. All of these works simulate the amorphous substance by Bethe lattice. The Bethe lattice is a mathematical construct that is rather artificial, but it illustrates a number of useful points[82,83]. It is a lattice that contains no closed loops. The Bethe lattice for coordination number 4 is shown in figure 10.

Joannopoulus[76] presented a general method for solving a Bethe lattice with arbitrary coordination for a tight-binding Hamiltonian with an arbitrary number of degrees of freedom per site and an arbitrary number of interaction integrals. We shall apply this method to the Bethe lattice of tetrahedrally bonded III-V compounds with sp^3s^* orbitals. Using the sp^3s^*

atomic orbitals as the basis, the Hamiltonian is given by

$$\widehat{H} = \sum_i |i\rangle\langle i|\widehat{H}|i\rangle\langle i| + \sum_{<i,j>} |i\rangle\langle i|\widehat{H}|j\rangle\langle j|, \tag{10}$$

where $< i, j >$ means that site i and site j are nearest-neighbors to each other, and

$$|i\rangle = (|si\rangle, |p_x i\rangle, |p_y i\rangle, |p_z i\rangle, |s^* i\rangle). \tag{11}$$

Here, $\langle i|\widehat{H}|i\rangle = H^a$ when $i = a$ (anion site), and $\langle i|\widehat{H}|i\rangle = H^c$ when $i = c$ (cation site). Namely,

$$\langle i|\widehat{H}|i\rangle = \begin{pmatrix} E(s_\pm) & 0 & 0 & 0 & 0 \\ 0 & E(p_\pm) & 0 & 0 & 0 \\ 0 & 0 & E(p_\pm) & 0 & 0 \\ 0 & 0 & 0 & E(p_\pm) & 0 \\ 0 & 0 & 0 & 0 & E(s^*_\pm) \end{pmatrix}, \tag{12}$$

where + (−) must be taken when $i = c$ (a). In equation 10, the nearest-neighbor interactions $\langle i|\widehat{H}|j\rangle$ are given by the same forms as H^{ac} or $\{H^{ac}\}^\dagger$ in equation 5 having $(g_0, g_1, g_2, g_3) = (1,1,1,1)/4,\ (1,1,-1,-1)/4,\ (1,-1,1,-1)/4$, or $(1,-1,-1,1)/4$. Namely, in the portion shown in figure 10,

$$\langle a_0|\widehat{H}|c_0\rangle \equiv V_0^a = \frac{1}{4}\begin{pmatrix} V(ss) & V(s_-p_+) & V(s_-p_+) & V(s_-p_+) & 0 \\ -V(s_+p_-) & V(xx) & V(xy) & V(xy) & -V(s^*_+p_-) \\ -V(s_+p_-) & V(xy) & V(xx) & V(xy) & -V(s^*_+p_-) \\ -V(s_+p_-) & V(xy) & V(xy) & V(xx) & -V(s^*_+p_-) \\ 0 & V(s^*_-p_+) & V(s^*_-p_+) & V(s^*_-p_+) & 0 \end{pmatrix} \tag{13}$$

$$\langle a_0|\widehat{H}|c_\nu\rangle \equiv V_\nu^a = T_\nu V_0^a T_\nu \tag{14}$$

$$\langle c_0|\widehat{H}|a_0\rangle \equiv V_0^c = \{V_0^a\}^\dagger \tag{15}$$

$$\langle c_0|\widehat{H}|a_\nu\rangle \equiv V_\nu^c = T_\nu V_0^c T_\nu \tag{16}$$

with $\nu = 1, 2, 3$, where

$$T_\nu = \begin{pmatrix} 1 & 0 & 0 & 0 & 0 \\ 0 & e_{1\nu} & 0 & 0 & 0 \\ 0 & 0 & e_{2\nu} & 0 & 0 \\ 0 & 0 & 0 & e_{3\nu} & 0 \\ 0 & 0 & 0 & 0 & 1 \end{pmatrix}, \quad (e_{\mu\nu} = 2\delta_{\mu\nu} - 1).$$

Now, we define the Green's function $\widehat{G}$ as a function of E by the equation $(E - \widehat{H})\widehat{G} = 1$. Then the Green's function matrix-element equations, in general, become

$$\{E - \langle i|\widehat{H}|i\rangle\}\langle i|\widehat{G}(E)|j\rangle = \delta_{ij} + \sum_{k\neq i}\langle i|\widehat{H}|k\rangle\langle k|\widehat{G}(E)|j\rangle. \tag{17}$$

For the Bethe lattice shown in figure 10, this equation can be rewritten as

$$\{E - H^a\}\langle a_0|\widehat{G}(E)|a_0\rangle = 1 + \sum_{\nu=0}^{3} V_\nu^a \langle c_\nu|\widehat{G}(E)|a_0\rangle, \tag{18}$$

$$\{E - H^c\}\langle c_0|\widehat{G}(E)|a_0\rangle = V_0^c\langle a_0|\widehat{G}(E)|a_0\rangle + \sum_{\nu=1}^{3} V_\nu^c \langle a_\nu|\widehat{G}(E)|a_0\rangle. \tag{19}$$

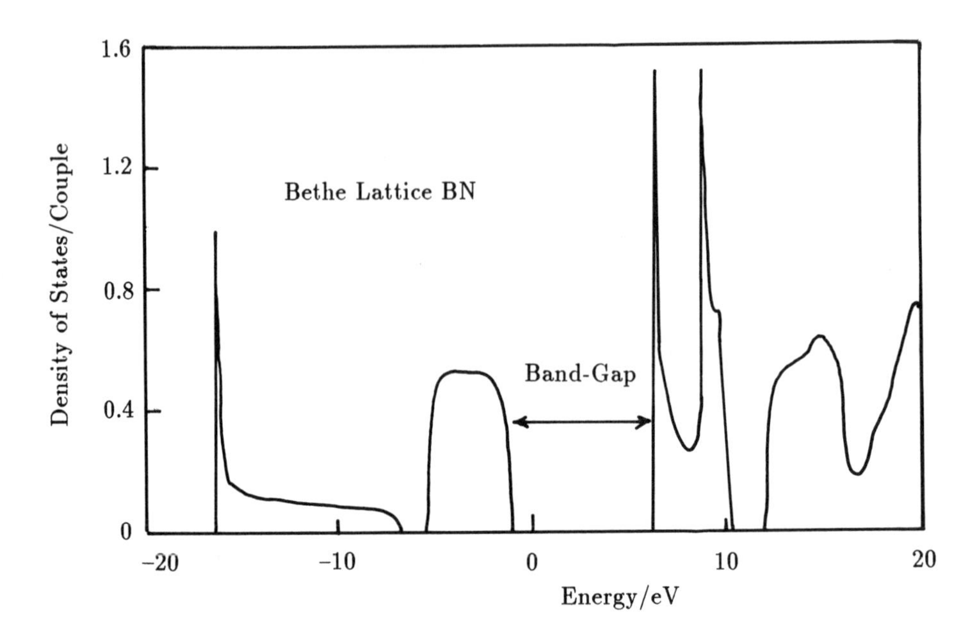

Figure 11: Calculated density of states in the BN Bethe lattice. The energy is measured from $E(\Gamma_{15}^{v})$ of cubic BN.

These equations of the infinite system can be solved by exploiting two symmetries of the Bethe-lattice structure; first, that every site can be transformed into any other site by a fixed set of transformations; and second, that any two nearest-neighbor sites are connected to each other by a single self-avoiding path. Therefore, we can define effective fields $\widehat{\Phi}_{\nu}^{a}$ and $\widehat{\Phi}_{\nu}^{c}$, for each equivalent bond, that contain all the information concerning the structure away from this bond, as shown in figure 10. In this figure, the effective fields are given by the following transfer matrices:

$$\Phi_{\nu}^{c} = \langle c_{\nu}|\widehat{G}(E)|a_{0}\rangle\langle a_{0}|\widehat{G}(E)|a_{0}\rangle^{-1} = \langle c_{\nu}|\widehat{G}(E)|c_{0}\rangle\langle a_{0}|\widehat{G}(E)|c_{0}\rangle^{-1}, \tag{20}$$

$$\Phi_{\nu}^{a} = \langle a_{\nu}|\widehat{G}(E)|c_{0}\rangle\langle c_{0}|\widehat{G}(E)|c_{0}\rangle^{-1} = \langle a_{\nu}|\widehat{G}(E)|a_{0}\rangle\langle c_{0}|\widehat{G}(E)|a_{0}\rangle^{-1}. \tag{21}$$

These fields are of the same symmetry as V_{ν}^{a} and V_{ν}^{c}; namely,

$$\Phi_{\nu}^{a(c)} = T_{\nu}\Phi_{0}^{a(c)}T_{\nu}. \tag{22}$$

Inserting equations 12–16 and 20–22 into equation 18 and 19, we have the equations for the effective fields,

$$\Phi_{0}^{a} = (E - H^{a} - \sum_{\nu=1}^{3} T_{\nu}V_{0}^{a}\Phi_{0}^{c}T_{\nu})^{-1}V_{0}^{a}, \tag{23}$$

$$\Phi_{0}^{c} = (E - H^{c} - \sum_{\nu=1}^{3} T_{\nu}V_{0}^{c}\Phi_{0}^{a}T_{\nu})^{-1}V_{0}^{c}, \tag{24}$$

and the equations for the diagonal elements of the Green's functions:

$$\langle a|\widehat{G}(E)|a\rangle = (E - H^{a} - \sum_{\nu=0}^{3} T_{\nu}V_{0}^{a}\Phi_{0}^{c}T_{\nu})^{-1}, \tag{25}$$

$$\langle c|\hat{G}(E)|c\rangle = (E - H^c - \sum_{\nu=0}^{3} T_\nu V_0^c \Phi_0^a T_\nu)^{-1}. \tag{26}$$

By solving equations 23 and 24 selfconsistently, we obtain the effective fields Φ_0^a and Φ_0^c. Thus, by substituting these fields into equations 25 and 26, we have the values of the Green's function diagonal elements. The density of states is then given as a function of energy by

$$N(E) = -\frac{1}{\pi} \text{ Im Tr } \{\langle a|\hat{G}(E)|a\rangle + \langle c|\hat{G}(E)|c\rangle\}. \tag{27}$$

In figure 11, the calculated density of states is shown as a function of energy for the BN Bethe lattice. The calculated band-gap is 7.4 eV. This value is, of course, larger than that for cubic BN.

4.3 Electronic Structures of BN/III-V Interfaces

It is well known that interface states with idiosyncratic density distribution in the band-gap have been generally found to dominate the electronic properties of insulator–semiconductor (I-S) systems on III-V semiconductors. Specifically, as shown in the previous section, InP I-S systems exhibit a high density of interface states in the lower half of the band-gap. The Fermi level without gate bias commonly lies 0–0.4 eV below the conduction band edge. On the other hand, GaAs I-S systems exhibit so high a density of interface states in the upper half of the band-gap that two-dimensional electron transport has not yet been achieved in these GaAs I-S systems. The surface Fermi level without gate bias lies 0.5–0.8 eV above the valence band maximum.

Several models have been proposed for the origin of the interface states, including the unified defect model of Spicer *et al.*[84,85], and the disorder-induced gap-state model of Hasegawa and Ohno[86]. Very recently, Yamaguchi[87] has theoretically investigated the electronic structures of III-V semiconductor interfaces, using the superlattice method, where the insulators were simulated with lattice-matched II-VI alloys. He concluded that the heterointerface bond relaxation strongly affects the formation of interface states in the band-gap. Even more recently, Yamaguchi[88] has also calculated the electronic structures of non-crystalline BN/III-V semiconductor interfaces, using the scattering-theoretic method[89,90], where the insulator BN films were simulated by tetrahedrally bonded Bethe lattices. The present final subsection reviews this theory and its calculated results.

The theoretical procedure for creating the heterojunction of the Bethe lattice and the crystal is schematically described in figure 12. First, the unperturbed system is taken to consist of the two infinite bulk materials. A perturbation that creates the interface, $\hat{V}$ is introduced. In the crystal, the perturbation of removing one monolayer (cation+anion) of index 0 can be represented by[89]

$$V_{\ell n} \equiv \langle \phi_\ell(\vec{q})|\hat{V}|\phi_n(\vec{q})\rangle = \lim_{u\to\infty} u\delta_{\ell 0}\delta_{n0}, \tag{28}$$

using the layer orbital basis

$$|\phi_\ell^{\alpha b}(\vec{q})\rangle = N_2^{-1/2} \sum_{\vec{\rho}} e^{i\vec{q}\cdot(\vec{\rho}+\vec{r}_b^\ell)}|\alpha b\ell;\vec{\rho}\rangle, \tag{29}$$

where ℓ labels different layers, $\vec{q}$ are the two-dimensional wave vectors, N_2 and $\vec{\rho}$ are the number and the lattice vectors of the two-dimensional Bravais lattice, and $\vec{r}_b^\ell$ are the position vectors of the atoms in the two-dimensional unit cell. In the above equation,

$$|\phi_\ell(\vec{q})\rangle = (|\phi_\ell^a\rangle, |\phi_\ell^c\rangle) = (|\phi_\ell^{sa}\rangle, |\phi_\ell^{p_x a}\rangle, \ldots, |\phi_\ell^{s^* c}\rangle). \tag{30}$$

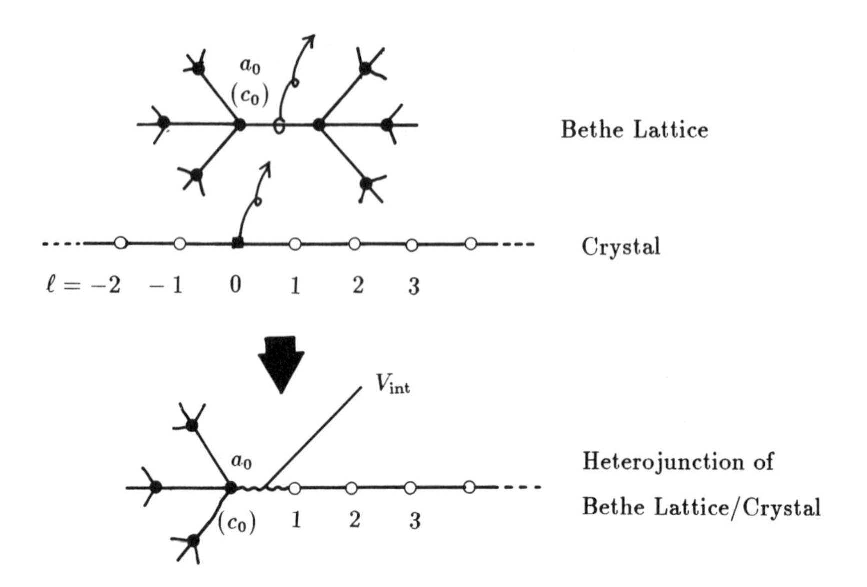

Figure 12: Schematic description of the heterointerface creation procedure.

The Green's function $\widehat{G}^{\rm crys} \equiv (E - \widehat{H} - \widehat{V})^{-1}$ in the layer orbital representation is given in terms of the perfect crystal Green's function $\widehat{G}^0 \equiv (E - \widehat{H})^{-1}$ by the following Dyson's equation:

$$G^{\rm crys}_{\ell n}(\vec{q}, E) \equiv \langle \phi_\ell(\vec{q}) | \widehat{G}^{\rm crys}(E) | \phi_n(\vec{q}) \rangle = G^0_{\ell n}(\vec{q}, E) + \sum_{mm'} G^0_{\ell m}(\vec{q}, E) V_{mm'} G^{\rm crys}_{m'n}(\vec{q}, E). \quad (31)$$

Substituting equation 28 into equation 31 gives

$$G^{\rm crys}_{\ell n}(\vec{q}, E) = G^0_{\ell n}(\vec{q}, E) - G^0_{\ell 0}(\vec{q}, E) G^0_{00}(\vec{q}, E)^{-1} G^0_{0n}(\vec{q}, E) \quad (\ell, n > 0). \quad (32)$$

By definition, the perfect crystal Green's function in the layer orbital representation, $G^0_{\ell n}(\vec{q}, E)$, can be given in terms of eigenvalues $E_{n\vec{k}}$ and eigenvectors $|n\vec{k}\rangle$ of the Hamiltonian, equation 2, by

$$G^0_{\ell n}(\vec{q}, E) = \sum_{n\vec{k}} \frac{\langle \phi_\ell(\vec{q}) | n\vec{k} \rangle \langle n\vec{k} | \phi_n(\vec{q}) \rangle}{E + i0^+ - E_{n\vec{k}}}, \quad (33)$$

where values of the tight-binding parameters for GaAs and InP crystals used in his work are described in reference [87].

On the other hand, the Bethe lattice having free surfaces is created by removing one of the four surrounding effective fields. Hence, the same procedure as in the previous section yields

$$\langle a_0 | \widehat{G}^{\rm amor}(E) | a_0 \rangle = (E - H^a - \sum_{\nu=1}^{3} T_\nu V_0^a \Phi_0^c T_\nu)^{-1}, \quad (34)$$

$$\langle c_0 | \widehat{G}^{\rm amor}(E) | c_0 \rangle = (E - H^c - \sum_{\nu=1}^{3} T_\nu V_0^c \Phi_0^a T_\nu)^{-1}, \quad (35)$$

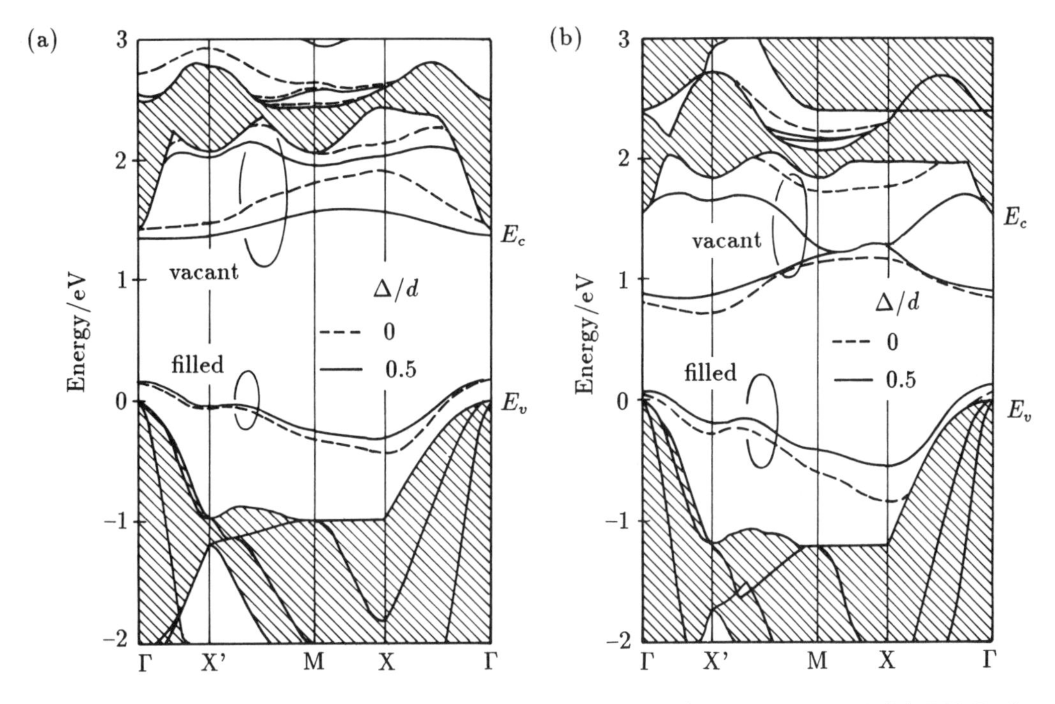

Figure 13: Electronic structures of (a) BN Bethe lattice/InP crystal and (b) BN Bethe lattice/GaAs crystal with (110) orientation.

for the Bethe lattice having a free surface; that is, a dangling bond. where a_0 represents an anion site with one dangling bond, and c_0 represents the cation site with one dangling bond.

Next, each bond at a free surface layer of the crystal is coupled with a dangling bond of the Bethe lattice. Then, the eigenvalues in the band-gap for this heterojunction system are given by the solutions E of the equation[90]

$$\det[1 - \widehat{G}^1(E)\widehat{V}^1] = 0, \tag{36}$$

where

$$\widehat{G}^1(E) = \begin{pmatrix} G^{\mathrm{amor}}(E) & 0 \\ 0 & G^{\mathrm{crys}}(E) \end{pmatrix}, \tag{37}$$

$$\widehat{V}^1 = \begin{pmatrix} 0 & V_{\mathrm{int}} \\ V_{\mathrm{int}}^{\dagger} & 0 \end{pmatrix}. \tag{38}$$

Here, V_{int} is the matrix element that couples the two interface atoms. In this work, V_{int} has been taken as the form

$$V_{\mathrm{int}} = \begin{matrix} \\ \langle\phi^{a_0}| \\ \langle\phi^{c_0}| \end{matrix} \begin{matrix} |\phi_1^a\rangle \quad |\phi_1^c\rangle \\ \begin{pmatrix} 0 & \bar{V}_0^a \\ \bar{V}_3^c & 0 \end{pmatrix} \end{matrix}, \tag{39}$$

for the (110) interface, where

$$\bar{V}_\nu^b = (1 + \Delta/d)^{-2}\{V_\nu^b(\mathrm{BN}) + V_\nu^b(\mathrm{GaAs})\}/2 \quad (b = a, c) \tag{40}$$

with the lattice relaxation parameter Δ. In equation 39, $\bar{V}_3^c$ or $\bar{V}_0^a$ are set to zero for the (111) interface having anion-terminated or cation-terminated crystals, respectively. Here, it must be noted that the layer orbital representations $\langle \phi^{b_0} | G^{\mathrm{amor}}(E) | \phi^{b_0} \rangle \equiv G_{00}^{\mathrm{amor}}(E)$ are equal to the atomic orbital representation $\langle b_0 | G^{\mathrm{amor}}(E) | b_0 \rangle$ due to the absence of connections between the atoms of each Bethe lattices. Equation 36 can readily be reduced to the following equation;

$$\det[1 - G_{00}^{\mathrm{amor}}(E) V_{\mathrm{int}} G_{11}^{\mathrm{crys}}(\vec{q}, E) V_{\mathrm{int}}{}^{\dagger}] = 0. \tag{41}$$

Thus, the interface electronic structures in the band-gap can be calculated as a function of two-dimensional wave vectors $\vec{q}$ by solving equation 41 with equations 32, 34, 35 and 39.

Figure 13 shows the interface electronic structures for InP and GaAs GaAs having Bethe lattice of BN as the overlayer insulators. Here, note that the lowest of vacant interface bands is induced by terminated B bonds. As shown in these figures, chemical trends of the interface bands in the band-gap are similar to the crystalline I-S interfaces; namely, GaAs I-S interface exhibits a vacant interface band deep below E_c, which yields a high density of localized states above the Fermi level. On the other hand, InP I-S interface exhibits no vacant interface band below E_c, when the interface bond relaxations are sufficiently small. There are also filled interface bands in the band-gap both in InP and in GaAs. These interface bands become deep with an increase in the bond relaxation. These trends are consistent with the experimental data.

References

[1] Mayer, F. and Zappner, R.: Ber., 1921, 54, 560.

[2] Patterson, R. J., Humphries, R. D. and Haberecht, R. R.: *Ext. Abst. National Electrochemical Society Meetings; Part 1, Pittsburgh, 1963, 103; Part 2, New York, 1963, 197.*

[3] Powell, C. F.: *Vapor Deposition (John Willey & Sons Inc., 1966), p. 381.*

[4] Baronian, W.: Mater. Res. Bull., 1972, 7, 119.

[5] Takahashi, T., Itoh, H. and Ohtake, A.: *Yogyo-kyokaishi*, 1981, 89, 63 (in Japanese).

[6] Sano, M. and Aoki, M.: Thin Solid Films, 1981, 83, 247.

[7] Motojima, S., Tamura, Y. and Sugiyama, K.: Thin Solid Films, 1982, 88, 269.

[8] Rand, M. J. and Roberts, J. F.: J. Electrochem. Soc., 1968, 115, 423.

[9] Hirayama, M. and Shohno, K.: J. Electrochem. Soc., 1975, 122, 1671.

[10] Murarka, S. P. Chang, C. C., Wang, D. N. K. and Smith, E. E.: J. Electrochem. Soc., 1979, 126, 1951.

[11] Adams, A. C. and Capio, C. D.: J. Electrochem. Soc., 1980, 127, 399.

[12] Yamaguchi, E. and Minakata, M.: *Ext. Abst. 15th Conf. Solid State Devices, Tokyo, 1983, p. 81.*

[13] Yamaguchi, E. and Minakata, M.: J. Appl. Phys., 1984, 55, 3098.

[14] Pierson, H. O.: J. Composite Mater., 1975, 9, 228.

[15] Furukawa, Y.: Jpn. J. Appl. Phys., 1984, 23, 376.

[16] Nakamura, K.: J. Electrochem. Soc., 1985, 132, 1757.

[17] Alexander, J. H., Joyce, R. J. and Sterling, H. F.: *Ext. Abst. The Electrochemical Society (Fall Meeting), Montreal, 1968, p. 343.*

[18] Hyder, S. B. and Yep, T. O.: J. Electrochem. Soc., 1976, 123, 1721.

[19] Gafri, O., Grill, A., Itzhak, D., Inspektor, A. and Avni, R.: Thin Solid Films, 1980, 72, 523.

[20] Sokołowski, M.: J. Crys. Growth, 1979, 46, 136.

[21] Sokołowski, M., Sokołowska, A., Rusek, A., Romanowski, Z., Gokieli, B. and Gajewska, M.: J. Crys. Growth, 1981, 52, 165.
[22] Szmidt, J., Jakubowski, A., Michalski, A. and Rusek, A.: Thin Solid Films, 1983, 110, 7.
[23] Schmolla, W. and Hartnagel, H. L.: J. Phys., 1982, D15, L95.
[24] Schmolla, W. and Hartnagel, H. L.: Solid State Electron., 1983, 26, 931.
[25] Miyamoto, H., Hirose, M. and Osaka, Y.: Jpn. J. Appl. Phys., 1983, 22, L216.
[26] Matsumoto, O., Sasaki, M., Suzuki, H., Seshimo, H. and Uyama, H.: *Proc. Int. Conf. CVD*, 1987, p. 552.
[27] Shanfield, S. and Wolfson, R.: J. Vac. Sci. Technol., 1983, A1, 323.
[28] Chorpa, K. L., Agarwal, V., Vankar, V. D., Deshpandey, C. V. and Bunshah, R. F.: Thin Solid Films, 1985, 126, 307.
[29] Chayahara, A., Yokohama, H., Imura, T., Osaka, Y. and Fujisawa, M.: Tech. Rep. Inst. Electron. & Commun. Eng. Jpn., 1986, SSD86-162, 37 (in Japanese).
[30] Noreika, A. J. and Frankombe, M. H.: J. Vac. Sci. Technol., 1969, 6, 722.
[31] Tanaka, K., Uemura, Y. and Iwata, M.: *Oyo-butsuri*, 1977, 46, 120 (in Japanese).
[32] Wiggins, M. D., Aita, C. R. and Hickernell, F. S.: J. Vac. Sci. Technol., 1984, A2, 322.
[33] Yasuda, K., Yoshida, A., Takeda, M., Masuda, H. and Akasaki, I.: Phys. Stat. Sol., 1985, a90, K7.
[34] Seidel, K. H., Reichelt, K., Schaal, W. and Dimigen, H.: Thin Solid Films, 1987, 151, 243.
[35] Weismantel, C., Bewilogua, K., Dietrich, D., Erler, H. J., Hinneberg, H. J., Klose, S., Nowick, W. and Reisse, G., Thin Solid Films, 1980, 72, 19.
[36] Weismantel, C.: J. Vac. Sci. Technol., 1981, 18, 19.
[37] Satou, M. and Fujimoto, F.: Jpn. J. Appl. Phys. Part 2, 1983, 22, L171.
[38] Zhou, P. F, Môri, T. and Namba, Y.: *Shinkuu*, 1985, 28, 581 (in Japanese).
[39] Yamaguchi, E., Minakata, M. and Furukawa, Y.: Jpn. J. Appl. Phys., 1984, 23, L49.
[40] Schmolla, W.: Int. J. Electron., 1985, 58, 35.
[41] Mishima, O., Yamaoka, S. and Fukunaga, O.: J. Appl. Phys., 1987, 61, 2822.
[42] Mishima, O., Tanaka, J., Yamaoka, S. and Fukunaga, O.: Science, 1987, 238, 181.
[43] Vick, G. L. and Whittle, K. M.: J. Electrochem. Soc., 1969, 116, 1142.
[44] Yamaguchi, E.: J. Appl. Phys., 1984, 56, 1722.
[45] Yamaguchi, E.: Phys. Rev., 1985, B32, 5280.
[46] Doni, E. and Pastori, G.: Nuovo Cimento, 1969, 64B, 117.
[47] Zunger, A.: J. Phys., 1974, C7, 76.
[48] Dovesi, R., Pisani, C. and Roetti, C.: Int. J. Quantum Chem., 1980, 17, 517.
[49] Zupan, J.: Phys. Rev., 1972, B6, 2477.
[50] Zupan, J. and Kolar, D.: J. Phys., 1972, C5, 3097.
[51] Zunger, A., Latzir, A. and Halperin, A.: Phys. Rev., 1976, B13, 5560.
[52] Robertson, J.: Phys. Rev., 1984, B29, 2131.
[53] Nakhmanson, M. S. and Smirnov: V. P.: Fiz. Tverd. Tela, 1971, 13, 3288 [Sov. Phys. Solid State, 1972, 13, 2763].
[54] Catellani, A., Posternak, M., Baldereshi, A., Jansen, H. J. F. and Freeman, A. J.: Phys. Rev., 1985, 32, 6997.
[55] Park, K. T., Terakura, K. and Hamada, N.: J. Phys., 1987, C20, 1241.
[56] Kurita, N. and Nakao, K.: J. Phys. Soc. Jpn., 1987, 56, 4442.
[57] Kleinman, L. and Philips, J. C.: Phys. Rev., 1960, 117, 460.
[58] Bassani, F. and Yoshimine, M.: Phys. Rev., 1963, 130, 20.

[59] Wiff, D. R. and Keown, R.: J. Chem. Phys., 1967, 47, 3113.
[60] Hamstreet, L. A. and Fong, C. Y.: Phys. Rev., 1972, B6, 1464.
[61] Zunger, A. and Freeman A. J.: Phys. Rev., 1978, B17, 2030.
[62] Tsay, Y. F., Vaidyanathan, A. and Mitra S. S.: Phys. Rev., 1979, B19, 5422.
[63] Dovesi, R., Pisani, C., Roetti, C. and Dellarole, P.: Phys. Rev., 1981, B24, 4170.
[64] Prasad, C. and Dubey, J. D.: Phys. Stat. Sol., 1984, 125, 629.
[65] Gielisse, P. J., Mitra, S. S., Plendl, J. N., Griffis, R. D., Mansur, L. C., Marshall, R. and Pascoe, E. A.: Phys. Rev., 1967, 155, 1039.
[66] Chrenko, R. M.: Solid State Commun., 1974, 14, 511.
[67] Fomichev, V. A. and Rumsch, M. A.: J. Phys. Chem. Solids, 1968, 29, 1015.
[68] Vogl, P., Hjalmarson, H. P. and Dow, J. D.: Phys. Rev., 1984, B30, 1929.
[69] Yamaguchi, E.: Jpn. J. Appl. Phys., 1986, 25, L643.
[70] Soma, T., Sawaoka, S. and Saito, S.: Mater. Res. Bull., 1974, 9, 755.
[71] Weaire, D. and Thorpe, M. F.: Phys. Rev., 1971, B4, 2508.
[72] Thorpe, M. f. and Weaire, D.: Phys. Rev., 1971, B4, 3518.
[73] Nagle, J. F., Bonner, J. C. and Thorpe, M. F.: Phys. Rev., 1972, B4, 2233.
[74] Joannopoulus, J. D. and Yndurain, F.: Phys. Rev., 1974, 10, 1974.
[75] Yndurain, F. and Joannopoulus, J. D.: Phys. Rev., 1975, B11, 2957.
[76] Joannopuolus, J. D.: Phys. Rev., 1977, B16, 2764.
[77] Laughlin, R. B., Joannopoulus, J. D. and Chadi, D. J.: Phys. Rev., 1980, B21, 5733.
[78] Martínez, E. and Ynduráin, F.: Phys. Rev., 1982, B25, 6511.
[79] Carrico, A. S., Elliott, R. J. and Barrio, R. A.: Phys. Rev., 1986, B34, 872.
[80] Barrio, R. A., Tagüeña-Martínez, J., Martínez, E. and Yuduráin, F.: J. Non-Cryst. Solids, 1985, 72, 181.
[81] Barrio, R. A., Elliott, R. J. and Carrico, A. S.: Phys. Rev., 1986, B34, 879.
[82] Bethe, H. A.: Proc. Roy. Soc. (London), 1935, A216, 45.
[83] Domb, C.: Adv. Phys., 1960, 9, 245.
[84] Spicer, W. E., Chye, P. W., Skeath, P. R., Su, C. Y. and Lindau, I.: J. Vac. Sci. Technol., 1979, 16, 1422.
[85] Spicer, W. E., Lindau, I., Skeath, P. R. and Su, C. Y.: J. Vac. Sci. Technol., 1980, 17, 1019.
[86] Hasegawa, H. and Ohno, H.: J. Vac. Sci. Technol., 1986, B4, 1130.
[87] Yamaguchi, E.: J. Phys. Soc. Jpn., 1988, 57, 2461.
[88] Yamaguchi, E.: *Proc. Int. Conf. Phys. Semicond.*, Warsaw (in press). WeP-103.
[89] Pollmann, J. and Pantelides, S. T.: Phys. Rev., 1978, B18, 5524.
[90] Pollmann, J. and Pantelides, S. T.: Phys. Rev., 1980, B21, 709.

Materials Science Forum Vols. 54 & 55 (1990) pp. 353-374

BORON NITRIDE THIN INSULATING FILMS ON GaAs COMPOUND SEMICONDUCTORS

V.J. Kapoor and G.G. Skebe
Electronic Devices and Materials Laboratory
Department of Electrical and Computer Engineering
University of Cincinnati, Cincinnati, OH 45221-0030, USA

ABSTRACT

Plasma enhanced chemical vapor deposition (PECVD) of high quality boron nitride films on GaAs compound semiconductor was investigated. Several processing variations of the ammonia and diborane reacton were employed, and the resulting chemical compostion of the films was analyzed. Processing parameters that were studied included: R.F. power, deposition pressure, ammonia to diborane ratio and deposition temperature. Deposition power exhibited the most pronounced effect on the composition of the films. The boron to nitrogen ratio in boron nitride films varied from 0.8 to 1.45 with decreasing R.F. power. For a deposition power of 30 watts, all other processing variations yielded approximately stoichiometric films. The oxygen content in the bulk of the boron nitride films was less than 2% for most conditions. The total hydrogen content in the boron nitride film was observed to decrease with increasing deposition temperature and power. The boron nitride film was studied as an active gate insulator for the metal-insulator-semiconductor field-effect transistor (MISFET) applications. Metal-insualtor-semicondutor capacitors were fabricated using boron nitride films. The capacitance-voltage measurements obtained exhibited a hysteresis type behavior which varied from 0.3 to 3.8 volts as measured at flatband conditions. Leakage current densities as low as 5×10^{-9} A/cm^2 were observed.

I. INTRODUCTION

Boron nitride (BN) is a material of high chemical inertness, thermal stability[1] and with a large bandgap[2]. Boron nitride can withstand high pressure and temperature[3]. As a result, it is used as a thin film dielectric, a protective coating, a diffusion barrier for alkali metals, a boron diffusion source and as a membrane for X-ray masks[4-6]. Rand and Roberts[4] have described the preparation and properties of boron nitride films deposited by chemical vapor deposition (CVD) of diborane and ammonia in a hydrogen carrier gas at temperatures of 600 C and 800 C. The dielectric quality and chemical stability reported was very good, with the exception of certain films which reacted with atmospheric moisture over long periods of time, extensively converting the films to orthoboric acid. Hirayama and Shohno[5] have reported the preparation of CVD boron nitride films for application as a boron source for diffusion in silicon. Diborane and ammonia reacted at temperatures of 700 C-1250 C with hydrogen acting as a carrier gas. They reported that deposition temperatures over 1000 C resulted in polycrystalline boron nitride which was thermally stable. At temperatures below 1000 C, amorphous boron nitride films were formed. These films readily decomposed during the heat treatment in nitorgen or hydrogen ambient. S. P. Murarka et al[6], deposited boron nitride films and phosphorus-doped boron nitride films. The films were chemically vapor deposited by a reaction between ammonia and diborane in an inert carrier gas at temperatures between 400 C and 700 C. They reported that some films were unstable in atmospheric ambients and that these films contained a relatively high amount of oxygen. Films that contained a low level of oxygen were stable at room temperature and humidity. W. Schmolla and H. L. Hartnagel[7] have reported the deposition of BN films in double-plasma reactor for semiconductor applications by using a plasma process which gives suitable electronically stimulated reactants under relatively low gas temperature conditions. The chemical and physical properties of the CVD-BN films are strongly dependent on the deposition parameters, such as the flow rates of the reactants, carrier gases and the substrate temperature during the deposition.

Boron nitride displays multiband electro-, photo- and cathode-ray luminescence. The crystal structure of boron nitride film is very similar to that of graphite-like hexagonal form or in the diamond-like cubic form. If crystalline thin film can be obtained with large carrier drift velocity, it can be a promising material for certain microwave applications[8,9]. To achieve the best BN film properties, a low oxygen content is essential, and the process parameters have to be optimized correspondingly. Good insulating properties occur with BN since the bandgap is very large, although the value depends upon the structure of the material ranging from 3.8 eV for hexagonal to 8 eV for cubic material. It is possible that boron and nitrogen do not behave as dopants in the semiconductor because they are of the same group in the periodic table as the semiconductor components.

High temperature thermal oxidation processess, resulting in high quality dielectrics in silicon technologies, are not compatible with GaAs materials. At temperatures greater that 450 C, dissociation of arsenic from the GaAs substrate occurs which degrades the semiconductors electrical characteristics. Because of this requirement, PECVD is an attractive method for thin film deposition on GaAs. In recent years, a number of techniques have been used to deposit dielectrics on III-V compound semiconductors, namely GaAs and InP. This compatibility has been suggested to lead to a lower surface state density and result in a suitable dielectric for MIS (metal-insulator-semiconductor) structures. Preliminary studies were conducted on boron nitride-gallium arsenide structures prepared by RF sputtering of a boron nitride target in a nitrogen ambient[10].

Although plasma enhanced chemical vapor deposition techniques have not previously been used to deposit boron nitride films, these techniques have been widely used for the deposition of silicon nitride passivating films. As mentioned earlier, the main advantage of using PECVD techniques is the low temperature at which high quality dielectric layers can be deposited. Reactions occuring only at high temperatures under classical thermal activation can occur at much lower temperatures in the presence of an energetic plasma field. Gas molecules entering the reactor are dissociated mainly by electron impact which results in very reactive radicals, atoms, and ions[11]. These reactive species then condense on the substrate surface, forming an amorphous film. Another significant advantage is that the substrate surfaces can be treated *in situ* in a gas plasma prior to deposition[12].

In this work we have investigated plasma enhanced chemical vapor deposition (PECVD) of high quality boron nitride films on GaAs compound semiconductor. Several processing variations of the ammonia and diborane reacton were employed, and the resulting chemical composition of the films was analyzed. Processing parameters that were studied included: R.F. power, deposition pressure, ammonia to diborane ratio and deposition temperature. The boron nitride film was then studied as an active gate insulator for the metal-insulator-semiconductor field-effect transistor (MISFET) applications.

II. THIN FILM DEPOSITION

Boron nitride thin films were deposited on n-type GaAs wafers with <100> orientation. The majority of substrates were doped with Sn to a concentration of 2×10^{17} atoms/cm^3 while the remaining substrates were Si doped to a concentration of 2×10^{17} atoms/cm^3. Prior to boron nitride deposition, the substrates were chemically cleaned in succession in the following electronic grade solvents: trichloroethylene, acetone and methonal. Following the chemical degrease, a mixture of 1:1 deionized water and hydrofluoric acid was used to etch any native oxide. The samples were then rinsed in deionized water for 30 seconds and blown dry with oxygen.

The deposition of boron nitride was performed in a Technics planaretch II PECVD system[13]. The plasma treatment chamber through which samples are inserted, processed and removed, consists of a 12 inch diameter by 4.5 inch high stainless steel vacuum chamber which is O-ring sealed to the base plate. The deposition plate and the plasma generating electrode capacitively couple the 30 KHz R.F. plasma power generator to the gas species. The pressure of the plasma treatment chamber is monitored by a MKS Baratron pressure gauge type 222A. Gas flow into the reaction chamber is controlled by mass flow controller. Gases used for deposition were: ammonia and diborane diluted in nitrogen carrier gas (0.649% B_2H_6 diluted in N_2). The overall chemical reaction is given below:

$$2NH_3 + B_2H_6 = 2BN + 6H_2 \quad (1)$$

The chemical composition of the deposited films depends on a number of processng parameters. The parameters investigated were: total pressure of the reacting ambient, R.F. Plasma power, deposition temperature, and diborane to ammonia gas ratios.

The deposition procedure began with etching the chamber of any previously deposited material. Carbon tetraflouride (CF_4) was used for this purpose. A CF_4 flow of 30 sccm and R.F. plasma power of 300 watts was used for 10 min. The etching procedure was necessary to achieve repeatable results. In order to reduce the amount of impurity species in the chamber during the plasma treatment, the chamber was baked out for 30 minutes prior to loading the samples. Impurities are thereby desorbed from the inner of the chamber.

Figure 1 depicts the deposition rate as a function of deposition pressure. The deposition rate decreased slightly from 7.2 Å/min at 230 mTorr to 5.8 Å/min at 450 mTorr. This slight decrease was considered to be negligible and therefore, the deposition pressure was deemed to be an insignificant factor affecting the deposition rate.

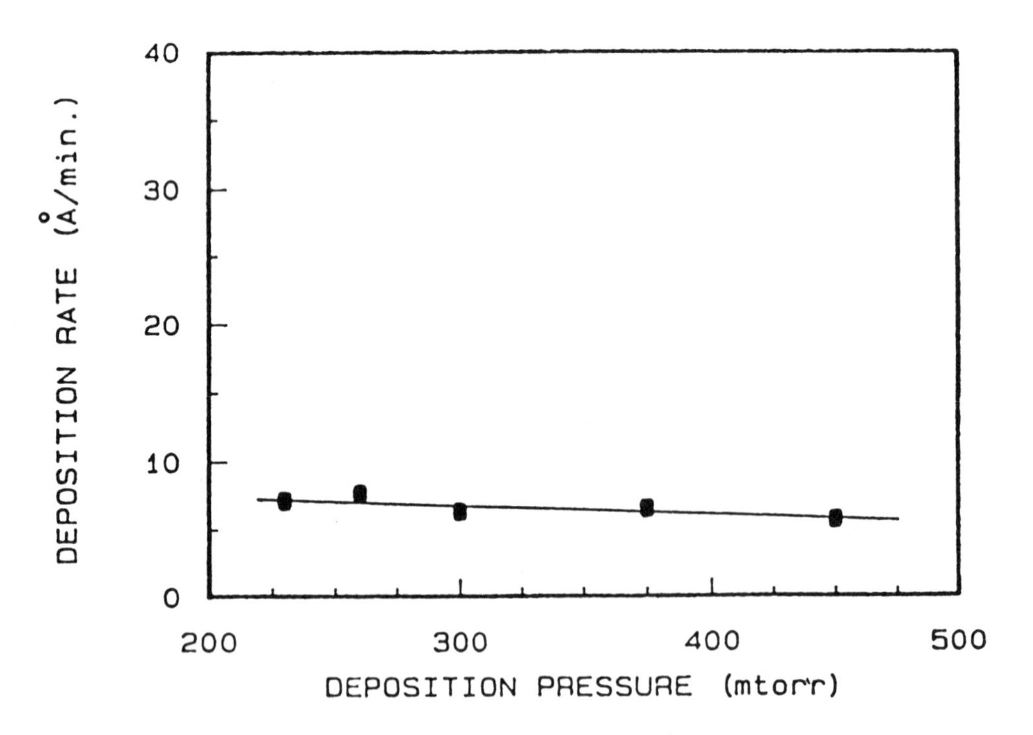

Figure 1 Deposition rate as a function of Deposition Pressure (P_w = 30 watts, NH_3/B_2H_6 = 10, T = 75°C)

Figure 2 depicts the deposition rate as a function of NH_3/B_2H_6 gas ratio. From an initial value of 11.2 Å/min at a ratio of 4.4, the deposition rate decreased as the NH_3/B_2H_6 ratio increased and resulted in a low of 2.3 Å/min at a ratio of 100. In this variation, the B_2H_6 gas flow was not held constant and decreased with increasing NH_3/B_2H_6 ratios in order to maintain a constant total gas flow. Therefore, the decrease in deposition rate coincided with a decrease in the percentage of boron in the reactant gases. Since there is a considerably greater quantity of nitrogen than boron present in the reactant gases, it is reasonable to conclude that the boron concentration is the limiting factor affecting the deposition rate under these conditions and that the deposition rate observed was due to a reduction in the boron reactant.

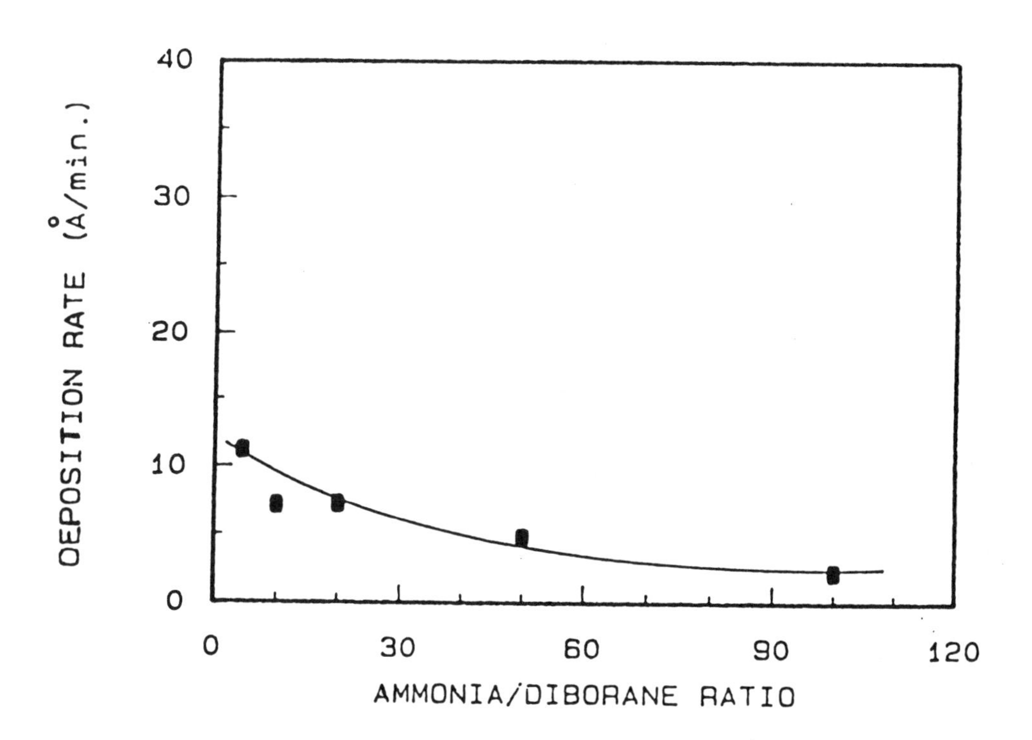

Figure 2 Deposition rate as a function of NH_3/B_2H_6 ratio. (P_w = 30 watts, P_r = 230 mTorr, T = 75°C)

Table 1 lists the refractive indices and deposition rates of the films resulting from the corresponding processing variations. The refractive indices of the films were observed to vary from 1.62 to 1.77. These values were found to be consistent with the findings of other researchers. Rand and Roberts[4] have reported a refractive index in the range of 1.7 - 1.8 for boron nitride deposited by chemical vapor deposition techniques; S. P. Murarka et al[6] have reported the refractive index for CVD boron nitride in the range of 1.7 to 2.8 for boron to nitrogen concentration ratios of 0.7 to 4.0.

Table 1 Boron nitirde processing conditions, refractive index and depostion rate.

Sample	Deposition Pressure	Plasma Power	Deposition Temp.	NH_3/B_2H_6 Ratio	Refractive Index	Deposition Rate(A/min)
1D	230 mT	30 W	75 C	10	1.72	7.2
2D	260 mT	30 W	75 C	10	1.75	7.7
3D	300 mT	30 W	75 C	10	1.67	6.4
AL2	375 mT	30 W	75 C	10	1.71	6.6
5D	450 mT	30 W	75 C	10	1.77	5.8
L10	230 mT	30 W	50 C	10	1.66	12.7
L11	230 mT	30 W	125 C	10	1.74	7.6
L12	230 mT	30 W	200 C	10	1.72	5.1
L13	230 mT	30 W	250 C	10	1.73	4.4
L14	230 mT	30 W	75 C	4.4	1.73	11.2
L15	230 mT	30 W	75 C	20	1.74	7.3
L16	230 mT	30 W	75 C	50	1.75	4.8
L17	230 mT	30 W	75 C	100	1.74	2.3
L18	230 mT	10 W	75 C	12	1.62	11.9
L19	230 mT	125 W	75 C	10	1.69	4.3
L20	230 mT	300 W	75 C	10	1.70	34.9

Figure 3 shows the deposition rate as a function of the deposition temperature. The deposition rate is seen to decrease with increasing temperature. This trend suggests that increasing temperatures cause the rate of reaction to increase near the gas distribution ring (the point at which the reactant gases enter the chamber), thereby resulting in an increased depletion of the reactant gases flowing over the surface of the samples. The depletion of the reactant gases as a function of distance from the gas distribution ring was noted in early depositions of the boron nitride films. The wafers were positioned at various locations on the deposition platen in order to investigate variations in deposition rate with respect to position. The results showed a decrease in the deposition rate (increased depletion of reactant gases) with increased distance from the gas distribution ring. As the rate of the reaction increases, so will the variation in deposition rate with respect to distance.

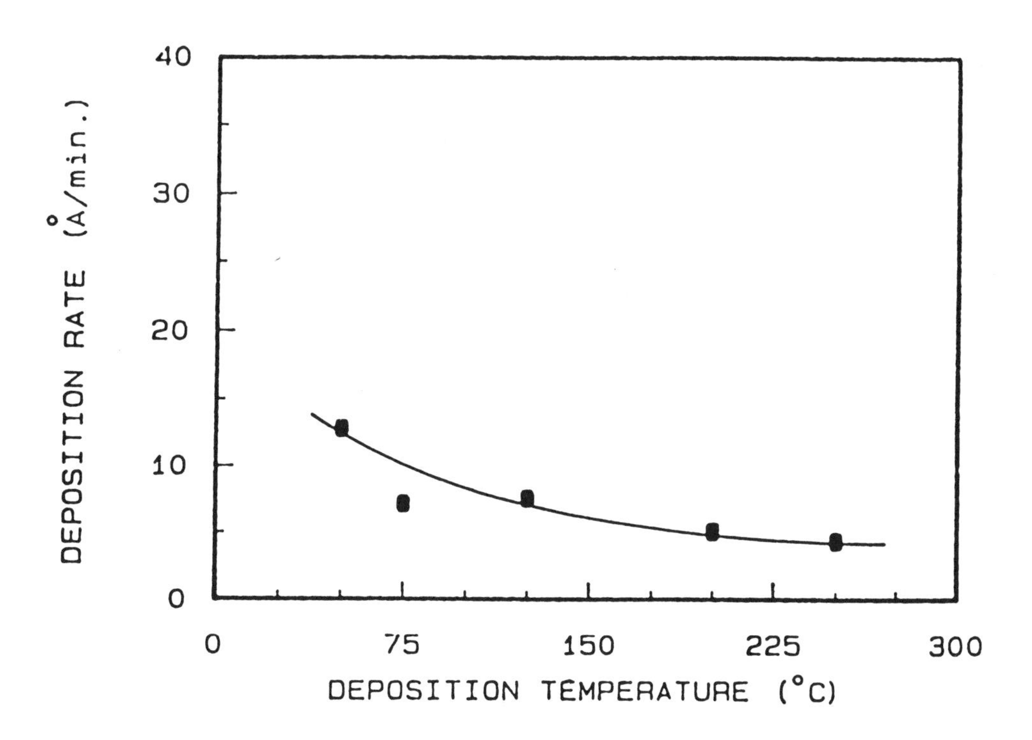

Figure 3 Deposition rate as a function of Deposition Temperature. (P_w = 30 watts, P_r = 230 mTorr, NH_3/B_2H_6 = 10)

Figure 4 shows the deposition rate as a function of R.F. plasma power. At R.F. powers from 10 to 125 watts, the deposition rate decreased from 11.9 to 4.3 Å/min. As the R.F. power increased from 125 watts, the deposition rate increased rapidly. This trend can be explained by the existence of two separate chemical reactions, both resulting in the formation of boron nitride. One reaction dominates at higher power levels and the other dominates at lower power levels as discussed below.

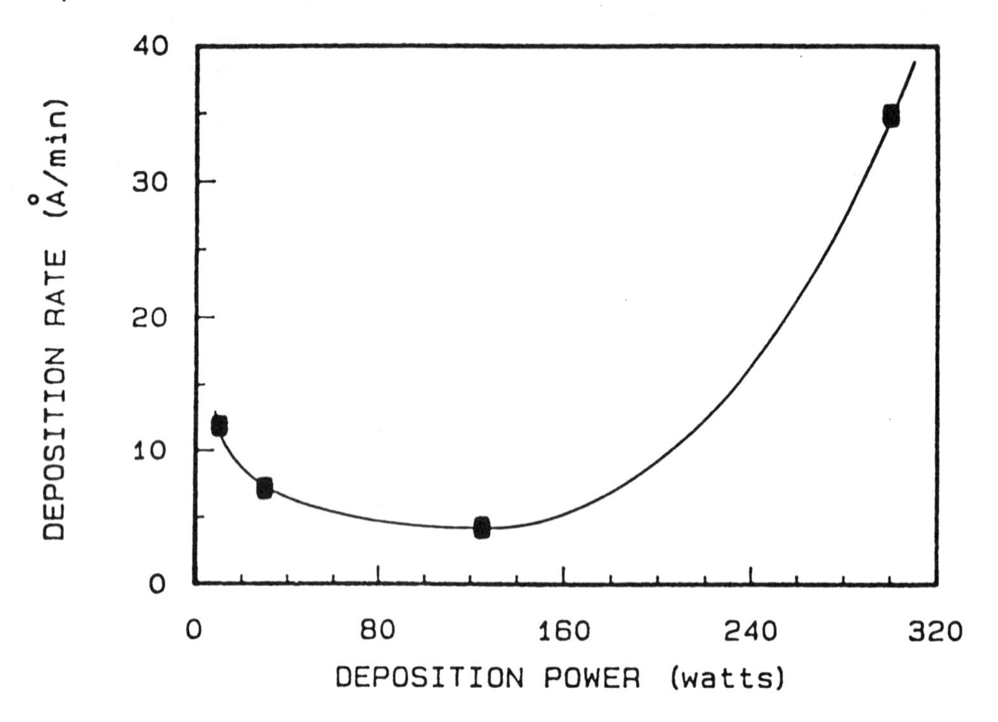

Figure 4 Deposition rate as a function of Deposition Power. (P_r = 230 mTorr, NH_3/B_2H_6 = 10, T = 75°C)

III. CHEMICAL COMPOSITION CHARACTERIZATION

Auger electron spectroscopy in conjunction with argon ion sputtering was used to determine the chemcial composition of BN thin films. We have used a Physical Electronics scanning Auger spectrometer model #545 with a cylindrical mirror analyzer with 0.6% energy resolution. The experiment involved bombardment of the specimen surface with a beam of 1 keV argon ions over a sample area of $3mm^2$ and simultaneously scanning with the primary electron beam of 4.0 μ A at 4 keV over a well defined sample area of 1.3×10^{-5} cm^2. The system argon pressure was 5×10^{-5} Torr whereas the sample chamber base pressure was less than 2×10^{-9} Torr. The argon ion beam parameters were selected to maintain a sputter rate of about 35 A/min for boron nitride insulating thin films[14].

Figure 5 shows the dependence of the atomic percent of boron (B), nitrogen (N) and oxygen (O_B) in the bulk of the film and the atomic percent of oxygen at the interface (O_I) on the deposition pressure. The atomic percent of boron and nitrogen remains relatively constant and very close to that of stoichiometric boron nitride for all values of pressure. The oxygen present in the bulk remained constant at 1.3% for pressures fo 375 mTorr and less, but increased slightly to 4.6% at 450 mTorr. The atomic % of oxygen at the interface also remained relatively constant over all pressures investigated.

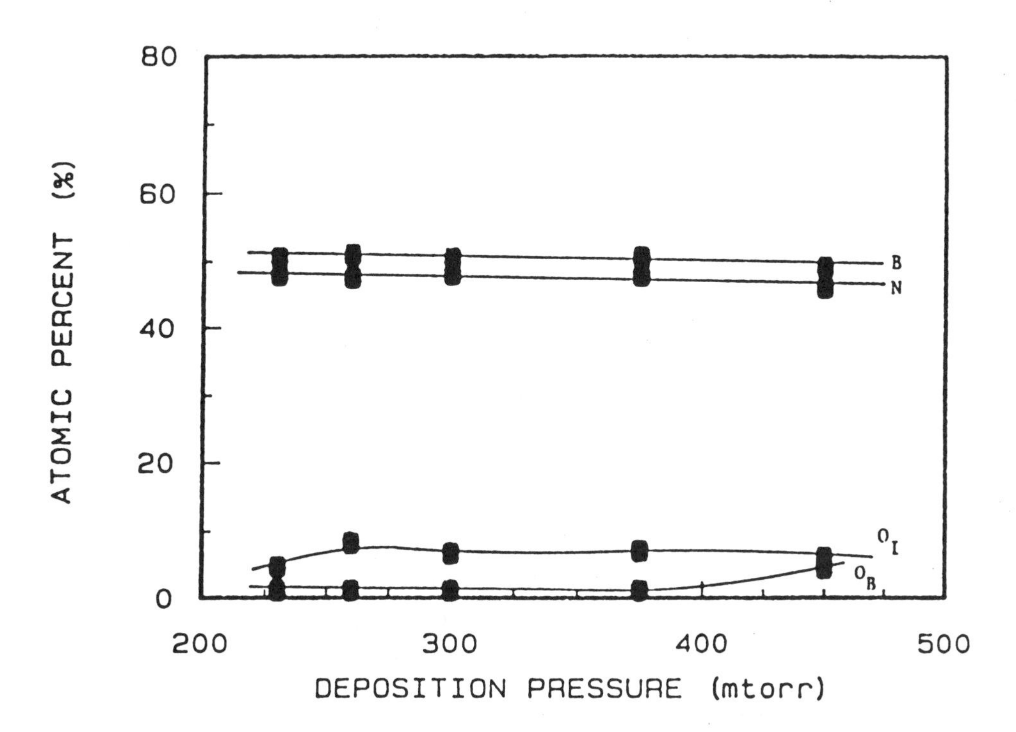

Figure 5 Atomic Percent of Boron (B), Nitrogen (N), Interfacial Oxygen (O_I), and Oxygen in the bulk (O_B) as a function of Deposition Pressure.

(P_w = 30 watts, NH_3/B_2H_6 = 10, T = 75°C)

Figure 6 shows the composition of the film as a function of deposition temperature. The atomic % of boron (B) and nitrogen (N) remain nearly constant and at approximately stoichiometric values. The atomic percent of oxygen in the bulk is approximately constant with temperature.

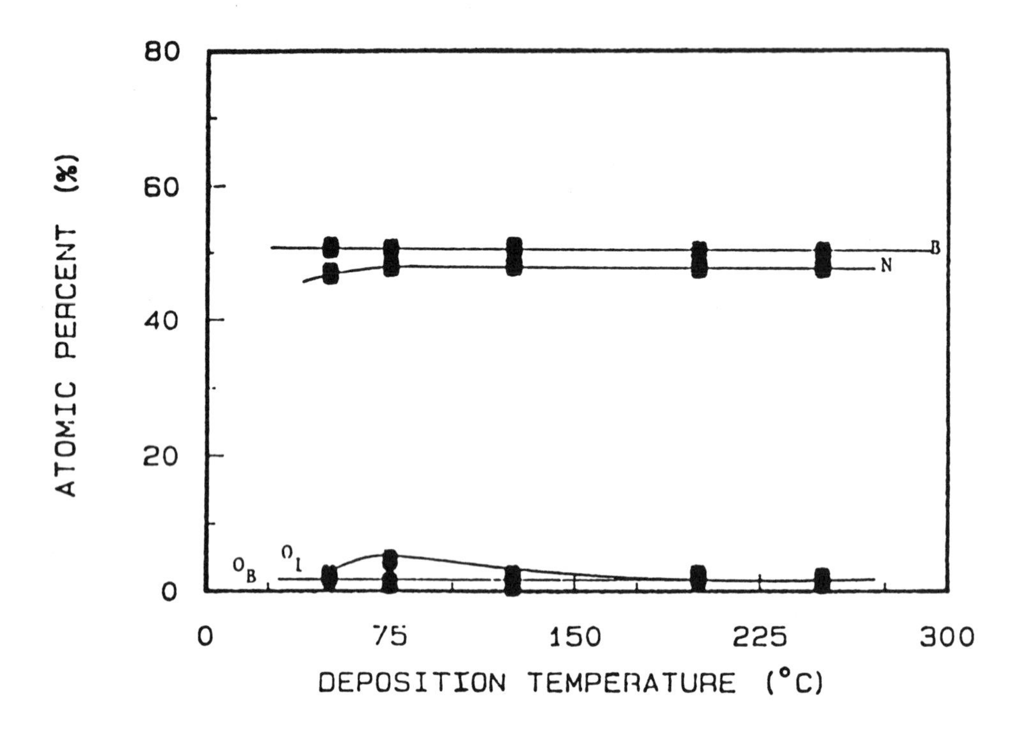

Figure 6 Atomic Percent of Boron (B), Nitrogen (N), Interfacial Oxygen (O_I), and Oxygen in the bulk (O_B) as a function of deposition temperature.
(P_w = 30 watts, P_r = 230 mTorr, NH_3/B_2H_6 = 10)

Figure 7 depicts the composition of the film as a function of NH_3/B_2H_6 gas ratio. Once again the atomic percent of boron (B) and nitrogen (N) remained approximately constant with ratios approaching the stoichiometric value. However, a slight decrease in both the boron and nitrogen concentrations and an increase in the atomic percent oxygen in the bulk is noted at an NH_3/B_2H_6 ratio of 20. The inerfacial oxygen (O_I) concentration increases with increasing NH_3/B_2H_6 ratios up to a ratio of 50. At higher NH_3/B_2H_6 ratios the atomic percent of interfacial oxygen decreases slightly to a value of 4.3% at a NH_3/B_2H_6 ratio of 100.

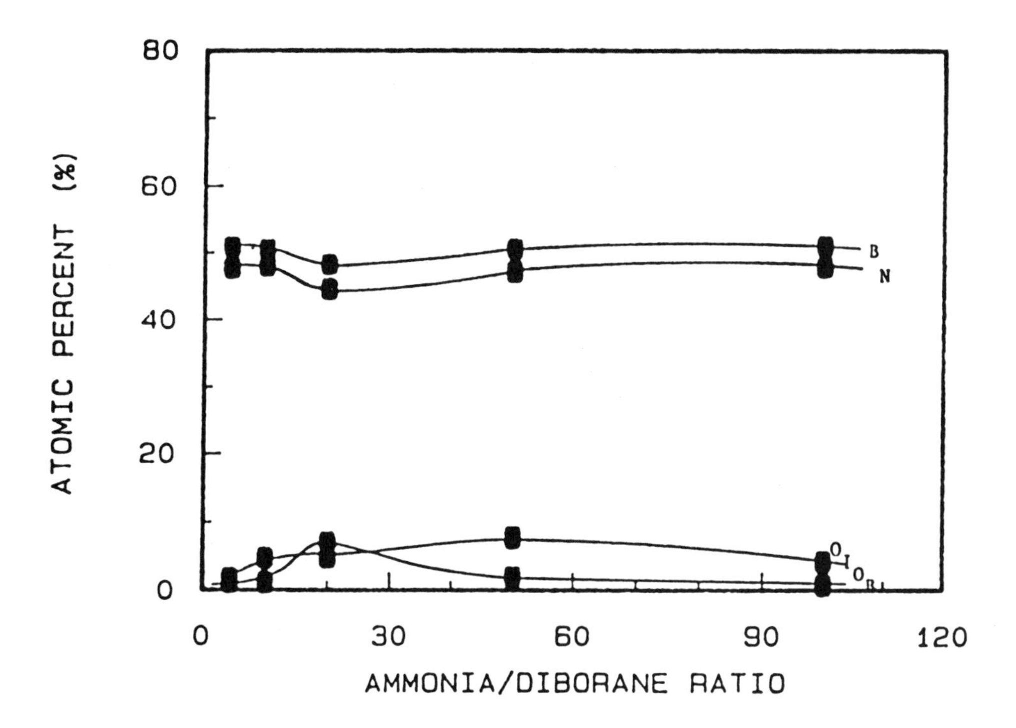

Figure 7 Atomic Percent of Boron (B), Hydrogen (H), Interfacial Oxygen (O_I) and Oxygen in the bulk (O_B) as a function of NH_3/B_2H_6 ratio.

(P_w = 30 watts, P_r = 230 mTorr, T = 75°C)

By far, the processing parameter having the greatest effect on the film composition was the deposition power. Figure 8 shows the atomic percent of boron (B), nitrogen (N), oxygen (O_B) and interfacial oxygen (O_I) as a function of deposition power. At an R.F. power of 10 watts, the atomic percent of nitrogen and boron were at their lowest values and both the atomic percent oxygen and interfacial oxygen were at their highest values. The boron to nitrogen ratio was also the highest value achieved. As the plasma power was increased to 30 watts, the oxygen levels decreased to less than 5%, and the boron to nitrogen ratio approached unity. Increasing the plasma power to 125 watts resulted in a composition similar to that of the 10 watt condition. At a plasma power of 300 watts, the oxygen levels decreased to a relatively low value (<7%), however, the boron to nitrogen ratio decreased to 0.80. This was the only processing variation which resulted in a nitrogen-rich film.

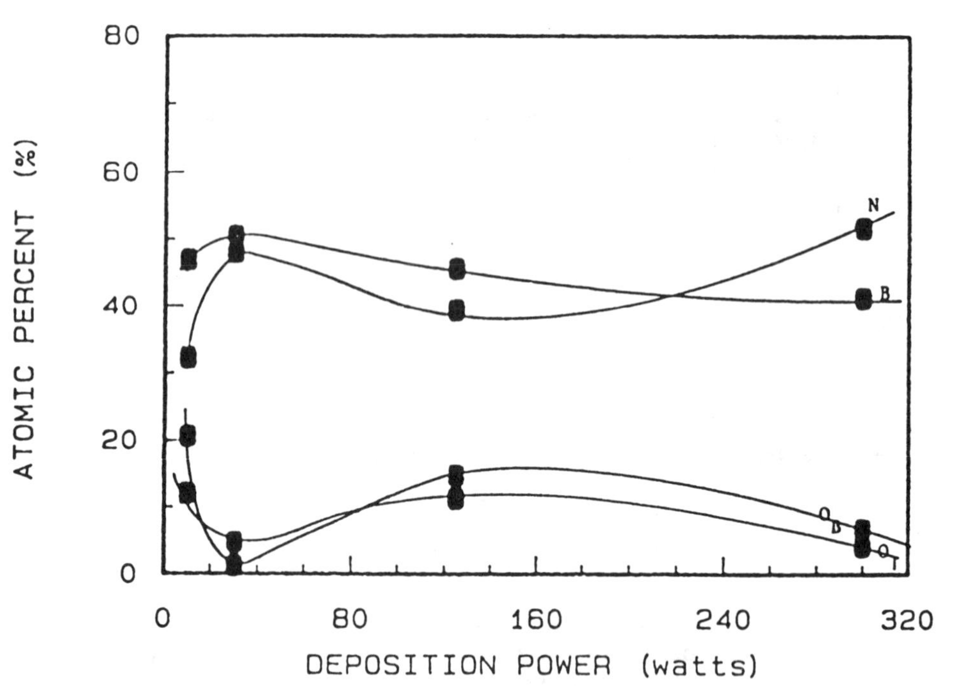

Figure 8 Atomic Percentage of Boron (B), Nitrogen (N), Interfacial Oxygen (O_I) and Oxygen in the Bulk (O_B) as a function of Deposition Power.
(P_r = 230 mTorr, NH_3/B_2H_6 = 10, T = 75°C)

S. P. Murarka et al [6] have described the preparation and properties of chemical vapor deposited boron nitride films. They have reported that the deposition rate of boron nitride films was proportional to the concentration of ammonia to the negative 0.2 power $[NH_3]^{-0.2}$ and to the concentration of diborane $[B_2H_6]$. They also have observed that higher deposition rates resulted in higher concentrations of boron and that in the absence of NH_3, pure boron films were obtained. Considering these observations, they have proposed that during the deposition, two nearly independent reactions occur which result in the simultaneous deposition of boron nitride and boron. The proposed reactions are:

$$B_2H_6 + 2\ NH_3 \rightleftarrows \begin{matrix}\text{complex}\\ \text{intermediate}\end{matrix}\ 2\ BN + 6\ H_2 \qquad (2)$$

and

$$B_2H_6 \rightleftarrows \begin{matrix}\text{complex}\\ \text{intermediate}\end{matrix}\ 2\ B + 3\ H_2 \qquad (3)$$

The deposition rate of pure boron (Equation 3) was assumed higher than that of boron nitride (Equation 2). This postulation was used to describe the decreasing deposition rate and the decreasing boron to nitrogen ratio with increasing NH_3/B_2H_6 ratio. The variation in deposition rate with respect to the NH_3/B_2H_6 ratio in this investigation (see Figure 2) showed a similar dependence. However, the chemical composition with respect to the NH_3/B_2H_6 ratio (see Figure 7) remained fairly constant and did not show the strong dependence of the boron to nitrogen ratio on the NH_3/B_2H_6 ratio found by Murarka. Therefore it is suggested here that the NH_3/B_2H_6 ratio dependence is the result of a boron-limited reaction.

The possible existence of two nearly independent chemical reactions could explain the strong dependence of the R.F. plasma power on both the film composition and deposition rate. The reaction in equation (3) suggests that deposited films would be boron-rich, at least under certain conditions. Chemical vapor deposited boron nitride films with boron to nitrogen ratios as high as 4.0 have been reported. However, boron nitride films deposited in this investigation have resulted in boron to nitrogen ratios near unity and at a high of only 1.45. This is probably the result of the nitrogen carrier gas ionizing and reacting directly with boron due to the decomposition of B_2H_6. With plasma enhanced deposition techniques, nitrogen becomes reactive in the plasma discharge (nitrogen is normally inert at temperatures less than 1000 C--typically CVD techniques utilize temperatures of 600-900 C). One would expect that the reaction of boron and nitrogen (originating from the carrier gas) would dominate at higher power levels since diatomic nitrogen requires a significant amount of energy to dissociate. The deposition performed at 300 wtts resulted in a nitrogen-rich film. This was the only processing variation which resulted in a nitrogen-rich film which further suggests that the reaction of equation (3) dominates at higher power levels. The reaction described in equation (2) is believed to dominate at lower plasma power. The existence of a single reaction at lower power could explain the minor variations in film composition with the other processing variations used.

Infrared absorption is a well-established technique for detecting light elements bonded in transparent solids. However, the sensitivity of this technique is not adequate for analysis of the BN thin films. The MIR technique, developed by Harrick[15] and further refined by other workers[16], makes it possible to increase the sensitivity of infrared absorption and then to measure the vibrational modes for hydrogen bonded to nitrogen (N-H) and for hydrogen bonded to boron (B-H) centers in BN. We have employed MIR spectroscopy to determine the amount of N-H and B-H and O-H bonds in the BN thin films. A boron nitride film of 2000 Å thickness was deposited on the surface of an optically polished GaAs plate of dimensions of 50X15X1 mm^3 with 45 entrance and exit faces. The light is reflected internally through the GaAs plate, with approximately 50 reflections before it leaves the opposite face. Each time the light reflects off the surface where the boron nitride is deposited some absorption occurs in the BN film at the point of reflection. The effect of the large number of reflections is to increase the sensitivity of infrared absorption and allow detection of N-H and B-H vibrational absorption modes in the boron nitride films. The multiple internal reflection spectroscopic measurements were performed at room temperature using a fixed-angle Perkin Elmer Model 0186-0382 MIR accessory and the specially formed MIR plates as discussed above.

The infrared spectrum was obtained from the boron nitride films and three distinct peaks related to the composition of the boron nitride films were identified. The most prominent of these is the narrow peak centered at 3420 cm^{-1} which has been identified as an N-H absorption peak. The broad peak centered at 3300 cm^{-1} and overlapping the N-H peak was an O-H absorption peak. The third peak observed was a B-H absorption peak centered at 2530 cm^{-1}.

Figure 9 shows the absorption coefficients for the hydrogen bonds as a function of deposition pressure. The concentration of N-H, B-H and O-H bonds follows the same trend over the entire range of deposition pressures indicating that the chemical structure of the boron nitride films remains nearly constant while the total hydrogen content varies slightly with deposition pressure. When considering these results in light of the Auger analysis results (Figure 5) which show approximately constant atomic concentrations fo boron, nitrogen and oxygen with deposition pressure, it appears that the deposition pressure has little effect, if any, on the chemical composition of the boron nitride films.

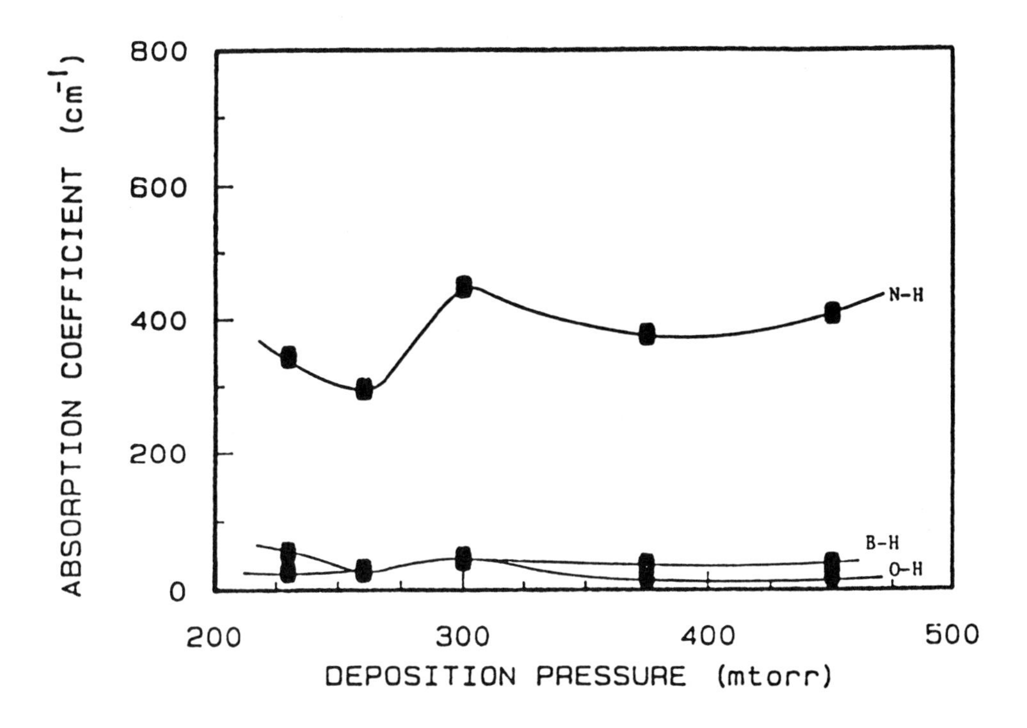

Figure 9 Absorption Coefficient for N-H, B-H and O-H bonds as a function of Deposition Pressure. (P_w = 30 watts, NH_3/B_2H_6 = 10, T = 75°C)

Figure 10 shows the absorption coefficients for hydrogen bonds as a function of the NH_3/B_2H_6 ratio. The sum of B-H and O-H bond concentrations are nearly constant with respect to NH_3/B_2H_6, while the N-H bond concentration varies considerably. This result implies that the boron, oxygen and hydrogen bonding in the film are dependent on one another, but the N-H bond is independent of the film composition and the concentration of N-H bonds is related only to the hydrogen content of the film.

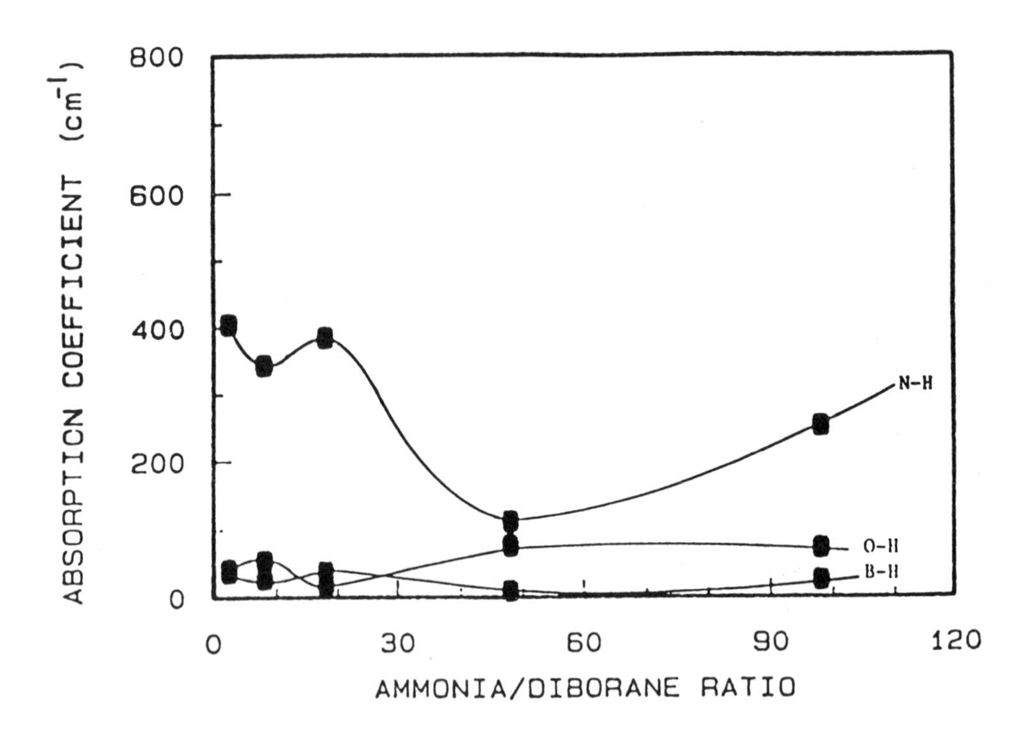

Figure 10 Absorbtion Coefficient for N-H, B-H and O-H bonds as a function of NH_3/B_2H_6 ratio. (P_w = 30 watts, P_r = 230 mTorr, T = 75°C)

Figure 11 shows the absorption coefficients of the B-H, N-H and O-H bonds as a function of deposition temperature. The concentration of B-H bonds remains approximately constant and the concentration of O-H bonds increases slightly with increasing temperature. However, the N-H bond concentration decreases dramatically with increasing temperature. The dependence of the deposition temperature on the N-H bond concentration can be explained as follows: The reaction of diborane and ammonia occurs in seperable steps. At the room temperature, there is an immediate formation of a stable white solid borohydride which is the diammoniate of doborane. The borohydride rearranges to the ring compound borazine ($B_3N_3H_6$) and some polymeric material $(BNH_2)n$ at temperatures in excess of 200 C. As the temperature is increased further, hydrogen is progressively driven off, forming polymeric

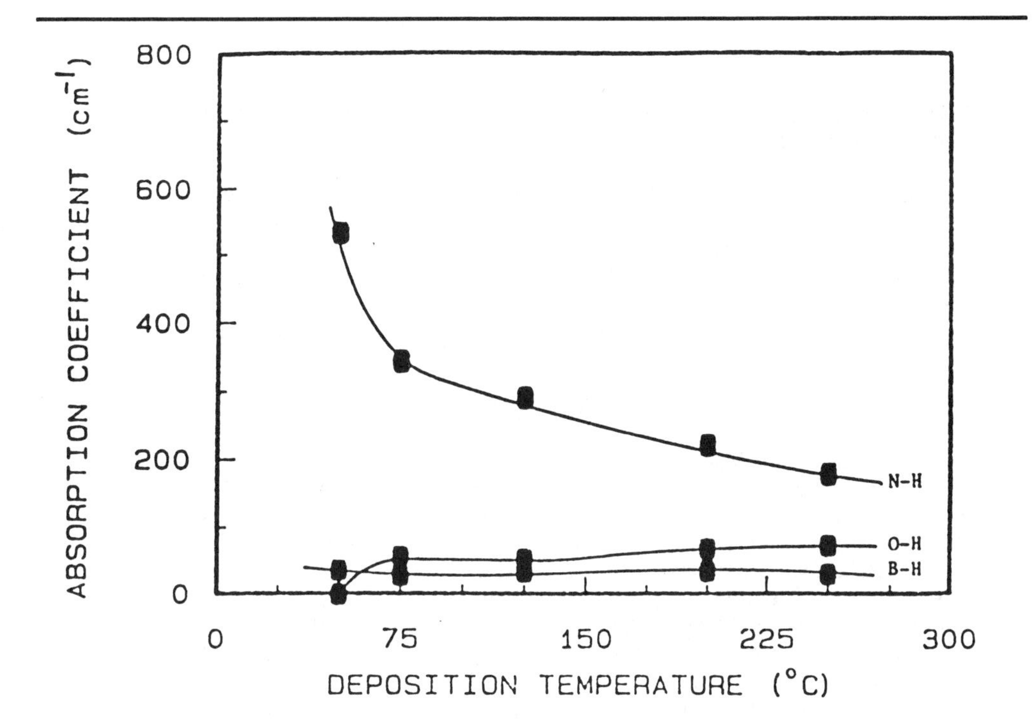

Figure 11 Absorbtion coefficient for N-H, B-H and O-H bonds as a function of Deposition Temperature.

(P_w = 30 watts, P_r = 230 mTorr, NH_3/B_2H_6 = 10)

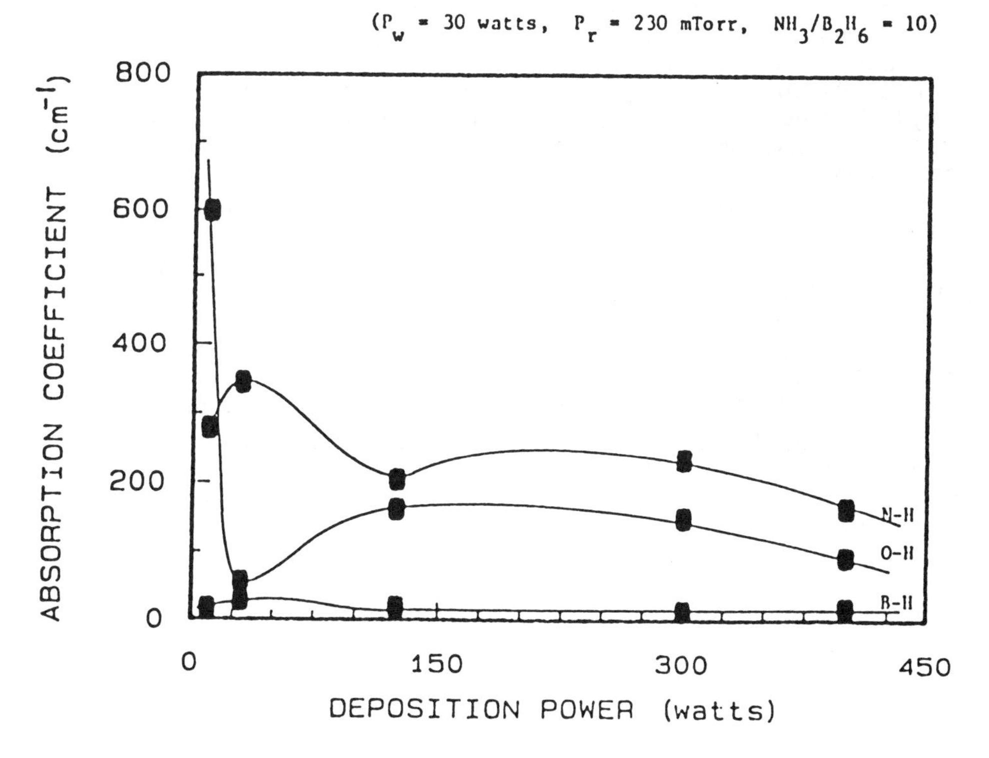

Figure 12 Absorbtion Coefficient for N-H, B-H and O-H as a function of Deposition Power.

(P_r = 230 mTorr, NH_3/B_2H_6 = 10, T = 75°C)

(BNH)n and finally BN. Therefore, one would expect the hydrogen content to decrease with increasing temperature.

The absorption coefficients fo the B-H, N-H and O-H bonds are shown in Figure 12 as a function of deposition power. The B-H concentration remains essentially nearly constant over the entire range of deposition power used. The total hydrogen bond concentration decreased with increasing power similar to the variation observed with increasing temperature. The O-H concentrations in these films were much higher than any of the other values measured in this investigation and varied with respect to the oxygen content of the films.

V. ELECTRICAL CHARACTERIZATION

Capacitance-Voltage (C-V) measurement techniques at 1 MHz were used to investigate the electrical properties of the boron nitride-GaAs substrate interface. The MIS (metal-insulator-semiconductors) structures were used as the vehicle by which the interfacial properties could be analyzed. The Al-boron nitride-GaAs (MIS) devices were prepared on <100>, n-type GaAs substrates. The boron nitride insulating film of thickness 900 Å was chemically vapor deposited on the GaAs substrate as discussed in Section II. An ohmic contact was formed on the back surface of the GaAs wafers before deposition by sequential thermal evaporation of 1000 Å of Au-Ge eutectic alloy, 1000 Å of Ni and 2000 Å of Au, followed by annealing at 400 C in a N_2 atmosphere for three minutes. The procedures for cleaning the substrate surface immediately prior to deposition has been reported previously[13]. The MIS structures for capacitance-voltage (C-V) measurement study were constructed by thermal evaporation of 6000 Å of Al on the dielectric BN film surface.

A typical plot of capacitance vs. voltage obtained for the Al-BN-GaAs structure is shown in Figure 13. The arrows on the curves indicate the direction in which the voltage was varied. The initial trace was obtained for voltages varying from negative to positive. The retrace was obtained for voltage varying from positive to negative and was measured immediately following the completion of the initial trace. The hysteresis voltage, which is the flatband voltage difference between the two traces, is the result of an interfacial charging phenomenon associated with most compound semiconductors.

The flatband voltage and the hysteresis voltage sift (Δv) were determined to be 0.88 votls and 1.1 volts respectively. This hysteresis is typical of majority carriers injection from the GaAs substrate into the boron nitride film and subsequent trapping at/or near the interface or in the bulk of the boron nitride film. In addition, the C-V curve is "stretched out" and true accumulation is not reached. The stretching may be due to a large density fo interface states. It was determined that Δv tended to vary with the atomic percent of oxygen present at the GaAs/BN interfacial region.

The dielectric breakdown of the film was determined to be 8×10^5 V/cm and the resistivity of the film was determined to be 10^{14} ohm-cm.

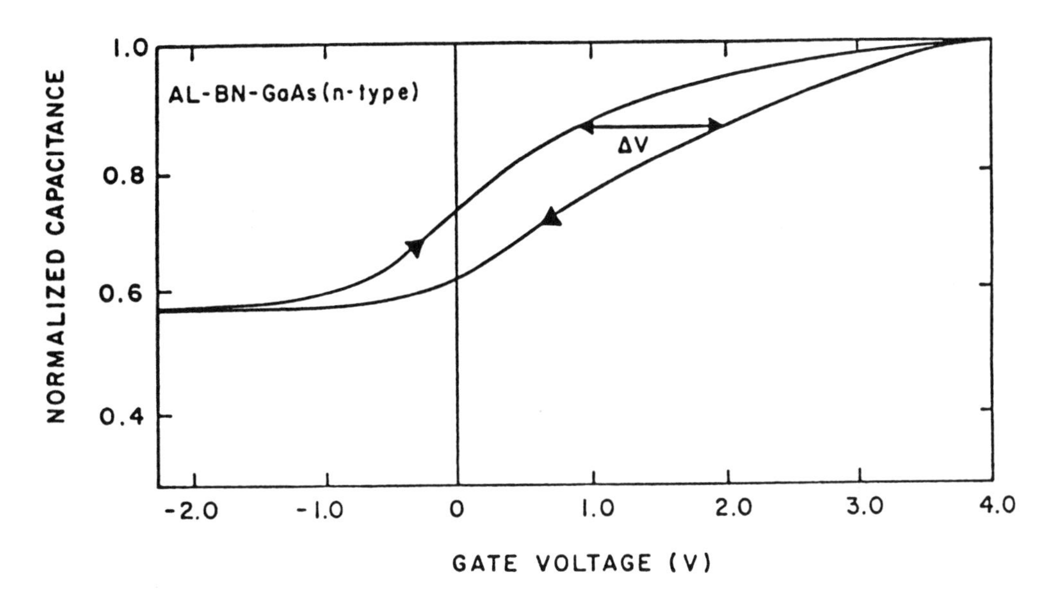

Figure 13 Capacitance-voltage plot for an Al/BN/GaAs structure.

IV. SUMMARY

Plasma enhanced chemical vapor deposition techniques have been employed to deposit boron nitride films by the reaction between ammonia and diborane. Variations in processing parameters resulted in films of various chemical composition. These processing variations were as follows: The R.F. power was varied from 10 to 400 watts, the deposition pressure was varied from 230 mTorr to 450 mTorr, the ammonia to diborane ratio was varied from 4.1:1 to 100:1 and the deposition temperature was varied from 50 to 250 C.

The boron to nitrogen ratio, as determined by Auger electron spectroscopy, varied from 0.8 to 1.45 although the majority of films were approximately stoichiometric. The refractive index and deposition rate, as determined by ellipsometry, varied from 1.62 to 1.77 and from 2.3 to 12.7 Å/min, respectively, over all processing conditions as shown in Table 1.

Multiple internal reflection infrared spectroscopy was used to determine the presence fo B-H, O-H and N-H bonds in the films. The total hydrogen content in the films was observed to decrease with increasing temperatures and increasing deposition power. The higher energy is believed to drive off hydrogen from polymeric $(BNH_2)n$, which is an intermediate step in the overall reaction.

The atomic percentage of oxygen in the bulk was less than 2% for most conditions. The atomic percentage of interfacial oxygen, however, was typically higher (between 5-10%).

The composition of the films and the deposition rate showed the greatest variation with respect to deposition power. The deposition rate decreased as the power increased from 10 to 125 watts and increased quickly with increasing powers above 125 watts. The atomic percent of oxygen, both in the bulk and the interface, increased to their highest values at low deposition power (10 watts). The only nitrogen-rich film was observed at a deposition power of 300 watts.

The current voltage (I-V) measurements and high frequency C-V measurements were performed on the MIS structures. The C-V measurements exhibited a memory or hysteresis type behavior. The flatband voltage measurements varied from 0.0 to 6.7 votls. Hysteresis voltages varied from 0.3 to 3.8 volts as a function of oxygen at the BN-GaAs interface. The lowest hysteresis voltage measured (0.3 volts) was from the sample which had the lowest concentration of oxygen and the lowest concentration of nitrogen. I-V measurements performed on the MIS capacitors exhibited leakage current densities as low as $5x10^{-9}$ Å/cm^2.

An investigation of the effect of an *in situ* plasma surface treatment on the interfacial properties of the boron nitride/gallium arsenide structure would be beneficial. This is of interest as several investigators have had some success in reducing the surface state density when this technique was employed. This is being further investigated.

V. REFERENCES

1) O. Gafri, A. Grill, D. Itzhak, A. Inspektor and R. Avni, Thin Solid Films, 1980, 72, 523.

2) T. H. Yuzuriha and D. W. Hess, Thin Solid Films, 1986, 140, 199.

3) S. Larach and R. E. Shrader, Phys. Rev., 1956, 104, 68.

4) M. J. Rand and J. F. Roberts, J. Electrochem. Soc., 1968, 115, 423.

5) M. Hirayama and K. Shohno, J. Electrochem. Soc., 1975, 122, 1671.

6) S. P. Murarka, C. C. Chang, D. N. K. Wang and T. E. Smith, J. Electrochem. Soc., 1979, 126, 1951.

7) W. Schmoll and H. L. Hatnagal, J. Electrochem. Soc., 1982, 129, 2636.

8) H. Miyamoto, M. Hirose and Y. Osaka, Jpn J. Appl. Phys., 1983, 22, L216.

9) S. B. Hyder and T. O. Yep, J. Electrochem. Soc., 1976, 123, 1721.

10) D. C. Liu, S. Nashino, V. J. Kapoor and J. A. Woolam, Workshop of Dielectric Systems for the III-V Compounds, San Diego, CA June 1982.

11) H. Dun, P. Pan, F. R. White and R. W. Douse, J. Electrochem. Soc., 1981, 128, 1556.

12) B. Bayraktaroglu and R. L. Johnson, J. Appl. Phys., 1981, 52, 3535.

13) G. J. Valco and V. J. Kapoor, J. Electrochem. Soc., 1987, 134, 685.

14) V. J. Kapoor, R. S. Bailey and S. R. Smith, J. Vac. Sci. Technol., 1981, 18(2), 305.

15) N. J. Harrick, Internal Reflection Spectroscopy, Interscience, New York, 1967.

16) V. J. Kapoor, R. S. Bailey and H. J. Stein, J. Vac. Sci. Technol., 1983, A1(2), 600.

Materials Science Forum Vols. 54 & 55 (1990) pp. 375-398

FUNDAMENTAL TRIBOLOGICAL PROPERTIES OF ION-BEAM-DEPOSITED BORON NITRIDE THIN FILMS

K. Miyoshi
National Aeronautics and Space Administration
Lewis Research Center
Cleveland, OH 44135, USA

ABSTRACT

The adhesion, friction, and micromechanical properties of ion-beam-deposited boron nitride (BN) films are reviewed in this chapter. The BN films are examined in contact with BN, metals, and other harder materials. For simplicity of discussion, the tribological properties of concern in the processes are separated into two parts. First, the pull-off force (adhesion) and the shear force required to break the interfacial junctions between contacting surfaces are discussed. The effects of surface films, hardness of metals, and temperature on tribological response with respect to adhesion and friction are considered. The second part deals with the abrasion of the BN films. Elastic, plastic, and fracture behavior of the BN films in solid-state contact are discussed. The scratch technique of determining the critical load needed to fracture interfacial adhesive bonds of BN films deposited on substrates is also addressed.

INTRODUCTION

The technical function of numerous engineering systems such as machines, instruments, and vehicles depends on processes of motion. According to its basic physical definition, the term ''motion'' denotes the change in the position of an object with time. Many processes in nature and technology depend on the motion and dynamic behavior of solids, liquids, and gases.

Tribology is ''the science and technology of interacting surfaces in relative motion and of associated subjects and practices'' [1]. Tribology deals with force transference between surfaces moving relative to each other and includes such subjects as adhesion, friction, wear, and lubrication. Mechanical systems such as bearings, gears, and seals are examples of components involving tribology. The function of tribological research is to bring about a reduction in the adhesion, friction, and wear of mechanical components, to prevent their failure and to provide long, reliable component life through the judicious selection of materials, operating parameters, and lubricants.

Ceramic materials are being used increasingly for machine elements in sliding or rolling contact. These elements include components of advanced engines, such as bearings, seals, and gears, and tools used in metal shaping, such as cutting tools and extrusion dies. The successful use of ceramics in these applications is limited more often by tribological problems than by material properties or processing deficiencies [2–5]. Clearly, there is a great need for a fundamental understanding of the surface interactions of ceramics with ceramics and other materials [6].

Monolithic ceramics continue to be of great interest for structural use in automobile and aerospace engines because of such properties as high-temperature strength, environmental resistance, and low density. However, the severity of operating conditions in engines has not favored the use of bulk ceramics because of their high sensitivity to microscopic flaws and catastrophic fracture behavior. This brittle nature translates into low reliability for ceramic components and, thus, limited application in engines [7–9].

For safety in critical automobile and aerospace applications, the components are required to display markedly improved toughness and noncatastrophic, or graceful, fracture [7–9]. Thus, the enhanced interest in ceramic materials has been further expanded to include high-performance ceramic-coated materials and fiber-reinforced ceramics.

Thin ceramic films are widely used in a variety of applications in which materials in monolithic form are not suitable because of diverse and special requirements. Also, thin ceramic films serve a variety of purposes: providing resistance to abrasion, erosion, corrosion, wear, radiation damage, or high-temperature oxidation; reducing adhesion and friction; and providing lubrication [10–14].

Boron nitride (BN) is a promising ceramic material for use as a high-temperature, wear resistant, hard solid lubricating film as well as a protective insulator film on semiconductors under a variety of environmental conditions [15–18].

Boron nitride can be deposited by a variety of techniques, including ion beam deposition [19,20], low-temperature chemical vapor deposition [21,22], plasma deposition [23,24], radiofrequency glow discharge [25], sputtering [26], borazine pyrolysis [27], and others [28]. The strength and durability of the films depend largely on the interfacial adhesive bond formed between the film and the substrate.

The objective of this chapter is to review the author's research of fundamental tribological properties and micromechanical properties of ion-beam-deposited BN films on metallic and nonmetallic substrates. Some earlier publications on this research are given in references 29 to 31.

First, the surface chemistry, microstructure, and bulk chemical composition of the BN films were investigated. Second, the BN films were brought into contact with BN, metals, and other harder materials to examine the tribological properties of the materials. For simplicity of discussion, the tribological properties of concern in the processes are separated into two parts.

First, the pull-off force (adhesion) and the shear force required to break the interfacial junctions between contacting surfaces are discussed. The effects of surface contaminant films, hardness of metals, and temperature on tribological response with respect to adhesion and friction are considered.

The second part deals with the abrasion of BN films. Elastic, plastic, and fracture behavior of BN films in solid-state contact are discussed. The scratch technique of determining the critical load needed to fracture interfacial adhesive bonds of ceramics deposited on substrates is also addressed.

ION-BEAM-DEPOSITED BORON NITRIDE

Thin films containing BN have been synthesized by using an ion beam extracted from a borazine ($B_3N_3H_6$) plasma [20]. The substrates used were both metallic and nonmetallic materials. The metallic substrates were 440C stainless steel, 304 stainless steel, and titanium (99.97 percent pure). The nonmetallic substrates were silicon, quartz (fused silica, SiO_2), gallium arsenide (GaAs), and indium phosphide (InP) (table I). The BN films on the metallic substrates were approximately 2 μm thick, and those on the nonmetallic substrates were 0.2 μm thick (table I).

Boron Nitride Surface

X-ray photoelectron spectroscopy (XPS) survey spectra of the BN film surfaces on the substrates obtained before sputter cleaning typically revealed a carbon peak as well as an adsorbed oxygen peak, as shown by curves (a) in figure 1 [29,30]. A layer of adsorbate on the surface consisted of water vapor and hydrocarbons from the atmospheric environment that may have condensed and become physically adsorbed to the BN film. After BN film surfaces had been sputter cleaned with argon ions for 45 min, small carbon and oxygen contamination peaks as well as the boron and nitrogen remained (curves (b) in fig. 1).

The B_{1s} photoelectron emission lines of the BN (fig. 1) peaked primarily at 190 eV, which is associated with BN. They also included a small amount of boron carbide (B_4C). The N_{1s} photoelectron lines for the BN peaked primarily at 397.9 eV, which is again associated with BN. The C_{1s} photoelectron lines taken from the as-received surface at 284.6 eV indicate the presence of adventitious adsorbed carbon contamination with a small amount of carbides. After sputter cleaning, the adsorbed carbon contamination peak disappeared from the spectrum, and the relatively small carbide peak could be seen. The O_{1s} photoelectron lines of the as-received BN surface peaked at 531.6 eV because of adsorbed oxygen contamination and oxides. After sputter cleaning, the adsorbed oxygen contamination peak disappeared from the spectrum, but the small oxide peak remained.

TABLE I.—NONMETALLIC SUBSTRATES AND BN FILM THICKNESS

(a) Nonmetallic substrates

Material	Doping		Resistivity, Ω-cm	Sliding		Primary cleavage plane
	Dopant	Concentration, cm^{-3}		Plane	Direction	
Si	---	-----	0.01 to 0.02	{100}	<110>	{111}
SiO_2 (quartz)	---	-----	---------	---	---	---
p-type GaAs	Zn	3×10^{17}	0.12	{100}	<110>	{110}
n-type GaAs	Te	3×10^{17}	---------	↓	↓	↓
InP	---	-----	---------			
n-type InP	Sn	3×10^{17}	0.010			

(b) BN film thickness for nonmetallic substrates

Substrate and film	BN film thickness,[a] nm
Si coated with BN at 200 °C	169.7 ± 4.4
Si coated with BN at 350 °C	118.3 ± 0.7
SiO_2 coated with BN at 200 °C	---------
SiO_2 coated with BN at 350 °C	---------
Zn-doped, *p*-type GaAs coated with BN	---------
Te-doped, *n*-type GaAs coated with BN	129.0
InP coated with BN	---------
Sn-doped, *n*-type InP coated with BN	116.0 ± 0.5

[a]Film thicknesses were determined by using a rotating analyzer ellipsometer with a helium-neon laser, a mercury arc lamp, or a xenon arc lamp as light sources.

The peak intensity for both boron and nitrogen associated with BN increased with argon ion sputter cleaning; that for carbon and oxygen decreased markedly. The BN film deposited on the 440C substrate was nonstoichiometric, with a boron-to-nitrogen (B/N) ratio of 1.6.

Microstructure and Bulk Composition

The ion-beam-deposited BN films were hard and semitransparent. For example, figure 2 presents a microstructural appearance of a 304 stainless steel substrate coated with the 2-μm-thick BN film under an optical microscope. The surface hardness of the ion-beam-deposited BN films is presented later in the section Micromechanical Properties.

A typical transmission electron photomicrograph (fig. 3(a)) indicates that the BN film lacked the macroscopic structural features common in both the monolithic cubic and hexagonal forms of BN [30]. In the absence of macroscopic crystallinity neither grains, grain boundaries, grain orientations, nor additional phases were found to exist. The electron diffraction pattern of the film (fig. 3(b)) indicates a diffused ring structure containing three rings that were identified as hexagonal (BN)4H. This evidence suggests the presence of, in addition to a structureless amorphous phase, a

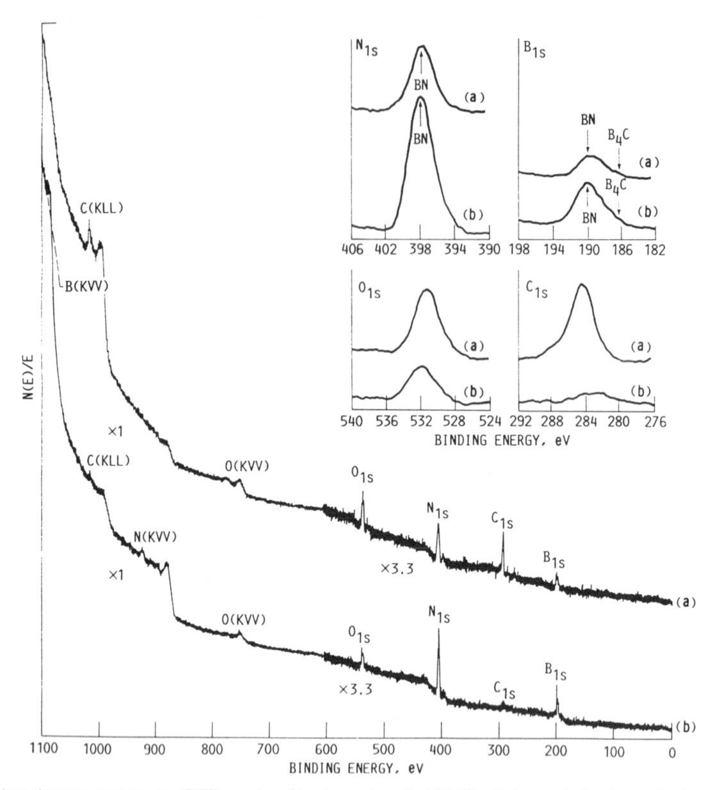

Figure 1.—X-ray photoelectron spectroscopy (XPS) spectra of ion-beam-deposited BN film before and after ion sputtering, curves (a) and (b), respectively (sputtering time, 45 min; substrate, 440C stainless steel).

Figure 2.—Microstructural appearance of 304 stainless steel coated with ion-beam-deposited BN films as observed through BN film under optical microscope.

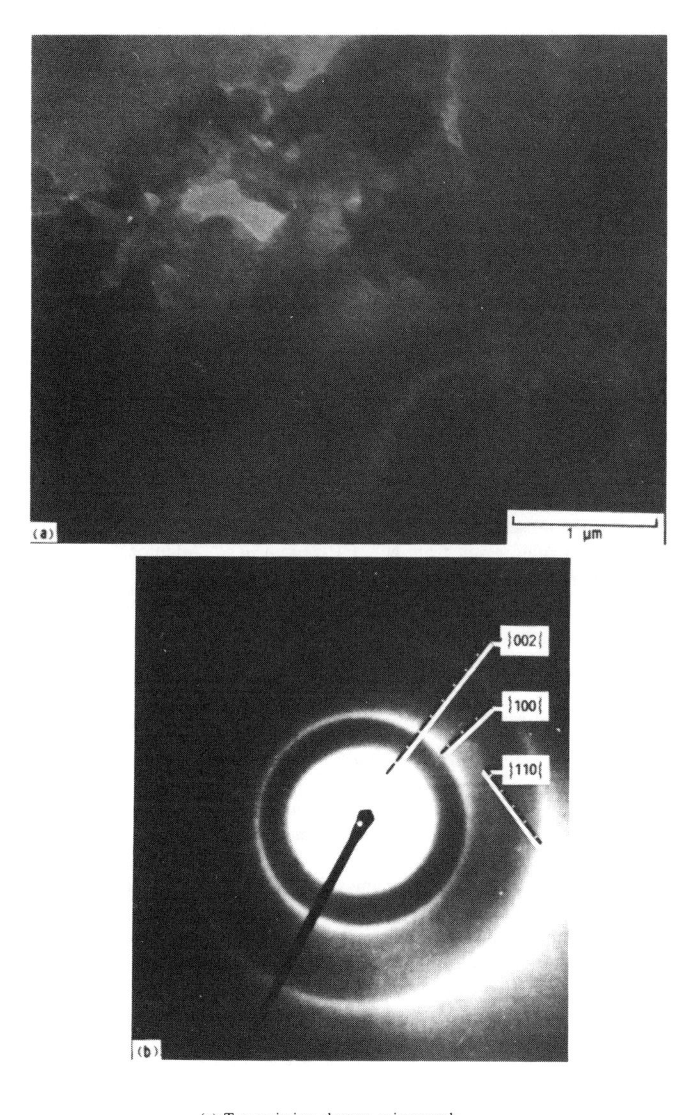

(a) Transmission electron micrograph.
(b) Electron diffraction pattern.

Figure 3.—Typical microstructure and electron diffraction pattern of BN coating film.

crystalline phase with a size range of 8 to 30 nm (i.e., the films were not completely amorphous; properties characteristic of hexagonal BN were detected). This finding agrees with electron microscopical and optical band gap observations [17,18].

Elemental depth profiles for the ion-beam-deposited BN film on silicon were obtained as a function of the sputtering time from Auger electron spectroscopy (AES) analyses by Pouch (fig. 4) [18]. The height of the boron and nitrogen peaks rapidly increased with an increase in the sputtering time. The oxygen and carbon peaks, however, decreased in the first 1 to 2 min and remained constant thereafter. The BN film deposited on silicon had a B/N ratio of about 2. Thus XPS and AES analyses clearly revealed that BN was nonstoichiometric and that small amounts of oxides and carbides were present on the surface and in the bulk of the BN film. Contaminants such as carbides (e.g., B_4C) and oxides may be introduced and absorbed in the BN film during ion beam deposition.

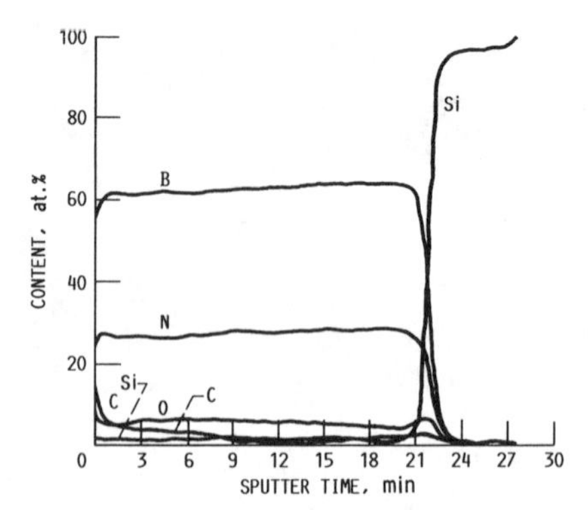

Figure 4.—Auger electron spectroscopy (AES) depth profile; atomic percent as function of sputter time for ion-beam-deposited BN film on Si (deposition temperature, 200 °C; ion beam energy, 150 eV).

The BN films deposited on the silicon substrates were probed by using secondary ion mass spectroscopy by Pouch [18]. The spectra indicated the presence of the following secondary ions: B^+, B_2^+, C^+, O^+, Si^+, Si_2^+, and SiO^+. The peak observed at 14 amu could result from N^+, CH_2^+, and SiO_2^+. Additional peaks at 24 and 25 amu were related to BN^+ and thus supported the XPS data. The SiO^+ signal was associated with the oxide present at the BN-to-silicon interface.

ADHESION AND FRICTION

Clean and Contaminated Surfaces

The surfaces of ceramics and metals usually contain, in addition to the constituent atoms, adsorbed films of water vapor or hydrocarbons that may have condensed from the environment.

In a vacuum environment, sputtering with rare gas ions or heating surfaces to very high temperatures can remove contaminants that are adsorbed on the surface of ceramics and metals. Removing adsorbed films from these surfaces results in very strong interfacial adhesion when two such solids are brought into contact. Typical adhesion results from ion-beam-deposited BN film in contact with a monolithic magnesia-doped silicon nitride (Si_3N_4) are presented in figure 5. The Si_3N_4 pins were sputter cleaned with argon ions before adhesion experiments. The surfaces of BN films were in two states: as-received and heat cleaned at 700 °C. The strength of adhesion is expressed as the force

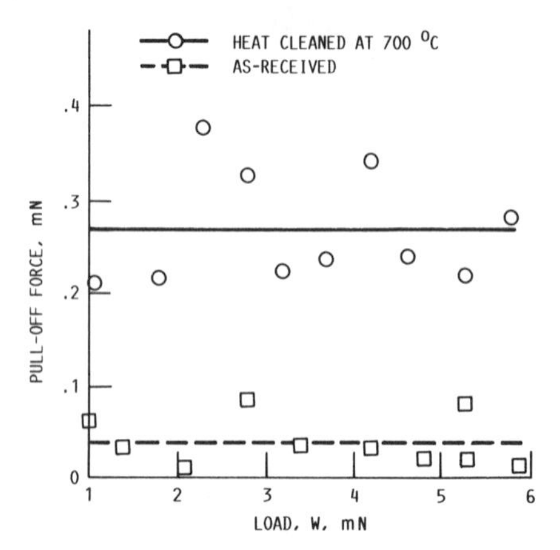

Figure 5.—Pull-off force (adhesion) as function of load for ion-beam-deposited BN films in contact with hemispherical monolithic Si_3N_4 pins at room temperature in vacuum.

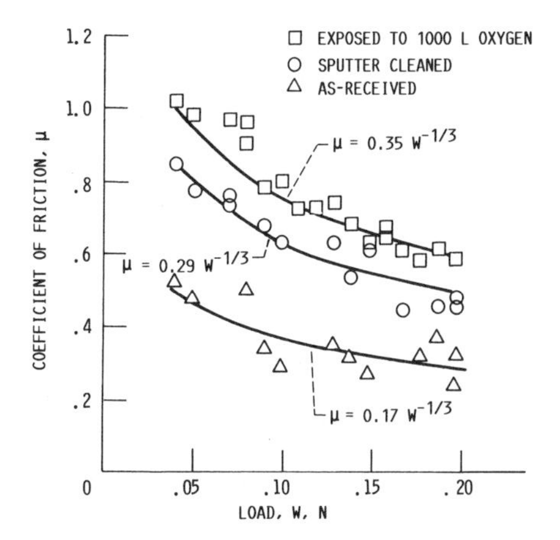

Figure 6.—Coefficient of friction as function of load for hemispherical BN pins in sliding contact with BN flats (single-pass sliding; sliding velocity, 3 mm/min; vacuum, 30 nPa; room temperature; L = 1×10^{-6} torr•sec).

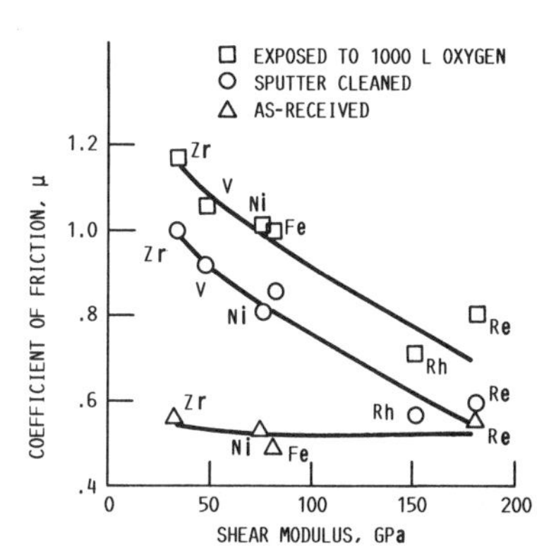

Figure 7.—Coefficient of friction as function of shear modulus of metals in contact with BN flats (single-pass sliding; sliding velocity, 3 mm/min; vacuum, 30 nPa; room temperature, L = 1×10^{-6} torr•sec).

necessary to pull the surfaces apart; it is called the pull-off force [32]. The plots indicate no significant change in the pull-off force with respect to load over the light load range of 1 to 6 mN. The data, however, indicate clearly that the contaminants on the BN surfaces greatly reduce the extent of adhesion (by about a factor of 7).

Figures 6 and 7 present typical plots of the coefficients of friction for the ion-beam-deposited BN films in contact with BN and with metals, respectively. The surfaces of BN films and metals were in three states: as received, argon ion sputter cleaned, and exposed to 1000 L (L = 1×10^{-6} torr•sec) of oxygen with an oxygen pressusre of 1×10^{-6} torr after sputter cleaning. The coefficients of friction are strongly affected by contaminants adsorbed on the as-received surfaces. The coefficients of friction for the sputter-cleaned surfaces are higher than those for the as-received surfaces. In other words, the presence of the contaminants on the as-received surface of the BN film reduces the adhesion and shear strength of the contact area.

In contrast, oxygen exposures to clean BN and metal surfaces strengthen the BN-to-BN adhesion and metal-to-BN adhesion (figs. 6 and 7). It is known that exposing both metal and ceramic surfaces to oxygen under carefully controlled conditions, after sputtering with argon ions or heating in vacuum, results in the adsorption of oxygen, which produces the following two effects: (1) The metal and ceramic oxidize and form an oxide surface layer, and (2) the oxide layers increase the shear strength of the contact and the coefficients of friction [33,34]. In these cases strong oxide-to-oxide bonding takes place at the interfaces, thereby raising the shear strength and the coefficient of friction.

Figure 6 also indicates that adhesion and friction for BN-to-BN contacts decrease as the load increases under all three surface conditions. Further, figure 7 indicates that adhesion and friction for sputter-cleaned metal-to-BN couples are smaller for metals with a large shear modulus. These subjects are discussed in somewhat greater detail in the following sections.

Area of Contact

Boron nitride films behave elastically up to a certain contact pressure. For example, when BN coated on a 440C bearing stainless steel flat is placed in contact with BN on a 440C stainless steel pin in a vacuum, the coefficient of friction is not constant. It decreases as the load increases, as shown in figure 6. To a first approximation for the load range investigated, the relation between coefficient of friction μ and load W on logarithmic coordinates is given by the expression

$$\mu = kW^{-1/3} \tag{1}$$

The exponent arises from an adhesion mechanism for the surfaces in solid-state contact. The area of elastic contact can be determined by the elastic deformation [35]. The friction is found to be a function of the shear strength of this elastic contact area. Similar contact and friction characteristics for diamond on diamond and diamond on silicon carbide were also found [35,36].

The data of figure 7, for surfaces sputter-cleaned with argon ions, indicate a decrease in coefficient of friction with an increase in shear modulus of the metals. Note that the magnitude of shear modulus, like Young's modulus, is generally dependent on the electron configuration of the metal [37]; its maximum value in a given period of the periodic table corresponds to the metal having the maximum number of unpaired *d* electrons. The minimum near the end of each period occurs for the element that has an s^2p^1 configuration. Shear and Young's moduli are also related to such physical, chemical, and mechanical properties as surface and cohesive energy, chemical stability, and tensile and shear strength [37–41]. Further, all the slidings in this investigation involve adhesion at the contact area between the metal and BN.

The morphology revealed that all the BN surfaces contacted by the metals contained transferred films of metal. On separation of the BN and metal in sliding contact, both the interfacial adhesive bonds between the metal and BN and the cohesive bonds in the metal were broken. In other words, the shear forces that broke both the interfacial adhesive bonds and the cohesive bonds in the metal were primarily responsible for the frictional force. The examined metal failed in shear or tension at some of the real areas of contact where the interfacial bonds were stronger than the cohesive bonds in the metal. Metals that have a low shear modulus exhibit larger areas of metal transfer than those with a high modulus. Such dependence of friction and metal transfer on shear modulus may arise from surface and cohesive energy as well as from ductility of the metals; therefore, it is interesting to compare the foregoing friction results, for metal-to-BN contacts, with surface energy and hardness of metals.

Figure 8 presents surface energy values at room temperature from recommended values suggested by Tyson and Miedema [38,40]. The surface energy for the metals increases with an increase in shear modulus. Note that Miedema [40] estimated the values at room temperature from values of the experimental surface energy and entropy by using temperature dependence factors. The surface energy is correlated with such thermochemical parameters for metals as electron density, electron negativity, and heat sublimation [39,40]. The calculated ideal shear strengths of the metals were also correlated with shear modulus [41]—the higher the shear modulus, the greater the shear strength.

The adhesion and friction of the metal-to-BN contacts were expected to increase as the surface energy (i.e., the bond energy of metals) increased. But figures 7 and 8 show that the adhesion and friction go in the opposite direction; they decreased with an increase in surface energy of metals. In other words, the friction was reduced with an increase in shear modulus, whereas the surface energy increased as the modulus value increased. Presumably, the ductility of metals, that is, the deformation of metals, has not been considered here [42,43].

Because of the marked difference in elastic and plastic deformation of ceramics and softer metals, solid-state contact between the two materials can result in considerable plastic deformation of the softer metal. This deformation can contribute to the adhesion and friction of the materials because it increases the real contact area. To gain an understanding of interface deformation under the action of a friction force, indentation experiments were conducted with the metal pin specimens. The hardness data (fig. 9(a)) indicate that at room temperature the Vickers hardness of metals increases as the shear modulus increases. Figure 9(b) presents areas of contact, calculated from the experimental data presented in figure 9(a), as a function of shear modulus of the metal. The area of contact was determined by the ratio of normal load to hardness. The calculated area of contact is very strongly dependent on the shear modulus of the metal; it decreases with increasing shear modulus. The decrease in the area of contact with an increase of shear modulus for these metals is greater than the corresponding increase in the surface energy (bond energy) with shear modulus. (See figs. 8 and 9(b).) Consequently, the shear force required to move the metal pin in a direction parallel to the surface of BN decreases with increasing shear modulus. This fact is consistent with the results shown in figure 7. Thus, such mechanical factors as hardness are of great importance [42,43]. During ceramic-to-metal contact, strong bonds form between the materials. These interfacial bonds are stronger than the cohesive bonds in the metal (as evidenced by the transferred metal) at the major part of the real area of contact. Hardness of metals plays an important role in adhesion and friction and exceeds that of the surface energy.

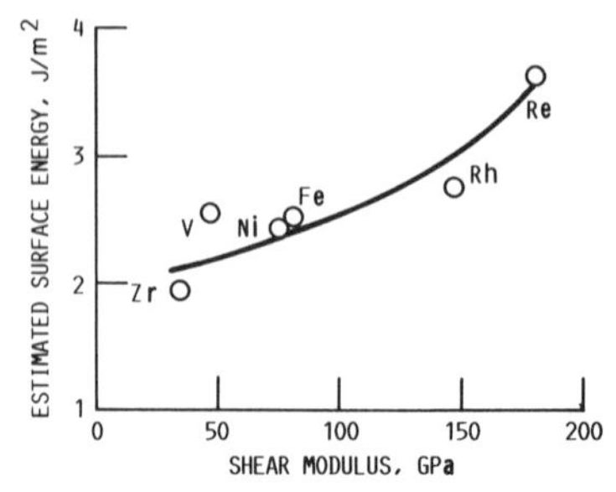

Figure 8.—Estimated surface energy of metal as function of shear modulus.

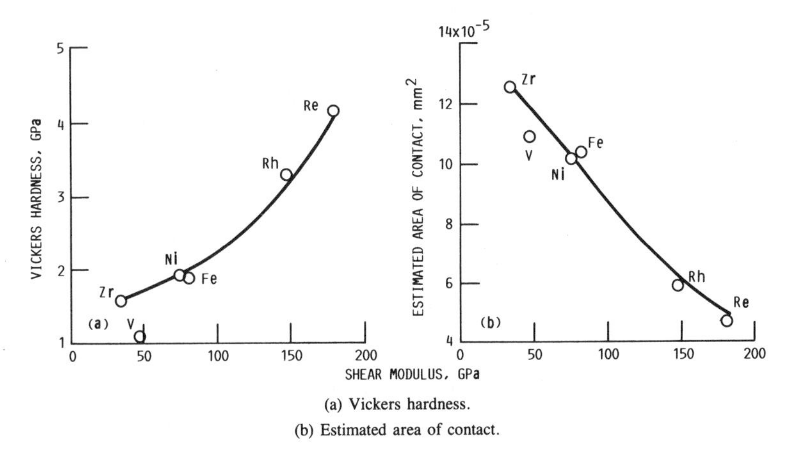

(a) Vickers hardness.

(b) Estimated area of contact.

Figure 9.—Vickers hardness of metals and estimated area of contact as functions of shear modulus.

Temperature Effects

An increase in surface temperature of a ceramic material tends to promote surface chemical reactions. These chemical reactions cause products to appear on the surface which can alter tribological behavior [44].

Figure 10 presents the average pull-off forces (adhesion) for the BN films at 700 °C in vacuum as a function of temperature. The BN films were heat cleaned at 700 °C in vacuum, and the monolithic Si_3N_4 pins were sputter cleaned with argon ions before adhesion experiments.

Although the pull-off force for the BN films increased slightly with temperatures up to 400 °C, it generally remained low at these temperatures. The pull-off force increased significantly at 500 °C and remained high, in the range of 500 to 700 °C. Strong adhesive bonding resulting from a strong chemical reaction between the BN film and monolithic Si_3N_4 can take place at the contacting interface at temperatures in the range of 500 to 700 °C.

In general, the behavior of pull-off force (adhesion) is primarily related to that of static and dynamic friction in vacuum [45].

Comparison of Coefficients of Friction

Figure 11 summarizes friction properties for ceramic-to-ceramic, metal-to-metal, and metal-to-ceramic contacts in vacuum environments. In addition to the BN coating, ceramics of concern include SiC, Si_3N_4, SiO_2, and ferrite

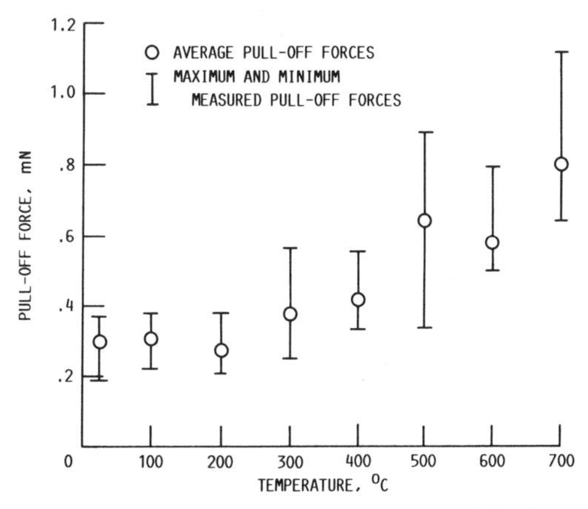

Figure 10.—Pull-off force (adhesion) as function of temperature for ion-beam-deposited BN films in contact with hemispherical monolithic Si_3N_4 pins in vacuum.

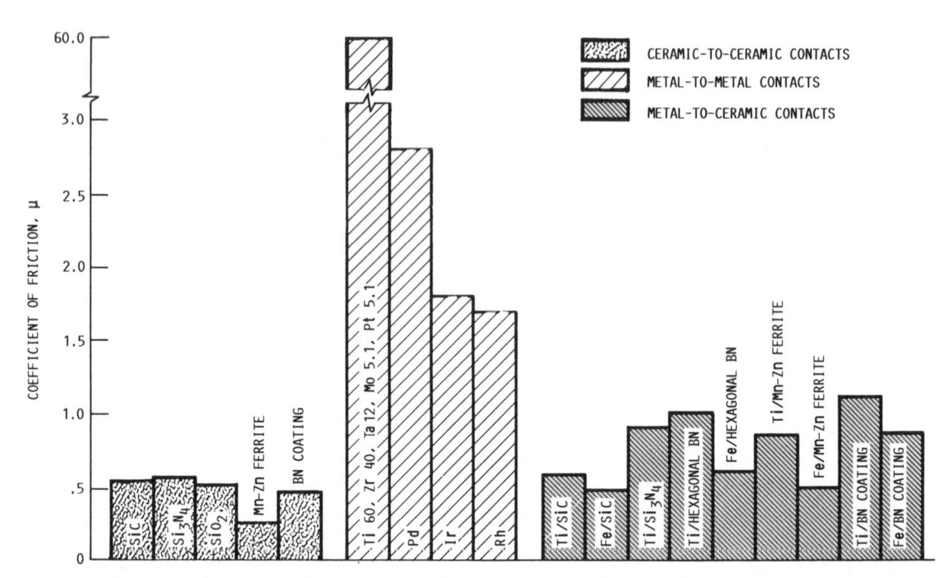

Figure 11.—Coefficients of friction for clean solid-to-solid interfaces (single-pass sliding; vacuum, 30 nPa).

in bulk form. Metals of concern include titanium, zirconium, tantalum, molybdenum, platinum, palladium, iridium, rhodium, and iron.

The data presented in figure 11 indicate the marked difference in friction for the combination of clean solids. The coefficients of friction due to adhesive bonding for ceramic-to-ceramic and metal-to-ceramic contacts were much lower than those for metal-to-metal contacts. The coefficients of friction for metal-to-metal contacts were extremely high. In general, the coefficient of friction for BN-to-BN contact can fit into the category of ceramic-to-ceramic contact. The coefficient of friction for metal-to-BN contact can fit into the category of metal-to-ceramic contact.

Figure 11 suggests that the use of ceramic materials, including bulk and coatinglike forms, in vacuum, spacelike environments is beneficial from tribological considerations.

MICROMECHANICAL PROPERTIES

Ceramics, in both coating and monolithic form, behave micromechanically in a ductile fashion to a certain contact stress when they are brought into contact with ceramics or other solids. Even at room temperature, ceramics such as BN and SiC behave elastically and plastically at low stresses under relatively modest conditions of rubbing contact; however, they microfracture under more highly concentrated contact stresses [30,31,36,46–50]. This microfracture, known as brittle fracture, is one of the most critical characteristics of a ceramic that must be considered in design for structural and tribological applications.

Single-pass scratch experiments were conducted to examine the deformation and fracture behavior of BN films sliding against a 0.2-mm-radius diamond pin in air at a relative humidity of 45 percent and at room temperature [30,31]. The BN films were deposited on both metallic and nonmetallic substrates.

Elasticity

When a BN film surface is brought into sliding contact with a diamond under relatively small load (<2 N), elastic deformation can occur locally in both the BN film and the diamond. Sliding occurs at the interface. The coefficient of friction was not constant and decreased as the load increased. The friction characteristics were similar to those for BN-to-BN contacts presented earlier in figure 6. In these experiments no permanent groove formation due to plastic flow and no cracking of the BN film with sliding were observed [31].

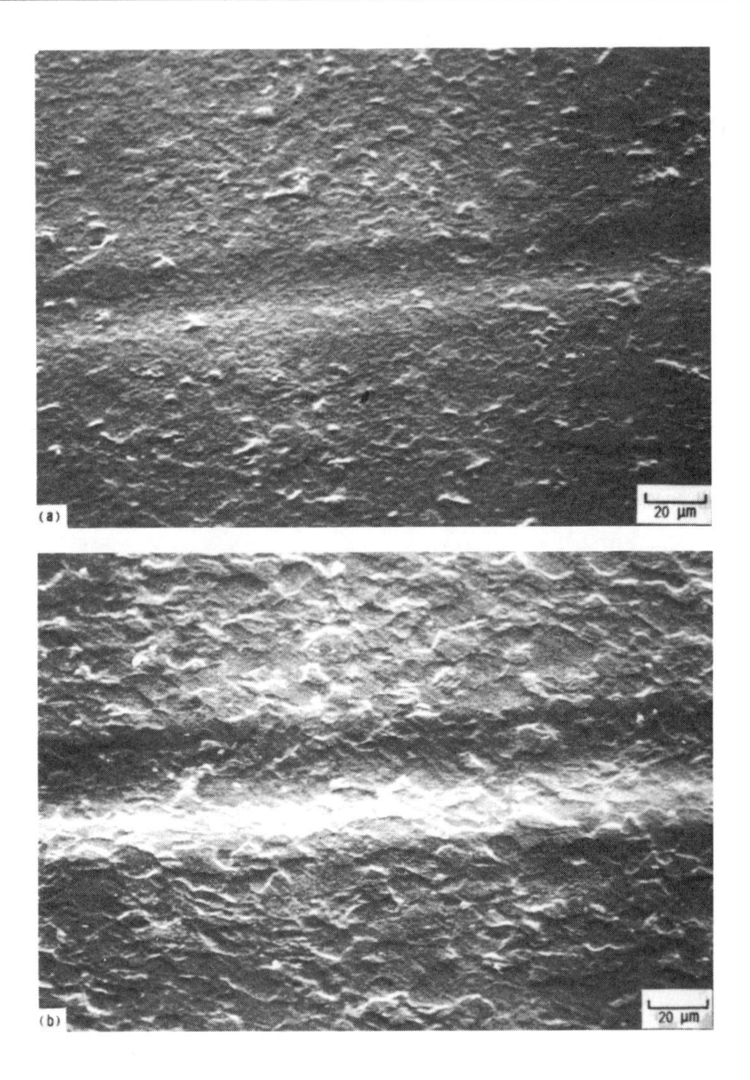

(a) 440C stainless steel substrate.
(b) Titanium substrate.

Figure 12.—Scanning electron photomicrographs of wear tracks on BN film surfaces generated by 0.2-mm-radius diamond pins (load, 4N; sliding velocity, 12 mm/min; laboratory air).

Plasticity

An increase (of a few newtons) in load results in plastic deformation of the BN film, as shown in figure 12. The spherical diamond indented and slid on the BN film without suffering permanent deformation, but it caused permanent grooves in the BN film during sliding. The BN film deformed plastically much like metallic films. Scratch measurements were therefore conducted with BN films deposited on the metallic and nonmetallic substrates, starting from the smallest loads at which the scratches were visible by optical and scanning electron microscopy and detectable by surface profilometry.

The width D and height H of a groove with some amount of deformed BN piled up along its sides are defined in figure 13. The widths of the grooves reported herein were obtained by averaging the widths from 10 or more measurements of surface profile traces, and by optical or scanning electron microscope examinations.

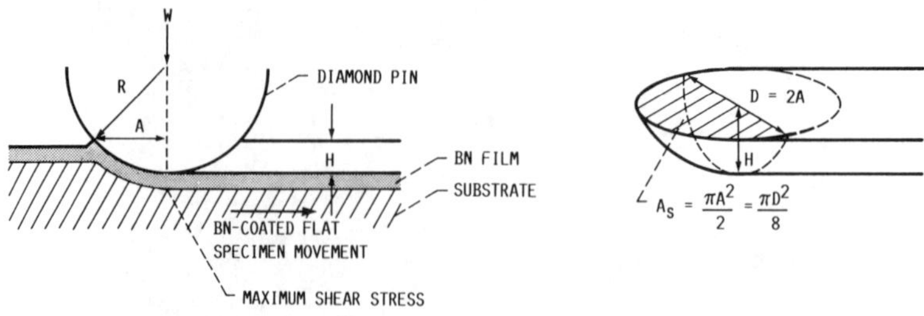

Figure 13.—Definition of mean contact pressure ($P = W/A_s$).

The relationship between the load and the width of the resulting scratch may be expressed with a number of empirical relations [51]. When the average scratch width D for BN films is plotted as a function of the load W on logarithmic coordinates, this is expressed as $W = kD^n$ (i.e., Meyer's law, as typically presented in figures 14 and 15).

Figures 14 and 15 present data for scratch widths obtained for BN films on 440C stainless steel, 304 stainless steel, and titanium substrates and on silicon, SiO_2, GaAs, and InP substrates, respectively. Comparative data for

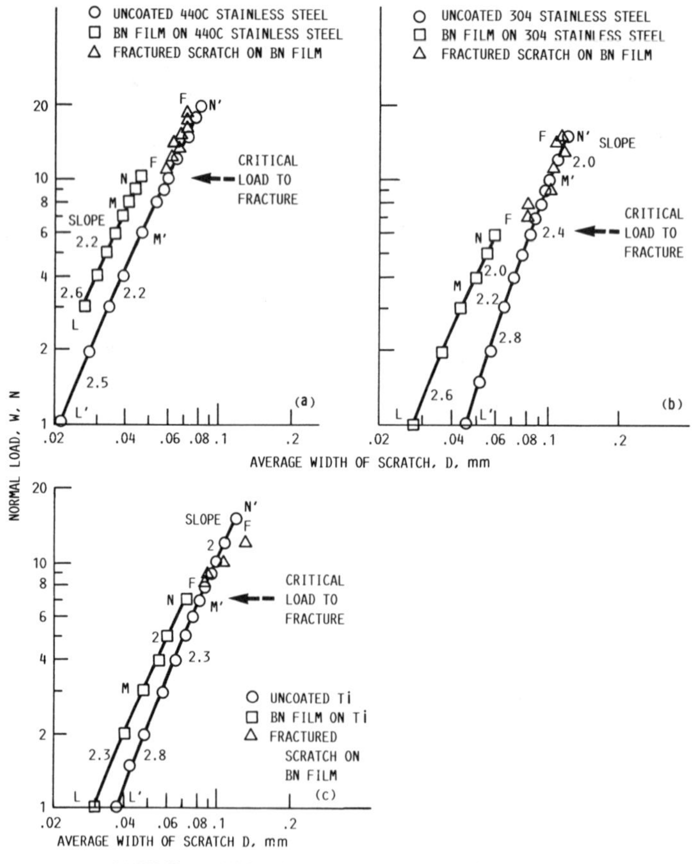

(a) BN film on 440C stainless steel and uncoated 440C stainless steel.
(b) BN film on 304 stainless steel and uncoated 304 stainless steel.
(c) BN film on Ti and uncoated Ti.

Figure 14.—Scratch width as function of load for metallic substrates (sliding velocity, 12 mm/min; laboratory air; portion FF represents fractured scratch on BN film).

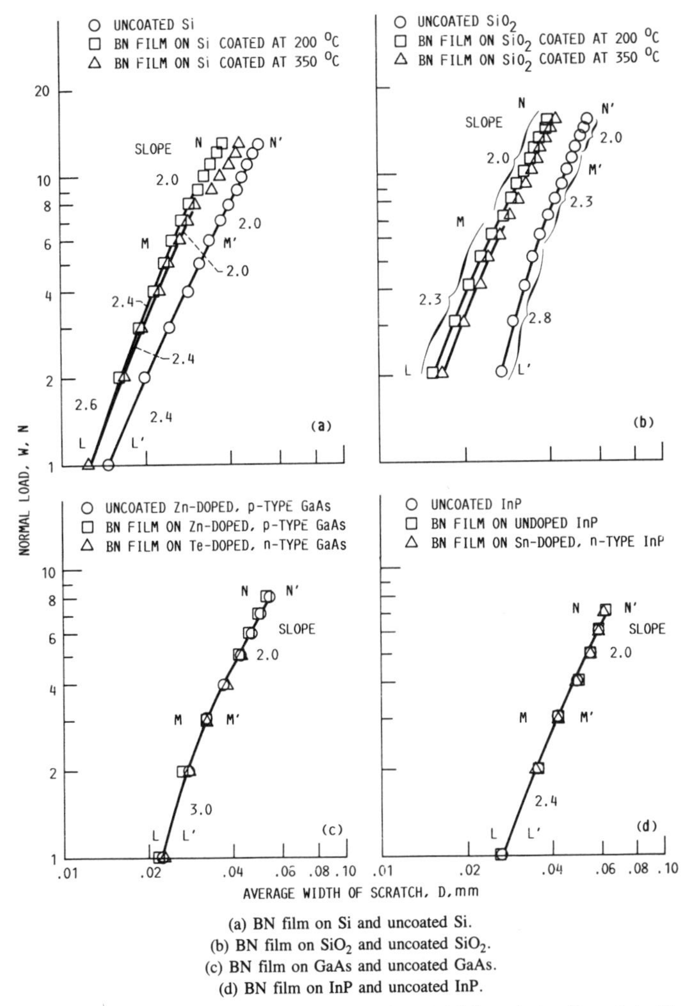

(a) BN film on Si and uncoated Si.
(b) BN film on SiO_2 and uncoated SiO_2.
(c) BN film on GaAs and uncoated GaAs.
(d) BN film on InP and uncoated InP.

Figure 15.—Scratch width as function of load for nonmetallic substrates (sliding velocity, 12 mm/min; laboratory air).

uncoated 440C stainless steel, 304 stainless steel, titanium, silicon, SiO_2, GaAs, and InP are also presented in figures 14 and 15. The portion LM for BN films or L′M′ for uncoated material is gradually curved but is considered to be composed of approximately straight portions of transitional slopes. For example, the transitional slopes are 2.6 and 2.2 for BN film on 440C stainless steel, and 2.6 and 2.4 for BN film on silicon.

The portion MN for BN films or M′N′ for uncoated materials is a straight line of slope 2. It is evident that MN or M′N′ is the range over which Meyer's law is valid for BN films on both metallic and nonmetallic substrates as well as for uncoated metallic and nonmetallic materials. Here the Meyer's index n is a constant and has a value of 2. Thus, the BN films on both metallic and nonmetallic substrates behave plastically much like metals when they are brought into sliding contact with hard solids such as diamond.

It is interesting that in figure 14 the portion FF, representing the condition of fracture where the load exceeds the critical load, is also roughly expressed by $W = kD^n$. The fractured scratch in the BN film on the substrate is almost as wide as the scratch in the uncoated metallic material used for the substrate. This evidence confirms that cracks are generated from the contact area rather than from the free surface of the film. It suggests that the substrate is responsible not only for controlling the critical load which will fracture the BN film but also for the extent of fracture. Furthermore, the critical load required to fracture a ceramic film on a substrate can be determined by measurements of the scratch width. The fracture and wear of BN films are discussed in greater detail in the next section.

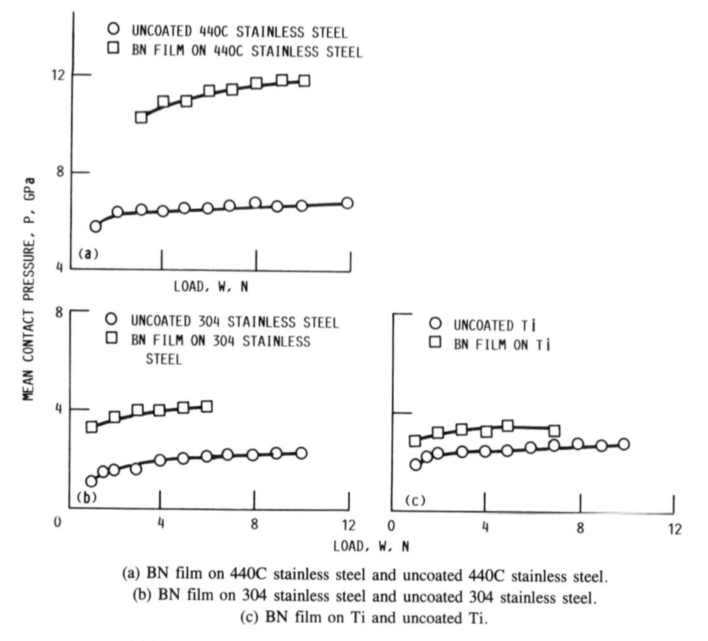

(a) BN film on 440C stainless steel and uncoated 440C stainless steel.
(b) BN film on 304 stainless steel and uncoated 304 stainless steel.
(c) BN film on Ti and uncoated Ti.

Figure 16.—Mean contact pressure (yield pressure) as function of load for metallic substrates (sliding velocity, 12 mm/min; laboratory air).

Mean contact pressure (yield pressure) P during sliding may then be defined by $P = W/A_s$, where W is the applied load and A_s is the projected contact area given by $A_s = \pi D^2/8$ (only the front half of the pin is in contact with the flat) [51,52].

The mean contact pressure over the contact area gradually increased until the deformation passed to a fully plastic state, as presented in figure 16 for metallic substrates and in figure 17 for nonmetallic substrates. The mean contact pressure at a fully plastic state P_m is higher than but proportional to the measured Vickers hardness (tables II–IV).

The mean contact pressure of 440C stainless steel, 304 stainless steel, titanium, silicon, and SiO_2 increased by a factor of up to 2 with the presence of BN films (figs. 16 and 17).

The mean contact pressures for BN films on doped GaAs and InP are the same as those for the uncoated surfaces of doped GaAs and InP (figs. 17(c) and (d)). Optical microscopic examination of the grooves on the BN films deposited on doped GaAs and InP substrates clearly revealed that the sliding action caused breakthrough of the film in the contact area. The breakthrough of the BN films was caused by poor interfacial adhesive bonds between the film and the substrate in and near the diamond contact region. The removal of the BN films induced direct contact of the diamond pin with the doped GaAs and InP substrates. It therefore resulted in no increase in the mean contact pressure of the doped GaAs and InP with the presence of BN films.

Fracture and Wear

When a much higher contact pressure (due to highly concentrated stress in the contact area between the diamond and BN film) is provided, the sliding action produces visible microscopic cracking in the BN film. Scanning electron photomicrographs of a wear track on a BN film deposited on a 304 stainless steel substrate (fig. 18) reveal that both plastic deformation and fracture occurred in the BN film. The fracturing resulted from brittle cracks being generated, propagated, and then intersected in the BN film and at the interface between the BN film and the substrate. The backscatter photomicrograph (fig. 18(b)) reveals two different materials: (1) The dark areas in the photomicrograph show where the BN film stayed on the substrate and (2) the light areas show where the BN film fragments came off, revealing the 304 stainless steel substrate.

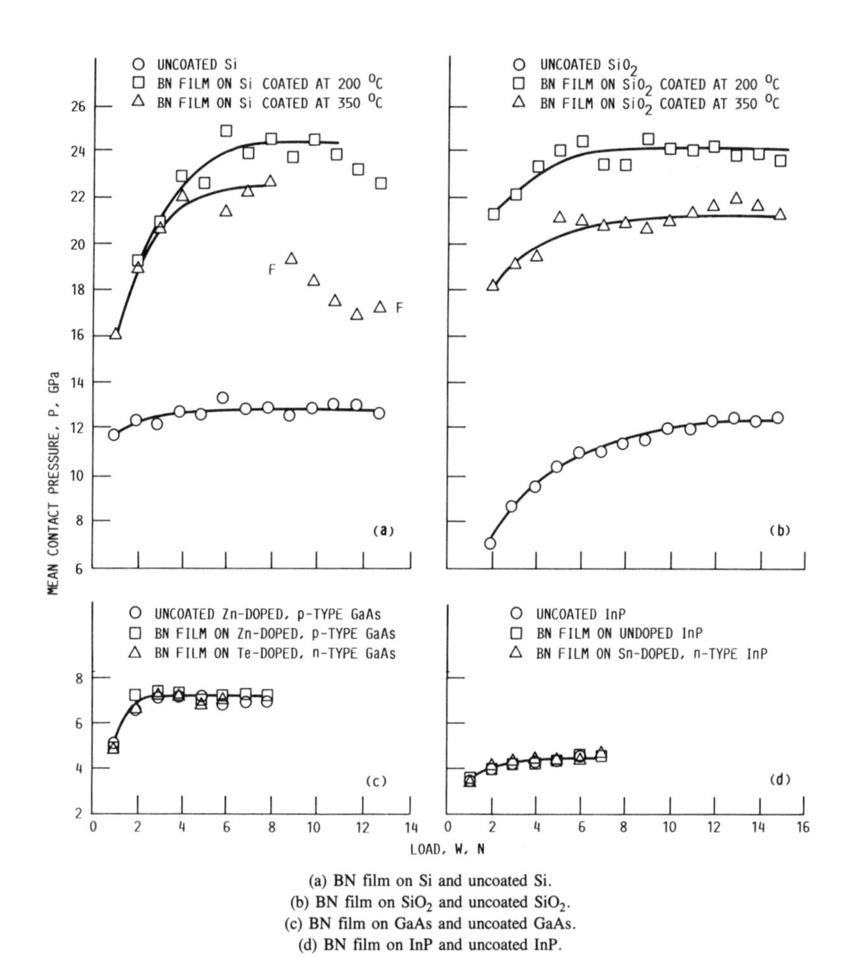

(a) BN film on Si and uncoated Si.
(b) BN film on SiO_2 and uncoated SiO_2.
(c) BN film on GaAs and uncoated GaAs.
(d) BN film on InP and uncoated InP.

Figure 17.—Mean contact pressure (yield pressure) as function of load for nonmetallic substrates (sliding velocity, 12 mm/min; laboratory air; portion FF represents fractured scratch on BN film).

TABLE II.—CRITICAL NORMAL LOAD TO FRACTURE BN FILM IN SLIDING CONTACT AND SHEAR STRENGTH OF INTERFACIAL ADHESIVE BONDS

Substrate	Vickers hardness, H_v, GPa		Mean contact pressure (yield pressure) at a fully plastic state, P_m, GPa	Critical normal load, W_c, N	Interfacial adhesive strength, S GPa	
	Substrate[a]	BN film[b]			$S = K\left(\frac{W_c P_m}{\pi R^2}\right)^{1/2}$	$S \simeq \frac{2W_c}{\pi DR}$
440C stainless steel	7.1	10.2	12	11	0.77	0.68
304 stainless steel	2.5	4.3	4.1	7	.37	.29
Ti	2.6	3.4	3.3	8	.40	.30

[a]Hardness measuring load, 2 N.
[b]Hardness measuring load, 0.05 N.

TABLE III.—MEAN CONTACT PRESSURE (YIELD PRESSURE) OF AND CRITICAL NORMAL LOAD TO FRACTURE BN FILM COATED ON NONMETALLIC SUBSTRATE

Substrate and film	Mean contact pressure (yield pressure) at fully plastic state P_m, GPa	Vickers hardness,[a] H_v, GPa	Ratio P_m/H_v	Critical normal load, W_c, N
Si coated with BN at 200 °C	24.0	18.8	1.28	7
Si coated with BN at 350 °C	21.9	17.5	1.25	7
SiO_2 coated with BN at 200 °C	23.9	---	---	3
SiO_2 coated with BN at 350 °C	21.2	17.2	1.23	3
Zn-doped, *p*-type GaAs coated with BN	7.15	5.72	1.25	3
Te-doped, *n*-type GaAs coated with BN	7.05	5.65	1.25	3
InP coated with BN	4.51	3.66	1.23	2
Sn-doped, *n*-type InP coated with BN	4.54	3.61	1.26	1

[a]Hardness measuring load, 0.05 N.

TABLE IV.—MEAN CONTACT PRESSURE (YIELD PRESSURE) OF AND CRITICAL NORMAL LOAD TO FRACTURE NONMETALLIC MATERIALS

Material	Mean contact pressure (yield pressure) at fully plastic state P_m, GPa	Vickers hardness,[a] H_v, GPa	Ratio P_m/H_v	Critical normal load, W_c, N
Si	12.6	9.89	1.27	5
SiO_2	12.4	---	---	1
Zn-doped, *p*-type GaAs	6.94	5.61	1.24	2
Te-doped, *n*-type GaAs	----	5.68	---	-
InP	----	---	---	1
Sn-doped, *n*-type InP	4.35	3.49	1.25	1

[a]Hardness measuring load, 0.05 N.

Scanning electron photomicrographs of a wear track on a BN film deposited on a 440C stainless steel substrate (fig. 19) reveal that the BN film deformed plastically during sliding at a load of 11 N. Permanent grooves were formed where the diamond began to slide. However, the sliding action also produced sudden gross flaking of the BN film. A large light area in the last half of the wear track (fig. 19(a)) is a fracture pit where the BN film was completely removed along the sliding direction of the diamond. This is confirmed by the backscatter photomicrograph (fig. 19(c)), which shows the light area to be the 440C stainless steel substrate. The fracturing and removal of the BN film from the 440C stainless steel substrate were caused by fracture of cohesive bonds in the BN film and interfacial adhesive bonds between the film and the substrate in and near the contact region with the diamond. The BN film was fractured and removed from the titanium substrate in a similar manner.

Investigators have detected released acoustic emissions when the intrinsic cohesive bonds in the ceramic coating film or the adhesive bonds between the film and substrate or both are broken and a new surface is created. The pattern and intensity of the acoustic emissions depend on the nature of the disturbance, that is, plastic flow, cracking, or flaking of fragments [53–56].

Figure 20 presents typical acoustic emission traces and friction force traces for a BN film deposited onto a nonmetallic substrate. When the BN film surface is brought into contact with a diamond pin under a small load (which is lower than the critical loads needed to fracture intrinsic cohesive bonds in the BN film and adhesive bonds between the film and substrate), no acoustic emission is detected (fig. 20(a)). The friction force trace at the same load (3 N) fluctuates slightly with no evidence of stick-slip behavior (fig. 20(b)). After the diamond has passed over the surface once, scanning electron microscopic examination of the wear track indicates that a permanent groove is formed in the BN film, much as occurs in metallic films under similar conditions [30,31]. However, no cracking of the BN film is observed with sliding.

An increase in load to or above the critical loads required to fracture the BN film and the interfacial adhesive bonds between the film and the substrate results in a small amount of cracking in and near the plastically deformed groove. The acoustic emission trace detected at a load of 9 N indicates evidence of a fluctuating acoustic emission signal output (fig. 20(c)). Acoustic emission is observed when the sliding appears to involve small amounts of cracking in addition to plastic flow. Such acoustic emission is due to the release of elastic energy when cracks propagate in the BN film. The friction force trace measured at the load of 9 N is characterized by randomly fluctuating behavior, but only occasional evidence of stick-slip behavior is observed (fig. 20(d)).

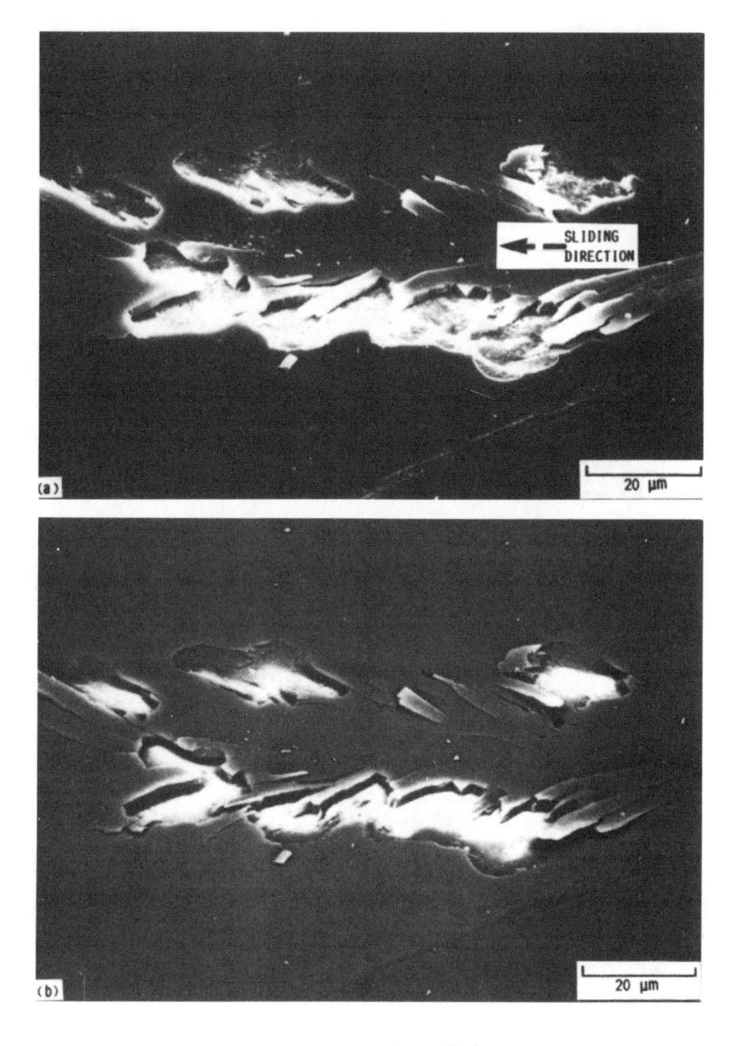

(a) Secondary electron image SEM.
(b) Backscatter electron image SEM.

Figure 18.—Scanning electron microscope (SEM) photomicrographs of wear track on BN film surface generated by 0.2-mm-radius diamond pin (substrate, 304 stainless steel; load, 7 N; sliding velocity, 12 mm/min; laboratory air).

When a much higher load is applied to the BN film, the sliding action produces, in addition to plastic flow, locally gross surface and subsurface fracturing in the film and at the interface between the BN film and the substrate. In such cases the acoustic emission traces are primarily characterized by chevron-shaped peaks (fig. 20(e)), whereas the friction force is primarily characterized by a continuous, marked stick-slip behavior (fig. 20(f)).

The behavior of acoustic emission is related to that of the friction force. For example, at point I in figures 20(e) and (f), the diamond pin comes to rest until point II is reached. At point II, the pin is set into motion and slips, and will continue to move until point III is reached. At point II, acoustic emission is released because the slip action produces fracturing at the interface between the BN film and the substrate. At point III, the pin comes to rest again. Thus, fracture in the film and at the interface between the BN film and the substrate is responsible for the observed acoustic emission signal output and friction behavior. Acoustic and friction measurements of the critical load required to fracture a ceramic film on a substrate agree well with the critical loads detected by optical and scanning electron microscopy of the scratches.

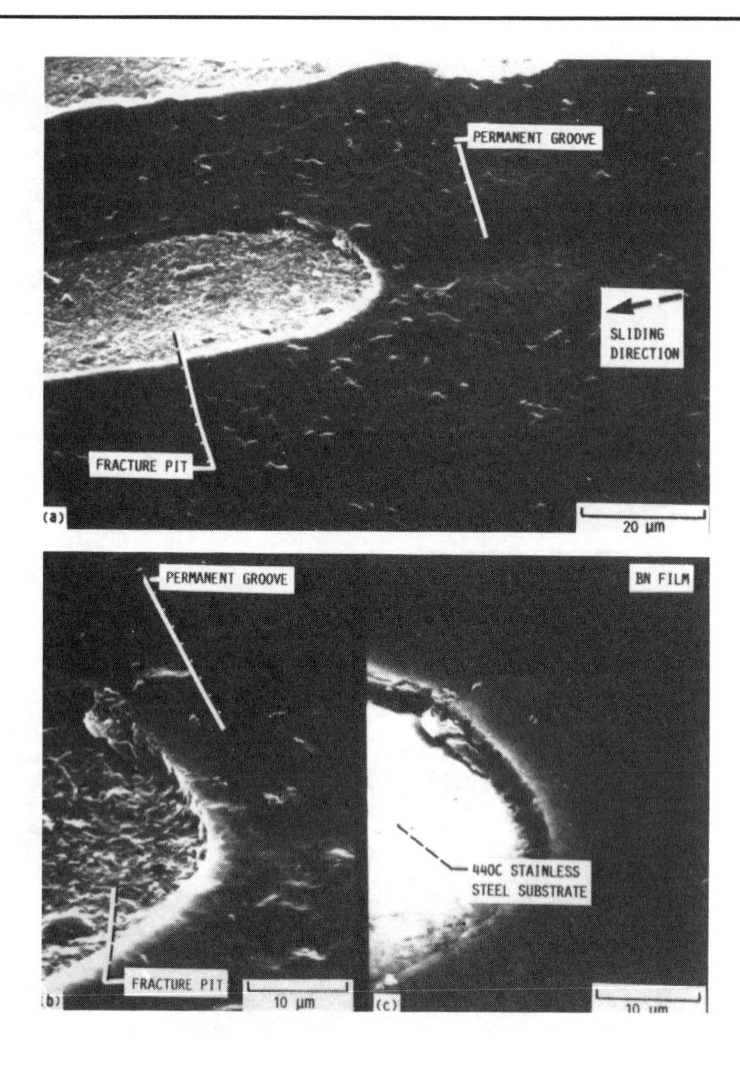

(a) Secondary electron image SEM (low magnification).
(b) Secondary electron image SEM (high magnification).
(c) Backscatter electron image SEM.

Figure 19.—Scanning electron microscope (SEM) photomicrographs of wear track on BN film surface generated by 0.2-mm-radius diamond pin (substrate, 440C stainless steel; load, 11 N; sliding velocity, 12 mm/min; laboratory air).

Figure 21 presents data on the critical loads required to fracture BN film and adhesive bonds between the film and metallic substrate as determined by acoustic emission and friction force measurements. The critical load to fracture is related to the hardness and strength of the metallic substrate. The harder the metallic substrate or the greater the strength of the substrate, the higher the critical load.

With uncoated silicon (table IV), the critical load to fracture cohesive bonds in the silicon was 5 N in sliding contact. With the BN film deposited on the silicon substrate, however, cracking occurred at a load of 7 N (table III). A similar trend was observed with good-quality BN films on SiO_2, as well as with poorly adhered films on zinc-doped GaAs and undoped InP (tables III and IV). Thus, the BN film on the brittle, nonmetallic materials plays an important role in surface fracture. The presence of BN film is very effective in increasing the critical load needed to initiate fracture in nonmetallic material.

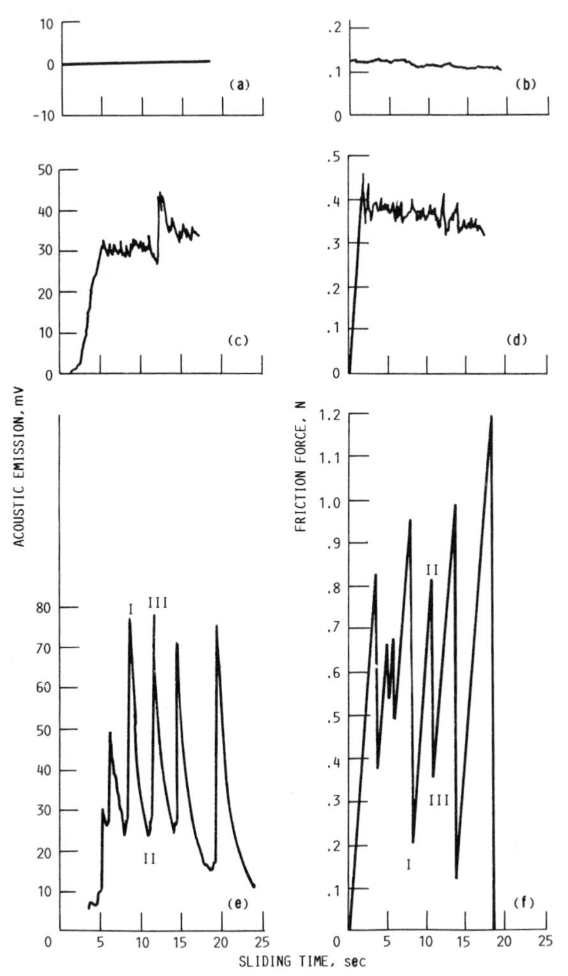

(a) No acoustic emission caused by diamond pin under small load, 3N.
(b) Friction force caused by diamond pin under small load, 3N.
(c) Acoustic emission caused by diamond pin under critical load, 9N.
(d) Friction force caused by diamond pin under critical load, 9N.
(e) Acoustic emission caused by diamond pin under much higher load, 12 N.
(f) Friction force caused by diamond pin under much higher load, 12 N.

Figure 20.—Typical acoustic emission traces and friction traces for BN film in contact with hemispherical diamond pin in laboratory air.

Adhesion and Shear Strength

Interfacial adhesive bonds between the coating and the substrate or cohesive bonds in the BN coating or both were broken when critically loaded. For example, with 440C stainless steel and titanium, some of the interfacial adhesive bonds between the BN film and the substrate were generally weaker than the cohesive bonds in the BN film. The failed surface revealed the complete removal of the BN film from the 440C stainless steel surface (fig. 19) or the titanium surface.

On the other hand, with the 304 stainless steel, silicon, and SiO_2 substrates, the interfacial adhesive bonds between the BN film and the substrate were generally stronger than the cohesive bonds in the BN film. The BN film first cracked when it was critically loaded. When a much higher load than the critical load was applied, the sliding action

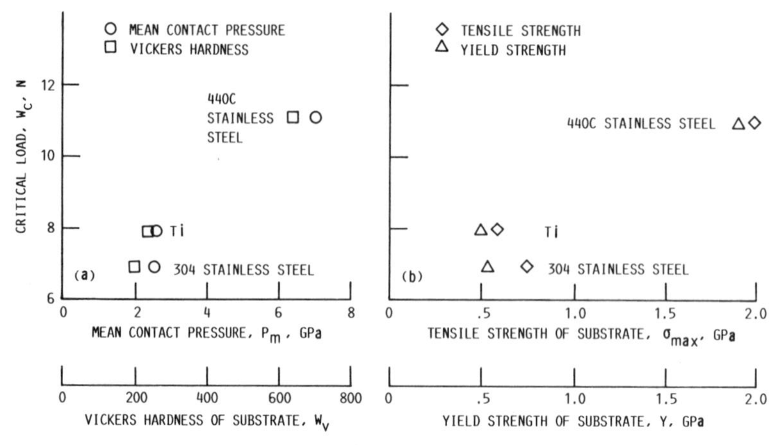

(a) Relation of critical load to hardness of substrate.
(b) Relation of critical load to strength of substrate.

Figure 21.—Critical load needed to fracture BN film and interfacial adhesive bonds between film and substrate in sliding contact with hemispherical diamond pin in laboratory air (mean contact pressure is scratch hardness of substrate).

induced fracture of interfacial adhesive bonds between the BN film and the substrate, and the BN film fragments were partially removed from the substrate (e.g., fig. 18).

Although the failure modes observed with the BN films on the metallic and nonmetallic substrates were different, it is interesting to compare the adhesion strength of the interfacial bonds between the BN film and the substrate.

Benjamin and Weaver [57] derived the following expressions for scratch adhesion in terms of shearing stress S produced at the coating/substrate interface by the plastic deformation, the hardness of the substrate (mean contact pressure at fully plastic state P_m), the critical load applied on the pin W_c, tip radius of the pin R, and the width of the resulting scratch D

$$S = K\left(\frac{W_c P_m}{\pi R^2}\right)^{1/2} \tag{2}$$

$$S \simeq \frac{2W_c}{\pi DR} \tag{3}$$

These relationships allow for the calculation of the shear strength (i.e., the adhesion strength of the interfacial bonds) [52,57]. The results are presented in tables II and V. The values of the critical loads were obtained and confirmed not only by optical and scanning electron microscopic examination of the scratches, but also by the acoustic emission technique. Table II reveals the strong correlation between the shear strength (i.e., adhesion strength) and the hardness of the substrate. The harder the metallic substrate, the greater the shear strength.

Lubricating Effects of Boron Nitride Film

The coefficients of friction measured in laboratory air for BN films on the metallic and nonmetallic substrates are presented in figures 22 and 23, with comparative data for the uncoated metallic and nonmetallic materials. The friction resulted primarily from abrasion (plowing) of the diamond. The data presented in figures 22 and 23 indicate the marked difference in friction for presence or absence of BN films. The presence of BN films decreases adhesion and plastic deformation, and accordingly, friction. Thus, the ion-beam-deposited BN film can be effectively used as a solid lubricating film.

TABLE V.—CRITICAL NORMAL LOAD TO FRACTURE INTERFACIAL ADHESIVE BONDS BETWEEN BN FILM AND SILICON SUBSTRATE AND SHEAR STRENGTH OF INTERFACIAL ADHESIVE BONDS

Coating condition,[a] °C	Critical normal load, W_c, N	Interfacial adhesive strength, S, GPa	
		$S = K\left(\frac{W_c P_m}{\pi R^2}\right)^{1/2}$	$S \simeq \frac{2W_c}{\pi DR}$
200	11	1.1	1.0
350	8	.91	.84

[a]Nominal temperature of substrate during ion-beam deposition.

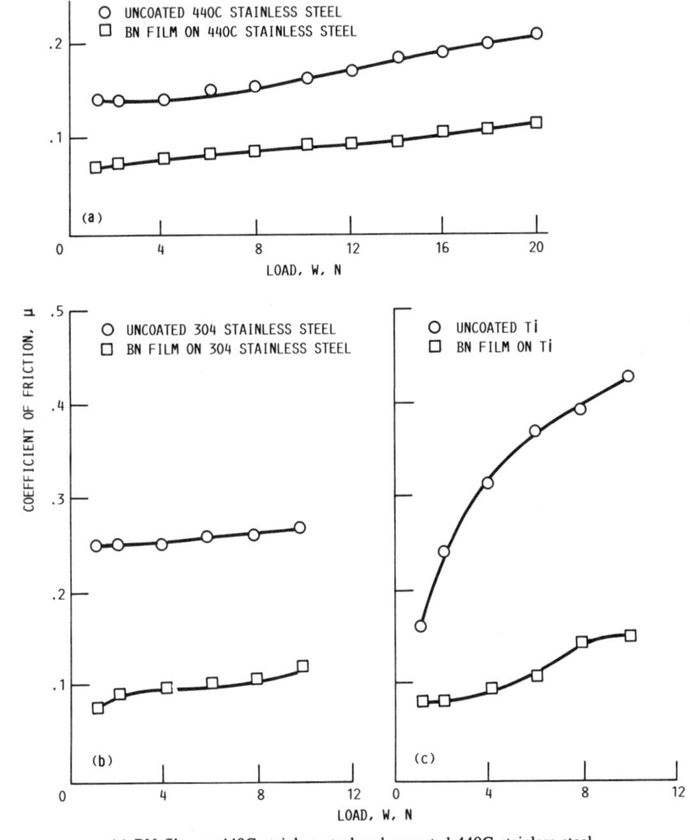

(a) BN film on 440C stainless steel and uncoated 440C stainless steel.
(b) BN film on 304 stainless steel and uncoated 304 stainless steel.
(c) BN film on Ti and uncoated Ti.

Figure 22.—Coefficient of friction as function of load for metallic substrates (sliding velocity, 12 mm/min; laboratory air).

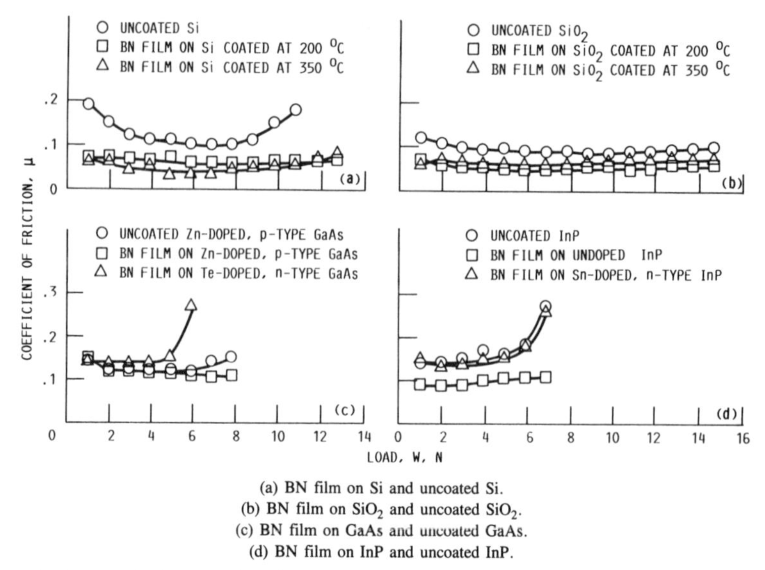

(a) BN film on Si and uncoated Si.
(b) BN film on SiO_2 and uncoated SiO_2.
(c) BN film on GaAs and uncoated GaAs.
(d) BN film on InP and uncoated InP.

Figure 23.—Coefficient of friction as function of load for nonmetallic substrates (sliding velocity, 12 mm/min; laboratory air).

CONCLUSIONS

Based on fundamental studies conducted on ion-beam-deposited boron nitride (BN) films grown on both metallic and nonmetallic substrates, the following comments can be made.

1. Ion-beam-deposited BN films are not completely amorphous but contain (to a small extent) hexagonal BN.
2. Both the sputter-cleaned surfaces and the bulk BN films are nonstoichiometric, with a boron-to-nitrogen (B/N) ratio of approximately 2, and contain small amounts of oxides and carbides in addition to BN.
3. Surface films on BN coatings affect their tribological behavior. For example, adsorbed carbon contaminants on a BN coating surface decrease both adhesion and friction, whereas oxygen, as a surface contaminant on metals in sliding contact with BN films, increases adhesion and friction.
4. For clean metal-to-BN contacts, strong bonds form between the materials, and the interfacial bonds are stronger than the cohesive bonds in the metal at a major part of the real area of contact. Adhesion and friction are strongly dependent on the ductility of the metals. Hardness of metals is of paramount importance to adhesion and friction, exceeding that of the surface energy of metals.

Adhesion, friction, surface energy, and hardness of a metal in contact with a BN film are all related to its shear modulus, which has a marked dependence on the electron configuration of the metal.

5. Boron nitride films, like metal films, deform elastically and plastically in the interfacial region between two solids in contact under load. Unlike metal films, however, when the contact stress exceeds a certain critical value, fracture can occur. The critical load to fracture is related to the hardness and strength of the substrates. The harder the substrate or the greater the strength of the substrate, the higher the critical load.

The critical load required to fracture a BN film on a substrate can be determined by measurements of the scratch width, acoustic emission, or friction as well as by optical or scanning electron microscopic examinations of the scratches.

6. The presence of BN films decreases adhesion and plastic deformation, and accordingly, friction. The BN films exhibited a capability for lubrication (low adhesion and friction) in both vacuum and air at atmospheric pressure.

ACKNOWLEDGMENT

The author would like to thank John J. Pouch and Samuel A. Alterovitz, NASA Lewis Research Center, for providing the BN films deposited on the nonmetallic substrates.

REFERENCES

1. Research Group on Wear of Engineering Materials: Friction, Wear, and Lubrication—Terms and Definition. O.E.C.D. Committee for Scientific Research, Delft, Netherland, 1966.
2. Sliney, H.E., Jacobson, T.P., Deadmore, D., and Miyoshi, K.: Ceram. Eng. Sci. Proc., 1986, **7**, 1039.
3. Quinn, T.F.J.: Wear, 1984, **100**, 399.
4. Sutor, P.: Ceram. Eng. Sci. Proc., 1985, **6**, 963.
5. Rice, R.W.: Ceram. Eng. Sci. Proc., 1985, **6**, 940.
6. Buckley, D.H., and Miyoshi, K.: Ceram. Eng. Sci. Proc., 1985, **6**, 919.
7. Kiser, J.D., Levine, S.R., and DiCarlo, J.A.: NASA CP-10003-SESS-1, 1987, **103**.
8. Sliney, H.E.: NASA CP-10003-SESS-1, 1987, 89.
9. Brindley, P.K.: NASA CP-10003-SESS-1, 1987, 73.
10. Buckley, D.H.: Surface Effects in Adhesion, Friction, Wear, and Lubrication. Elsevier, 1981, 245.
11. Czichos, H.: Wear, 1984, **100**, 579.
12. Halling, J, and Arnell, R.D.: Wear, 1984, **100**, 367.
13. Hintermann, H.E.: Wear, 1984, **100**, 381.
14. Schintlmeister, W., Wallgram, W., Kanz, J., and Gigl, K.: Wear, 1984, **100**, 153.
15. Miyoshi, K., and Buckley, D.H.: Wear, 1986, **110**, 295.
16. Buckley, D.H.: ASLE Trans., **21**, 1978, 118.
17. Alterovitz, S.A., Warner, J.D., Liu, D.C., and Pouch, J.J.: Proceedings of the Symposium on Dielectric Films on Compound Semiconductors. V.J. Kapoor, D.J. Connolly, and Y.N. Wong, eds. Electrochemical Society, Pennington, 1986, 59.
18. Pouch, J.J., Alterovitz, S.A., and Warner, J.D.: NASA TM-87258, 1986.
19. Weissmantel, C.: J. Vac. Sci. Technol., 1981, **18**, 179.
20. Shanfield, S., and Wolfson, R.: J. Vac. Sci. Technol. A, 1983, **1**, 323.
21. Motojima, S., Tamura, Y., and Sugiyama, K: Thin Solid Films, 1982, **88**, 269.
22. Muraka, S.P., Cheng, C.C., Wang, D.N.K., and Smith, T.E.: J. Electrochem. Soc., 1979, **126**, 1951.
23. Hyder, S.B., and Yep, T.O.: J. Electrochem. Soc., 1976, **123**, 1721.
24. Liu, D.C., Valco, G.J., Skebe, G.G., and Kapoor, V.J.: Proceedings of the Symposium on Silicon Nitride Thin Insulating Films, V.J. Kapoor and H.J. Stein, eds., Electrochemical Society, Pennington, 1983, 141.
25. Miyamoto, H., Hirose, M., and Osaka, Y.: Jpn. J. Appl. Phys. Part 2, 1983, **22**, L216.
26. Wiggins, M.D., and Aita, C.R.: J. Vac. Sci. Technol. A, 1984, **2**, 322.
27. Adams, A.C.: J. Electrochem. Soc., 1981, **128**, 1378.
28. Nakamura, K.: J. Electrochem. Soc., 1985, **132**, 1757.
29. Miyoshi, K., Buckley, D.H., and Spalvins, T.: J. Vac. Sci. Technol. A, 1985, **3**, 2340.
30. Miyoshi, K., Buckley, D.H., Pouch, J.J., Alterovitz, S.A., and Sliney, H.E.: Tribology—Friction, Lubrication and Wear, **2**, Mechanical Engineering Publications, London, 1987, 621.
31. Miyoshi, K., et al.: Surf. Coat. Technol., 1987, **33**, 221.
32. Tabor, D.: NASA CP-2300, 1983, **1**, 1.
33. Pepper, S.V.: J. Appl. Phys., 1976, **47**, 2579.
34. Miyoshi, K., and Buckley, D.H.: Ceram. Eng. Sci. Proc., 1983, **4**, 674.
35. Bowden, F.P., and Tabor, D.: The Friction and Lubrication of Solids—Part II. Clarendon Press, Oxford, 1964, 158.
36. Miyoshi, K., and Buckley, D.H.: ASLE Trans., 1979, **22**, 79.
37. Gschneidner, K.A., Jr.: Solid State Physics, **16**, F. Seitz and D. Turnbull, eds., Academic Press, 1964, 275.
38. Tyson, W.R.: Can. Metall. Q., 1975, **14**, 307.
39. Jones, H.: Met. Sci. J., 1971, **5**, 15.
40. Miedema, A.R.: Z. Metllk., 1978, **69**, 287.
41. Miyoshi, K., and Buckley, D.H.: ASLE Trans., 1984, **27**, 15.
42. Tabor, D.: J. Lubr. Technol., 1977, **99**, 387.
43. Rabinowicz, E.: Friction and Wear of Materials. John Wiley and Sons, 1965.
44. Miyoshi, K., Buckley, D.H., and Srinivasan, M.: Am. Ceram. Soc. Bull., 1983, **62**, 494.
45. Miyoshi, K.: Surf. Coat. Technol., 1988, **36**, 487.
46. Steijn, R.P.: J. Appl. Phys., 1961, **32**, 1951.
47. Steijn, R.P.: J. Appl. Phys., 1963, **34**, 419.
48. Steijn, R.P.: Wear, 1964, **7**, 48.
49. Steijn, R.P.: ASLE Trans., 1969, **12**, 21.
50. DuFrane, K.F., and Glaeser, W.A.: NASA CR-72530, 1969.
51. Tabor, D.: The Hardness of Metals. Clarendon Press, Oxford, 1951.
52. Ahn, J., Mittal, K.L., and MacQueen, R.H.: Adhesion Measurements of Thin Films, Thick Films, and Bulk Coatings, K.L. Mittal, ed., ASTM Special Technical Publication 640, 1978, 134.
53. Hintermann, H.E.: Wear, 1984, **100**, 381.
54. Hintermann, H.E.: Science of Hard Materials, R.K. Viswanadham, D.J. Rowcliffe, and J. Gorland, eds., Plenum, 1983, 357.
55. Sekler, J., Steinmann, P.A., and Hintermann, H.E.: Surf. Coat. Technol., 1988, **36**, 519.
56. Rickerby, D.S.: Surf. Coat. Technol., 1988, **36**, 541.
57. Benjamin, P., and Weaver, C.: Proc. R. Soc. London A, 1960, **254**, 163.

Materials Science Forum Vols. 54 & 55 (1990) pp. 399-416

FAST ATOM BEAM TECHNIQUES FOR BN AND OTHER HARD FILM FORMATIONS AND APPLICATIONS TO FRICTION-REDUCING COATINGS

H. Kuwano

NTT Applied Electronics Laboratories, Nippon Telegraph and Telephone Corporation
Musashino-shi, Tokyo 180, Japan

ABSTRACT

This paper describes fast atom beam techniques and friction-reducing solid lubricant coatings. The characteristics of a modified McIlraith-type source having a charge-exchange cell and the calculations of fast atom formation based on resonance charge transfer are shown.

Frictin-reducing coatings are proposed using fast atom beam techniques, consisting of a hard bottom layer film with a solid lubricant top layer film. A dual fast atom beam technique incorporating two fast atom beam sources improves film crystallinity and adhesion between the films and their respective substrates. Solid lubricant coatings are newly formed by an Mo-S complex target.

Friction-reducing coatings were accomplished by bi-layer films consisting of hard bottom layer, such as BN, B, TiB_2 and B_4C, with an added top layer MoS_2 solid lubricant. The bi-layer solid lubricant, especially BN bottom layer film, had 1/5 of MoS_2 solid film lubricant friction coefficient, less than 0.01 and had high durability. The hard films obtained by accelerated fast atom bombardment of their growth surface had either a single or poly-crystalline structure, while those without bombardment had an amorphous structure.

Feasibility of the bi-layer solid lubricants are demonstrated by fabrications bi-layer solid lubricant ball bearings which were tested up to 10^6 oscillatory cycles and showed low motive torque.

§1. INTRODUCTION

This paper presents 1) fast atom beam characteristics, 2) hard coatings, such as BN, B, TiB_2 and B_4C using dual fast atom beam technique, 3) MoS_2 solid lubricant coatings using an Mo-S complex target, 4) friction-reducing coatings consisting of a hard bottom layer film with an added solid lubricant top layer film, 5) friction characteristics.

A fast atom beam is free from any electric fields for targets and other vacuum devices. Fast atom beam techniques for sputter deposition, etching, analysis and implantation are very easily applied to insulators and composite materials because charging of the specimen surface is not a serious problems in this process. The techniques are applied to solid lubricant film formations[1,2], insulator etching[3] and secondary ion mass spectrometry[4].

When temperature, vacuum atmosphere, or other factors preclude the use of liquid lubricants, a dry lubricant is necessary. Solid lubricants such as MoS_2 and WS_2 are therefore very important for vacuum instruments or for spacecraft [1, 5-9]. Since the life expectancies of the most recently developed communications satellites are more than 10 years under severe conditions. Frictin-reducing solid lubricants for communications satellite mechanisms requiring long life and high reliability represent one important application.

In this paper, friction-reducing bi-layer solid lubricant coatings consisting of a hard bottom layer film such as BN, B, TiB_2, and B_4C with an added MoS_2 solid lubricant is shown by using fast atom beam techniques. Hard coatings are an important engineering advance for improving the resistance of tools to wear. However, some problems in adhesion exist, as well as a lack of toughness.

Dual fast atom beam technique is studied to improve crystallinity of hard coatings and its adhesion problems. An MoS_2 deposit is newly controlled by an Mo-S complex target. The feasibility of bi-layer friction-reducing coatings was demonstrated through preparation and testing of bi-layer-coated ball bearings.

§2. FAST ATOM BEAM CHARACTERISTICS

A fast atom is defined as an energetic uncharged atom having significantly enough above thermal energy, namely above 10 eV, in the paper. To get a fast atom beam, it is easy to neutralize energetic ions emitted from an ion source. The neutralization of ions is the most effectively achieved in the mechanism of resonant charge transfer.

The probability of a resonant charge transfer of an ion with a neutral atom of the same species (symmetrical resonance charge transfer) is given by Eq. (1):

$$P = l^{-1}dx = nsdx, \tag{1}$$

where n is the number of neutrals per cm^{-3}, s cross section in cm^2 for symmetrical resonant charge transfer, dx is the differential path length in cm, and l is the mean free path in cm. The values of s in Eq. (1) are derived from Fig. 1, which demonstrates the relation between the cross section s and the particle velocity v for different rare gases, as quoted in Ref. 10. The probability of ion charge transfer to a neutral particle, R_{n0}, is calculated by

$$R_{n0} = 1 - e^{-nsl}, \tag{2}$$

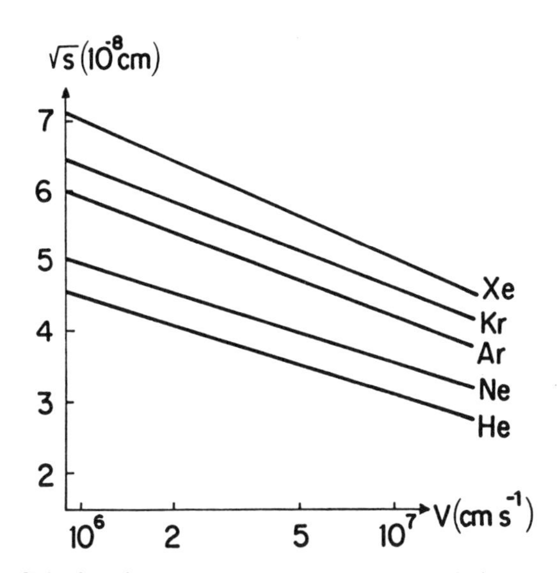

Figure 1: Calculated cross sections for symmetrical charge transfer.

The proportion of energetic neutral particles, R_{n0} in each energy Ar^+ ion beam is shown in Fig. 2.

In this paper, a modified McIlraith cold cathode source (saddle field source) [1-4,11,12,18] with a charge-exhange cell is applied as a fast atom source as shown in Fig. 3. The plasma chamber, anode rods, cathode grid, and charge-exchange cell are all made of graphite. Fast atom beams are extracted through mesh holes and the charge-exchange cell in the grounded graphite. Using Ar gas, the discharge of the source generally operates at a pressure of about 10^{-1}-10 Pa. The discharge current is controlled between 0-500 mA at the power supply by varying the anode potential from 0 to 5 kV. Pressure in the sample chamber is maintained in the $5X10^{-2}$-$1X10^{-1}$ Pa range.

The proportion of high-energy neutral particles is studied by measuring the secondary electron current produced by neutral particles and ions, and by measuring the ion beam current at the collector (Fig. 4).

The relationship between the deflection voltage and the Mo collector current is plotted in Fig. 4. The residual collector current in the beam is attributable to fast atoms because the residual collector currents are almost constant over 0.5-kV deflection voltage.

If the beam does not contain electrons, the proportion of high-energy neutral particles in the beam, R_{n0} is represented by the equation

$$R_{n0} = \frac{N_0}{N_0 + N_+} \tag{3}$$

$$= [\frac{1 + e_+ + e_0(I_i/I_d - 1)}{1 + e_+}]^{-1}, \tag{4}$$

where N_0 is the number of high-energy neutral particles per second, N_+ is the number of ions per second, e_0 is the efficiency of secondary electron emission caused by high-energy neutral particle bombardment in the collector, e_+ is the efficiency of secondary electron emission caused by ion bombardment in the collector, and $I_i = N_+ + e_0N_0 + e_+N_+$ is the

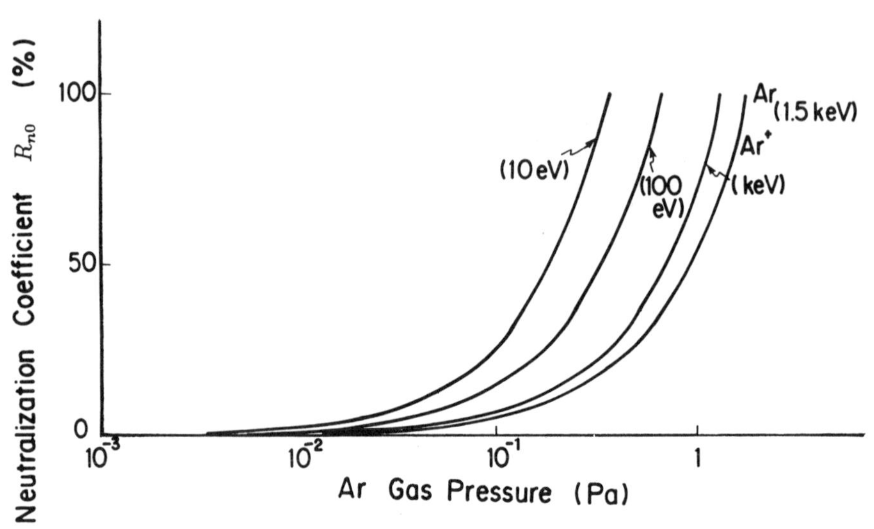

Figure 2: Calculated R_{n0} in each energy Ar^+.

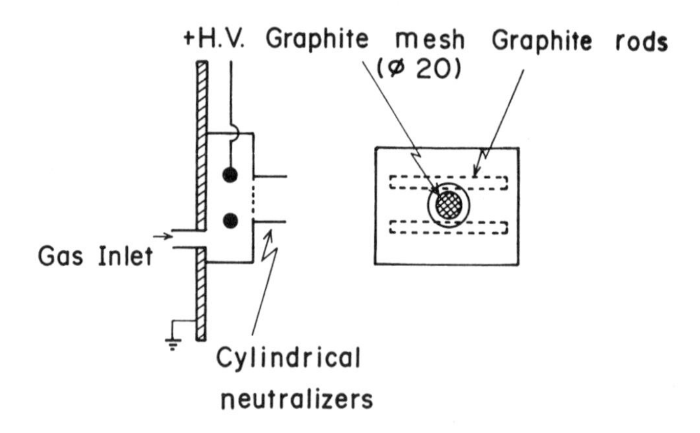

Figure 3: Schematic diagram of modified McIlraith cold cathode source.

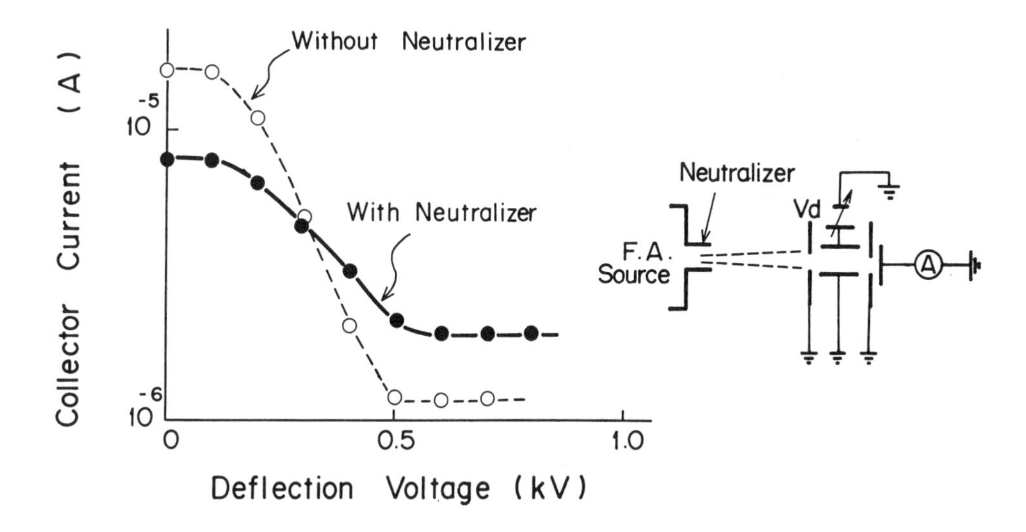

Figure 4: Measurement of fast atom beam. Discharge conditions: Ar gas pressure 5 Pa, discharge voltage 1.6 kV, and discharge current 100 mA.

current in the Mo collector at a deflection voltage of 0 V, and $I_d = e_0 N_0$ is the saturation current of the high deflection voltage. This equation determines the difference between the secondary electron emission efficiency caused by neutral particle bombardment and that caused by ion bombardment. The dependence of the coefficient of secondary emission on the kinetic energy of Ar, Ne and He atoms and ions is shown in Fig. 5 [13].

The experimental results for the source characteristics are discussed using Eqs. (2) and (4) [3]. The equations are applied in Figs. 6 and 7, employing Ar gas discharge. Figure 6 shows the dependence of neutralization coefficient on the discharge current. R_{n0} values with and without the charge exchange cell, which range from 90 % to 95 %, are approximately 20 % greater than those without the cell.

The R_{n0} dependence on the Ar gas pressure is shown in Fig. 7. As the Ar gas pressure increases, the R_{n0} values become greater with and without the charge exchange cell. For example, the R_{n0} value at 5 Pa in a source with the charge exchange cell is 98 %. In Figs. 6 and 7, the calculated values from Eq. (2) are in accordance with the experimental values for a source without a charge-exchange cell. These results indicate that the neutralization mechanism in the source is mainly a resonant charge transfer. There would be a tendency to have lower energy neutrals than the ions because the charge exchange cross section decreases with energy.

The neutral atom energy distribution of a modified McIlraith type source was estimated by measuring the residual ion energy with a retarding electrodes and a Faraday cup. The principal kinetic energy for fast atoms emitted from the source appears to be almost 95 % of the anode potential [3,14].

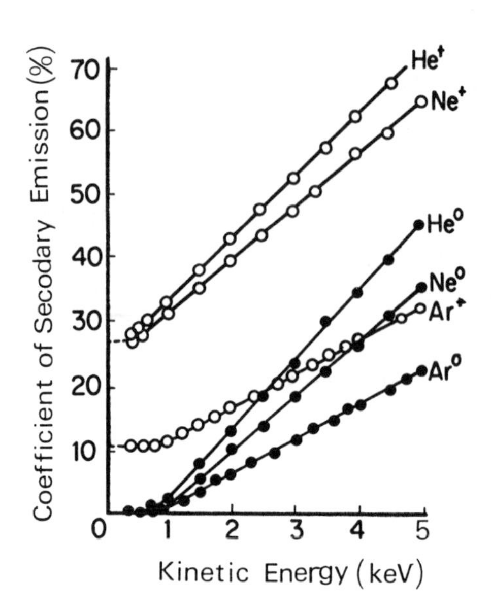

Figure 5: Dependence of the coefficient of secondary emission on the kinetic energy of Ar, Ne, and He atoms and ions for Mo target.

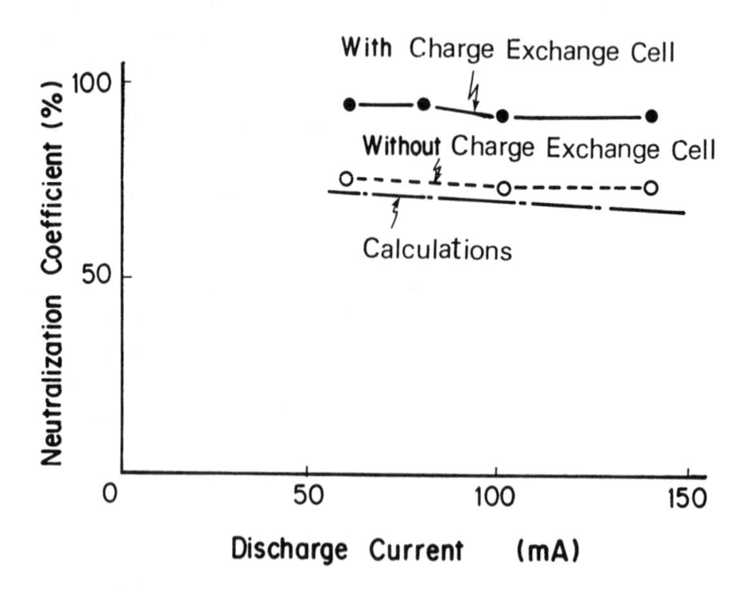

Figure 6: R_{n0} dependence on discharge current. Discharge conditions: Ar gas pressure 5 Pa, discharge voltage 1.2-1.8 kV, and discharge current 60-140 mA.

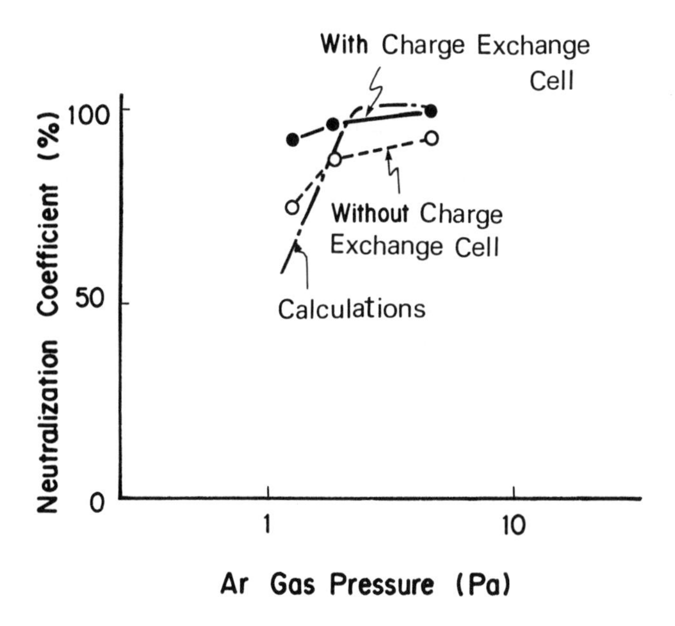

Figure 7: R_{n0} dependence on Ar gas pressure. Discharge conditions: Ar gas discharge voltage 1.1-1.6 kV, discharge current 100 mA.

§3. FRICTION-REDUCING BI-LAYER COATINGS

The concept of bi-layer frictin-reducing coatings is to provide a hard bottom layer coating (BN, B, TiB_2 and B_4C) as well as a solid lubricant top layer coating (MoS_2 or WS_2) as shown in the model in Fig. 8 [2]. The hard coating protects the substrate from mechanical damage. The hard materials noted above were selected boron compounds for this study because they possess an unusual combination of properties including great hardness, low density, high melting point and resistance to chemical erosion.

§4. HARD COATING USING DUAL FAST ATOM BEAM TECHNIQUE

In Fig. 9, a schematic diagram of the dual fast atom beam system[2] for hard coatings is shown. The system is similar in concept to several previously described dual ion beam systems [15-18]. One of the fast atom sources is used to sputter a target, while the other is used to bombard the substrate during film growth. The substrate was bombarded principally to improve the crystalline structure and the adhesion between the substrate and films as well as to clean the substrate surface. Both fast atom sources were operated with argon. The substrates were placed in different positions on a rotating turntable to form the coatings on the balls, the inner races and the outer races of the ball bearings.

The discharge in the fast atom beam source was generally operated at an argon pressure of about 10^{-1}-1 Pa. Sample chamber pressure was maintained in the $5.3X10^{-2}$-$1.1X10^{-1}$ Pa range. The proportion of high-energy neutral particles in the beam was above 90 %.

Electron diffraction patterns for BN films deposited on silicon substrates using BN target are shown in Fig. 10 [19]. Figure 10 (b), which obtained with fast atom bombardment of their growth surface shows a halo indicating amorphous structure, while (a) without fast atom bombardment shows partly a series circular rings. The film was partly identified as a poly-crystalline cubic BN including amorphous.

Electron diffraction patterns for B, TiB_2 and B_4C coatings on silicon substrates using each compound target are shown in Fig. 11 [2]. The coatings also obtained with fast atom bombardment of their growth surfaces at room temperature showed Laue spots and diffraction lines indicative of single or poly-crystalline structures. On the other hand, all of these coatings obtained without fast atom bombardment at room temperature showed halos indicating an amorphous structure. In each material, the discharge voltage for the films exhibiting the Laue spots was smaller than that for the films exhibiting diffraction lines. This suggests that there are optimum fast atom beam energies for single-crystalline growth and the excessive energies damage the film resulting in poly-crystalline or amorphous structures.

§5. SOLID LUBRICANT COATINGS USING COMPOSITE TARGETS

An Mo-S complex target was examined for the purpose of stoichiometric control in the sputtering [1]. The Mo-S complex target consists of two parts: a circular target of sulfur (S) with two fan-shaped target of molybdenum (Mo) placed on top of the S target.

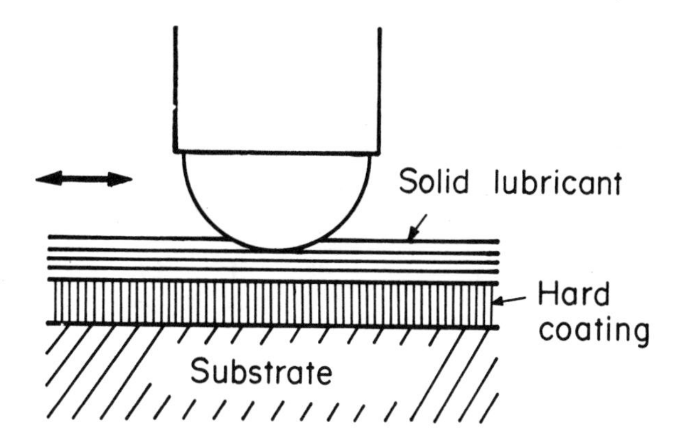

Figure 8: Friction-reducing bi-layer coating model.

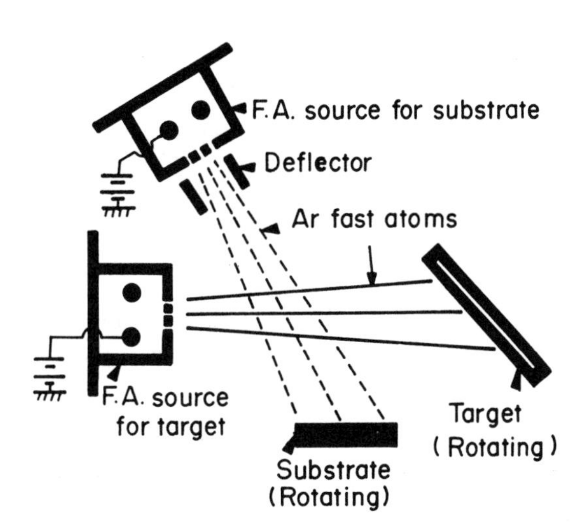

Figure 9: Schematic diagram of dual fast atom beam system.

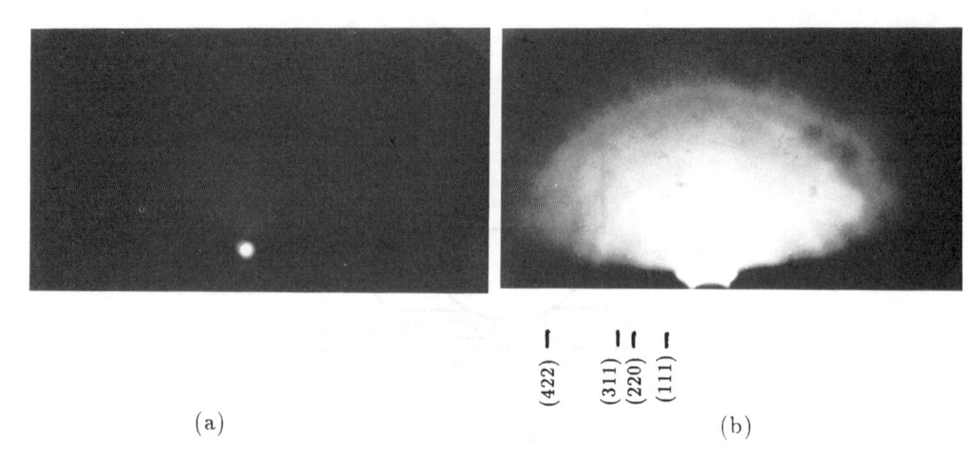

Figure 10: Electron diffraction (RHEED) patterns for BN on Si substrates. (a)Coating without Ar fast atom bombardment at room temperature. (b)Coating with bombardment at a discharge voltage of 1.5 kV at room temperature.

The ratio of the fast atom-bombarded area of the S target to that of the Mo target changes the composition of the deposited film. The fast atom beam technique allows simultaneous irradiation of the complex target, consisting of the insulator (S) and the conductor (Mo), without incurring serious problems. Details of the stoichiometric control of an Mo-S complex target is show in the inset in Fig. 12.

Some experiments were performed using a conventional MoS_2 compound target [19]. The deposition films using this target are rich in Mo and are remarkably different stoichiometrically from MoS_2 as shown in Fig. 13. The chemical composition of the films is $MoS_{1.4}$-$MoS_{1.6}$. The main reasons this film is rich in Mo are (1) the different sputtering yields of S and Mo, (2)the reaction of S atoms sputtered from the target with residual gas in the vacuum chamber to form compounds such as H_2S and SO_2.

Electron diffraction patterns of stoichiometric MoS_2 film using an Mo-S complex target were examined [1]. The patterns indicate that nonannealed film exhibits an amorphous structure and that the MoS_2 film annealed at 400 °C for 2 h exhibits a poly-crystalline structure.

§6. FRICTION CHARACTERISTICS

The friction characteristics of the deposited films were measured with a sliding frictions resistance testing system, in which a 500 g weighted steel ball (6 mm in diameter) was moved back and forth on the specimen surface at room temperature in air.

6.1. Friction characteristics of hard coatings

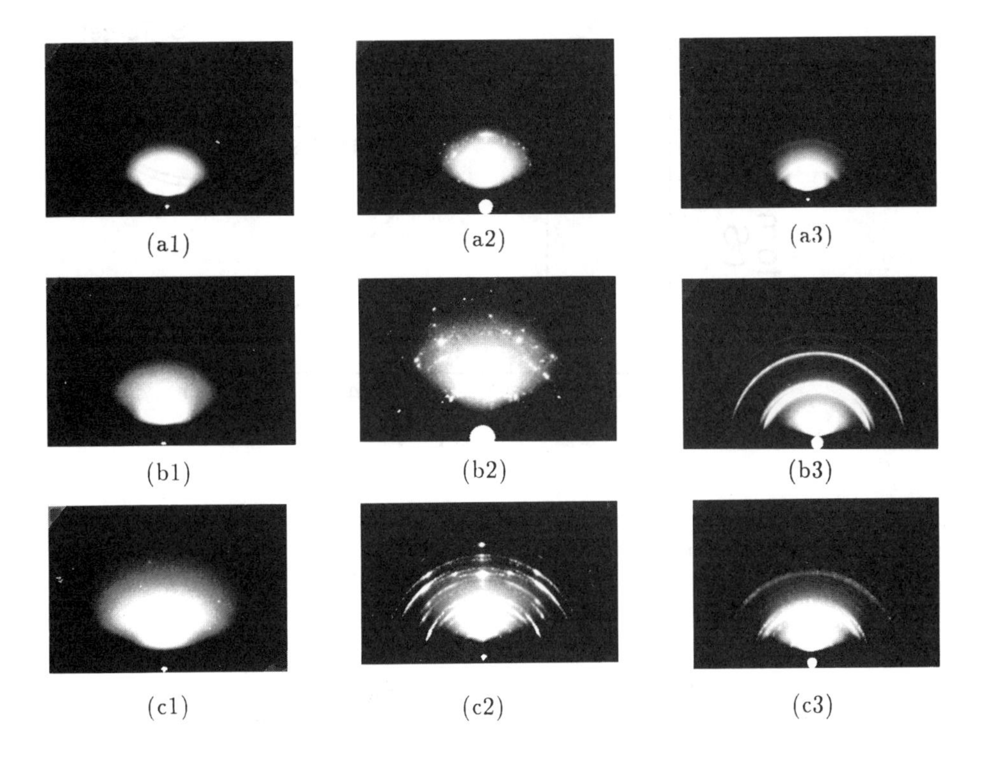

Figure 11: Electron diffraction (RHEED) patterns of the hard coatings. (a) B coatings at room temperature. (a1) Coating without Ar fast atom bombardment of the growing surface. (a2) Coating with bombardment at a discharge voltage of 1 kV. (a3) Coating with bombardment at 1.5 kV. (b) TiB_2 coatings at room temperature. (b1) Coating without bombardment. (b2) Coating bombarded at 1.5 kV. (b3) Coating bombarded at 2.0 kV. (c) B_4C coatings at room temperature. (c1) Coating without bombardment. (c2) Coating bombarded at 1.5 kV. (c3) Coating bombarded at 2.0 kV.

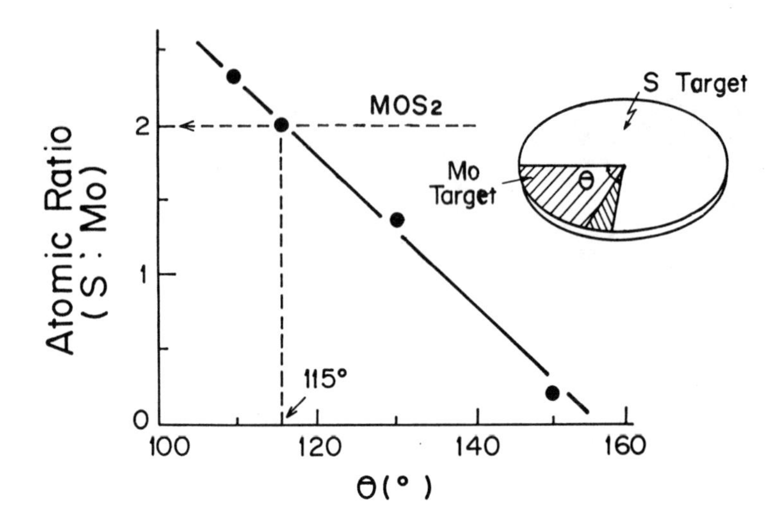

Figure 12: Relationship between vertical angle of Mo target and chemical composition ratio (S/Mo) obtained using Mo-S complex target.

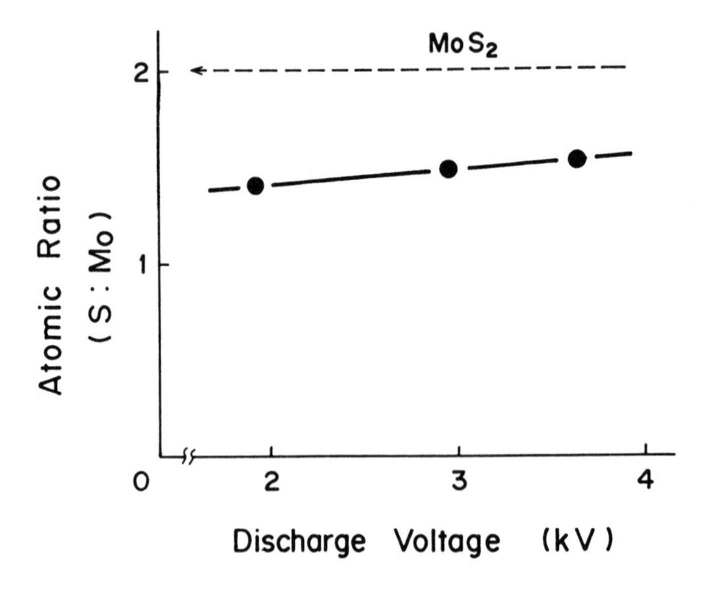

Figure 13: Chemical composition of Mo-S films obtained using MoS_2 compound target.

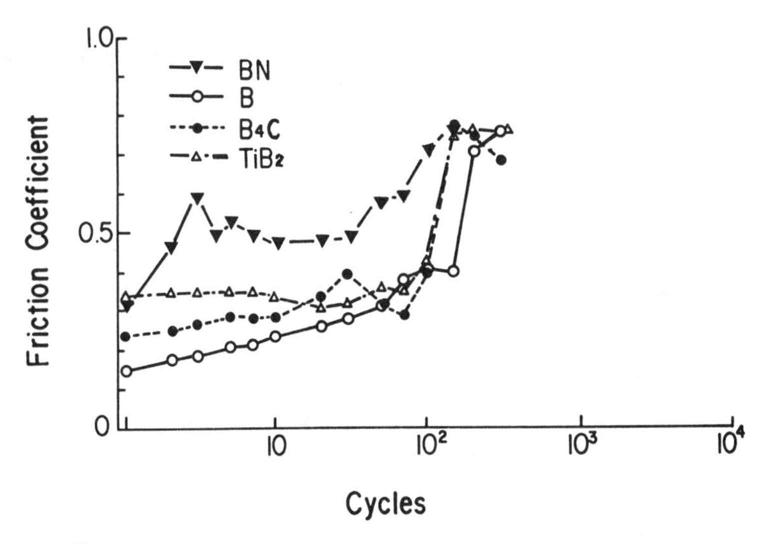

Figure 14: Friction characteristics of hard coatings.

In Fig. 14, Friction characteristics of BN, B, TiB_2 and B_4C hard coatings formed on the substrates of SUS440C are shown. Each film was bombarded by fast atom during its growing surface. The film thickness was about 50 nm. Each of the friction characteristics of these hard coatings shows clearly that of typical hard materials. The friction coefficients are very high, namely above 0.5 after 100 cycles.

6.2. Friction characteristics of MoS_2 solid lubricant

The friction characteristics of films of various ($MoS_{0.5}$-MoS_8) chemical compositions deposited at the Si wafer substrate temperature of 30 °C are shown in Fig. 15. The film thickness was about 100 nm. The films were annealed in a vacuum atmosphere (10^{-1}-1 Pa) at 300 °C for 2 h. It is clear that friction coefficient of the MoS_2 film is better.

The friction characteristics of the stoichiometric MoS_2 films deposited at the substrate temperature of 30 and 70 °C are plotted in Figs. 16(a) and 16(b), respectively. Films were annealed at various temperatures (60-400 °C) for 2 h in a vacuum atmosphere. The higher the annealing temperature, the better the lubricating properties of the film became. The friction coefficients and durability values of the films deposited using the Mo-S complex target are superior to those deposited using a conventional MoS_2 compound target [20]. The friction coefficient is below 0.1 for films deposited at the substrate temperature of 30 °C and subsequently annealed at 300 or 400 °C.

As also seen for films deposited at 70 °C the higher the annealing temperature, the smaller the friction coefficient becomes. Comparatively speaking, however, the lubricating properties of the nonannealed films deposited at 70 °C are much better than those deposited at 30 °C. Additionally, the friction coefficient of the annealed film at 300 °C is less than 0.05.

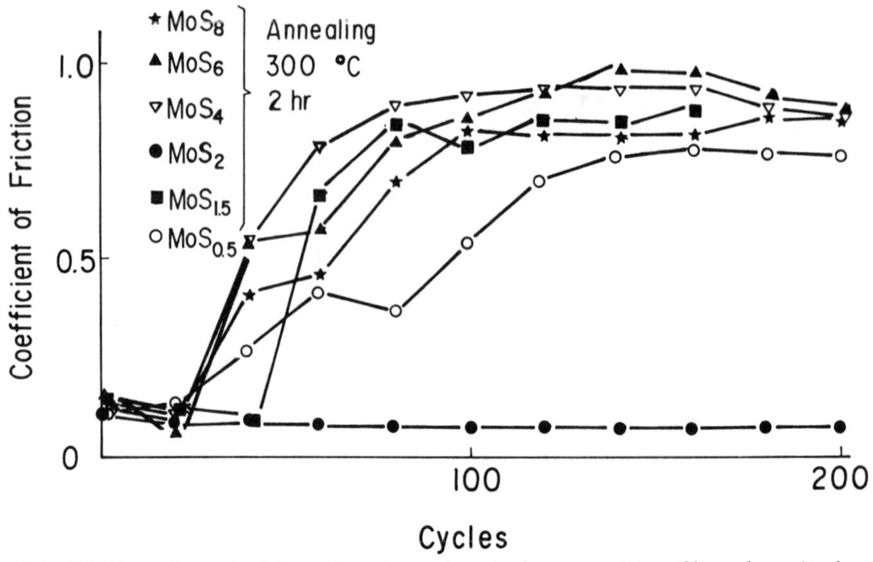

Figure 15: Friction characteristics of various chemical composition films deposited using Mo-S complex target at substrate temperature of 30 °C.

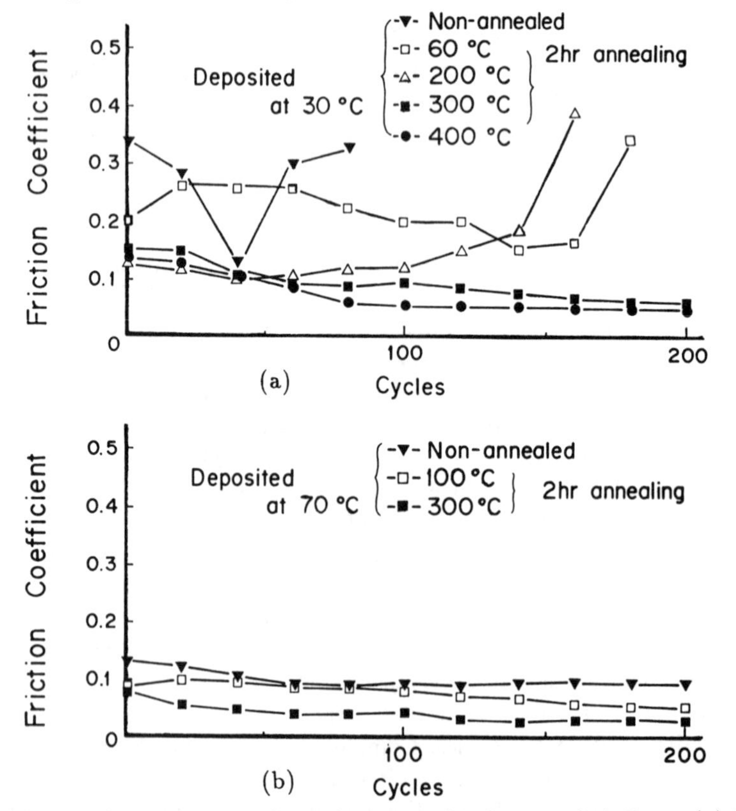

Figure 16: Friction characteristics of stoichiometric MoS_2 annealed films. (a) Deposited at substrate temperature of 30 °C; (b) deposited at substrate temperature of 70 °C.

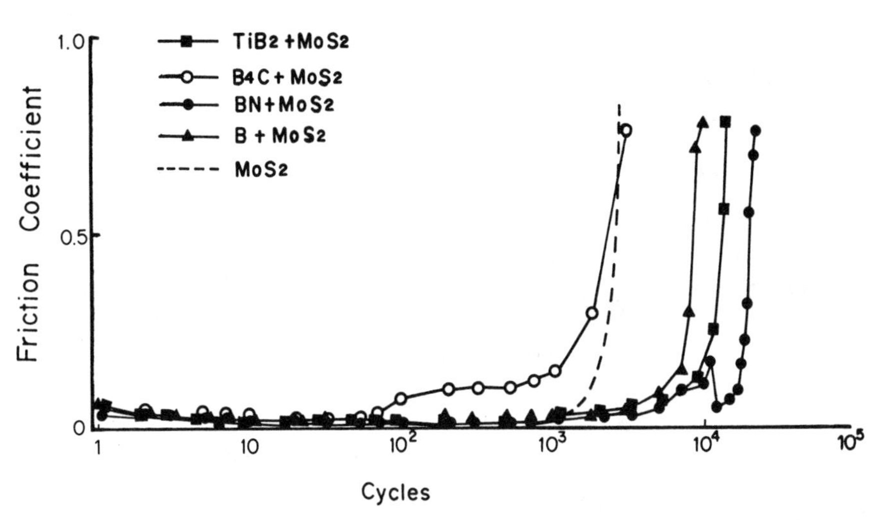

Figure 17: Friction characteristics of friction-reducing bi-layer coatings on SUS440C stainless steel.

6.3. Friction characteristics of bi-layer solid lubricant films consisting of hard bottom layer with an added top layer of MoS_2 solid lubricant

The friction characteristics of bi-layer solid lubricants consisting of MoS_2 over the hard coatings (BN, B, TiB_2 and B_4C) on SUS440C stainless steel as well as the MoS_2 solid lubricant (MoS_2) alone are compared in Fig. 17. The hard coating and MoS_2 thicknesses were about 50 and 100 nm, respectively.

It is clear that the friction coefficients of the bi-layer solid lubricants (MoS_2/BN, MoS_2/B, MoS_2/TiB_2) change from 0.01 to 0.02 or from 1/5 to 2/5 of that of the MoS_2 single-layer lubricant alone. Moreover these stable low friction coefficients were maintained near 10^4 cycles. However, the friction coefficient of the MoS_2/B_4C system increased because of peeloff.

The motive torque of the bi-layer (MoS_2/TiB_2) solid lubricant bearing pair is shown in Fig. 18. In the figure, the bi-layer solid lubricant bearing pair is compared with the single-layer MoS_2 bearing pair. These were tested using a moving-coil type ball bearing oscillatory test equipment [21]. In the test, a 20 mm diameter ball-bearing pair was given a rigid axial prowled of 3 kg. The running tests were performed by steady-cycle oscillation of the ball bearing outer races. The oscillatory angle and speed of the housing and outer race were set at 3.7 degree at a frequency of 17 Hz, respectively. The motive torque of the bearing pair was determined by using spring deformation.

The motive torque measured within the oscillatory angle in the bi-layer bearings was small as that in MoS_2 single-layer bearings. On the other hand, all of measured values were larger in over swing test. Accordingly, values are given in the figures for both the normal swing (within oscillatory angle) and the overswing (out of oscillatory angle). Up to $3X10^6$ cycles, the torque for the bi-layer was about 1/4 of the single-layer MoS_2 lubricant torque. In Fig. 18, the motive torques of bi-layer solid lubricant were generally lower than for single-layer MoS_2 films, in agreement with the friction test results in Fig. 17.

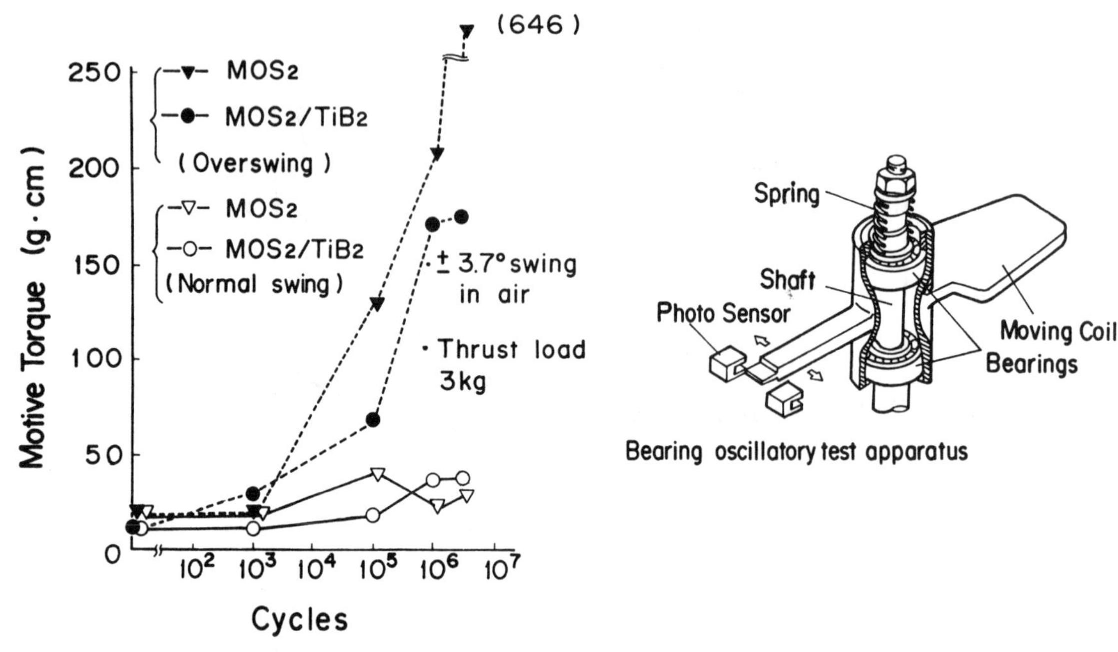

Figure 18: Motive torque of MoS_2/TiB_2 bi-layer coated ball bearings.

§7. CONCLUSION

Fast atom beam techniques and friction-reducing solid lubricant coatings were presented. The tendencies of a modified McIlraith-type source characteristics are very close to the calculated resonant charge transfer values. The highest proportion of high-energy neutral atoms in the beam is about 98 %.

The bi-layer solid lubricant consisting of hard bottom layer with an added MoS_2 solid lubricant, especially MoS_2/BN and MoS_2/B systems, provided low friction coefficients of less than 0.01. In addition, the hard films obtained by the fast atom bombardment of their growth surface were found to have either single or poly-crystalline structures, whereas the films not subject to surface bombardment had an amorphous structure. In MoS_2 film formations, a new chemical composition control method using an Mo-S complex target allows for the sufficient control of the film composition.

Bi-layer solid lubricant coating ball bearings were produced and tested, which demonstrated the feasibility of applying the bi-layer solid lubricant to practical bearing surfaces.

ACKNOWLEDGMENTS

The author wish to his appreciation to Drs. Hisao Takata and Kiyoshi Itao for their encouragement. Thanks are also due to Messrs. Kazutoshi Nagai , Fusao Shimokawa and Yoshimitu Asanuma for their valuable discussions and helps in experiments.

REFERENCES

1) Kuwano, H. and Nagai, K.: J.Vac.Sci.Technol. A3, 1809(1985).
2) Kuwano, H. and Nagai, K.: J.Vac.Sci.Technol. A4, 2993(1986).
3) Kuwano, H. and Shimokawa, F.: J.Vac.Sci.Technol. B6, 1565(1988).
4) Nagai, K. and Kuwano, H.: in *Ion Mass Spectrometry, SIMS 4*, edited by A.Benninghoven, J.Okano, R.Shimizu and H.W.Werner (Springer, Berlin, 1984). p448.
5) Spalvins, T.: ASLE Trans. 12, 36(1969).
6) Spalvins, T.: J.Vac.Sci.Technol. A3, 2329(1985).
7) Gardos, M.N.: Lubr. Eng. 32, 463(1976).
8) Nishimura, M., Nosaka, M., Suzuki, M., and Miyazawa, Y.: ASLE Proc. of the 2nd Int.Conf.on Solid Lubrication, 128(1978).
9) Christy, R.I.: Thin Solid Films 80, 289(1981).
10) Hasted, J.B.: *Physics of Atomic Collisions* (Butterworths, London, 1964), p143.
11) McIlraith, A.M.: Nature 212, 1422(1966).
12) Franks, J.: J.Vac.Sci.Technol 16, 181(1979).
13) Arifov, V.A.: *Interaction of Atomic Particles with a Solid Surface* (Consultants Bureau, New York, 1969), p235.
14) Shimokawa, F., Kuwano, H., and Nagai, K.: in *proceedings of the Tenth Symposium on Ion Source and Ion-Assisted Technology,* edited by T.Takagi (Institute of Electrical Engineers of Japan, Tokyo, 1986), p101.
15) Weissmantel, C., Fiedler, O., Hecht, G. and Reisse, G.: Thin Solid Films 13, 359(1972).
16) Harper, J.M.E. and Gambino, R.J.: J.Vac.Sci.Technol. 16, 1901(1979).
17) Weissmantel, C., Bewilogua, K. and Breuer, K.: Thin Solid Films 96, 31(1982).
18) Harper, J.M.E., Cuomo, J.J., and Hentzell, H.T.G.: J.Appl. Phys., 58, 550(1985).
19) Shimokawa, F., Kuwano, H., and Nagai, K.: in *Proceedings of the Ninth Symposium on Ion Source and Ion-Assisted Technology*, edited by T.Takagi (Institute of Electrical Engineers of Japan, Tokyo, 1985), p467.
20) Kuwano, H. and Nagai, K.: in *Proceedings of the International Ion Engineering Congress of ISIAT and IPAT '83*, edited by T.Takagi (Institution of Electrical Engineers of Japan, Tokyo, 1983) Vol.2, p977.
21) Miyake, S. and Takahashi, T.: ASLE Trans. 30, 248(1987).

AUTHOR INDEX